Thi:

Cl

Parallel Processing in the Visual System

THE CLASSIFICATION OF
RETINAL GANGLION CELLS AND
ITS IMPACT ON THE
NEUROBIOLOGY OF VISION

PERSPECTIVES IN VISION RESEARCH

Series Editor: Colin Blakemore
University of Oxford
Oxford, England

Parallel Processing in the Visual System
THE CLASSIFICATION OF RETINAL GANGLION CELLS AND
ITS IMPACT ON THE NEUROBIOLOGY OF VISION
Jonathan Stone

A Continuation Order Plan is available for this series. A continuation order will bring delivery of each new volume immediately upon publication. Volumes are billed only upon actual shipment. For further information please contact the publisher.

Parallel Processing in the Visual System

THE CLASSIFICATION OF RETINAL GANGLION CELLS AND ITS IMPACT ON THE NEUROBIOLOGY OF VISION

JONATHAN STONE

School of Anatomy
University of New South Wales
Sydney, Australia

PLENUM PRESS • NEW YORK AND LONDON

Library of Congress Cataloging in Publication Data

Stone, Jonathan, 1942–
  Parallel processing in the visual system.

  (Perspectives in vision research)
  Bibliography: p.
  Includes index.
  1. Retinal ganglion cells — Classification. 2. Vision. 3. Mammals — Physiology. I.
Title. II. Series. [DNLM: 1. Retina. 2. Neurons — Classification. 3. Classifica-
tion — Methods. 4. Visual perception — Physiology. WW 270 S878p]
QP479,S78 1983   599′.01823   83-13676
ISBN 0-306-41220-9

For Margaret

Foreword

In the mid-sixties, John Robson and Christina Enroth-Cugell, without realizing what they were doing, set off a virtual revolution in the study of the visual system. They were trying to apply the methods of linear systems analysis (which were already being used to describe the optics of the eye and the psychophysical performance of the human visual system) to the properties of retinal ganglion cells in the cat. Their idea was to stimulate the retina with patterns of stripes and to look at the way that the signals from the center and the antagonistic surround of the respective field of each ganglion cell (first described by Stephen Kuffler) interact to generate the cell's responses. Many of the ganglion cells behaved themselves very nicely and John and Christina got into the habit (they now say) of calling them I (interesting) cells. However. to their annoyance, the majority of neurons they recorded had nasty, nonlinear properties that couldn't be predicted on the basis of simple summation of light within the center and the surround. These uncoop-erative ganglion cells, which Enroth-Cugell and Robson at first called D (dull) cells, produced transient bursts of impulses every time the distribution of light falling on the receptive field was changed, even if the total light flux was unaltered.

From this chance discovery of two major classes of ganglion cells (now called X and Y cells) has grown a whole new approach to the anatomy, physiology, and devel-opment of the visual pathway, as well as to human psychophysics. Jonathan Stone has made a number of important contributions to this study of parallel analysis within the visual system, including the first full description of the third (rather motley) class of ganglion cells, the W cells. In this monograph he shows how influential this way of thinking has been in visual science as well as in other aspects of sensory physiology. What better start could there be to this new series of monographs on vision, *Perspectives in Vision Research?* The series will provide up-to-the-minute authoritative accounts of all aspects of visual science, still perhaps the broadest and most active area within neuroscience.

Colin Blakemore
Oxford

Preface

This monograph began as an account of the classification of retinal ganglion cells in the cat and other mammals, and its scope could well have been limited to that. But although the classification of ganglion cells is a complex and intriguing problem in itself, its major importance lies, I believe, in the impact that it has had on our understanding of the visual pathways. Once it was established that retinal ganglion cells form a number of functionally distinct groups, the visual centers of the brain were analyzed and reanalyzed in terms of those groups. From this work there emerged a new understanding of these centers, leading to the idea of "parallel processing" in the visual system, i.e., that the visual pathways comprise parallel-wired sets of neurons that code and transmit different aspects of the visual image. It has become a challenge to trace this parallel organization and to ascertain both the value and the limitations of the concept of parallel processing in the analysis of the visual pathways. Historically, the idea of parallel processing was first developed in the study of the somatosensory pathways, and it is currently being extended to audition and olfaction, as well as vision. Its value and limitations in all these contexts need exploration and assessment.

As a consequence, only the first of the three parts of this monograph is concerned with the classification of retinal ganglion cells. Much the longest part is Part III, which concerns the impact that the classification has had on our understanding of the lateral geniculate nucleus, superior colliculus, and visual cortex, and on the analyses of retinal topography, of the influence of deprivation on the visual pathways, and of visual perception. Part III ends with a survey of parametric processing in sensory systems other than vision, and a proposal for a "parametric systematics" of neuronal classification.

Part II concerns the methodology of classification. I argue there that although classification is a fundamental process in the conceptual organization of scientific knowledge, many visual neurobiologists (myself included) have paid too little attention to the methodologies we have used in classifying nerve cells. A case is argued for a particular approach to classification, in the context of (1) historical and contemporary approaches to classification, (2) the epistemological issues involved, and (3) the biological context of the problem.

I have incurred many debts in writing this monograph. Some are old and intangible. From my father, I learned from childhood the value for scholarship of erudition, simplicity of analysis, and self-reliance. My scientific mentor, P. O. Bishop, brought me into this field, many of whose horizons he had pioneered or was about to explore; from him I learned the value of constant recourse to experimentation and of tolerance for others' interpretations. Their influence on my work has been strong and abiding. Other of the debts are more recent and tangible. I owe much to colleagues with whom several sections of the monograph have been developed as separate essays, in particular Michael

ix

H. Rowe, Bogdan Dreher, and Audie G. Leventhal. Their contributions to Chapters 1, 2, 5, 8, 10, and 11 were fundamental. I owe warm gratitude to Daniele Dubois (who typed the first draft) and to Trudy Wiedeman (who typed the second draft); to Sharon McDonald and Peter Wells for their help with the illustrations; and to Paul Halasz for his patience and engineering skills in developing and helping me with a computer-based storage of the text. Many colleagues, including Colin Blakemore, Bogdan Dreher, Michael Cooper, Audie Leventhal, James McIlwain, Marilee Ogren, David Rapaport, Michael Rowe, and Mark Rowe, read and improved the manuscript, and for their valuable ideas and suggestions I owe my thanks and appreciation. I am grateful also to the many scientists and writers who gave ready permission for the reproduction of illustrations and text from their papers. My wife has borne and parried my recurring frustrations with the task with an affection and intelligence on which I have come much to rely.

Jonathan Stone
Sydney

Contents

I. THE CLASSIFICATION OF RETINAL GANGLION CELLS

II. ON THE METHODOLOGY OF CLASSIFICATION

III. THE IMPACT OF GANGLION CELL CLASSIFICATION

The Classification of Retinal Ganglion Cells

I

From the Beginning

Ganglion Cell Classification

to 1966

1

In 1933, G. H. Bishop published a report entitled "Fibre Groups in the Optic Nerve." This is perhaps the earliest study, morphological or physiological, to which present concepts of the parallel organization of the visual pathways can be traced. It was an electrophysiological study in which Bishop obtained oscillographic records of compound action potentials, generated in the excised optic nerve of the frog, and in the exposed but still-attached nerve of the rabbit, by a brief electrical shock applied to the nerve some distance from the recording lead. In both species, the compound action potential showed early and later components (Fig. 1.1), suggesting that the axons of the nerve were not homogeneous in the velocity at which they conduct action potentials, but rather fell into two or more groups (Bishop suggested three), with distinct conduction velocities and therefore distinct calibers. In the frog, the three groups had conduction velocities of 10, 3 and 0.4 m/sec; in the rabbit, the approximate values were 20–50, 7–17, and 4 m/sec. The report concluded with an optimistic prediction that the presence of conduction velocity groupings in the optic nerve would be a useful starting point for the further investigation of the visual pathways.

Bishop's optimism no doubt stemmed from the striking correlations between axon caliber and sensory modality described in the somatosensory system; the evidence for these correlations came from clinical observations reinforced by neurophysiological studies (Chapter 12, Section 12.1), in many of which Bishop had taken part. It is remarkable, therefore, that in 1959 (thus, 26 years after his original prediction), Bishop reviewed the evidence concerning the functional correlates of conduction velocity in the somatosensory and visual systems, and came to the more pessimistic conclusion that none was apparent in the visual pathways, and that the significance of conduction velocity was more phylogenetic than functional. Since 1959, however, reports from a number of laboratories, including an important contribution from Bishop et al. (1969), have sub-

3

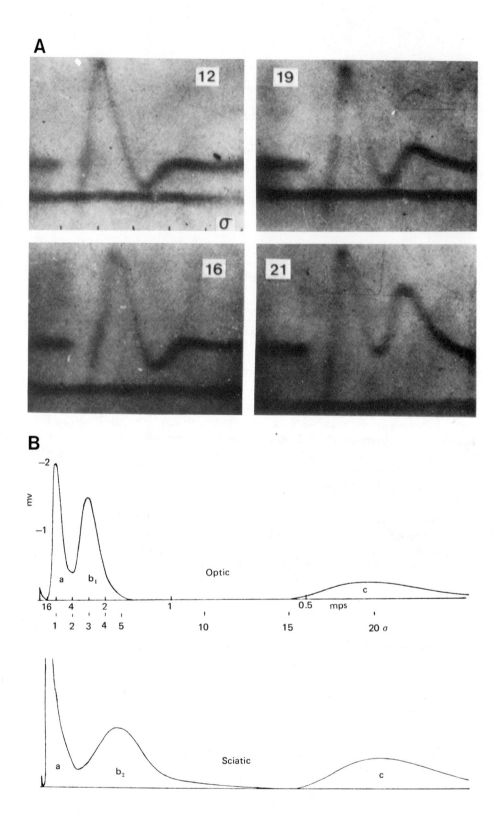

stantially borne out Bishop's original suggestion. What led Bishop to change his hypothesis between 1933 and 1959? Why did it take so long for the functional correlates of conduction velocity to be established for the visual pathways, after the early recognition of striking correlates in the somatosensory pathways, and after Bishop's 1933 prediction? By what steps did present classifications of retinal ganglion cells develop? This chapter attempts to summarize the major studies relevant to those questions, published between 1933 and 1966.

The work is conveniently, and not too arbitrarily, traced in three strands, which were successfully interwoven in 1966 and subsequent years, but which were pursued largely independently before then: (1) studies of conduction velocity groupings among optic nerve axons, (2) studies of the receptive fields of retinal ganglion cells, and (3) studies of the morphology of retinal ganglion cells.

1.1. CONDUCTION VELOCITY GROUPINGS IN THE OPTIC NERVE

In the 1920s and 1930s, G. H. Bishop was a collaborator in a series of studies (e.g., Erlanger *et al.*, 1926; Heinbecker *et al.*, 1933, 1934) that concerned the fiber groups found in peripheral somatosensory nerves. One of the starting points for these studies had been clinical observations in the late 19th century (reviewed by Gasser and Erlanger, 1929) of differential sensory effects of pressure block on peripheral nerves. It was observed that when a nerve was subjected to pressure, the sense of touch was lost before temperature sense, and that pain was the most persistent sensation. These observations led to the idea that these sensations were served by different nerve fibers, with different susceptibility to pressure. Ranson (1921) reviewed evidence that the fibers subserving pain are small, and Erlanger (1927) showed that the fastest sensory fibers are found only in muscle nerves and, therefore, presumably must subserve muscle sense. Gasser and Erlanger (1929) showed that fast (and, therefore, large) fibers are the most susceptible, and small (slow) fibers the most resistant, to pressure block, providing a physiological basis for the earlier clinical observations. Bishop and Heinbecker (1930, 1932) studied fiber groups in visceral and cervical sympathetic nerves, Heinbecker *et al.* (1934) reviewed this and earlier evidence for the division of function among somatosensory nerves, and Bishop (1959) reviewed the question again, with the perspective provided by a further generation of scientists.

This stream of work formed the context for Bishop's attempt in 1933 to ascertain

---

Figure 1.1. First evidence of conduction velocity groupings in the optic nerve. (A) Oscilloscope traces showing evidence of conduction velocity groupings in frog optic nerve (from Bishop, 1933). The numbers on the traces (12, 16, 19, 21) indicate the relative strengths of the electrical stimulus used. Note the emergence of a second deflection (presumably representing a second conduction velocity group) with stronger stimuli (19, 21). Time base in milliseconds. (B) Bishop's (1933) comparison of frog optic nerve and sciatic nerve. Bishop commented: "Diagrammatic plot of conducted action potential of frog optic nerve . . . and for comparison (below) plot of potentials of the frog sciatic for same conduction distance." In each nerve he identified three deflections (a, b, c), representing three conduction velocity groups. [Reproduced with kind permission of the American Physiological Society.]

whether conduction velocity groups are to be found in the optic nerve. His starting point was the correlation between size and function established for somatosensory nerves:

> It has been found in all nerves so far studied that the nerve fibers occur in groups, with relatively vacant spaces between them, and that these groups are related to function. For instance in the saphenous nerve, the group of fibers with fastest conduction mediates touch and pressure, a slower conducting group mediates pain and temperature and a still slower group is motor.

His conclusion was a specific prediction for the visual system:

> In peripheral nerves a group of larger fibers mediates sensations of touch, including those permitting spatial discrimination, while a group of smaller fibers mediates pain and temperature. . . . By analogy one might anticipate that the larger fibers of the optic nerve would also mediate that aspect of vision concerned with spatial discrimination of form, while the smaller fibers would be concerned with the quantitative factor of intensity. The best that can be said of such a speculation is that there seems to be no serious objection to be made to it, and perhaps, that it suggests a point of attack for the further analysis of vision.

Bishop's paper did not arouse immediate or widespread interest however. The conduction velocity groupings in frog optic nerve have not been reinvestigated, although Maturana *et al.* (1960) confirmed the range of velocities that Bishop reported; the groupings in rabbit nerve were confirmed only 35 years later, by Lederman and Noell (1968), who observed fast and slow-conducting components in the potential recorded in the optic tract following photic stimulation. In the meantime, studies of conduction velocity groupings in the optic nerve began to concentrate on the cat, the first descriptions being provided by Bishop and O'Leary (1938, 1940, 1942). Bishop and O'Leary (1938; Fig. 1.2) described early and late components in the field potential elicited in cat optic tract by stimulation of the optic nerve; their latencies indicated the presence of distinct groups of axons with velocities of 60 and 25 m/sec. Relying on von Gudden's (1886) observation that the small fibers of the optic tract pass to the superior colliculus (SC) rather than to the lateral geniculate nucleus (LGN), and on their own observation that the response of the visual cortex to optic nerve stimulation seems determined by the fast-conducting group, Bishop and O'Leary concluded:

> . . . the division of the optic tract into two size groups, as indicated by two discrete potential waves, represents a functional division, as has been proved true for other nerves.

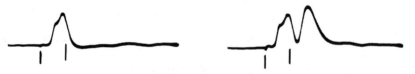

Figure 1.2. First evidence of conduction velocity groupings in cat optic nerve and tract (from Bishop and O'Leary, 1938). Field potentials recorded in the optic tract following stimulation of the optic nerve at stimulus strengths just maximal for an early potential (left) and for a second, later potential (right). Upward deflection, negative in tract. Time base shows millisecond divisions. [Reproduced from the *Journal of Neurophysiology* with kind permission of the American Physiological Society.]

Moreover, they calculated that, despite the prominence of the fast-conducting potential, "there are probably five times more fibres in the slower conducting group than in the fast."

Bishop and O'Leary's conclusion that only large fibers of the optic tract project to the LGN and cerebral cortex seems to have proven incorrect; the slower (25 m/sec) axons carry the activity of a functional class of retinal ganglion cells called *X cells* and their predominant projection is to the A laminae of the LGN and thence to area 17 of the visual cortex (Chapters 6 and 8). Bishop and O'Leary's equation of von Gudden's (1886) small fibers with their 25 m/sec group also appears incorrect, as the 25 m/sec (X-cell) group does not appear to project substantially to the SC (Chapter 2, Section 2.3.3.7, Table 2.1; also Chapter 7, Section 7.2.2). Nevertheless, the outlines of present understanding were beginning to emerge.

Bishop and O'Leary (1942) provided two new observations relevant to the present discussion. First, they argued for the presence of two conduction velocity groups slower than the 25 m/sec they had seen earlier. The faster of these was only a little slower than the 25 m/sec group and its present status is still unclear. The 25 m/sec group is the modern t_2 group comprising X-cell axons; and the slightly slower group could comprise the axons of the area centralis X cells (Stone and Freeman, 1971), or it could comprise the faster of the axons of W cells. The slower of Bishop and O'Leary's (1942) slower-than-25 m/sec groups seems clearly identifiable, however. Bishop and O'Leary noted that the potential wave generated by this group is "temporally dispersed . . . causing a large response from the superior colliculus." This appears to be the first description of the activity of the axons of W-class ganglion cells and of their strong projection to the SC.

Second, Bishop and O'Leary developed their analysis of the distribution of optic tract axons between the SC and the LGN (specifically its major dorsal component, the dLGN). Distinguishing three conduction velocity groups now, rather than two, they still concluded that the retinal input to the LGN is predominantly fast-fiber. However, they now noted evidence that some medium-velocity fibers also reach the LGN [present understanding is that the medium-caliber (X-cell) axons are numerically dominant] and they noted further than "some of the smallest (therefore slowest) fibers of the nerve" also go to the LGN; perhaps they observed the W-cell input to the C laminae of the LGN, not recognized in terms of single cells until 1975 (Chapter 6, Section 6.1.2). They also concluded that the major termination site of the slowest-conducting optic tract axons was in the region of the SC, a conclusion well corroborated by recent work.

The first report from another laboratory to take up the question of conduction velocity groups in cat optic nerve seems to have been Chang's (1951) suggestion of three conduction velocity pathways in the retinogeniculocortical system of the cat, each component related to a component of color vision, a suggestion that has not been corroborated. Subsequently, the Australian group P. O. Bishop, Jeremy, and Lance (1953) studied the optic nerve of the cat in some detail and described two conduction velocity groups (one with conduction velocity of 30–40 m/sec, the other approximately 20 m/sec); some of their records are shown in Fig. 1.3A. P. O. Bishop and MacLeod (1954) traced the same groups into the LGN and gave them the still-used labels, t_1 (fast) and t_2 (slow) (Fig. 1.3B; the label t was used because the potentials were recorded in the

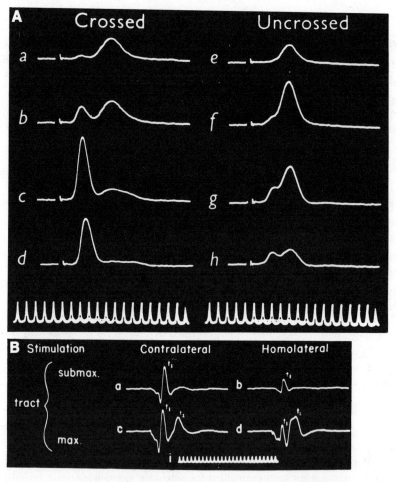

Figure 1.3. Labeling of t_1 and t_2 groups of cat optic nerve. (A) Antidromic field potentials recorded from the crossed and uncrossed optic nerves after stimulation of one optic tract. Two deflections are apparent in each nerve (e.g., in b and h), suggesting the presence of two conduction velocity groups. The depth of the stimulating electrodes was increased from a through d and e through h, causing changes in the relative amplitudes of the two deflections. Time intervals are 0.2 msec. [From Bishop *et al.* (1953). Reproduced with kind permission of the *Journal of Physiology.*] (B) Field potentials recorded in the optic tract in response to electrical stimulation of the contralateral and homo- (ipsi-) lateral optic nerves (from Bishop and MacLeod, 1954). At maximal or supramaximal stimulus strengths, two components are apparent in each response, labeled t_1 and t_2. Time intervals are 0.2 msec. [Reproduced from the *Journal of Neurophysiology* with kind permission of the American Physiological Society.]

optic tract). Chang (1956) observed the same two groups (Fig. 1.4) and argued for the presence of a "very fast" (70 m/sec) group, whose presence has not subsequently been confirmed.

In 1955, G. H. Bishop resumed his studies of cat optic nerve (Bishop and Clare, 1955), providing evidence for the presence of four conduction velocity groups, of which

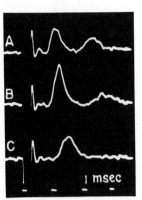

Figure 1.4. Evidence that the t_2 group does not project to the SC (from Chang, 1956). Field potentials recorded in cat optic nerve following electrical stimulation of: (A) contralateral optic tract, (B) contralateral LGN, (C) contralateral brachium of the SC. Chang drew attention to the absence of a second potential in (C), even at high stimulus strengths; this indicated, he suggested, that the t_2 fibers do not project to the SC. [Reproduced from the *Journal of Neurophysiology* with kind permission of the American Physiological Society.]

two correspond to t_1 and t_2 groups of P. O. Bishop and MacLeod (1954), the other two groups being, as previously, slower-conducting (6 and 3.4 m/sec). As previously, however, the potentials attributed to these slow-conducting axons were small and, arguably, unconvincing, presumably because of the small caliber of the axons and of variation in their conduction velocities (which means that their activities reach a recording lead quite asynchronously, even after a single, brief electrical volley). The most compelling descriptions of the conduction velocities of these very-slow-conducting axons have, therefore, come from single-unit studies (e.g., Stone and Hoffmann, 1972; Stone and Fukuda, 1974a; Cleland and Levick, 1974a; Kirk *et al.*, 1975; and see Chapter 2, Sections 2.1.3 and 2.3.3.4). Indeed, it is probably fair comment that, despite the accuracy of Bishop and O'Leary's (1942) conclusions concerning the presence and destination of such fibers, and the corroborative reports of Bishop and Clare (1955) and Spehlmann (1967), their presence was not widely accepted before the descriptions of single cells with very slow axons.

Lennox (1957) argued for the presence of three conduction velocity groups in cat optic nerve and tract, on the basis of a single-unit study. As with the three groups proposed by Chang, two of Lennox's groups corresponded to the t_1 and t_2 groups proposed by Bishop *et al.* (1953). The other group was faster-conducting (56 m/sec); its existence has not subsequently been confirmed.

All the above studies of optic nerve axons concerned their extraretinal segments in the optic nerve and tract, which are myelinated. Granit (1955), Dodt (1956), and Motokawa *et al.* (1957) studied conduction velocities among cat optic nerve axons along their intraretinal course, where they are unmyelinated. Dodt (1956) in particular studied the groups present in the retina and recorded the field potentials they generated at many retinal locations. Dodt's field potential records indicate the presence of two groups within the retina with conduction velocities of 2.5 and 1.8 m/sec (Fig. 1.5). These groups presumably are the retinal counterparts of t_1 and t_2, their relatively slow velocities resulting (Stone and Freeman, 1971) from the intraretinal axons being unmyelinated.

Sefton and Swinburn (1964) described three groupings among axons in the optic nerve and tract of the rat. By analogy with the cat they designated them t_1 (mean conduction velocity 13.5 m/sec), t_2 (5.5 m/sec), and t_3 (3.0 m/sec) (Figure 6.23).

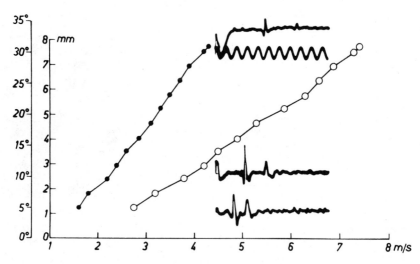

Figure 1.5. Early measurements of intraretinal conduction velocities of t_1 and t_2 (from Dodt, 1956). Plot of the latencies of the early (●) and late (○) components of the field potential recorded in cat retina following stimulation of the contralateral lateral geniculate body. The traces show the potentials recorded (from above down) 7.75, 3.5 and 1.2 mm nasal to the optic disc. Each trace shows two potentials, suggesting that two conduction velocity groups are present. The slopes of the two curves indicate intraretinal conduction velocities of 2.4 and 1.4 m/sec for the fast and slower groups. The units on the abcissa are presumably milliseconds. [Reproduced with kind permission of Birkhauser-Verlag.]

Evidence of Functional Correlates

A number of the groups of workers mentioned above speculated on the significance of the conduction velocity groups they observed. Chang (1951) and Lennox (1957) presented evidence that conduction velocity is related to the color-coding properties of ganglion cells, an idea that has not been confirmed, although a rare color-sensitive cell with a slow-conducting axon has been described (Cleland and Levick, 1974b). Bishop and Clare (1955) discussed the possibility first raised by Bishop and O'Leary (1940) that the different fiber groupings are more related to differences in phylogenetic history than to functional differences, an idea pursued further in Bishop's (1959) review. Thus (Bishop and Clare, 1955):

> A further attack on this problem of the significance of fiber size might be shifted from . . . sensory physiology . . . to . . . comparative anatomy. Once original patterns of nervous system organization have been developed, Nature has been extremely conservative in maintaining them, but has not hesitated to add successive complications to any primitive scheme. Still there can be recognized many instances where fibers of large size serve more recently acquired and more highly differentiated structures.

This suggestion is, of course, so generally stated that it must prove at least partially true, but it raised a valuable consideration and the study of the phylogenetic history of the visual pathways is still in its infancy. Several more specific indications of the functional importance of conduction velocity groupings can be detected in this earlier work, however, principally in terms of the different destinations of the various groups of axons, and the pioneering studies of Bishop and O'Leary (1938, 1940, 1942) on projections of the different conduction velocity groups were soon expanded in several ways. For example, Bishop and MacLeod (1954) described separate negative postsynaptic field potentials generated in cat LGN by t_1 and t_2 fibers. The implications of this observation are three: first, that both t_1 and t_2 fibers project to the LGN; second, that both excite (rather than inhibit) the cells of the LGN; and third, that they may excite different populations of geniculate cells. These three implications are now established as important components of the circuitry of the LGN. Conversely, Chang (1956) noted evidence that the fast axons of cat optic nerve project to the SC, while the slow axons do not. Chang's slow group conducted at 30 m/sec, equivalent to Bishop and O'Leary's (1940) medium-velocity group, so this evidence is not in conflict with their conclusions. Some of Chang's evidence is shown in Fig. 1.4; one of his conclusions is particularly germane:

> It is remarkable that there is only one peak in record C (recorded at the disc following a stimulus to the superior colliculus) and no other peaks can be produced no matter how strong is the stimulus.

The short latency of this single peak led him to conclude that

> The optic nerve contributes mainly, if not exclusively, large fibres to the superior colliculus and pretectal area. . . .

In fact, the small caliber W-cell axons have been shown subsequently to project massively to the SC, as Bishop and O'Leary (1940) and Bishop and Clare (1955) had earlier concluded, but Chang's report appears to be the first to argue for the present view that the large axons of Y cells do project there, and that the medium-caliber axons (of X cells) apparently do not, at least in substantial numbers (see Chapter 7, Section 7.2, for more detailed discussion).

Bishop and Clare (1955) confirmed the earlier conclusions of Bishop and O'Leary (1940) concerning axonal destination, including two apparently incorrect conclusions (they also did not detect the projection of t_1 fibers to the SC or of t_2 fibers to the A laminae of the LGN), but also added two elements that have been amply corroborated. First, they confirmed that the slower-than-t_2 fibers project to the SC and pretectum in considerable numbers. Second, expanding on a possibility raised by Chang (1951), they inferred the maintenance of cell size relationships along the visual pathways:

> It would appear that in general [in the spinal cord] a large presynaptic fiber connects with a larger postsynaptic one, a small one with a small one. Likewise across the geniculate synapse large fibers relay to radiation fibers with approximately the same conduction rate, and therefore presumably the same size range. As a first approximation the size groupings may be inferred to apply to the whole extent of pathways as well as to those unit segments . . . so far . . . studied, and to apply to cell size as well as to fiber size.

Despite these clear early statements of a tendency for large and small neurons at one stage in the visual pathways to connect with large and small neurons respectively at the next stage, the problem was not approached again until 1967 in the rat and 1971 in the cat (Chapter 2, Section 2.1.2).

Altman and Malis (1962) also found physiological evidence of a small fiber projection to cat SC, but perhaps the most remarkable evidence of functional correlates of conduction velocity to emerge during this period was the correlation reported by Maturana *et al.* (1960) between the receptive field properties of ganglion cells in frog retina and the conduction velocity of their axons. They noted that the cells with slowest-conducting (<0.5 m/sec) axons had small receptive fields with ON- or OFF-centers; cells with somewhat faster axons (2 m/sec) had ON-OFF receptive fields, while cells with the fastest axons (8 m/sec) had very large OFF-center receptive fields. Further, these workers concluded (see Section 1.2.2) that the different cell types perform "discrete and invariant" types of complex analysis on the visual image, a conclusion that would seem to reinforce the functional significance of the conduction velocity differences between them. They further noted a correlation between cell type and the level of termination of the cell's axon in the optic tectum. These seem precisely the sorts of correlates of conduction velocity that Bishop had anticipated in 1933, and it is remarkable that this finding had little impact on subsequent work on conduction velocity groupings in the optic nerve. For example, in none of the papers (e.g., Gouras, 1969; Fukada, 1971; Cleland *et.al.,* 1971; Hoffmann *et al.,* 1972; Stone and Hoffmann, 1972) that contributed to present understanding of the functional correlates of the axonal conduction velocity of ganglion cells in the cat or monkey is Maturana and co-workers' evidence of correlates in the frog even referred to. The reason for this may be that Maturana and co-workers themselves placed little emphasis on the finding. They note the correlation without illustrating their data, and they made the central point of their paper the use of a "naturalistic" approach to the study of ganglion cell receptive fields, by which they discovered the "natural invariants" of function. They characterized the "natural" operations of these cells in terms of their responses to visual stimuli and at one point in discussion they seem to suggest that properties of the cell that are not relevant to this operation are "accidental properties," whose further investigation would be of little value.

1.2. RECEPTIVE FIELD STUDIES OF RETINAL GANGLION CELLS

Hartline (1938) was the first to describe the receptive fields of single ganglion cells of a vertebrate retina. He recorded from single axons dissected from the axon bundle layer of the retina of the bullfrog and explored the retina with a spot of light, looking for regions of the retina from which the spot could modulate the action spike activity of the particular fiber under study. He reported two basic findings in this study: first, the range of response properties to be found among ganglion cells, and second, the limited region of the retina from which the activity of a cell could be modulated.

Hartline's understanding of the importance of the first of these findings can be sensed in his description of the different responses he saw in different cells.

> It is not until the bundles have been dissected down until one, or only a few fibres remain active that a new and striking property of the vertebrate optic response is revealed. For such experiments show conclusively that not all of the optic nerve fibres give the same kind of response to light. This diversity of response among fibers from closely adjacent regions of the same retina is extreme and unmistakable; it does not depend upon local conditions of stimulation or adaptation, but appears to be an inherent property of the individual ganglion cells themselves.

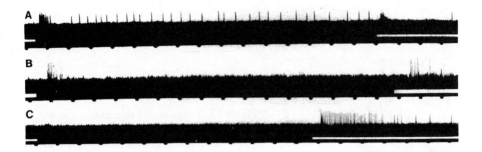

Figure 1.6. First records of ON-, OFF-, and ON-OFF-responses of single cells (from Hartline, 1938). The original legend reads:

"Oscillographic records of the action potentials in three single intraocular optic nerve fibers of the frog's eye, showing three characteristic response types.

A: Response to illumination of the retina consisting of an initial burst of impulses, followed by a maintained discharge lasting throughout illumination. There is no response to cessation of illumination in this fiber (the off response in this record is partly due to retinal potential, partly to another fiber which discharged several small impulses).

B: Response only to onset and cessation of light.

C: Response only to cessation of illumination.

The time is marked in ⅕ second, and the signal marking the period of illumination fills the white line immediately above the time marker."

[Reproduced with kind permission of the American Physiological Society.]

Hartline described cells that respond when the light spot turns on, others that respond when it turns off, and others that respond at both on and off (Fig. 1.6). He noted that the region of the retina to which a cell is sensitive is limited, and introduced the term *receptive field:*

> *Spatial effects:* No description of the optic responses in single fibres would be complete without a description of the region of the retina which must be illuminated in order to obtain a response in a given fiber. This region will be termed the receptive field of the fiber. The location of the receptive field of a given fiber is fixed; *its extent, however, depends upon the intensity and size of the spot of light used to explore it, and upon the condition of adaptation; these factors must, therefore, be specified in identifying it.* [italics added]

In two subsequent papers (1940a,b), Hartline showed that a ganglion cell's sensitivity to a light spot stimulus is not uniform over its receptive field, but is maximal near the center of the receptive field and decreases toward its edge. As a consequence, the size of receptive field depends on the intensity and size of the stimulus used to plot it (Fig. 1.7). He noted that over certain ranges of spot intensity and size, the intensity and area of a spot eliciting a threshold response from the cell, were interchangeable, and he also presented the first stimulus–response relationship for single ganglion cells. Further, he extended several of these observations to the alligator and turtle. Hartline did not publish again on vertebrate ganglion cells, and 13 years elapsed before the next studies of vertebrate receptive fields were published by Barlow (1953), who also studied the frog, and by Kuffler (1953), who studied the cat (which was, therefore, the first mammal to be so studied). These two studies can now be seen, together with Hartline's papers, as

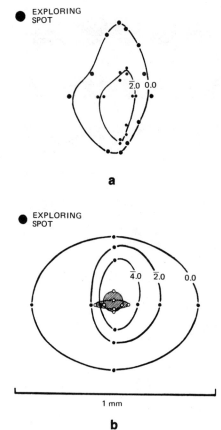

a

b

1 mm

Figure 1.7. First receptive field plots (from Hartline, 1940a). The original legend reads:

"Charts of the retinal region supplying single optic nerve fibers (eye of the frog).

a: Determination of the contours of the receptive field of a fiber at two levels of intensity of exploring spot. Dots mark positions at which exploring spot (50 micron diameter) would just elicit discharges of impulses, at the intensity whose logarithm is given on the respective curve (unit intensity = 20,000 meter candles). No responses at log I = −3.0, for any location of exploring spot. This fiber responded only at "on" and "off".

b: Contours (determined by four points on perpendicular diameters) of receptive field of a fiber, at three levels of intensity (value of log I given on respective contours). In this fiber steady illumination (log I = 0.0 and −2.0) produced a maintained discharge of impulses for locations of exploring spot within central shaded area; elsewhere discharge subsided in 1–2 seconds. No maintained discharge in response to intensities less than log I = −2.0; no responses at all to an intensity log I = −4.6."

[Reproduced with kind permission of the American Physiological Society.]

the start of a still-continuing series of studies on the receptive field properties of vertebrate retinal ganglion cells.

The technical contributions of Barlow's and Kuffler's studies were quite distinct, Barlow developing the area/threshold technique used by Hartline on the excised, exposed frog retina, Kuffler developing a technique for recording from the intact eye of a mammal and employing the animal's optics in the presentation of stimuli; but the principal developments they proposed in the understanding of receptive fields were very similar. Both showed that a ganglion cell's receptive field may be heterogeneous, comprising a center region and a concentric "surround" with very different properties. In the frog, the influence of this surround region was evident in the weakening of the cell's response to a spot stimulus when the spot was enlarged beyond the apparent borders of what Hartline had termed the *receptive field*, but subsequently has come to be termed the *receptive field center*. This inhibitory surround was present only in cells with ON-OFF-center regions. Some evidence of it was noted by Hartline (1940b); it was clearly described by Barlow (Fig. 1.8). Kuffler's (1953) description of cat ganglion cell receptive fields has for many years formed the starting point for any account of visual receptive

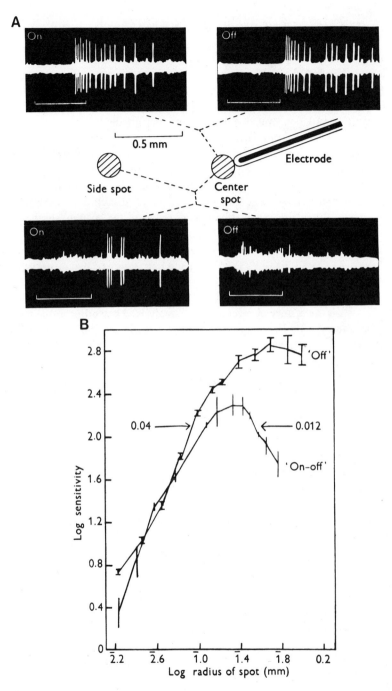

Figure 1.8. Evidence of surround component of ON-OFF receptive fields in the frog (from Barlow, 1953). (A) the ON-response (left) and OFF-response (right) of an ON-OFF-cell, elicited by a spot in the receptive field center (top traces), are inhibited by a side spot turned on and off simultaneously (lower traces). Time marks represent 0.2 sec. (B) Radius/sensitivity curves for an OFF-cell and an ON-OFF-cell in frog retina. As the radius of the stimulus spot increases, the cell becomes responsive to successively lower intensities of light. The "fall-off" in the curve for the ON-OFF unit at larger radii represents the influence of an inhibitory surround. [Reproduced with kind permission of the *Journal of Physiology (London)*.]

fields in mammals. He showed that, as in the frog, ganglion cells in the cat are most sensitive to stimuli at the center of their receptive field, sensitivity decreasing with distance from the center, over distances of 0.5 to 1.0 mm (2–4°). Kuffler showed further that receptive fields of most (in his data, all) cat ganglion cells comprise a center region where photic stimulation evokes either an ON- or OFF-discharge from the cell and a concentric region where stimulation elicits the opposite response (i.e., an OFF- or ON-discharge) (Fig. 1.9). The influences of these two receptive field components on the cell are antagonistic (Fig. 1.10) so that illumination of both regions by a large spot causes a weaker response than illumination of just the center region, by a small spot. That is, cat ganglion cells tend to be most sensitive to small visual stimuli, an observation that has been the starting point for many subsequent studies of the spatial selectivity of mammalian ganglion cells.

From these findings, the analysis of the receptive field properties of ganglion cells over the next 13 years (i.e., until 1966) seems to have followed two distinct lines, which in the following paragraphs are termed *parametric analysis* and *analysis in terms of feature extraction*. These two lines of analysis were suggested in Rowe and Stone's (1980a) essay on the history of receptive field analysis. They differ in their technology, terminology, and emphasis; most fundamentally, however, they differ in the underlying assumptions of the investigators pursuing them.

1.2.1. Parametric Analyses of Receptive Fields

One line of studies was pursued principally in the cat, the major papers including Kuffler *et al.* (1957), Barlow *et al.* (1958), Wiesel (1960), McIlwain (1964), Rodieck and Stone (1965a,b), and Rodieck (1965). These papers explored the parameters of the center/surround receptive field described by Kuffler (1953) without proposing that any particular properties were of primary significance. This was also true of earlier studies

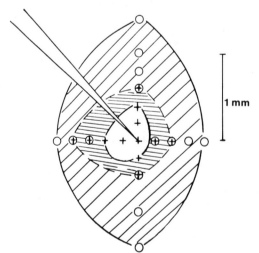

Figure 1.9. First plot of the receptive field of a mammalian ganglion cell (from Kuffler, 1953). The original legend reads:

"Distribution of discharge patterns within receptive field of ganglion cell (located at tip of electrode). Exploring spot was 0.2 mm in diameter, about 100 times threshold at centre of field. Background illumination approximately 25 m.c. In central region (crosses) "on" discharges were found, while in diagonally hatched part only "off" discharges occurred (circles). In intermediary zone (horizontally hatched) discharges were "on-off". Note that change in conditions of illumination (background, etc.) also altered discharge pattern distribution."

[Reproduced from the *Journal of Neurophysiology* with kind permission of the American Physiological Society.]

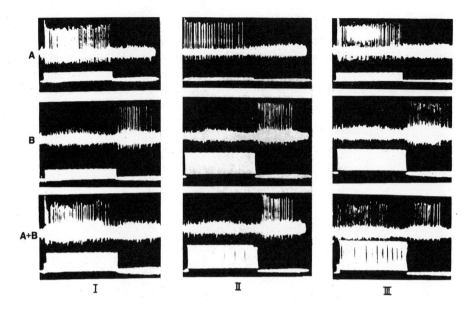

Figure 1.10. Center–surround antagonism in cat retina (from Kuffler, 1953). Interaction of two separate light spots. Spot A, 0.1 mm in radius, was placed in the center of the receptive field. Spot B, 0.2 mm in radius, was 0.6 mm away, in the surround. Flashed separately, they generated ON- (upper) and OFF- (middle) responses. When the two spots were flashed simultaneously, both responses were reduced (bottom). Flash duration was 0.33 sec, potentials were 0.3 mV. In II the intensity of spot B has been increased and in III the intensity of both spots has been increased. [Reproduced from the *Journal of Neurophysiology* with kind permission of the American Physiological Society.]

of Hartline, Barlow, and Kuffler so that the studies described here as parametric can be viewed as a steady continuation of the earlier approach. Kuffler *et al.* (1957) studied the maintained discharge of retinal ganglion cells in the absence of any localized visual stimulus. Barlow *et al.* (1958) presented evidence that the organization of these fields changes with dark adaptation, the surround influence diminishing as the retina adapts to low ambient illumination. Wiesel (1960) reported two major observations. First, he showed that the antagonistic effect of the surround varies from cell to cell, being usually weaker when the center region is large. Second, he showed that ganglion cells at the area centralis have markedly smaller receptive field centers than those in peripheral retina. This finding supported the idea that receptive field center size may be an important determinant of visual acuity. Hubel and Wiesel (1960) extended these observations to the retina of the monkey, reporting the same pattern of receptive field organization as in the cat, but noting the presence of color sensitivity in some cells, and also the generally smaller receptive fields of monkey ganglion cells. These properties match the high spatial resolution of which the monkey is capable, and the monkey's ability to discriminate color. Hubel and Wiesel (1961) extended these observations to relay cells of the LGN of the cat, showing that relay cells have the same pattern of receptive field organization as retinal ganglion cells, but with a general increase in surround antagonism.

Similarly, Rodieck and Stone (1965a,b) reported two principal findings. First, they showed that the responses of cat retinal ganglion cells to stationary and to moving visual stimuli could be regarded as the product of a single receptive field mechanism, with a "Mexican hat"-shaped sensitivity profile, as shown in Fig. 1.11. They presented evidence that, assuming (1) a constant time course for the cell's responses to a flashing

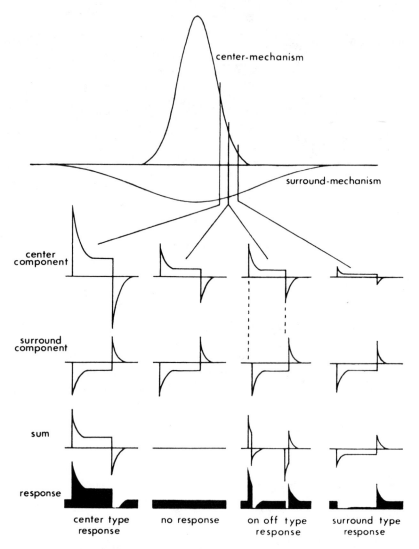

Figure 1.11. Model of cat ganglion cell receptive field (from Rodieck and Stone, 1965a). The model assumed separate center and surround mechanism with spatial distributions shown at the top. The model also assumed that a small stimulus spot elicited responses whose time courses were invariant with spot position, but whose amplitude varied following the functions shown at the top. A slight latency difference between center and surround influences was postulated to account for ON-OFF-responses. The model was used by Rodieck (1965) to develop a mathematical model of receptive field function. [Reproduced from the *Journal of Neurophysiology* with kind permission of the American Physiological Society.]

stationary spot stimulus, whatever its position in the receptive field, and (2) linear summation by the ganglion cell of influences reaching it from different parts of its receptive field, then the cell's responses to various moving stimuli were predictable from knowledge of its sensitivity profile. Rodieck (1965) gave this model formal mathematical treatment and explored much of its potential. Rodieck and Stone's second point is shown in Fig. 1.12. They argued that a population of ganglion cells distributed over an area of the retina can signal far more information to the brain than any one ganglion cell. The pattern of firing at *any given moment* in a distributed set of ganglion cells can signal the contrast, length, breadth, velocity, and direction of any moving stimulus. Of these parameters, a single ganglion call can signal only the contrast, breadth, and velocity of a stimulus (but not its length or direction), and then only if time is available for the cell's full response to the stimulus to be encoded (e.g., > 500 msec for a 0.5-deg-wide object moving a 1 deg/sec).

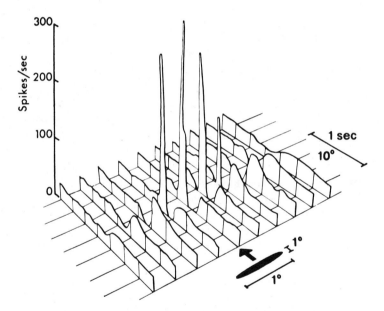

Figure 1.12. Visual coding by an array of ganglion cells (from Rodieck and Stone, 1965*b*). The original legend reads:

"Response of an off-center unit to different chordal movements of a 1° black disk. The disk moved through 20° (10° each side of the receptive field of the unit) with a velocity of 10°/sec. The responses to different paths are shown in perspective view. For simplicity of illustration the histograms are shown as smooth lines obtained by tracings from the plotted histograms. The arrow points to the histogram obtained after we had set the disk to move horizontally through the center of the receptive field. The next histogram to the upper right was obtained for the disk moving horizontally half a degree higher, etc. Time in each histogram is directed to the lower right. As discussed in the text, this plot may alternatively be viewed as a map of the local response, at a certain instant, of similar off-center units at corresponding positions in this region of the retina to the movement of the disk as shown. The angular distance between histograms is much larger than the equivalent angular distance along them in order that each can be seen. The disk illustrated is therefore drawn with a similar linear distortion."

[Reproduced from the *Journal of Neurophysiology* with kind permission of the American Physiological Society.]

McIlwain (1964) described a curious "periphery effect" in cat retinal ganglion cells, showing that the thresholds to visual stimuli of some cat retinal ganglion cells are lowered by stimulation of the retina at very considerable distances (8–10 mm) away. The effect has been extensively confirmed (e.g., Levick *et al.,* 1967) and shown to be particularly prominent in Y-class cells (Cleland *et al.,* 1971). It indicates the operation of a facilitatory mechanism that spreads widely across the retina and, interestingly, it requires a reexamination of what is meant by a receptive field. Hartline used the term *receptive field* to refer to "the area of retina which must be illuminated in order to obtain a response from a given fiber." Kuffler (1953) observed the surround components of cat ganglion cell receptive fields and "enlarged" Hartline's definition

> . . . to include all areas in *functional* connection with a ganglion cell. . . . Not only the areas from which responses can actually be set up by retinal illumination may be included . . . but also all areas which show a functional connection, by an inhibitory or excitatory effect on a ganglion cell. This may well involve areas which are somewhat remote from a ganglion cell and by themselves do not set up discharges.

By either definition, areas of the retina that give rise to McIlwain's periphery effect would be included in a receptive field, so that Y-class ganglion cells could be regarded as having receptive fields of very large size [up to 40° radius (McIlwain, 1966)]. Perhaps the most useful answer to the question "how big is the receptive field of a Y cell in cat retina?" is Hartline's original comment that the extent of a receptive field depends on the stimulus used to plot it. When a receptive field is plotted with a small (say 1° or smaller) flashing spot stimulus, the periphery effect is often undetectable and the receptive fields of Y cells have center regions of 1–2° in diameter and surround regions up to 6° in diameter. With stimuli better suited to elicit the periphery effect (a large moving object, for example) the receptive field of the cell is as much as 10 times wider. A receptive field is, therefore, the area of the retina (or visual space) whose illumination by a particular stimulus causes modulation of a cell's activity. The extent of a receptive field depends on the stimulus used to plot it.

All the above studies, with which must be included those of Wagner *et al.* 1960, 1963) on receptive fields of ganglion cells in goldfish retina, pursued the analysis of receptive fields in a piecemeal way, exploring the variability in receptive field properties along parameters such as dark/light adaptation, retinal topography, responses to moving stimuli, and others. Several studies considered the problem of how these cells might code visual information; as, for example, in the "array of ganglion cells" argument in Fig. 1.12, but none seems committed to viewing particular properties of the cells as of overriding importance. Enroth-Cugell and Robson's (1966) study, which began the development of the Y/X/W classification, was designed to extend this stream of analysis by measuring yet another parameter, the responses of individual ganglion cells to grating stimuli of threshold and suprathreshold contrast; the classification of ganglion cells into X and Y groups reported there was an unexpected observation shown only several years later to be of fundamental importance.

It could be argued that most of the studies of receptive fields just characterized as parametric do not belong in an account of ganglion cell classification since, between the pioneering studies of Hartline (1938, 1940*a,b*) and of Enroth-Cugell and Robson's

(1966) study, none of these studies proposed, or sought to propose, a classification of ganglion cells. They have been discussed because it seems clear that it was from this series of studies that the Y/X/W classification emerged; its emergence is traced in the following chapter (see especially Section 2.1).

1.2.2. Feature Extraction Analyses of Receptive Fields

The second major stream of work on ganglion cell receptive fields between 1953 and 1966 constituted an attempt to establish a new paradigm for the understanding of ganglion cells and, by using natural stimuli for activating the cells instead of artificial spots and moving geometrical figures, to discover the true operations of individual ganglion cells. The true operation performed by a cell could (it was argued) be characterized by some feature of the visual world to which the cell was uniquely sensitive; for example, a small object moving in a particular direction. This stimulus was called the cell's *trigger feature* (Barlow, 1961). The function of the cell was to detect the occurrence of that feature in a particular part of the animal's visual field (i.e., in the receptive field of the cell) and signal that occurrence to the brain. The work was done in the frog, pigeon, and rabbit, and has added greatly to knowledge about ganglion cell receptive fields.

In his study of the receptive fields of frog retinal ganglion cells, Barlow (1953) used the spot stimuli first used by Hartline (1938, 1940*a,b*), developing the area/threshold technique for plotting receptive fields and demonstrating the inhibitory surrounds of ON-OFF-center cells. Thus, Barlow's experimental approach in this paper seems parametric in the sense used above. In discussion, however, Barlow included the following passage that (as Hughes, 1971, has noted) seems to be the starting point for the idea that ganglion cells can be meaningfully classified in terms of their trigger features:

> . . . an optic nerve fibre is the final common path for activity aroused in a considerable region of the retina, and if some purposive integration has taken place, it should be possible to relate this to the visual behaviour of the frog. According to Yerkes (1903), the frog uses its eyes mainly in feeding; it also escapes from large moving objects. . . . When feeding, its attention is attracted by its prey, which it will approach, and finally strike at and swallow. Any small moving object will evoke this behaviour, and there is no indication of any form discrimination. In fact, 'on-off' units seem to possess the whole of the discriminatory mechanism needed to account for this rather simple behaviour . . . it is difficult to avoid the conclusion that the 'on-off' units are matched to this stimulus and act as 'fly-detectors.'

Barlow extended the idea to include "OFF-center" units, and concluded that "the retina is acting as a filter, rejecting unwanted formation and passing useful information."

The idea that different ganglion cell groups have discrete functional roles that can be determined by studying their receptive field properties was the central point of the reports by Maturana *et al.* (1960) and Lettvin *et al.* (1961) on the receptive fields of frog retinal ganglion cells. These workers described five "natural groups" among frog retinal ganglion cells, concluding that cells of each group perform a discrete and invariant analysis of the visual image and code a particular feature of the image whose occurrence it signaled to the brain. The operations were defined in terms of the stimulus most effective for activating the cell, and the cells were subsequently named after its particular operation. The five operations that Maturana *et al.*(1960) proposed were:

1. Sustained edge detection
2. Convex edge detection
3. Changing contrast detection
4. Dimming detection
5. Dark detection

Subsequently, Lettvin *et al.* (1961) named the cell groups as follows:

Group I: Boundary detectors
Group II: Movement gated, dark convex boundary detectors
Group III: Movement or changing contrast detectors
Group IV: Dimming detectors
Group V: "Unclassified" (this group was *not* labeled dark detectors)

An example of the analysis of one of these operations is shown in Fig. 1.13. Maturana *et al.* (1960) presented evidence that these different operations could be related to different conduction velocities of the ganglion cell axons concerned, to different laminae of termination of axons in the optic tectum, and to different morphologies of the retinal ganglion cells involved.

This approach attracted great interest, for example in the Symposium on Principles of Sensory Communication held in Boston in 1959 [W. Rosenblith (ed.), 1961]. Some years later, Bishop and Henry (1972) commented that

> Lettvin and his colleagues brought about a major revolution in visual neurophysiology. . . . It was a shift in attention from image space to object space. It is as though the investigators had turned their backs on the retina so as to put themselves . . . in the same place as the animal, and to see for themselves what the single neuron was reporting.

The possibility of solving the mysteries of sensory coding by the use of very simple experimental techniques combined with the imaginative use of the intellect, was accompanied by a considerable impatience with the use of "unnatural stimuli" such as had typically been used in parametric analyses of receptive fields.

> Spots of light are not natural stimuli for the frog in the way that a fly or worm is. . . . As the studies performed by four of the five classes are essentially independent of change in illumination, to study luminosity responses cannot contribute further insight into their natural functions and can only inform about what we would consider *accidental properties* of these cells. [italics added; Maturana *et al.,* 1960]

These workers therefore recommended that the investigator undertake nonquantitative (qualitative) experiments to identify those aspects of a cell's response which are its true operation and those which are only "accidental." When that true operation has been identified, meaningful quantitative questions can be asked about it, and answered. No suggestion was made, however, as to how the true operation could be reliably identified, except that natural stimuli should be tried in a nonquantitative way. Lettvin *et al.* (1961) argued the ineffectiveness of any more artificial approach, expressing the view

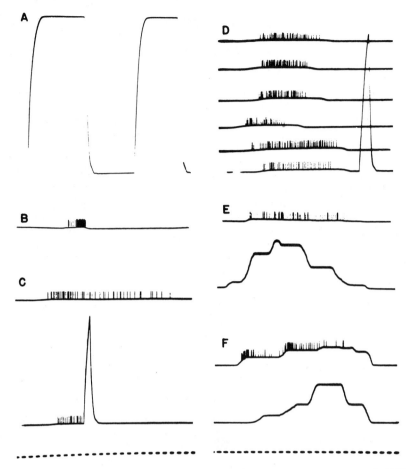

Figure 1.13. Analysis of a "convex boundary detector" in frog retina (from Maturana *et al.*, 1960). The original legend reads:

"Class 2. Convex edge detectors. Single-fibre recording from the tectum (shaped spikes). A photomultiplier monitored the same sweep in which the spikes are registered; an upward deflection of the base line indicates a reduction of illumination.

A: OFF and ON of the general illumination; no response.

B: Burst of activity in response to a small dark moving object (1 degree in diameter).

C: Upper trace, sustained response to the same object stopped in the RRF. Lower trace, the sustained response was elicited by the small object stopped in the RRF but was erased by a transient darkening of the general visual field.

D: Invariance of the response to movement under changes of illumination. The small object was moved slowly through the RRF shortly after the level of illumination was set. The records are not aligned because the movement did not start at the same instant after the beginning of the sweep. Most of the differences in the response are due to slight changes in the speed and path followed by the moving object. Lower trace, bright light. Upper trace, dim light (100:1 ratio).

E: Absence of response to a straight edge. Upper trace, response to the small object (disc, 1 degree in diameter) moved in steps through the RRF. Lower trace, absence of response to a dark band 7 degrees × 20 degrees moved long-edge first. The greater upward deflection of the second trace indicates the greater darkening of the visual field that it produced, as compared with the small object.

F: Response to a corner. Upper trace, response to a corner of the dark band (7 degrees × 20 degrees) moved across the RRF. Lower trace, absence of response to the straight edge. The darkening produced by the corner is almost as large as that produced by the band. Time marks, 5/second."

[Reproduced with kind permission of the Rockefeller University Press.]

pungently as they introduced two very "complex" types of neurons in frog optic tectum ("sameness" neurons and "newness" neurons):

> The descriptions are provisional and may be too naturalistic in character. However, we have examined well over a hundred cells and suspect that what they do will not seem any simpler or less startling with further study. . . . Of course, if one were to perform the standard gestures, such as flashing a light at the eye, probably the cells could be classified and described more easily.

A similar emphasis on the "true" operation performed by a ganglion cell can be sensed in the metaphor used by Barlow *et al.* (1964) at the end of their important report of speed and direction-selectivity among retinal ganglion cells in the rabbit (Fig. 1.14).

> It would be ridiculous to analyse the sound output of a motor car without being aware that cars are . . . a means of transport. Would it be any less absurd to investigate the spectral sensitivity of a retinal unit without realizing that it signalled direction of motion? Different classes of unit convey (different) information . . . and this fact must be taken into account when planning the observations that are to be made on them.

Indeed, a general feature of the feature extraction approach is that a distinction is always drawn between the feature considered important [variously called a *natural invariant* (Maturana *et al.,* 1960), or *trigger feature* (Barlow, 1961; Barlow *et al.,* 1964), or *key feature* (Levick, 1975)] and features of the cell considered unimportant, *accidental* (Maturana *et al.,* 1960) or *secondary* (Hughes, 1979). Arguably, this distinction is fundamental to the feature extraction approach and was developed so that experimental emphasis could be placed on the important feature of the cells. It is nevertheless true that the assumption that the true operation of a cell can be determined readily and with certainty places the feature extraction analysis at odds methodologically with the parametric approach. In the latter approach, this assumption is avoided and the classifications developed from parametric studies were based on a wide range of the cells' properties. A considerable debate now exists in the literature on the relative merits of these different approaches to cell classification (Levick, 1975; Rowe and Stone, 1977, 1979, 1980*a,b;* Hughes, 1979), and the reader is referred to these papers for a full account. Briefly, the issues of the debate include:

Terminology: Workers in the parametric approach have used nondescriptive (usually alphanumeric) terms to name the groups of cells distinguished, to avoid the implication that certain features of the cell are of particular importance; arguably, this was necessary for the use of a hypothetico-deductive methodology (Rowe and Stone, 1977, 1979). The feature extraction classifications have generally used descriptive terminologies, naming the groups of cells after the feature thought to determine their functional role.

Testability and rigor: It has been argued that parametric classifications are testable, in that the properties of cell groups and their proposed functions are independently characterized and are related to each other by hypothesis; whereas the typologoical thinking and reliance on receptive field properties to define function involved in the feature extraction approach make the functional significance of groupings in such classifications untestable. Conversely, it has been argued that unambiguous definitions of cell groups, such as are employed in feature extraction analyses, are essential if a clearly defined and rigorous classification is to be established.

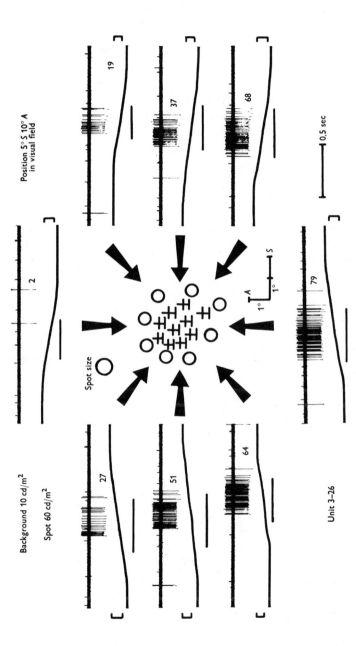

Figure 1.14. Directionally selective properties of a rabbit ganglion cell (from Barlow *et al.*, 1964). The original legend reads:

"Responses of a directionally selective unit (axon recording) to motion in different directions. Map of receptive field in centre. Each pair of records shows (lower trace) movement of a spot of light right through the receptive field in the direction of the adjacent arrow, and (upper trace) the response elicited. Conventions as follows: ±, response to stationary spot at both on and off; O, no response; there were no responses outside the ring of Os. Anterior (A) and superior (S) meridians in the visual field are shown together with 1° calibration marks. All records read from left to right. In upper trace, electrode positivity is downward; the number of spikes is shown immediately after each response. For lower trace, vertical calibration bar shows 5° displacement of light spot; horizontal bar indicates approximately when the spot was within the receptive field."

[Reproduced with kind permission of the *Journal of Physiology (London)*.]

Explanatory power: It has been argued that the heuristic power of the parametric approach is provided by the testability of such classifications; that only by testing can incorrect elements of the classification be identified and better formulations substituted. Conversely, it has been argued that the development of clear definitions allows the ready application of feature extraction classifications to other parts of the visual pathway and to other species, without the uncertainty and ambiguity common in the parametric approach.

It is fairly easy, in fact, to characterize most modern studies of ganglion cell classification as either parametric or feature extraction in approach; this point was developed in some detail by Rowe and Stone (1980*a*). Much of the variety of classification schemes and names currently applied to ganglion cells may be attributed to the differences between scientists in their methodology of classification.

1.3. MORPHOLOGICAL CLASSIFICATIONS OF GANGLION CELLS

Up to 1966, classifications of vertebrate ganglion cells on the basis of their morphology had been proposed by Cajal (1893) for a wide variety of vertebrates, by Polyak (1941) for the monkey, and by Brown (1965) for the rat.

Cajal's descriptions include the retinas of fish, amphibians, reptiles, birds, and mammals, and are based on Golgi-impregnated sections. His observations include many features of ganglion cell morphology, but his classification turned particularly on the sublamina of the inner plexiform layer in which the cells' dendrites spread (Fig. 1.15). Thus, he described ganglion cells of the first, or second, or third, etc. sublamina, diffuse ganglion cells (whose dendrites spread through all laminae), and bistratified ganglion cells. However, dendritic lamination is proving an important feature of ganglion cell morphology (Famiglietti and Kolb, 1976; Nelson *et al.,* 1978). Cajal does not propose a correlation between these features of a cell and other potentially important features of its morphology (such as the number or spread of its dendrites, the size of its soma, its connections with the brain, or the thickness of its axon).

Cajal's descriptions were drawn upon by Lettvin *et al.* (1961) to propose a morphological basis for the five types of ganglion cells they distinguished physiologically. Pomeranz and Chung (1970) provided support for this correlation reporting that one physiological type (class 1, edge detectors) and one morphological type (cells with a single densely branching dendrite) are both absent in the tadpole.

Polyak's (1941, 1957) classification of primate retinal ganglion cells rested principally on the shape of the cell's dendritic tree (hence "parasol" or "shrub" ganglion cells), or on its size ("giant" and "midget" cells) (Fig. 1.16). In the case of one cell class, the "midget" ganglion cell, Polyak correlated the size of the cell with its synaptic connections and retinal distribution. He showed that a midget ganglion cell has a single dendrite that contacts one cone bipolar cell and that cells with these characteristics are most highly developed and numerous near the fovea. All these features suggest that this cell group is of great importance in high-resolution vision. Perhaps as a result of these several correlations, the midget ganglion cell group is still widely used.

Brown (1965) examined ganglion cells in whole mounts of rat retina stained *in*

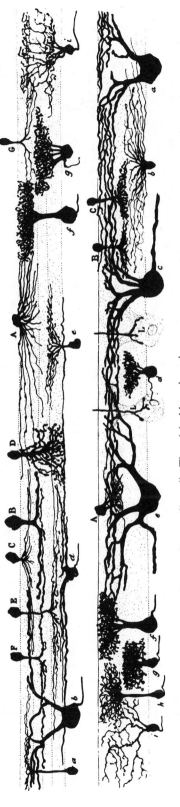

Figure 1.15. Cajal's (1893) classification of mammalian ganglion cells. The original legends read:

Upper: "amacrine cells and ganglion cells from the dog retina. A, stellate amacrine cell destined for the first sublayer; B, giant amacrine cell of the third sublayer; G, C, stellate amacrine cells destined for the second sublayer; E, amacrine cell destined for the third sublayer; F, small amacrine cell destined for the fourth sublayer; *a*, ganglion cell whose upper cluster spreads in the second sublayer; *b*, giant ganglion cell destined for the fifth sublayer; *e*, small ganglion cell whose cluster spreads in the fourth sublayer; *f*, middle-sized ganglion cell which arborizes in the first and in a portion of the second sublayers; *g*, ganglion cell which arborizes in the third and in a portion of the fourth sublayers; *i*, two-layered cell ["cellule bistratifée"]."

Lower: "Ganglion cells from the dog retina: *a*, giant ganglion cell whose cluster spreads in the first and a portion of the second sublayer; *b*, small ganglion cell whose multiple processes disappear in the fifth sublayer; *c*, giant cell whose cluster seems to spread mainly in the second sublayer; *d*, *g*, small ganglion cells with clusters in the fourth sublayer; *e*, giant ganglion cell of the second sublayer; *h*, another ganglion cell destined for the second and partially for the first sublayer; *f*, middle-sized ganglion cell destined for the fourth sublayer; *i*, unstratified ganglion cell. A, B, C, spongioblasts. L, lower terminal arborization of a bipolar cell." [Reproduced from *La Cellule*, Vol. 9; translation by Thorpe and Glickstein (1972).]

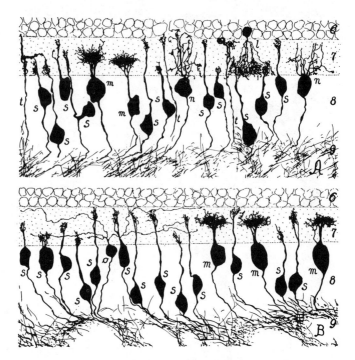

Figure 1.16. Polyak's (1941, 1957) classification of primate ganglion cells. The original legend reads:
"Types of nerve cells, mostly ganglions, from central area of the retina of a chimpanzee, stained with the Golgi method. Labelling: *l* so-called amacrine cell that in this instance does not possess an axon fiber; *m*, parasol ganglions; *n*, shrub ganglions; *o*, small ganglion with long, loose dendritic branches; *s*, midget ganglions; *t*, terminations of nerve fibers in inner plexiform layer, probably of extraretinal origin (so-called exogenous or centrifugal fibers). Note generally reduced dimensions of all varieties in comparison with the same varieties found in extra-areal periphery of retina; characteristic compact appearance of the treetops in *m*, and the loose appearance of same in *n* and *o* varieties; the minute dimensions and delicate structure of the treetops in the *s* variety located either close to the inner nuclear layer (*6*) or close to the ganglion layer (*8*); descent of all axis cylinders, the actual "optic nerve fibres", down to the fiber layer of retina (*9*), where they pass over to optic disc, eventually to form the optic nerve."

The relationship between Polyak's descriptions and more recent classifications is taken up in Chapter 3 (Section 3.1.4). Many of the ganglion cells classified here as "parasol" (*m*), shrub (*n*), or midget (*s*) may comprise the X-like cells of monkey retina. The small cell with loose dendrites (*o*) may represent the W-like system. [Reproduced from *The Vertebrate Visual System* by S. Polyak with permission of the University of Chicago Press. Copyright 1957 by the University of Chicago.]

vivo with methylene blue. He distinguished two groups of cell, "loose" and "tight." The terms refer to the density of branching of the cells' dendrites (Fig. 3.11). The "loose" cells had somewhat larger dendritic fields (mean 400 μm as against 300 μm) and slightly larger cell bodies. The "loose" cells ramified more superficially in the inner plexiform layer, perhaps receiving connections from a different set of bipolar cell terminals. Brown suggested that the loose and tight cells may correspond to two classes of receptive field (those with and those without surround components, respectively) described by Brown and Rojas (1965). Bunt (1976) has subsequently provided evidence from Golgi-stained material of greater variety in the morphology of rat ganglion cells.

Despite the obvious importance of the morphological features on which these workers based their classifications, it did not prove possible to draw compelling correlations between them and physiologically based classifications of the same cells. The most widely used morphological classification of ganglion cells [the $\alpha/\beta/\gamma$ grouping proposed by Boycott and Wassle (1974) for cat ganglion cells] does allow such a correlation to be drawn. That classification is based not on particular features of the cells, but on correlations between several features (soma size, dendritic branching, axon size) and, importantly, on a strong correlation with the physiologically based Y/X/W classification of the same cells (see Chapter 2, Sections 2.1.4 and 2.3.6). It is still not a comprehensive classification; as discussed in Section 2.3.3, it continues to develop and change as new evidence accumulates. Its success stems, I believe, from the correlations it establishes between widely different properties of the ganglion cells concerned.

1.4. FUNCTION OR PHYLOGENY AS A BASIS FOR GANGLION CELL CLASSIFICATIONS?

Finally in this chapter I would return to two questions raised at its beginning. First, why did G. H. Bishop change his hypothesis about the significance of conduction velocity groupings, from his suggestion in 1933 that they are related to functional roles performed by different ganglion cell groups, to his suggestion in 1959 that they relate to the phylogenetic history of the cell groups concerned, and not to their function? The reason was fairly simple; in the intervening 26 years, no evidence had accumulated at all of function/velocity correlations for the optic nerve. Thus, Bishop noted in 1959 that

> In the somaesthetic system . . . all the fiber size components have retained sensory correlates, with the possible exception of the C fiber system of man.

There was some evidence of duplication of function in different fiber groups with, for example, touch being mediated by both β and γ groups, the larger-caliber β system being capable of better spatial discrimination. He suggests that

> The more recent duplication of primitive functions within a given sensory modality accomplishes an advance in the functional competence of the whole apparatus.

In vision, he notes,

> Even if such visual functions as perception of form, color and frequency . . . are thought of as separate modalities, these visual functions . . . do not differ so far as is known with respect to sizes of fibers serving them.

He concludes that

> Although fiber groups corresponding to and presumably analogous to those of the somaesthetic system all serve one function, visual, the relation is rather obviously to the phylogeny of the visual system rather than its functional differentiation into modalities.

Bishop's pessimism was, we now know, premature; striking correlates between cell function and axonal conduction velocity were about to emerge. Indeed, Lettvin and co-workers' (1961) paper is part of the proceedings of a symposium held in 1959, and these workers tested conduction velocity correlates of receptive field properties explicitly because of Bishop's early report. Nevertheless, the second question raised earlier remains: why did it take as long as it did (30–40 years) for function/velocity correlates to be established in the visual system, after their clear enunciation for the somatosensory system in the 1920s and 1930s?

In retrospect, five factors seem to have contributed to the delay. First, physiologists of vision were largely inattentive to the problem. In the 20 years between G. H. Bishop's 1933 report and the report of P. O. Bishop *et al.* (1953), there were only three reports of physiological studies of the optic nerve or tract, all from one laboratory (Bishop and O'Leary, 1938, 1940, 1942). Second, once receptive field and conduction velocity studies of ganglion cells picked up momentum in the 1950s, the studies were done largely by separate groups of workers, who were inattentive to each other's work (the present writer was for a long time among the inattentive). Third, the feature extraction analysis of receptive fields tended to focus attention on a cell's trigger features and distract attention from the analysis of any "accidental" property of the cell, such as the conduction velocity of its axon. Fourth, the optic nerve is less accessible than somatosensory nerves; and fifth, there was for a long time no basis in visual psychophysics or in receptor morphology for expecting functional subgroupings within the optic nerve. The distinctions between pain, temperature, touch, vibratory and proprioceptive sensation are part of common experience and formed a ready starting point for the physiological analysis of somatic sensation. It is only recently that the psychophysicists of vision have provided evidence, for example, of distinct spatial and temporal "channels" in the visual system (Chapter 11). On this point, visual physiologists and anatomists may have been a little ahead of the psychophysicists. Again, distinct receptors were described for the submodalities and separate pathways of somatic sensation, but the rod/cone distinction among visual receptors bears no apparent relation to the Y/X/W classification of ganglion cells. These are submodalities developed from the same receptors.

An important conclusion seems to follow from the last paragraph. It is that the tempo at which ganglion cell classification has developed has depended on the readiness of investigators to draw together widely different properties of the cells; to seek correlations between receptive field properties, morphology, retinal distribution, central projections, and visual behavior. Bishop's (1959) paper made an important contribution by raising another major consideration, the phylogenetic history of cell populations, and his particular suggestion of phylogenetic correlates of conduction velocity seems likely to prove both right and wrong. Much of the data base on which he relied has subsequently proved incorrect or incomplete, and we have seen striking evidence accumulate of correlations between a cell's receptive field properties, retinal distribution, dendritic morphology, and central projections and its axonal conduction velocity and caliber, just as Bishop had originally (1933) anticipated. Yet, it is hard to deny that the phylogenetic history of retinal ganglion cells is likely to prove an important factor in the understanding of their properties. From the present vantage point, it seems simply unnecessary to consider phylogeny and function as alternative primary correlates of fiber caliber. Since (it is commonly held) evolution occurs in response to environmental pressure, i.e., because an animal that develops certain functional capabilities survives better, then any pathway that develops in phylogeny must presumably serve a valuable new function, or serve an old function better. Any classification of ganglion cells should, and no doubt will eventually have to, take into account both the functional properties of a cell class and its phylogenetic history. In Sections 2.3 and 2.4 of Chapter 2, developments in ganglion cell classification are proposed that attempt to do just that.

The Y/X/W Classification of Cat Retinal Ganglion Cells

2

Since 1966, our understanding of ganglion cells in cat retina has been transformed by the development of classifications of those cells into groups that are distinct in their receptive field physiology, morphology, axonal caliber and conduction velocity, relative numbers, retinal distribution, central projection, and, perhaps, in their phylogenetic history. Evidence has gathered that the different ganglion cell classes subserve substantially different functional roles within the visual system (Chapter 11), determine much of the circuitry of central visual nuclei (Chapters 6, 7, and 8), and are closely related to the topographical specializations of the retina (Chapter 9).

Some degree of controversy has accompanied the development of these classifications. The controversy has not concerned whether groupings exist, or whether they are important; rather it has concerned the methodology of classification and, as a consequence, the number of groups considered to be present, the names given to them, and the functional significance attributed to them. In practice, current schemes of ganglion cell classification seem readily identified (in terms discussed in Chapter 1, Section 1.2) as "parametric" (such as the Y/X/W and $\alpha/\beta/\gamma$ schemes), as "feature extraction" classifications (such as the brisk/sluggish/sustained/transient/concentric/nonconcentric scheme), or as amalgams of the two. Examples of such amalgams are Caldwell and Daw's (1978) Y/X/sluggish classification of rabbit ganglion cells and Rodieck's (1979) Y/X/phasic/tonic/suppressed-by-contrast/direction-selective/local edge detector/ color-coding classification of cat ganglion cells. Much of this chapter is presented using the Y/X/W terminology. This choice is not meant to imply that there is, or even should be, general agreement concerning classification and terminology. Clearly, however, an investigator or reviewer has to make some choice, and perhaps the best way to make that choice is in terms of the underlying methodology of classification involved. Grounds for the present choice are discussed in Rowe and Stone (1977, 1979) and in Chapters 4 and 5.

The Y/X/W classification has grown out of the parametric studies of retinal receptive fields discussed in Chapter 1, and a number of papers particularly important to its emergence were published in 1966. Principal among these is the report of Enroth-Cugell and Robson, which described the distinction between X and Y cells, and also provided their names. Other relevant studies published that year included that of Fukada and co-workers, which was one of the first consciously to seek correlates of conduction velocity in the visual responses of cat ganglion cells; and the report of this writer and Fabian, which provided early descriptions of the receptive field properties of the class presently termed *W cells*. [The very earliest description of a cell class now included in the W-cell group in fact appears in a footnote to the report of Rodieck and Stone (1965a); Rodieck (1967) soon after published a fuller description of these cells, the "suppressed-by-contrast" cells.]

In retrospect, four principal steps in the development of the Y/X/W Classification can be recognized:

1. The description of the X/Y difference (Enroth-Cugell and Robson, 1966).
2. The establishment of the conduction velocity correlates of the X/Y difference (Fukada, 1971; Cleland *et al.*, 1971).
3. The description of the W-cell group.
4. The description of corresponding morphological classes of ganglion cells (Boycott and Wässle, 1974).

These four steps are reviewed in Section 2.1, and in Section 2.2 a formal exposition of the Y/X/W classification is set out in terms of categories and taxa. Sections 2.3 and 2.4 pursue these matters further. Since Y, X and W cells were first described, their properties have been intensively explored and, this later experimentation has had a considerable impact on the original classification. That impact is traced in Section 2.3, which also concerns the different interpretations to which the variety of ganglion cell properties have been subject. Section 2.4 follows some implications of the range of ganglion cell properties now recognized, proposing a new development in ganglion cell classification, and Section 2.5 summarizes some of the problems of classification that emerge in Sections 2.2 and 2.4.

2.1. THE DEVELOPMENT OF THE Y/X/W CLASSIFICATION

2.1.1. Description of the X/Y Difference

Enroth-Cugell and Robson (1966) set out to measure a contrast-sensitivity function for individual ganglion cells in cat retina. They employed an oscilloscope-based visual display with which they could generate grating patterns of alternating bright and dark bars. The intensities of the bars could be varied, so that the contrast between light and dark bars could be high or low. The width of the bars could also be varied (i.e., the spatial frequency of the grating could be changed). The grating pattern could also be made to drift across the oscilloscope screen at a controllable speed. Enroth-Cugell and Robson chose the grating form of stimulus because it had been used in psychophysical studies of human vision, for example to characterize the human contrast-sensitivity function. In brief, the human ability to detect the presence of a grating stimulus depends on the spatial frequency of the grating (very fine gratings becoming indistinguishable from a uniform background even if of high contrast), and on the contrast between white and dark bars (the greater the contrast, the finer the grating that can be resolved). A plot of the minimum contrast needed for the detection of a grating against the spatial frequency of the grating is known as a contrast-sensitivity function (see Chapter 11, Section 11.3.1 and Fig. 11.4).

Enroth-Cugell and Robson succeeded in obtaining contrast-sensitivity functions for individual ganglion cells (Fig. 2.1). They placed the oscilloscope screen so that the cell's receptive field was approximately at its center, and then caused the grating pattern to drift across the receptive field at a constant velocity. The actual velocity was varied with the spatial frequency of the grating, so that, at any particular spatial frequency, four black–white cycles of the grating crossed the receptive field every second. Typically, the grating would evoke a discharge in an ON-center cell each time a light bar crossed the center of the receptive field, and from an OFF-center cell each time a dark bar crossed the center. The cell's discharge was therefore modulated in synchrony with the cycle-by-cycle passage of the grating, and the modulation could be reduced to zero by either reducing the contrast or increasing the spatial frequency of the grating. The cell then fired at a rate and in a pattern indistinguishable from its firing when the screen was quite uniform in luminance; the cell could not "see" the grating. For each cell examined, Enroth-Cugell and Robson plotted the minimum contrast required to cause modulation of the cell's firing, as a function of the spatial frequency of the grating. This plot represents the ability of the cell to resolve grating from background, and seems clearly analogous to the psychophysically determined human function discussed above.

A striking and unexpected finding emerged, however. Enroth-Cugell and Robson reported that this contrast-sensitivity function could be obtained in a simple way for only a proportion of the cells encountered; in their data, for only a minority of cells (approximately 25%). The majority of cells behaved in a "nonlinear" way that made it impos-

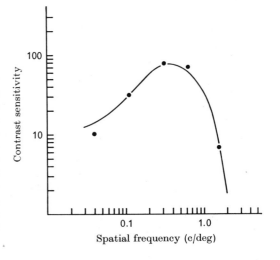

Figure 2.1. Contrast-sensitivity function for an X cell. The function is based on the responses of the cell to sinusoidal gratings drifting at 4 Hz. The ordinate represents reciprocals of the contrast required to evoke a response. For the closed circles, the criterion response was objectively determined as 10 spikes/sec. For the continuous line, the criterion was judged subjectively as barely audible when played over a loudspeaker. [From Enroth-Cugell and Robson (1966). Reproduced with kind permission of the *Journal of Physiology (London)*.]

sible to plot a threshold contrast/spatial frequency function for them without an additional assumption. In these cells, when the spatial frequency of the grating was reduced, the modulated pattern of firing caused by lower-frequency gratings was replaced by an *unmodulated increase* in firing rate. Two contrast-sensitivity functions could, in fact, be obtained for such cells, one for the modulated response and another, extending to higher spatial frequencies, for the unmodulated response (Fig. 2.2). The ganglion cells encountered fell unambiguously into the "linear" and "nonlinear" subgroups that, after further testing with other stimuli, were termed X and Y *cells,* respectively.*

Enroth-Cugell and Robson made several important observations on the two cell groups. To test the idea that the unmodulated firing of Y cells to fine gratings was due to some nonlinearity in the cells' behavior, they presented cells of both classes with a stationary grating stimulus, which could be turned on and off. (More accurately, the grating was presented and withdrawn, i.e., the contrast between black and white bars was reduced to zero by bringing both to the mean level of illumination of the grating pattern.) When a bright bar of the grating was centered on an OFF-center receptive field (whether of an X or Y cell), it generated a discharge from the cell when the grating was withdrawn, and an inhibition of firing when it was presented (Fig. 2.3, top row of histograms). The grating generated a similar pattern of discharge, but shifted 180 deg in phase, when a dark bar was centered on an OFF-center receptive field (Fig. 2.3, third-row of histograms). When the border between bright and dark bars was centered on the receptive field of an X cell (ON-center or OFF-center), the cell gave no response (Fig. 2.3, left histogram in second and bottom rows); this, it was argued, indicated that the X cell is "linear." The two halves of its receptive field were receiving opposite stimuli (i.e., one half was brightened and one half darkened); the net change in luminance across the receptive field was zero, since the two halves of the field were equally and oppositely affected. The cell showed "linearity" by summing the influences reaching it from the different parts of its receptive field to zero. This interpretation has been substantially

*Dr. Enroth-Cugell relates that during their experiments, the two groups of cells were called I (interesting) and D (dull). These designations referred to the fact that only for I cells (later, X cells) could the contrast-sensitivity measurements, which were the original object of the work, be made in a straightforward way.

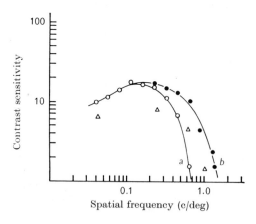

Figure 2.2. Contrast-sensitivity functions for a Y cell. The open circles represent the function obtained when, as for X cells, the response being judged was the modulated response of the cell to a drifting grating. The closed circles represent values obtained when the criterion response was a response, modulated or unmodulated, to the grating. The cell's unmodulated response is the more sensitive at higher spatial frequencies (> 0.6 cycle/deg). [From Enroth-Cugell and Robson (1966). Reproduced with kind permission of the *Journal of Physiology (London)*.]

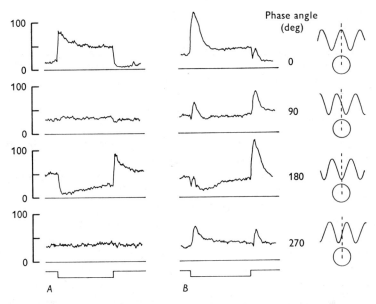

Figure 2.3. The linearity test of Enroth-Cugell and Robson (1966). The original caption reads:

"Responses of an off-centre X-cell (*A*) and an off-centre Y-cell (*B*) to the introduction and withdrawal of a stationary sinusoidal grating pattern. The contrast (0.32) was turned on and off at 0.45 Hz. Downward deflexion of the lowest trace in both *A* and *B* indicates withdrawal of the pattern (contrast turned off), upward deflexion indicates introduction of the pattern (contrast turned on). The upper line in each pair is the pulse density of the ganglion cell discharge (scale at left: pulses/sec); the length of the zero line represents a duration of 2 sec. The 'phase angle of the pattern', i.e. the angular position (in degrees) of the (cosine) grating relative to the mid point of the receptive field centre, is given at the right of the figure and is illustrated by the sketches. *A*: X-cell (no.84); spatial frequency 0.13 c/deg. *B*: Y-cell (no. 13); spatial frequency 0.16 c/deg."

Compare the two histograms in the second row; the left-hand histogram shows almost no response while the right-hand histogram shows two peaks. The same difference is apparent between the bottom two histograms. The two-peaked response obtained from the Y cell is the evidence of its nonlinearity. [Reproduced with kind permission of the *Journal of Physiology (London)*.]

confirmed, although nonlinearities are now recognized as commonly present in X cells (see Section 2.3.3.3). With Y cells, on the other hand, no such "null position" for the stimulus could be found; even when two halves of the field were stimulated with symmetrically opposite stimuli, the cell always responded to the beginning and end of the stimulus. The cell summed the influences reaching it from different parts of its receptive field "nonlinearly."

Enroth-Cugell and Robson noted two other important differences between X and Y cells, and guessed a third. They noted, first, that the receptive field centers of Y cells are generally larger than those of X cells; and second, that X cells are encountered more frequently (relative to Y cells) near the area centralis of cat retina. These two findings have been widely confirmed and have been interpreted (see below) as evidence that X cells subserve high-resolution pattern vision, while Y cells subserve movement vision or perhaps low-resolution pattern vision. Further, Enroth-Cugell and Robson guessed that

Y cells had larger-caliber axons than X cells, but did not test the point. It has, of course, been confirmed in several reports that Y-cell axons do comprise the stoutest and, therefore, the fastest-conducting group of optic nerve axons, but 5 years elapsed (1966–1971) before these reports appeared; indeed, before any report appeared that looked further at the X/Y classification of cat retinal ganglion cells.

2.1.2. Conduction Velocity Correlates of X and Y Cells

In 1966, Fukada and co-workers reported experiments to measure the flicker fusion frequency of individual ganglion cells (in cat retina) and to relate that measure to the axonal conduction velocity; ganglion cells with faster-conducting axons tended to be able to follow flickering light stimuli to higher rates than cells with slow-conducting axons. The finding has been modified and expanded subsequently (Fukada and Saito, 1971). Fukada and co-workers' study was one of the first to seek to relate the visual responses of cat retinal ganglion cells to the conduction velocities of their axons.

Indeed, it is remarkable, considering the intensity of subsequent work on the cat, that correlates between axonal conduction velocity and the receptive field properties of ganglion cells were first clearly established for two other species, the frog (Maturana *et al.*, 1960) and the monkey (Gouras, 1969). Maturana and co-workers noted (see Chapter 1, Section 1.2.2) that the different classes of ganglion cells they distinguished in frog retina had different conduction velocities. However, because they defined those groups in terms of the coding operations the cells performed, as indicated by certain of their receptive field properties, and suggested that other properties are accidental, their paper seemed to dismiss rather than bring out any possible significance of this correlation. In 1969, Gouras distinguished two groups of ganglion cells in monkey retina that, in retrospect, seem in many ways akin to X and Y cells. Cells of one group were color-coding (see Chapter 3, Section 3.1.2) and gave tonic responses to flashing spot stimuli; cells of the other group were less sensitive to color and gave phasic responses to flashing stimuli. Gouras showed that the conduction velocities of the axons of the two groups were quite distinct, the phasic cells having fast-conducting axons, and the tonic cells having slower-conducting axons (Fig. 2.4). Gouras' finding has since been confirmed and expanded in the monkey (see Chapter 3, Section 3.1); it was an important step in the development of ganglion cell classification.

In 1971, Fukada (from Tokyo) and Cleland and co-workers (from Canberra) independently reported correlations between receptive field properties and axonal conduction velocities of cat retinal ganglion cells, and both suggested that Y cells have faster axons than X cells. Fukada's (1971) paper followed a paper by Saito *et al.* (1970) that distinguished two groups of cat retinal ganglion cells: cells of one group (type I) gave phasic responses to stationary flashing stimuli, cells of the other group (type II) gave tonic responses. These response differences are shown in Fig. 2.5 (the legend to that figure describes the use of the terms *tonic* and *phasic*). The type I/type II distinction initially depended on the tonic/phasic criterion, but in 1971, Fukada added several observations, noting that type I cells tended to have faster axons (Fig. 2.6A) and larger receptive fields than type II cells. He commented that type I and II cells may correspond, respectively, to Y and X cells.

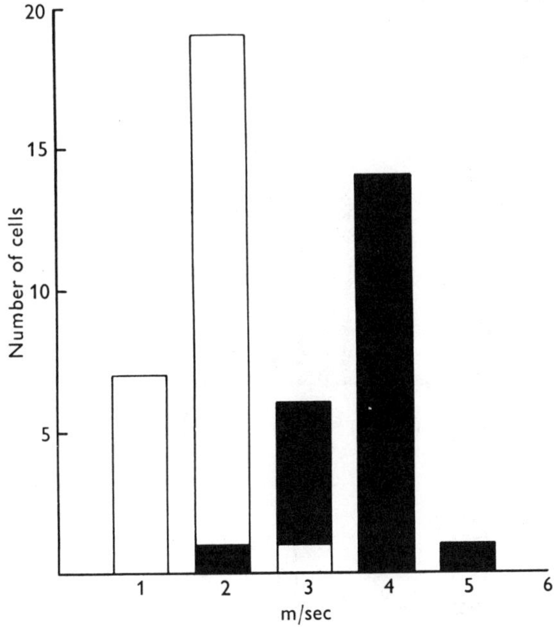

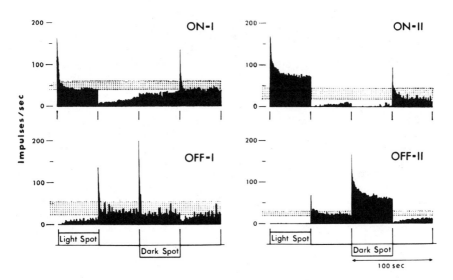

Figure 2.4. Conduction velocity differences between ganglion cell classes in the monkey. Gouras's (1969) evidence of a conduction velocity difference between "phasic" and "tonic" classes of monkey retinal ganglion cells. These frequency/conduction velocity histograms show that phasic cells (closed blocks) have on the average faster axonal conduction velocities than tonic cells (open blocks). [Reproduced with kind permission of the *Journal of Physiology (London)*.]

Figure 2.5. "Tonic" and "phasic" types of cat ganglion cells. Saito and co-workers' (1970) evidence for two types (I and II) of cat retinal ganglion cells, type II giving tonic responses to stationary flashing stimuli, type I giving phasic responses. The original caption reads:

"Response patterns of four types of units to light and dark spot stimuli shown in the form of PST histogram. . . . Dotted range in each pattern shows the background discharge rate of the unit. Spot size: 1° in diameter for the on-center units, 4° in diameter for the off-center units."

Note that for ON-II and OFF-II cells, the firing rate of the unit stays above the background discharge rate (represented by the hatching) while the appropriate stimulus (light spot for ON-II, dark spot for OFF-II) is maintained. Thus, their response is tonic. For ON-I and OFF-I cells, the firing rate falls back to the background rate even while the stimulus is maintained.

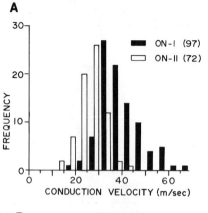

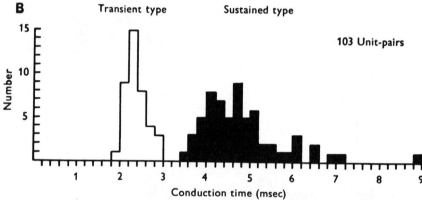

Figure 2.6. Conduction velocity correlates of cat X and Y cells. (A) Fukada's (1971) evidence of a conduction velocity difference between type II (X) and type I (Y) cells. This frequency/conduction velocity histogram shows that type I cells have on the average higher axonal conduction velocities than type II cells. The ordinate shows the numbers of cells encountered at each velocity; the abscissa shows conduction velocity in meters per second. [Redrawn from the original and published with kind permission of *Vision Research.*] (B) Cleland and co-workers' (1971) measurement of retinogeniculate conduction times for pairs of retinal ganglion cells and geniculate relay cells, for "transient" (Y) and "sustained" (X) cells. The data are shown as a frequency/conduction time histogram. Conduction times are 2–3 msec for Y cells and 3.5 msec for more X cells. The longest conduction times may have been recorded from W cells, whose presence was recognized only subsequently. [Reproduced with kind permission of the *Journal of Physiology (London).*]

Cleland *et al.* (1971) were interested in the transformation of visual information that occurs in the LGN of the thalamus (that part of the thalamus specialized to relay retinal information to the visual cortex). To study this transformation, they recorded simultaneously from a relay cell in the LGN and from the retinal ganglion cell or cells that provided its excitatory drive. They saw the distinction between X- and Y-class ganglion cells (which they termed *sustained* and *transient* cells, respectively) and drew two important conclusions. First, each relay cell receives excitatory drive from either X or Y cells, but usually not from both. Second, the time taken for the action spike to reach its relay cell in the LGN, after leaving the soma of its ganglion cell, is about twice as long for an X cell as for a Y cell (typically 4–6 msec as against 2–3 msec, Fig. 2.6B). Their calculations suggested that the axons of Y cells comprise the fast (30–40 m/sec) t_1 group described by Bishop *et al.* (1953) and Bishop and MacLeod (1954), while the X cells comprise the t_2 (17–23 m/sec) group. These were striking findings that, with the earlier work of Noda and Iwama (1967) on the rat (see Chapter 3), and the independent studies of the LGN reported by Stone and Hoffmann (1971), Fukada and Saito (1972), and Hoffmann *et al.* (1972), led to intensive study of the LGN in terms of Y, X, and (subsequently) W cells (see Chapter 6).

Both Fukada (1971) and Cleland *et al.* (1971) included several other observations on parameters of X and Y cells; Fukada, for example, showed an "adapting" response in X cells but not Y cells, while Cleland *et al.* (1971) noted that McIlwain's periphery effect is strong in most Y cells, but weak or absent in X cells, and also noted the particular responsiveness of Y cells to fast-moving stimuli. These observations began a long and continuing series of studies of the properties of X and Y cells, summarized in Section 2.3.3.

2.1.3. The W-Cell Grouping

Scattered among a number of reports of cat retinal ganglion cells between 1965 and 1971 are mentions of receptive fields quite distinct from those described by Kuffler, and from the X- and Y-cell receptive fields described by Enroth-Cugell and Robson (1966). The earliest is a footnote to the report of Rodieck and Stone (1965*a*). Subsequently, Stone and Fabian (1966) and Rodieck (1967) described small numbers of cells with very distinct receptive fields (Fig. 2.7). Barlow and Levick (1969) described an infrequently encountered "luminance unit" (Fig. 2.8). Fukada (1971) mentioned a small number of units that could not be classified as type I or II and noted that their axons were relatively slow-conducting; and Cleland *et al.* (1971) described one cell with a distinct receptive field organization. Also mentioned in conduction velocity studies, particularly those of Bishop and O'Leary (1942), Bishop and Clare (1955), Spehlmann (1967), and Bishop *et al.* (1969), is evidence of axons with conduction velocities slower than those of the t_2 axons (see Chapter 1, Section 1.1). Indeed, Bishop *et al.* (1969) suggested that such fibers might form up to 60% of the total number of axons in cat optic nerve.

These two findings ("different" receptive fields and slow-conducting axons) were brought together by Stone and Hoffmann (1972), whose report stemmed from experiments that were originally planned to be yet another parametric exploration of Y and X cells. We were concerned to study variations in the properties of these two cell groups as a function of their retinal location. Gradations in their properties along this parameter have been described subsequently and are reviewed in Section 2.3.3, but we were distracted from that study by the appearance in our recordings of cells with two distinctive features: their receptive fields differed from those of X and Y cells, and their axons were slower-conducting than those of X cells. The receptive field "types" we saw included fields with ON-OFF-centers [Fig. 2.9, first described in the cat by Stone and Fabian (1966)] and suppressed-by-contrast receptive fields [first described by Rodieck (1967)], both of which are obviously distinct from those of X and Y cells (Fig. 2.10). Also

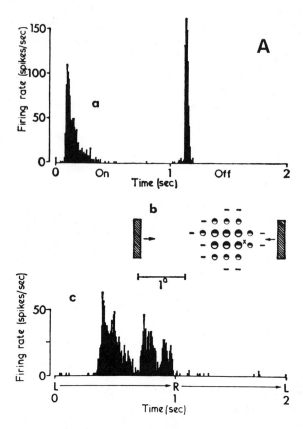

Figure 2.7. Early evidence of W cells. (A) Stone and Fabian's (1966) description of a direction-selective cell with an ON-OFF receptive field from cat retina. (a) The response of the cell to a small (¼-deg diameter) spot flashing on and off in the receptive field center. (b) A plot of the receptive field, the half filled circles showing positions at which an ON-OFF-response was obtained. The dashes represent an absence of response. (c) The response of the cell to a bar of light [represented in (b)] moved across the field. The cell responded only to movement from left-to-right. [Reproduced from *Science*, Vol. 152, with kind permission of the American Association for the Advancement of Science; copyright 1966 by the American Association for the Advancement of Science.] (B) Rodieck's (1967) description of a suppressed-by-contrast cell in cat retina.

described by Stone and Fabian (1966), Fukada (1971), and Stone and Hoffman (1972) were receptive fields that had the antagonistic center-surround organization first described by Kuffler (1953), yet still seemed quite distinct from X and Y cells. Stone and Hoffman (1972), Stone and Fukuda (1974*a,b*); Fukuda and Stone (1974), and Cleland and Levick (1974*a,b*), showed that these also have slower-than-t_2 axons, and the many observations reported in these papers are incorporated into Section 2.3.3.

The idea that the ganglion cells with these "different" receptive fields and slow-conducting axons had small cell bodies was first suggested by Stone and Fabian (1966), then again by Fukada (1971) and Stone and Hoffmann (1972). The latter workers also noted Bishop and co-workers' (1969) conclusion that the axons of these cells project to the SC of the midbrain [this was first suggested by Bishop and O'Leary (1942)], and

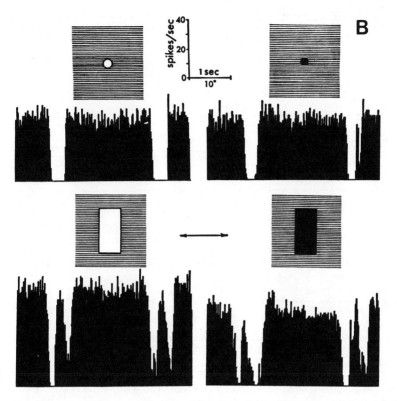

The upper two histograms show averaged response histograms of the cell's activity as light and dark spots (2-deg diameter) moved across the receptive field. The cell responds with a drop in firing rate to both directions of movement. The characteristic to which Rodieck drew attention is that the cell's firing is suppressed by both dark and light spots. Similarly, (lower histograms) both light and dark bars cause suppression of firing, which seems particularly marked when their edges cross the receptive field. [Reproduced from *Science*, Vol. 157, with kind permission of the American Association for the Advancement of Science; copyright 1967 by the American Association for the Advancement of Science.]

not to the diencephalon. Stone and Hoffmann suggested that these cells, despite their various receptive field properties, had sufficient properties in common to be regarded as a single functional group, which they termed *W cells*.

The term *W cell* was chosen by Rodieck [while preparing his major book (Rodieck, 1973)], so that the alphabetical sequence W/X/Y followed the increasing axonal velocities of the three cell groups; it was then adopted by Stone and Hoffman (1972) and Stone and Fukuda (1974*a,b*).

2.1.4. Morphological Classes of Cat Ganglion Cells

Brown and Major (1966) described a bimodality in the dendritic field sizes of cat ganglion cells, but did not propose a formal classification. Leicester and Stone (1967) and Shkolnik-Yarros (1971) proposed groupings of cat ganglion cells based principally

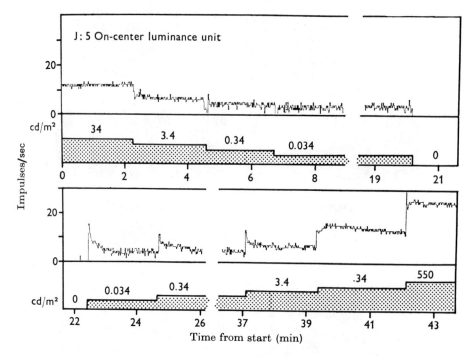

Figure 2.8. Barlow and Levick's (1969) evidence for "luminance units," i.e., ganglion cells in cat retina that seemed well suited to code ambient luminance. The original caption reads:

"Maintained discharge of a luminance unit at different adaptation levels. Impulses were counted for 1 sec periods in successive channels of the analyser. . . . Notice the regular decline with decreasing luminance and the return to nearly the same values when the luminance was increased."

[Reproduced with kind permission of the *Journal of Physiology (London)*.]

on dendritic morphology. The most successful morphological classification has, however, been that proposed by Boycott and Wässle (1974), who distinguished three principal cell groups. Termed α, β, and γ cells (and illustrated in Fig. 2.11), these cell groups were distinct in soma size, dendritic morphology, and axonal caliber. Boycott and Wassle suggested that α cells correspond to Y cells, β cells to X cells, and γ cells to W cells. They noted considerable variety within the γ-cell grouping, and suggested the possibility of a distinct cell group, the δ cell. Their suggestion of a general correlation between α and Y cells, between β and X cells, and between γ and W cells has been extensively confirmed and developed.

Since Boycott and Wässle's report, several studies have appeared that develop the classification they suggest in two principal ways. First, Famiglietti and Kolb (1976) and Nelson *et al.* (1978) have presented evidence that ON-center and OFF-center ganglion cells differ morphologically in that their dendrites spread in different sublaminae of the inner plexiform layer of the retina. Thus, two subgroups might be distinguished among each of the α, β, and γ classes, corresponding to the ON-center and OFF-center varieties of receptive fields found among Y, X, and W cells. Indeed, Kolb (1979) distinguishes

Figure 2.9. Responses of an ON-OFF-center W cell to visual stimuli. The visual stimuli used are represented in the diagram at the right of each response trace. Under each response trace is a second trace that shows the timing of the stimulus; it moved up when the luminance at the center of the receptive field increased, down when it decreased. (A–C) When a light spot was flashed on and off at each of a number of positions within the receptive field center, the cell responded with a phasic burst of firing at both the onset and the end of the flash. (D) A large flashing spot did not change the cell's firing rate from the near-zero rate observed without any localized stimulus. This is presumably because illumination of the region around the receptive field center inhibits both ON- and OFF-responses; apparently the cell has an inhibitory surround. (E, F) The cell

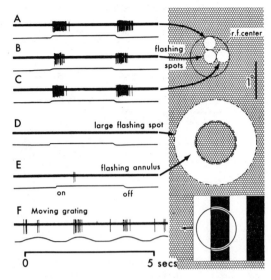

shows no clear response to an annulus that illuminates just the surround region of the receptive field (E); and only a weak response to a coarse grating (F).

These results suggest that this cell has an ON-OFF-center region and a purely inhibitory surround mechanism. [From Stone and Fukuda (1974a). Reproduced from the *Journal of Neurophysiology* with kind permission of the American Physiological Society.]

three classes of ganglion cells (I, II, and III) that correspond closely to α, β, and γ cells, respectively, and within each class two subgroups, a and b (see Section 2.3.3 and Fig. 2.18). Second, a group of cells has been recognized that resembles β cells in soma size, but differs from them in dendritic processes, resembling γ cells. Early evidence of these cells came from Rowe and Dreher's (1979, 1982b) study of ganglion cells labeled from forebrain sites identified by physiological work as receiving W-cell input. The correlation between W and γ cells suggested by Boycott and Wässle led to the prediction that many small-bodied ganglion cells would be filled by injections of horseradish peroxidase (HRP) into the medial interlaminar nucleus (MIN), deep C laminae, and ventral (vLGN) components of the LGN. In fact, few small and many medium-sized somas were labeled, and Rowe and Dreher suggested that many forebrain-projecting W cells have medium-sized somas. Their suggestion has received support from three subsequent studies. Stone and Clarke (1980) described, from Golgi-impregnated retinal whole mounts like those that Boycott and Wässle had used, the existence of many cells with medium-sized somas typical of β cells, yet widely branching dendritic fields more characteristic of γ cells (Figs. 2.12 and 2.13A). Second, Kawamura *et al.* (1979) and Leventhal *et al.* (1979, 1980) reported that HRP injections into the retinal-recipient zone of the pulvinar labeled medium-sized somas in the ganglion cell layer of the retina. Moreover, Leventhal *et al.* (1980) obtained HRP labeling of the cells' dendrites, showing the dendrites to be wide-spreading and very different from those of β cells (Fig.

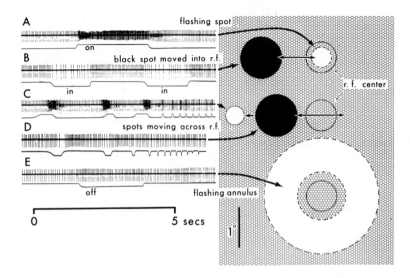

Figure 2.10. Response to visual stimuli of a tonic ON-center receptive field. (A) The cell responded to the presentation of a light spot in its center region with a sustained burst of firing, with little inhibition occurring at the end of the stimulus. (B) The cell's firing rate was reduced by a black spot moved into the receptive field center, but showed little excitation (typical of an ON-center X cell) when the spot left the receptive field. (C, D) The cell responded well to a slowly moving light (C) or dark (D) spot, with an increase and decrease in firing rate, respectively. However, the responses seem clear only with slow stimulus movements, indicated by the relatively long-lasting deflections toward the left-hand end of these traces. (E) A flashing annulus produced a weak increase in firing rate when it turned off, and decrease when it turned on; i.e., the receptive field had an OFF-surround.

In general, the cell lacked the transient component generally found in X cells; cells of this group also have larger receptive fields, slower-conducting axons, and distinct central projections from those of X cells. [From Stone and Fukuda (1974a). Reproduced from the *Journal of Neurophysiology* with kind permission of the American Physiological Society.]

2.13B). Third, Leventhal, Rodieck, and Dreher (personal communication) have traced the central projections of these cells, to the pulvinar, MIN, vLGN, and SC. They showed further that, as Stone *et al.* (1980) had predicted, these cells have axons as thin as the small-soma γ cells, despite their relatively large cell bodies.*

Recently, Kolb *et al.* (1981) suggested a quite radical development of the $\alpha/\beta/\gamma$ classification. From observations on Golgi-impregnated whole mounts of cat retina, they

---

*When classifying these cells, Stone and Clarke (1980), noting that the medium-soma cells concerned resemble small-soma γ cells in their dendritic morphology, receptive field properties, in some of their central projections, in the caliber and velocity of their axons, and in showing little systematic variation in their properties with retinal eccentricity, considered them a subgroup of the previously recognized γ-cell class. Leventhal and co-workers, on the other hand, noting that the cells at issue project to the pulvinar, MIN, and perhaps the C laminae of the LGN quite distinctly from small-soma γ cells, classified them as a separate group, ϵ cells. In the sense discussed in Chapter 5 (Section 5.2), this designation, though handy, seems typological in effect: it distinguishes the cells too sharply from the small-soma γ cells. They

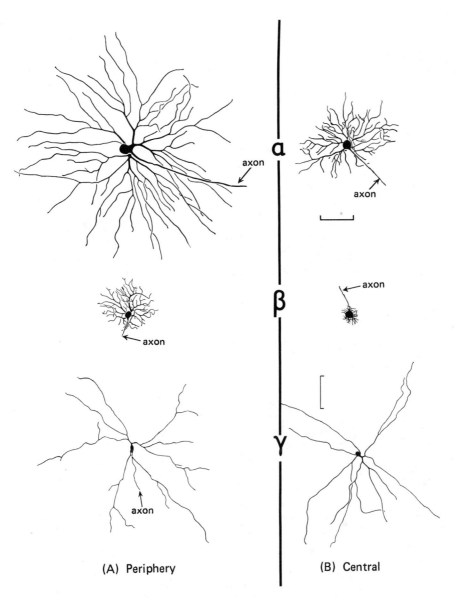

Figure 2.11. Morphological classes of cat ganglion cells: Boycott and Wässle's (1974) drawings of α, β, and γ cells, arranged to group central or peripheral cells together. (A) α, β, and γ cells from peripheral retina; (B) α, β, and γ cells from near the area centralis. Boycott and Wässle draw attention to the similarity between the peripheral β and central α classes. Note also that the central α and β cells are much smaller than their counterparts in peripheral retina, but that the central γ cell differs little in size from the peripheral γ cell. [Reproduced with kind permission of the *Journal of Physiology (London)*.]

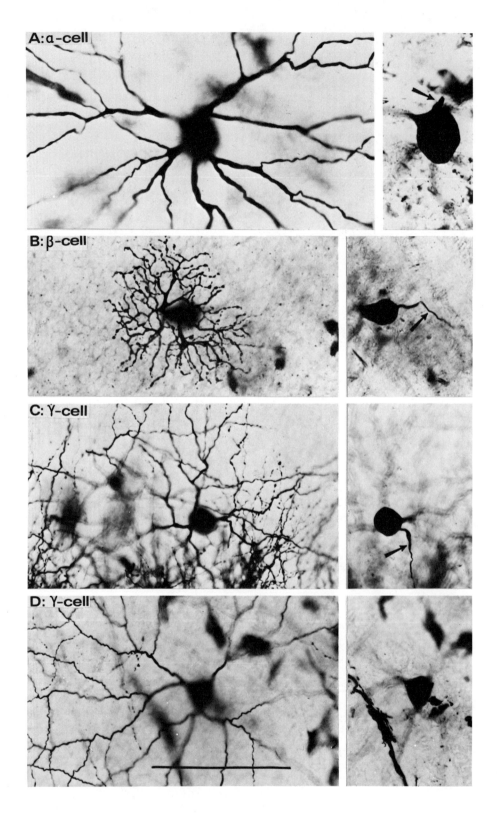

conclude that 23 different types of ganglion cells can be recognized. Two of the classes were the α and β classes of Boycott and Wässle (1974), and were given the same names. These workers suggest that one of the remaining 21 types (termed *G3*) corresponds to Boycott and Wässle's γ cells, and they also recognize the δ cell described briefly by Boycott and Wässle. The other 19 classes are thus being proposed for the first time.

Kolb *et al.*'s (1981) report provides valuable descriptions of previously unrecognized variation in ganglion cell morphology, but I suspect that the difference between their 23-group classification and the $\alpha/\beta/\gamma$ grouping stems more from the approach to classification taken by different workers than from the variety of cellular structures observed. That is, some workers decide to define entirely new cell types on the basis of one or a few cell properties, and therefore conclude that the overall population comprises many "types" with little variation between cells of one type; while other workers choose to propose a smaller number of groupings, each encompassing somewhat greater variation in properties. The issues involved in such choices are discussed in Sections 2.3 and 2.5 and in Chapter 5 (Section 5.2), where I argue that, following the experience of zoologists, the latter approach is the more powerful for the understanding of neurons. These issues have been discussed previously (Tyner, 1975; Rowe and Stone, 1977, 1979, 1980*b*).

Further detail of the morphology of cat ganglion cells is included in Section 2.3.3.6, in the context of an analysis of morphological variation. Two points can be made to summarize this more historical section. First, the recognition of the morphological parameters of ganglion cell groupings gave considerable impetus to the study of these groupings and of their implications for central visual pathways. Second, the morphological classification, like the physiologically based W/X/Y classification, is not static, but still developing as new evidence accumulates. Since the physiological and morphological classifications are based on physical properties of the same cells, many authors are beginning to consider them interchangeable ways of naming and grouping the same cells. It seems likely that before long a single classification of these cells will be adopted that is based on both physiological and morphological parameters. In Table 2.1, for example, the properties of Y, X, and W cells are shown as including the α, β, or γ morphologies.

2.2. THE Y/X/W CLASSIFICATION: CATEGORIES AND TAXA

This section comprises a formal statement of the Y/X/W classification. As discussed previously (Rowe and Stone, 1977, 1979), the classification is formulated as a set

---

Figure 2.12. Further variation in morphology of cat ganglion cells. Photomicrographs from Stone and Clarke (1980) illustrating medium-soma γ cells. (A) The dendritic field of a typical α cell is shown at the left, and its soma at the right. (B) Dendritic field and soma of a typical β cell, as found in peripheral retina. (C, D) Somas and dendrites of two cells with medium-sized somas (20–25 μm in diameter, a size typical of a β-cell soma), but with dendritic trees more typical of the γ cells of Boycott and Wässle (1974). [Reproduced with kind permission of the *Journal of Comparative Neurology*.]

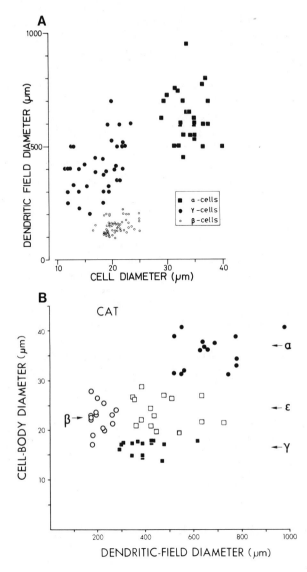

Figure 2.13. Graphical evidence of a medium-soma γ-cell group. (A) From the Golgi study of Stone and Clarke (1980), a plot of the dendritic field diameter of cat retinal ganglion cells against the diameter of their somas. Note that many cells represented as γ cells (closed circles) have somas as large as β cells (open circles), but much larger dendritic fields. These are the medium-soma γ cells illustrated in Figs. 2.12C and D. (B) Similar data from the work of Leventhal, Rodieck, and Dreher (personal communication). These workers also observed the cells termed *medium-soma γ cells* by Stone and Clarke (1980), naming them ε *cells*.

Note that abscissa and ordinate are interchanged between the two graphs. The agreement between the two sets of data is close. Leventhal and co-workers' data indicate that medium-soma γ cells project principally to forebrain sites (the RRZ of the pulvinar, the deep C laminae of the dLGN, the vLGN), and Rowe and Dreher (1979, 1982*b*) have shown that they also project to the MIN. Medium-soma γ cells appear to form the major W-cell input to the forebrain. Some medium-soma γ cells project to the pretectum and SC of the midbrain, along with small-soma γ cells, but they form only a minor part of the W-cell projection to the midbrain.

of testable hypotheses. Many recent studies have tested these hypotheses, the result being that several developments have already been incorporated into the classification and others such as those discussed in Sections 2.3 and 2.4 seem necessary. To ensure the testability of the classification and also relate it to the function of the visual system, it seems necessary for the classification to comprise both taxa (the groups of cells being delineated) and categories (the functional roles being performed by different cell groups). The

Table 2.1. Some Properties of Cat Retinal Ganglion Cells[a]

	Y cells	X cells	W cells
Receptive field center size (1)	Large, 0.5–2.5 deg	Small, 0.15–1.0 deg	Large, 0.4–2.5 deg
Linearity of center–surround summation (2)	Nonlinear	Linear or weakly nonlinear	Not tested
Periphery effect	Present (3)	Usually absent (3)	Absent (4)
Axonal velocity	Fast, 30–40 m/sec (5)	Slow, 15–23 m/sec (5)	Very slow, 2–18 m/sec (6)
Soma size, peripheral retina	Large, $>$ 22 μ diameter	Medium, 14–22 μ	Small–medium, 8–24 μ (7)
Proportion of population (8)	$<$ 10%	$\sim$ 40%	50–55%
Retinal distribution	Concentrate around area centralis, more numerous relatively in peripheral retina (10)	Concentrate at area centralis (9)	Concentrate at area centralis and in streak (10)
Central projections	To laminae A, A1 (11) and C (12) of LGN, to MIN (13) and via branching axon to SC (14). From A-laminae of LGN to cortical areas 17 and 18 (15), also by branching axon; and from MIN to areas 17, 18 and perhaps 19 (16)	To laminae A, A1 (11), and C (12) of LGN; thence to area 17 (15); to midbrain (a minority), but probably not to SC (17)	To SC (18), to C laminae of LGN and cortical areas 17, perhaps 18 (19), and strongly to 19 (20); to vLGN (21), MIN (22), and pulvinar (23)
Responses to standing contrast (25–28)	Phasic (transient) in most cells (25), some are tonic (sustained) especially near area centralis; all tonic when dark-adapted (27)	Most give tonic responses in mesopic conditions (26), many are transient when light-adapted (27)	Either tonic or phasic (28)
Nasotemporal division (24)	Nasal cells project contralat.; most temporal cells ipsilat.; strip of intermingling centered just temporal to area centralis	Nasal cells project contralaterally, temporal cells project ipsilaterally; narrow strip of intermingling centered on area centralis	Nasal cells project contralaterally; most temporal cells also project contralaterally; about 40% of temporal cells project ipsilaterally
Receptive field "layout" (29)	ON-center/OFF-surround or OFF-center/ON-surround	As for Y cells	Some have same layout as Y and X cells; others have ON-OFF-centers, some have purely inhibitory centers, are directionally selective or color-coded

(continued)

Table 2.1. Some Properties of Cat Retinal Ganglion Cells[a] (Continued)

	Y cells	X cells	W cells
Morphological correlates (30)	α cells, with a and b subgroups related to ON-center/OFF-center difference (31)	β cells, also with a and b subgroups (31)	γ cells, including small-soma γ cells and δ cells (30), also medium-soma γ (32) or ϵ (33) cells; a and b subgroups probable (31)

[a]The following are references relevant to the bracketed numbers in the table. I have tried to include all the pioneering studies in each area, but the list is not comprehensive for all studies that have examined each property.

(1) Cleland and Levick (1974a), Enroth-Cugell and Robson (1966), Stone and Fukuda (1974a).

(2) Enroth-Cugell and Robson (1966), Derrington et al. (1979).

(3) Cleland et al. (1971), Cleland and Levick (1974a).

(4) Cleland and Levick (1974a), Stone and Fukuda (1974a).

(5) Bishop et al. (1953), Cleland et al. (1971), Fukuda (1971).

(6) Bishop et al. (1969), Stone and Hoffmann (1972), Stone and Fukuda (1974a), Cleland and Levick (1974a).

(7) Boycott and Wässle (1974), Cleland and Levick (1974a), Cleland et al. (1973, 1975), Fukuda and Stone (1974, 1976), Stone and Fukuda (1974b), Levick (1975), Stone and Clarke (1980), Leventhal et al. (1980, 1981).

(8) Cleland and Levick (1974b), Fukuda and Stone (1974), Rowe and Stone (1976a).

(9) Cleland et al. (1973), Enroth-Cugell and Robson (1966), Fukuda and Stone (1974), Stone (1978).

(10) Fukuda and Stone (1974), Rowe and Stone (1976a), Stone and Keens (1980).

(11) Cleland et al. (1971, 1976), Fukada and Saito (1972), Hoffmann et al. (1972), Wilson et al. (1976), Leventhal, Rodieck, and Dreher (personal communication).

(12) Cleland et al. (1971), Hoffman et al. (1972), Cleland et al. (1976), Wilson et al. (1976).

(13) Dreher and Sefton (1975, 1979), Mason (1975).

(14) Hayashi et al. (1967), Hoffmann (1973), Kelly and Gilbert (1975), Wässle and Illing (1980).

(15) Stone and Dreher (1973), Dreher et al. (1980).

(16) Burrows and Hayhow (1971), Garey and Powell (1967), Stone and Dreher (1973), Ferster and LeVay (1978), Leventhal (1979), Kimura et al. (1980), Dreher et al. (1980).

(17) Cleland and Levick (1974a), Fukuda and Stone (1974), Leventhal et al. (personal communication) but see Wassle and Illing (1980).

(18) Hoffmann (1973).

(19) Cleland et al. (1976), Wilson et al. (1976), Holländer and Vanegas (1977).

(20) Maciewicz (1975), LeVay and Gilbert (1976), Holländer and Vanegas (1977), Kimura et al. (1980), Dreher et al. (1980).

(21) Spear et al. (1977), Leventhal et al. (personal communication).

(22) Dreher and Sefton (1975, 1979), Fukuda and Stone (1974), Leventhal et al. (personal communication) but see Wässle and Illing (1980).

(23) Dreher and Sefton (1975, 1979), Rowe and Dreher (1979, 1982b).

(24) Stone and Fukuda (1974b), Kirk et al. (1976a,b).

(25) Cleland et al. (1971), Fukada (1971).

(26) Cleland et al. (1973), Enroth-Cugell and Shapley (1973), Jakiela et al. (1976).

(27) Jakiela et al. (1976).

(28) Cleland and Levick (1974a,b), Stone and Fukuda (1974a).

(29) Cleland et al. (1971), Cleland and Levick (1974a,b), Enroth-Cugell and Robson (1966), Stone and Fukuda (1974a).

(30) Boycott and Wässle (1974).

(31) Famiglietti and Kolb (1976), Nelson et al. (1978), Kolb (1979).

(32) Stone and Clarke (1980).

(33) Leventhal et al. (1980, 1981).

classification then embodies at least two sets of hypotheses: first, that the categories proposed reflect real subdivisions of visual function, and second, that the taxa suggested subserve those functional roles (Bock, 1974; Rowe and Stone, 1977, 1979).

In the case of cat retinal ganglion cells, the following three categories are suggested: *high-resolution pattern vision, movement vision,* and *ambient vision.* That is, it is proposed that (at least) three relatively independent functions can be differentiated within mammalian vision. The following three taxa (groups of cells) are suggested among cat retinal ganglion cells: *Y cells, X cells,* and *W cells.* Finally, it is suggested that Y cells subserve movement vision, that X cells subserve high-resolution pattern vision, and that W cells subserve ambient vision.

2.2.1. Evidence for Choice of Categories

The categories of movement detection and high-resolution pattern vision are drawn from recent studies of the psychophysics of human vision, discussed in Chapter 11 (Section 11.3.1). These studies present evidence that within conscious visual perception, two independent mechanisms can be detected, one whose stimulation gives rise to the perception of movement (or of flicker of a stationary pattern), and a second whose stimulation gives rise to the perception of patterns. Several workers have suggested that the former function may be subserved by Y-class ganglion cells and the latter function by X cells.

The category "ambient vision" is drawn from studies of the visual capacity which remains in humans, monkeys, and cats after destruction of the areas of cerebral cortex to which retinal X and Y cells project. In the cat, for example, the activities of X and Y cells are relayed through the dLGN to cortical areas 17 and 18 (Chapter 6), yet destruction of these two areas has little effect on the animal's ability to relearn a variety of pattern discriminations and figure/ground discriminations (Sprague *et al.,* 1977), causing only a reduction in performance on tasks requiring high spatial resolution (Berkley and Sprague, 1979). Clearly, the surviving pattern vision might be subserved by Y cells, and indeed Lehmkuhle *et al.* (1980a,b) have recently argued for the involvement of the Y-cell system in low-frequency pattern vision (Chapter 11, Section 11.3.2). Alternatively, it might be served by the substantial part of the W-cell system that survives this lesion. In the monkey, removal of the striate cortex, which is the site of termination of both X- and Y-cell projections (Chapter 6), causes a profound loss of visual performance, followed by a slow recovery of a considerable level of spatial vision (Kluver, 1942; Weiskrantz, 1972, 1978; Humphrey, 1974; Keating, 1980). That residual vision includes the ability to differentiate visual figures from background (Humphrey, 1974), and to discriminate speed of movement (Keating, 1980). The monkey's ability to discriminate objects from each other is reduced, but far from eliminated, and the animal recovers the ability to use vision to move freely around objects. Humans with lesions to the visual cortex suffer partial blindness; areas of their normal visual field (called *scotomas*) are nonfunctional and, at a conscious level, the patient is blind to stimuli presented in them. Yet reports (Poppel *et al.,* 1973; Sanders *et al.,* 1974; Weiskrantz *et al.,* 1974) suggest that such subjects are capable of directing their eyes or hands toward stimuli presented in scotomas; and of discriminating shapes and orientations of large (but not small) objects or patterns.

In the human, such residual vision is dramatically different in character from normal, conscious vision, and Humphrey (1974) suggests that there is a similar qualitative difference in the monkey as well. The cat seems less affected than primates by destruction of areas 17 and 18, and it is difficult to assess whether a change in visual experience occurs in this species. What does seem clear, however, is that the principal remaining visual pathway in such animals is the projection from the retina to the midbrain and thence, via the extrageniculate thalamus, to the circumstriate cortex and, in the cat, from the retina to area 19 of the cortex, via the dLGN. In the cat, and probably in the monkey, the principal component of those pathways is formed by W-class ganglion cells, with a lesser component formed by Y cells and, in the cat, X cells (Chapters 6 and 7). Furthermore, no other psychophysical phenomena have yet been described that might be attributed to W cells; and the pathways that survive destruction of the cortical termination of X- and Y-cell projections via the dLGN, are the principal pathways into which retinal W cells enter. Further, the destruction of area 19 in the cat, i.e., of the principal area of the cortex to which W cells project, results in a sharp loss of visual function (Sprague *et al.*, 1977).

For these reasons, the visual capability that survives destruction of the cortical terminations of X and Y cells may be principally mediated by W-group ganglion cells. Trevarthen (1968), and subsequent workers, including Weiskrantz (1972, 1978), Humphrey (1974), Rowe and Stone (1980*b*), and Stone *et al.* (1979), have termed that residual capacity *ambient vision*. It includes functions such as the perception of visual space, some low-resolution pattern vision, and the reflex direction of gaze.

2.2.2. Evidence for Choice of Taxa

The taxa Y cells, X cells, and W cells have been widely discussed in the literature. The features relied on for the delineation of these taxa have not been constant, but have expanded in number as data have accumulated about them. The delineation of the three groups is presently made on the basis of the properties listed in Table 2.1. These properties include some of those used in the earlier papers on cell classification, e.g., the differences in receptive field size, linearity of summation, and retinal distribution described between X and Y cells by Enroth-Cugell and Robson (1966), and the slow axonal velocity and distinct receptive field properties of W cells described by Stone and Hoffmann (1972). They also include more recently described properties, such as the morphological features of the cells described by Boycott and Wässle (1974), Leventhal *et al.* (1980), and Stone and Clarke (1980), and the pattern of retinal distribution of W cells described by Rowe and Stone (1976*a*).

One element of the classification discussed above that distinguishes it from feature extraction classifications is the proposing of subgroups, e.g., the W1 and W2 subgroups of W cells. (Table 2.2). The suggestion of this subgrouping began with Stone and Fukuda's (1974*a*) tentative conclusion that most W cells seemed to fall into "tonic" and "phasic" subgroups; Rowe and Stone (1977) proposed a more formal subgrouping of

Table 2.2. **A Classification of Cat Retinal Ganglion Cells**

Higher taxa:	XY cells		W cells	
Lower taxa:	Y cells	X cells	W1 cells	W2 cells
Higher categories	Conscious or focal vision		Ambient vision: includes some movement and pattern vision, and spatial localization within the visual field	
Lower categories	Movement vision	Pattern vision (includes color vision in primates)	Distinct lower categories not yet clear	

W1 (equivalent to tonic) and W2 (equivalent to phasic) groups. Although this seemed an unremarkable development at the time, the suggestion of different taxonomic levels within a classification raises fundamental issues of interpretation and has been subject to two criticisms. First, it has been argued that ganglion cell groupings should be, and indeed have been, rigorously defined in terms of particular properties, and that those definitions have a simplifying influence on work in the field. Second, it has been argued that groups in a classification should be homogeneous, i.e., that the members of a group should be identical in the defining properties. A group that contains cells with a range of properties (such as the range of receptive field properties found in the W-cell grouping) is then a contradiction in terms, and should be abandoned, with several more homogeneous groups set up in its place.

The two arguments are related; they amount, I believe, to an essentialist or typological thrust in classification, a critical consideration of which can be found in Rowe and Stone (1977, 1979) and in Chapters 4 and 5. Two elements of that consideration are usefully foreshadowed here. First, despite their heterogeneity in receptive field properties and in certain other properties, W cells have important features in common, in particular their shared projections to the SC and to the C Laminae of the dLGN and thence to area 19 of the visual cortex, their common tendency to concentrate in the visual streak, and their morphology. The substitution of five, seven, or nine physically more homogeneous groups for the more general W-cell grouping loses that information about common properties. Consequently, and second, it seems important for classifications of nerve cells to be multiple-level, i.e., to comprise higher- and lower-order groupings. This allows the properties common to cells in the higher groups to be expressed, as well as the differences between cells, which provide the basis for subgroupings; and, as is argued in Sections 2.3 and 2.4, it allows development of a biological context for classification to supplement the purely functional criteria often relied upon. If the use of higher- and lower-level groups is valid (and it has been employed in most biological classifications since the time of Aristotle), then it follows that some groups in a classification (the higher-level groups) will contain a greater variety of properties than other (lower-level) groups, the level of variation of properties within a grouping increasing with taxonomic level. The problem of determining higher- and lower-level taxonomic groupings is considerable. A multiple-level classification has been employed not because it is simpler than a single-level. typological classification, but because it allows a more meaningful interpretation of the relationships between biological entities, such as animals or cells.*

In summary, therefore, I would stress four features of the Y/X/W classification just discussed:

1. Its hypothetico-deductive formulation, as a set of hypotheses of the relationship between the proposed taxa (Y, X, and W cells) and categories (movement vision, high-resolution pattern vision, and ambient vision).
2. The use of nondescriptive terminology for the taxa, necessary to preserve the testability of the scheme.
3. The use of multiple levels of taxa and (though still unspecified) categories, introducing the possibility of richer, more biological interpretations of differences between ganglion cells.
4. The tendency, consequent to (1) and (2) above, of the classification to generate its own development and eventual replacement.

In Section 2.3, the third of the above points is discussed in some detail, and in Section 2.4, the last of the points is illustrated. The Y/X/W classification is not a static codification of ganglion cell types, but a developing system of ideas about the functional groupings present among ganglion cells.

2.3. THE INTERPRETATION OF VARIATION†: A CENTRAL PROBLEM IN CELL CLASSIFICATION

The exploration of the properties of Y, X, and W cells has proceeded intensively since the first descriptions of these groups and has led to the recognition of a great deal of variation of properties within as well as between all three groups. The interpretation of that variation is central to any classification.

The number of physical attributes of a neuron or population of neurons is probably

---

*As an example from animal taxonomy, the domestic cat is distinguishable from its close carnivore relatives (such as the tiger) by certain physical features, such as size and coloring; carnivores are distinguishable from herbivorous mammals by still other features, such as the shape of their teeth and the organization of the digestive tract; mammals as a group are distinguishable from nonmammalian vertebrates by other features, such as skin specializations and the mode of feeding their young; while vertebrates as a group are distinguishable from invertebrates by other features, such as the vertebral column, and so on. Each of these distinctions, which together form the basis of the evolutionary classification of animals, rests on a different set of properties, depending on whether distinction is being made between two species of carnivores, between two orders of mammals, between two classes of vertebrates, or between vertebrates and invertebrates. The concepts central to the phylogenetic understanding of animals cannot be encompassed in a single-level classification.

†This section of Chapter 2 was initially drafted quite differently, as a "straightforward" account of the exploration of the properties of Y, X, and W cells, setting out the findings of different workers under ad hoc headings. The account seemed unorganized conceptually, however, and I am indebted to M. H. Rowe for pointing out the central issue of the interpretation of variation. We have written on the topic separately (Rowe and Stone, 1980b), and the present discussion follows the argument of that paper.

infinite in principle (Pratt, 1972) and certainly very high in practice. Classifying therefore involves the interpretation of some physical features as of greater importance than others for understanding the functional role of the cell.* The assessment of the importance of particular physical features of cells is not a self-evident matter, a distinguishing between the clearly important and clearly unimportant properties of a cell. Nor is it a matter of recognizing large variations in physical properties that presumably indicate substantial variations in their functional significance. For example, the differences in receptive field size found among X cells as a function of eccentricity are as great as the differences between X and Y cells at any given retinal location, yet may have, as argued below (Section 2.3.2), a quite different functional significance. Interpreting variation seems to require not only measurement of the magnitude of variation but also consideration of its significance for the cell's function and also, it is argued, for the cell's evolutionary history and potential. Three points concerning the interpretation of cell properties deserve emphasis:

1. A first step in the interpretation of variation is the recognition that among populations of nerve cells, as among animals, more than one type of variation can be recognized. Three types of variation are distinguished below; two of these, termed *role-indicating* and *systematic variation,* seem related to visual function or mechanisms while the third, *residual variation,* seems related to mechanisms of evolution. These three types of variation are probably not exhaustive; others may be recognized. However, the distinction between them allows some of the complex factors that seem to determine the properties of neurons to be recognized separately.
2. A second step is the recognition that any functionally distinct group of nerve cells will contain substantial variation within it. Groups can be distinguished, however, that encompass greater variation than others (as in animal taxonomy, the group *mammals* encompasses more variation than *carnivores*). This recognition leads to the development of multiple-level classifications in which higher-order groups (such as an Order in animal taxonomy) contain far more variation than a lower-level group (such as a species).
3. Differences in the interpretation of variation underlie the difference between the approaches to the classifying of visual neurons termed *parametric* and *feature extraction* in Chapter 1, and above.

2.3.1. "Single" and "Multiple" Interpretations of Variation

Two distinct interpretations of variation seem to underlie the "feature extraction" and "parametric" approaches to ganglion cell classification; they are here termed *single* and *multiple* interpretations of variation, following Rowe and Stone (1980*a,b*).

---

*One theoretical approach that has sought to deal with the high number of properties of an organism is numerical taxonomy. Its potential for classifying neuronal populations is discussed in Chapter 4, Section 4.1.2.

In the "single" interpretation of variation, a physical difference between cells either is taken as of primary or key importance, indicating a functional difference between them; or it is considered, usually implicitly, to be unimportant or secondary. Groupings of cells are then based on variations in the key property or properties. As an example of this approach, visual physiologists have often classified neurons according to certain of their receptive field properties (Section 1.2.2), so that if two cells differ in a key receptive field property they are put into different groups in the classification, however similar they might be in other properties. Conversely, if they are similar in that receptive field property, they are put in the same group no matter how much they differ in other properties. The groups in such classifications are designed for homogeneity of certain receptive field properties; only one level of significance is recognized in interpreting variation (important/not important), and the groups in the classification exist at a single taxonomic level. This seems to have been the interpretation of variation that underlies the feature extraction classifications of visual neurons (Section 1.2.2).

In the "multiple" interpretation of variation, attention is given to the different possible sources of variation in the physical properties of cells. Some variation in the properties of nerve cells is doubtless related to differences in the cells' functional roles; other variation may reflect variation in a particular functional role with varying conditions (for example, eccentricity-related variations in ganglion cell properties); yet other variation may reflect the different phylogenetic histories of various cell populations. In addition, it is recognized that some variation may indicate broad groupings of cells, while other instances of variation may indicate subgroups within these broader groups. As a consequence, a multilevel classification can be developed with higher- and lower-order taxa and categories; in the general case, variation from all three sources can be recognized within each group and subgroup. This is the interpretation of variation that seems most appropriate for the Y/X/W classification.

How are these different types of variation to be recognized and distinguished? The interpretation of variation is part of the inductive generalization involved in constructing a classification, for which, as argued in Chapter 5 (Sections 5.1–5.3) there is no established methodology. However, a rigorous methodology is equally lacking for the single interpretation of variation; the lack is not a problem, provided that any particular interpretation of variation is formulated as a testable hypothesis about the functional roles of the cells being studied.

2.3.2. Sources of Variation in the Properties of Ganglion Cells

The variation in physical properties found among retinal ganglion cells (and other neurons as well) seems attributable to at least three sources:

1. *Variation related to functional role:* Such variation between cells seems to indicate that different populations of cells perform different functional roles and should, in a function-oriented classification, be placed in separate groups or subgroups. An example is the variation in central projections of the different groups of ganglion cells distinguished in cat retina, which results in the different

areas of the visual cortex (areas 17, 18, and 19) receiving inputs from different ganglion cell classes (see Chapter 8, Section 8.1.1). Variation of this sort is herein termed *role-indicating variation*.

2. *Variation in properties related to systematic variation in the conditions in which a population of cells performs its function:* For example, several properties of X and Y cells vary with retinal eccentricity, and with the level of light/dark adaptation, presumably reflecting a difference in the performance of these cells with changes in one or other of these parameters. This sort of variation will be termed *systematic variation*.

3. *Variation related not to function, but to the evolutionary history and potential of the cell population:* This will be termed *residual variation*. Variation within biological populations that seems unrelated to their function is well recognized at both the whole animal (Ayala, 1978) and macromolecular (Lewontin, 1974, 1978) levels, and is thought to result from diversity in the genetic pool of the population. Its significance seems to be that it provides part of the genetic adaptability of the population to environmental pressure, and hence forms part of its evolutionary potential.

A number of striking examples of residual variation have been reported among populations of nerve cells. van Buren (1963) noted that the number of layers formed by the concentration of ganglion cells around the human foveola can vary considerably between apparently normal eyes; Stone (1978) described variation in the number of ganglion cells in the cat retina, in the number of large ganglion cells, and in the distribution of large cells around the area centralis; and Webb and Kaas (1976) described variation between individual owl monkeys in the development of a fovea centralis. Hickey and Guillery (1979) described substantial variation in the pattern of lamination of the human LGN, for which a functional correlate has still to be suggested; and similar variation was reported in the marmoset by Spatz (1978). Similarly, Haight and Neylon (1978) described extensive variation in brain morphology in the marsupial brush-tailed possum. Such variation, if substantial, should also result in residual variation in the functional capabilities of different individuals; indeed, it must do so if it is to provide a substrate for natural selection. Although seldom investigated specifically, such variation presumably accounts for much of the "noise" or error variance in the results of behavioral experiments. One study has described such variation even in the relatively stereotyped behavior pattern of fish and other nonmammalian vertebrates (Barlow, 1977).

2.3.3. A Multiple Interpretation of Variation in the Properties of Ganglion Cells

This section attempts an account of the properties of ganglion cells that combines a cataloging of properties with a multiple interpretation of variation in those properties. The account follows Table 2.1. All the interpretations suggested are themselves hypotheses, part of the inductive generalization of proposing taxa for a classification. The literature review is not comprehensive; the section is particularly concerned with variation in properties and its significance for cell classification, and mention is not made of many studies important in other contexts.

2.3.3.1. Receptive Field Size

Role-indicating variation: There is general agreement that at any retinal location, X cells have the smallest receptive field center regions of the three classes, while Y and W cells have larger receptive fields. This relationship is shown in Fig. 2.14; these data are from the study of Stone and Fukuda (1974a) but is in substantial agreement with other reports, such as Cleland *et al.* (1971), Fukada (1971), Ikeda and Wright (1972a), Cleland and Levick (1974a,b), and Cleland *et al.* (1979). The small receptive field size of X cells has been widely interpreted as evidence that these cells subserve high resolution, while the larger receptive field size of Y and W cells is taken to indicate that high spatial resolution is not an important feature of their function.

Systematic variation: In all three cell groups, there is a trend for receptive field size to increase with retinal eccentricity, indicating some gradient in spatial resolution with eccentricity. The trend is less marked among W cells than among either X or Y cells. As a result of it, however, when data from different eccentricities are pooled, the receptive field size ranges of the three cell classes overlap substantially.

Residual variation: At any eccentricity (Fig. 2.14), there is some range of receptive field size within all three cell groups that has yet to be related to any functional specialization. Bullier and Norton (1979a) have described similar variations in receptive field size, as well as in several other parameters of receptive fields at a particular retinal location.

2.3.3.2. Time Course of Center Response

Role-indicating variation: Between X and Y cells, a consistent difference has been described between the time courses of the responses of X and Y cells to standing contrast stimuli, i.e., to a small dark (or light) spot presented in the center region of an OFF- (or ON-) center receptive field. Specifically, although cells in both groups respond to the presentation of such a stimulus with a phasic burst of firing, only in X cells is the firing rate sustained above resting level as long as the stimulus continues. Some classifiers (e.g., Cleland *et al.*, 1971; Ikeda and Wright, 1972a) have made this a defining feature in their classification of ganglion cells. The sustained/transient difference attracted the interest of psychophysicists studying "pattern"- and "movement"-detecting mechanisms in human vision, since some psychophysical experiments suggested that cells subserving pattern detection should give sustained responses to standing contrast stimuli, while the responses of movement detectors should be transient. However, Kulikowski and Tolhurst (1973) noted that the sustained detector mechanism of human psychophysics seems far more sustained than the responses of cat X cells; and conversely, Rowe and Stone (1977) argued that, because the sustained/transient difference is not apparent between cat X and Y cells until typically 500 msec after the onset of a stimulus, eye movements that occur during normal fixation would obscure the difference. This suggests that, despite its consistency in the experimental situation and its usefulness as an identifier of

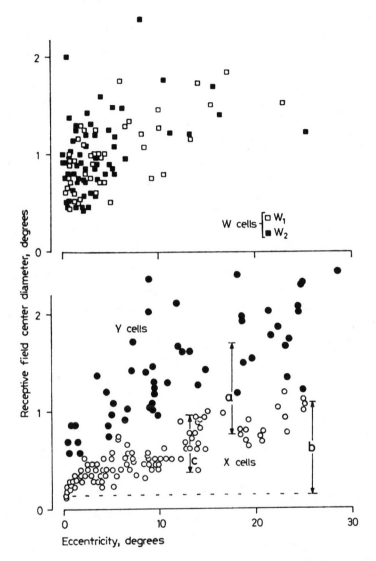

Figure 2.14. Example of types of variation in ganglion cell properties. Relationship between the diameter of the center region of a receptive field and the retinal eccentricity of the cells; for Y, X, and W cells from cat retina. The data are from Stone and Fukuda (1974a), redrawn by Rowe and Stone (1980b) to illustrate various types of variation in receptive field size. The arrows on the lower graph indicate: *role-indicating variation* in receptive field size between X cells and Y cells (a); *systematic variation* in the receptive field size of X cells with eccentricity (b) (comparable increases are also apparent for Y cells and, less markedly, for W cells); and *residual variation* in the receptive field size of X cells (c). Residual variation is also apparent among Y cells and among both subgroups of W cells. [Reproduced with kind permission of Karger AG.]

X and Y cells, the functional significance of the sustained/transient difference is not well understood. An analogous, but sharper, sustained/transient difference exists between the major subgroups of W cells (Stone and Fukuda, 1974a; Rowe and Stone, 1977). Here also its functional significance is unclear.

One group of cells, termed *luminance units* by Barlow and Levick (1969) and included by Stone and Fukuda (1974a,b) in the W-cell group, give extremely tonic responses to standing contrast stimuli. Moreover, there is a monotonic relationship between their ongoing firing rate and ambient illumination (Fig. 2.8) that may indicate their role in the monitoring of ambient illumination; for example, for the control of pupil size.

Systematic variation: Among X and Y cells, but not among W cells, the sustained-ness and transientness of the cells' responses vary consistently with two parameters: light/dark adaptation and eccentricity. Following dark adaptation, the responses of both X and Y cells become increasingly sustained or tonic (Cleland *et al.*, 1973; Jakiela *et al.*, 1976). Indeed, the time course of the response of a Y cell can be almost identical with that of an X cell adapted to a lower level of ambient illumination (Fig. 2.15). Conversely, the responses of cells in both classes become transient at high levels of light adaptation. Second, there is some still-scattered evidence that the responses of both X and Y cells may become more tonic at the area centralis (e.g., Hochstein and Shapley, 1976), the effect being graded with eccentricity. As a result, many Y cells at the area centralis are as sustained in their responses as peripheral X cells. No comparable trend in time course of response have been described for W cells.

Residual variation: Even controlling for eccentricity and adaptation, there is considerable variability in the "tonicity" of all major classes of ganglion cells (Bullier and Norton, 1979a), which may prove to be unrelated to a functional specialization, and hence residual in the present sense.

2.3.3.3. Other Receptive Field Properties

By contrast with a parameter such as central projections, which almost all authors interpret as of major importance (for an opposing view, see Hughes, 1979), many receptive field properties have been subject to widely differing interpretations. The interpretation of receptive field size is discussed above; all workers who have commented seem to consider it of importance. The differing interpretations of other properties have made this section most difficult to write.

Role-indicating variation: Most disagreement centers around which receptive field properties are of central importance. For example, Kuffler (1953) observed two sorts of receptive fields, those with ON-centers and OFF-surrounds and those with the opposite organization (OFF-center, ON-surround). Subsequently, some workers have emphasized this ON-center/OFF-center difference. Jung (1973), for example, proposed that the retinal input to the visual cortex comprises two systems, a B (brightness-coding) system mediated by cells with ON-center receptive fields, and a D (darkness-coding) system mediated by OFF-center cells. Other authors, finding the ON-center/OFF-center difference among all classes of ganglion cells distinguished on other grounds (e.g., among W, X, and Y cells), have not attributed much functional importance to it. Similarly,

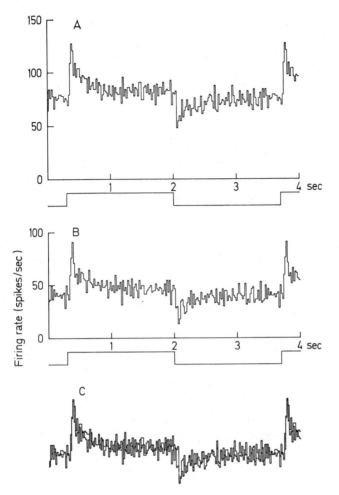

Figure 2.15. Evidence that the sustained/transient difference between X and Y cells is quantitative rather than qualitative (from Jakiela *et al.,* 1976). The stimulus was a spot of light 0.1° in diameter centered on the receptive field. The response of a Y cell (A) light-adapted at 1.7 log units and of an X cell (B) light-adapted at 3.0 log units have very similar time courses, as shown by their superimposition in (C). Thus, the tonic or phasic feature of a cell's response depends on the level of illumination to which it is adapted. [Reproduced with kind permission of Springer-Verlag.]

some authors have attributed "key importance" to the distinction between "concentric" receptive fields (those described by Kuffler, 1953) and "nonconcentric" fields (such as direction-selective or suppressed-by-contrast fields of some W cells); while other workers, finding the former organization among all major classes of ganglion cells (i.e., Y, X, and some W cells), have not attributed much significance to it and have noted some continuity of properties between certain concentric and nonconcentric "types" (e.g., Fig. 7 in Stone and Fukuda, 1974a).

Linearity of summation was one of the Y/X differences first noted by Enroth-

Cugell and Robson (1966), and some authors (e.g., Levick, 1975; Hochstein and Shapley, 1976) have considered it as of definitional importance for the Y/X distinction. Other authors (e.g., Stone and Fukuda, 1974a) have placed less emphasis on the property. Its functional significance is not understood, although it remains a strong criterion for distinguishing X cells from Y cells. Some support for a reduction in emphasis on linearity in distinguishing X from Y cells comes from the suggestion by Derrington *et al.* (1979) that the nonlinearity of Y cells stems from the receptive field mechanism responsible for McIlwain's periphery effect (see Chapter 1, Section 1.2.1) and for the "shift effect" described by Kruger and Fischer (1973). Derrington and co-workers' (1979) results indicate that the shift effect is commonly present to some degree in X cells, whose receptive field summation is presumably to some degree nonlinear as a result. The difference in linearity between X and Y cells thus seems relative, rather than absolute. Victor and Shapley (1979) also concluded that Y cells receive input from a nonlinear receptive field mechanism that does not substantially impinge on X cells; however, they consider that the center and surround mechanisms of X and Y cells are distinct in properties. In their view, the nonlinearity that characterizes Y cells derives partly from the extra, nonlinear input to them, and partly from the properties of their center and surround mechanisms.

The variety of receptive field properties found among W cells has also been variously interpreted. For example, Cleland and Levick (1974b) argue that cells with ON-OFF-center receptive fields seem specialized to respond to edge stimuli within a small region of the visual field, and called them *local edge detectors*. By contrast, Stone and Fukuda (1974a) gave less emphasis to this property, partly because all ganglion cells are specialized for the local detection of edges, in that they have receptive fields that are small relative to the size of the visual fields and many have an antagonistic surround component to their receptive fields; and partly because they saw a continuity of properties between ON-OFF-center and phasic ON-center and OFF-center receptive fields found among W cells. Stone and Fukuda (1974a) tentatively class cells with the latter three receptive field characteristics as a major subgroup of W cells [which they called *phasic* and Rowe and Stone (1977) subsequently termed *W2 cells*]; while Cleland and Levick (1974a,b) place the same cells in two distinct classes of cells (local edge detector and sluggish-transient).

Similarly, the color-coding properties of a small percentage of ganglion cells, first described by Cleland and Levick (1974b), and the "suppressed-by-contrast" properties of a small fraction, first described by Rodieck (1967), have also been variously interpreted. Cleland and Levick (1974a,b) and Rodieck (1979) considered these properties sufficient basis for setting up two more groups of ganglion cells, at the same taxonomic level as X and Y cells. Stone and Hoffmann (1972), Stone and Fukuda (1974a), and Rowe and Stone (1977), by contrast, noted the properties that these two groups have in common with W-class ganglion cells, and include them as subgroups of this grouping.

Some evidence is available that the visual responses of X and Y cells involve different neurochemical transmitters. Kirby and Enroth-Cugell (1976) studied the effects of picrotoxin and bicuculline on the visual responses of X and Y cells. These two drugs are antagonists of GABA (γ-aminobutyric acid), which is considered a likely retinal transmitter. Kirby and Enroth-Cugell found that both drugs consistently reduced the responsiveness of Y cells, but had little effect on X cells. This implicates GABA as a transmitter for Y cells, especially for the surround mechanism, which was more strongly affected than the center. Kirby (1979) reported an inverse result with strychnine, whose action is antagonistic to that of glycine, another suspected retinal neurotransmitter. Strychnine reduced the responsiveness of X cells but not of Y cells, and again the surround mechanism was the more affected. Glycine may then be important to the retinal circuitry of X cells, GABA for Y cells. Kirby noted that these pharmacological differences between X and Y cells were relative rather than absolute: a few Y cells were sensitive to strychnine and a few X cells to picrotoxin and bicuculline. Much remains to be learned about the neurotransmitters of all ganglion cell types; this evidence of different transmitters seems clearly to be role-indicating variation in the present sense.

The responsiveness of Y cells to fast-moving visual stimuli was first noted by Cleland *et al.* (1971). Most authors have interpreted this as of considerable functional significance. It forms part of the basis, for example, for the suggestion that Y cells may subserve movement vision (Chapter 11); that Y cells may subserve binocular vision (Levick, 1977); and that area 18 (which receives a strong Y-cell input) may be principally concerned with the perception of movement (e.g., Orban and Callens, 1978).

A functionally important difference between X and Y cells was suggested by Ikeda and Wright (1972*b*) in terms of the effect of defocusing of the retinal image. The responsiveness of Y cells is less dependent on sharp focusing of the retinal image than is the responsiveness of X cells. This may provide a retinal component for the neural mechanism of amblyopia; Ikeda and Tremain (1978, 1979) have argued, for example, that X cells (especially those at the area centralis) are more sensitive than Y cells to blurring of the retinal image during postnatal development, so that X cells are the more affected by optical defects of the eye or strabismus during the first few weeks of life (see Chapter 10, Section 10.3, for further discussion).

Systematic variation: A considerable degree of systematic variation has been described in the receptive field properties of cat ganglion cells, particularly of X and Y cells. The variation with eccentricity in receptive field size and in the tonicity of response of both X and Y cells has already been noted. In addition, the velocity-selectivity of X cells appears to shift with eccentricity, responsiveness to fast stimulus movements being greater among peripherally located cells than in those at the area centralis (Dreher and Sanderson, 1973). Ikeda and Tremain (1978) report that the effect of defocusing of the retinal image on the responses of X cells is more marked at of the area centralis than in the periphery of the retina. They do not comment as to whether there is a comparable gradient for Y cells. The presence of systematic variation in linearity of summation has yet to be investigated in detail. Hochstein and Shapley (1976) noted that a Y cell at 15° eccentricity was more nonlinear than an X cell at 60°, but the linearity of area centralis X and Y cells has still to be assessed.

Residual variation: In given conditions, e.g., at a particular retinal eccentricity and level of ambient illumination, some residual variation in receptive field properties persists within all the major cell groups. Residual variation in several properties of ganglion cells is described in the various parts of this section. Bullier and Norton (1979*a*) present measurements of what appears to be residual variation in several properties of X and Y cells, noting that it is more prominent in X cells; this led them in an earlier study to suggest the presence of "intermediate" cells, with properties of both X and Y cells. Residual variation seems particularly prominent among W cells, at least in their receptive field size, axonal conduction velocity, and morphology, and it is possible that some of the variety of "layouts" of ON- and OFF-areas of a receptive field found within the W-cell grouping may also prove to be residual in the present sense, rather than related to functional specializations.

2.3.3.4. Axonal Conduction Velocity

Role-indicating variation: The axonal velocities of Y, X, and W cells are reliable (but not perfect) identifiers of the three cell groups (Fig. 2.16). Y cells have fast-con-

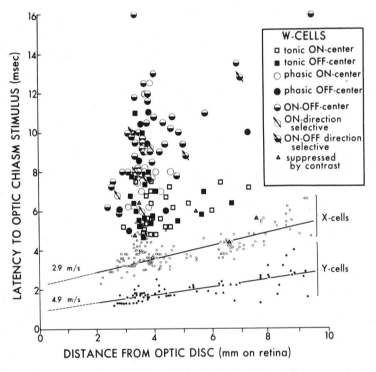

Figure 2.16. Evidence of distinct conduction velocities of Y, X, and W cells from cat retina (from Stone and Fukuda, 1974a). The graph plots the latency of the cells' antidromic responses to stimulation of the optic chiasm against the distance of the cells from the optic disc. Effectively, this is a plot of latency against conduction distance. The latencies of the three major classes are largely separated along the ordinate, and Rowe and Stone (1976b) have argued that much of the overlap of latencies between X and W cells is due to the pooling of data from different animals.

For X and Y cells, latency increases with distance, as shown by the regression lines. The inverse of their slopes suggests intraretinal velocities of 4.9 m/sec for Y cells and 2.9 m/sec for X cells. The latencies of the W cells are widely spread, indicating a wide range in the axonal conduction velocities of W cells, with some tendency for "tonic" W cells to have shorter latencies that "phasic" W cells. [Reproduced from the *Journal of Neurophysiology* with kind permission of the American Physiological Society.]

ducting axons (30–40 m/sec along their myelinated segment in the optic nerve and tract), equivalent to the t_1 group of Bishop and MacLeod (1954); X cells have slow-conducting axons (17–23 m/sec, the t_2 group); while most W cells have distinctly slower axons than X cells. There is also a less marked difference in conduction velocity between the major subgroups of W cells, W1 cells having faster-conducting axons than W2 cells (mean velocity 11.6 m/sec as against 6.6 m/sec). The evidence of these conduction velocity relationships has emerged from the reports of Fukada (1971), Cleland et al. (1971), Stone and Hoffmann (1972), Stone and Fukuda (1974a), Cleland and Levick (1974a,b), Kirk et al. (1975, 1976a,b), and Rowe and Stone (1976a,b).

Three distinct arguments have been canvassed that suggest that axonal velocity is of functional importance. Ikeda and Wright (1972a) and Singer and Bedworth (1973, 1974) suggested that the fast velocity of Y-cell axons enables Y-cell activity generated by an eye movement to reach the dLGN before X-cell activity; and, by activating the inhibitory interneurons of the dLGN, to "wipe clean" prior activity in the X-cell system, ready for a new burst of X-cell activity following the eye movement. The suggestion seems to be weakened by the consideration that the difference in retinogeniculate conduction times between X and Y cells (about 2–4 msec depending on retinal eccentricity) is small with respect to the photic latency of X- and Y-cell responses [both about 30–40 msec to 80 msec (Ikeda and Wright, 1972a)]. Second, it is likely that conduction velocity is simply related to axonal caliber and caliber may be related to the degree of branching of an axon. The stout caliber and fast conduction of the axons of Y cells may simply reflect the known branching of Y-cell axons between forebrain and midbrain structures (Hayashi et al., 1967; Hoffmann, 1973; Singer and Bedworth, 1973; Kelly and Gilbert, 1975), and within the LGN (Bowling and Michael, 1980). The suggestion implies, of course, that W cells have particularly restricted axonal terminations, and little evidence is available to support this implication. Third, Bishop (1959) argued that large axonal caliber reflects the recent phylogenetic origin of a fiber system. However, although several authors (including myself, 1966) have suggested that W cells are a phylogenetically old system, there is only limited evidence in favor of this idea, or of the idea that Y cells are phylogenetically more recent in origin than the thinner-axon X or W cells. Both Y- and W-like ganglion cells have in fact been described in all mammals so far examined in these terms, suggesting that both classes may have appeared fairly early in mammalian history. The X-cell system, on the other hand, seems more labile, being weak or absent in the rat (Hale et al., 1979), difficult to identify in the possum (Rowe et al., 1977), apparently somewhat better developed in the rabbit (Caldwell and Daw, 1978), clearly present in the cat, and very highly developed in certain primates (Dreher et al., 1976; Sherman et al., 1976).

Despite its value as an identifier of a ganglion cell as Y, X, or W class, therefore, the functional significance of axonal conduction velocity is not well understood. The above considerations provide only a limited basis for considering conduction velocity a role-indicating property of ganglion cells.

Systematic variation: Among X- and Y-cell axons, but apparently not among W-cell axons, the velocity of axonal conduction varies with eccentricity. Specifically, between cells located at the center of the area centralis and regions 1–2 mm away, there is a distinct gradient in the axonal velocity of X- and Y-cell axons, in both cases velocity

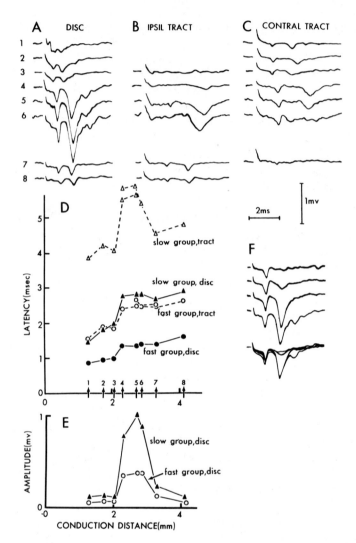

Figure 2.17. Early evidence of systematic variation in axonal conduction velocities of cat ganglion cells (from Stone and Freeman, 1971). (A) Field potentials generated by stimulation of the optic disc and recorded at a series of locations (1–8) extending from near the optic disc (1) across the area centralis (4–6) and beyond into temporal retina (7, 8). Note that the field potential is large in amplitude at positions 4–6, suggesting that this is a region of high ganglion cell concentration, the area centralis. (B, C) Potentials generated at the same sites by stimulation of, respectively, the ipsilateral and contralateral optic tracts. (D) Plot of the relationship between latency and position (effectively conduction distance) for disc-activated fast and slow groups, and for tract-activated fast and slow groups. In each plot, latency values obtained at positions 1–3, 7, and 8 (i.e., outside the area centralis) lie approximately along a straight line. Values obtained at positions 4–6 (in the area centralis) are relatively long for both fast and slow groups, indicating that the axons concerned (i.e., the axons of area centralis Y and X cells) are relatively slow-conducting. (E) Plot of the amplitude of fast- and slow-conducting groups, as a function of retinal position. The large amplitudes at positions 4–6 mark the area centralis. [Reproduced with kind permission of Springer-Verlag.]

being less at the area centralis (Stone and Freeman, 1971; Kirk *et al.*, 1975; Fig. 2.17). Beyond 1–2 mm from the area centralis, there is little evidence of any eccentricity-related variation in axonal conduction velocity (Rowe and Stone, 1976*b*), but within the central 1–2 mm, Kirk *et al.* (1975) have argued that the decrease in t_1 and t_2 velocities is graded with eccentricity. As a consequence, t_1 velocity reduces to 25 m/sec at the area centralis (Stone and Freeman, 1971), close to the value for peripheral t_2 axons (17–23 m/sec), and t_2 velocity drops to 10–15 m/sec, into the range attributed to W cells. No eccentricity-related gradient in axonal conduction velocity has been reported for W cells.

Bishop *et al.* (1953) reported evidence that t_1 and t_2 groups differ in axonal conduction velocity between temporal and nasal retina. The presence of this gradient has been questioned (e.g., Kirk *et al.*, 1975; Rowe and Stone, 1976*b*), but more recent evidence has tended to confirm the original observations (Stone *et al.*, 1980). The significance of the nasotemporal gradient is not well understood, but evidence of it has been reported in species other than the cat [e.g., the opossum (Rowe *et al.*, 1978), and the possum (Tancred and Rowe, 1979)]. It may reflect some difference in the phylogenetic origins of nasal and temporal retina. A comparable gradient has not yet been described for W cells.

Residual variation: At any eccentricity, there appears to be limited variation in the axonal conduction velocities of X and Y cells, and a wide variation in the axonal velocities of W cells of both major subgroups (Fig. 2.16).

2.3.3.5. Retinal Distribution

Role-indicating variation: The different retinal distributions of Y, X, and W cells have been interpreted as strong evidence of functional differences between them. The tendency for X cells to concentrate at the area centralis (Enroth-Cugell and Robson, 1966) has, for example, been interpreted as support for the suggestion that these cells subserve high-resolution vision. The concentration of W cells in the visual streak (Rowe and Stone, 1976*a;* see Chapter 9) raises the possibility that this feature of retinal topography is particularly highly developed in the retinal projection to the midbrain. Y cells are relatively (though not absolutely) most frequent in peripheral retina, suggesting that they are concerned with detection of images in peripheral visual field. Fukuda and Stone (1974), Wässle *et al.* (1975), and Stone (1978) reported that Y cells increase in absolute frequency toward the area centralis. Fukuda and Stone (1974) and Stone (1978) also provided evidence of a localized minimum in Y-cell density at the area centralis. Stone *et al.* (1978*b*) suggested that this feature may reflect a strong retinal specialization at the area centralis, since it is absent or weakened in the Siamese cat, in which the area centralis is poorly developed.

Systematic variation: Stone *et al.* (1977) and Stone and Keens (1978, 1980) have reported evidence of a difference in the relative numbers of medium-sized and small cells between temporal and nasal areas of retina. Specifically, medium-sized cells seem relatively more numerous in temporal retina than in nasal retina. The functional significance of this trend is not understood.

Residual variation: No evidence has yet been provided.

2.3.3.6. Morphology of Ganglion Cells

A striking correlation was reported by Boycott and Wässle (1974) between three morphological classes of ganglion cells that they observed in cat retina (and termed α, β, and γ cells) and, respectively, the Y, X, and W cells described by the physiologists (Section 2.1.4). Other studies have provided substantial support for the correlation; cells characteristic of the groups Boycott and Wässle distinguished are illustrated in Fig. 2.11, and a more recently recognized subgroup of the γ-cell class is discussed in Section 2.1.4 and illustrated in Fig. 2.12.

Role-indicating variation: In functional terms, the small dendritic fields of β cells presumably contribute to determining the small receptive fields of X cells and their role in high-resolution vision; conversely, the larger dendritic fields of α and γ cells may determine their large receptive field centers. The significance of differences in receptive field center size is discussed above. The axonal caliber of the three classes is presumably closely correlated with their axonal conduction velocity, whose significance is also discussed above. Famiglietti and Kolb (1976) and Nelson *et al.* (1978) have described a morphological basis for the ON-center/OFF-center difference in receptive field organization found among W, X, and Y cells. Their finding is discussed in Section 2.1.4 and illustrated in Fig. 2.18. Briefly, they suggest that ON- and OFF-center ganglion cells differ in the sublamina of the inner plexiform layer in which their dendrites spread. The dendrites of ON-center cells spread in a sublamina nearer the ganglion cell layer and are contacted by a particular class of cone bipolar cells, the "invaginating" bipolars; while the dendrites of OFF-center cells spread in a sublamina near the inner nuclear layer and are contacted by "flat" cone bipolar cells. However, the significance of the ON-center/OFF-center difference (discussed in Section 2.3.3.3) is not well established.

Kolb (1979) has taken this analysis a step further. Working from serial sections of the inner plexiform layer of cat retina examined in the electron microscope, she distinguished three classes of ganglion cells (I, II, and III), which seem to correspond closely to α, β, and γ cells, respectively. Further, she distinguished a and b subgroups of each class, by the level of the inner plexiform layer at which their dendrites spread. The additional step she took was to identify the synapses formed onto the dendrites of cells of each subgroup as formed by bipolar or by amacrine cells. As illustrated in Fig. 2.18B and C, the class II (β) cells receive predominantly synapses from cone bipolar cells, while class I (α) and III (γ) cells receive predominantly amacrine cell input. Thus, the afferent circuitry of the ganglion cell classes differs significantly between the W-, X-, and Y-cell groups.

The functional significance of the relatively "loose" pattern of dendritic branching seen in γ cells or of the denser pattern of branching seen in α and β cells is also not clearly established. It may be related to the "brisk"/"sluggish" difference in firing rates reported by Cleland and Levick (1974a) between X and Y cells on the one hand and W cells on the other.

Kolb *et al.* (1981) have recently proposed the existence of 23 morphological groups of ganglion cells. They do not discuss the problem of how variation in one property of a cell is to be interpreted, but their classification implies that they consider the variants on which they based their groupings to be "role-indicating." The evidence is not available, but to me it seems likely that much of the variation on which they base the 23 groups (principally variation in dendritic branching, dendritic field size, and soma size) will prove better understood as residual variation in the present sense. A number of their groups comprise only one or two cells from their sample, and differ from other groups in only one feature. For example, the one cell in their G9 class differs from β cells only in soma size.

Systematic variation: Two examples of systematic variation in ganglion cell morphology can be suggested. First, both α and β cells vary in morphology with eccentricity. Towards the area centralis, for example, α cells become smaller in soma size, axonal caliber, and dendritic field size. Similarly, many β cells in peripheral retina have multiple-stem dendrites (in contrast to the single-stem dendrite that characterizes them in central retina), and a distinctly larger cell body and dendritic spread than β cells at the area centralis. As a consequence, as Boycott and Wässle note, it is difficult to distinguish a centrally located α cell from a peripherally located β cell, although they remain distinct from neighboring cells of the other classes. The γ cells seem to form a distinct class of dendritic fields at all eccentricities; their dendritic field spread does become smaller at the area centralis, but the reduction is much smaller than in α or β cells. Second, evidence has begun to emerge of a difference in soma size of α and β cells (but not of γ cells) between temporal and nasal areas of retina. Specifically, the mean soma size of α and β cells seems smaller in nasal than temporal retina (Stone et al., 1977, 1980). Similar differences have been described in several mammalian species (see Chapter 9, Section 9.6.1); they may reflect the different developmental and/or phylogenetic histories of nasal and temporal retina.

Residual variation: At any particular eccentricity, the morphologies of α and β cells are fairly constant. Among γ cells, however, Boycott and Wässle (1974) noted considerably greater variety, particularly in dendritic branching patterns, which has still to be related to any functional differentiation among W cells, although it does match well the general variety apparent among W cells in receptive field size, receptive field layout, and axonal conduction velocity. Boycott and Wassle described small numbers of a cell group they termed δ cells, which might be a distinct subclass of γ cells. They seem too few in number to be related to the major subgroupings of W cells proposed, for example, by Rowe and Stone (1977). They may prove, however, to represent a distinct functional subgroup. Similarly, some of the 23 "types" of ganglion cells described by Kolb et al. (1981) may be better viewed as variations of previously recognized classes than as new, separate groupings.

2.3.3.7. Central Projections

Role-indicating variation: All workers seem agreed that the differing central projections of the different ganglion cell classes are evidence of their distinct functions.

X cells project principally to laminae A, A1, and C of the dLGN (Cleland et al., 1971; Hoffmann et al., 1972; Figs. 2.19 and 2.20) and thence to area 17 of the visual cortex (Stone and Dreher, 1973; Singer et al., 1975; Dreher et al., 1978, 1980); a minority of X cells projects to the midbrain but apparently not to the SC (Fukuda and Stone, 1974; Cleland and Levick, 1974a). Although Wässle and Illing (1980) have recently

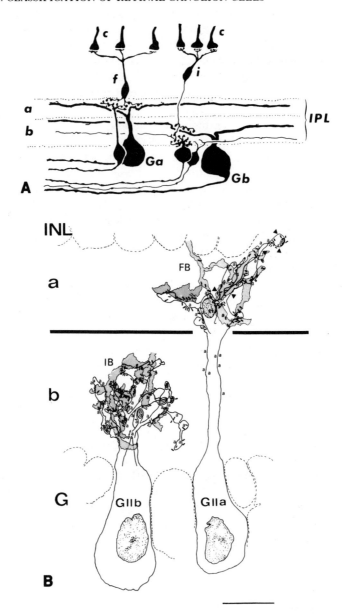

Figure 2.18. Afferent circuitry of different ganglion cell classes. (A) Summary figure from Nelson *et al.* (1978) of the morphological correlates of the ON/OFF difference among cat ganglion cells. The dendrites of some ganglion cells spread in the region of the inner plexiform layer (IPL) nearest the bipolar cells (sublamina a); these ganglion cells are labeled Ga. The dendrites of other ganglion cells (labeled Gb) spread in the region of the IPL near the ganglion cells (sublamina b). The Gb cells illustrated resemble α, β, and γ cells; the two Ga cells resemble α and β cells. The original caption reads:

"Organization of cone bipolar cells and ganglion cells in the IPL of the cat retina. Flat cone bipolar cells (f) have axon terminals ending in sublamina a, contacting the dendrites of a-type ganglion cells (Ga).

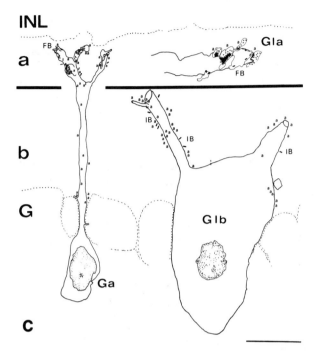

Invaginating cone bipolar cells (i) have axon terminals which ramify lower in the IPL, in sublamina b where they contact b-type ganglion cell dendrites (Gb). Ganglion cells of various morphologies branch either in sublamina a or sublamina b; these prove to be off-center and on-center, respectively. c, cones."

[Reproduced from the *Journal of Neurophysiology* with kind permission of the American Physiological Society.]

(B) Kolb's (1979) representation of synapses formed on class II (β) cells. The original legend reads:

"Reconstruction from serial sections of two medium-bodied ganglion cells in the central area of cat retina (GIIa, GIIb). The IIa ganglion cell receives *no* bipolar input until it branches in sublamina a. Flat cone bipolar terminals (FB, shaded profiles) provide the majority of synapses to its spines and dendrites, with amacrine synapses (a) representing the remaining 30% of the input. AII amacrines also provide input to IIa ganglion cell dendrites in sublamina a (black triangles). The class IIb ganglion cell receives mainly invaginating cone bipolar (IB) input (70%) (shaded profiles) in sublamina b of the IPL. Neighbouring ganglion cell profiles are indicated below (dotted lines, G). Incomplete dendrites on the reconstructed cell are represented by stippled cut surfaces. Scale bar: 12 μm."

(C) Kolb's (1979) representation of synapses formed on class I (α) and a probable class III (γ) cell. The original legend reads:

"Reconstructions from serial sections of two ganglion cells in the central area of cat retina. The small Ga cell (left) has an 8 μm body and a slender apical dendrite which does not branch until reaching sublamina a of the neuropil where three simple dendrites arise. The majority of synapses on the dendrites of the class Ga cell are from amacrines (a) but some flat cone bipolar synapses occur in sublamina a. The GIb ganglion cell has a large cell body (28 μm) and two of the main dendrites received predominantly amacrine input (a) but also some invaginating cone bipolar (IB) input in sublamina b of the IPL. A portion of a large dendrite (GIa) restricted to sublamina a receives patches of amacrine (a) or flat cone bipolar synapses (FB, ribbons). Dotted lines indicate amacrines in INL and ganglion cell bodies (G). Scale bar: 15 μm."

An important conclusion from (B) and (C) for the present context is that cone bipolar cells form the major input to class II (β) cells, while amacrine cells form the major input to the other two classes. [Reproduced with kind permission of Chapman & Hall.]

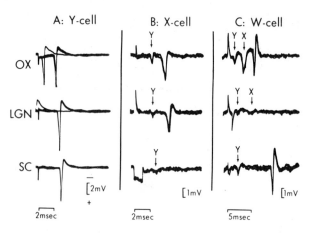

Figure 2.19. Evidence of different central projections of Y, X, and W cells (from Fukuda and Stone, 1974). These oscilloscope traces show antidromic responses of individual ganglion cells to stimulation of their axons by electrodes placed along their length, i.e., at various points along the retinal projections to central visual nuclei. The Y cell (left column) responds to stimulation of the optic chiasm (OX), of the SC, and of the A laminae of the LGN. The X cell does not respond to SC stimulation, nor the W cell to LGN stimulation. Thus, the X cell seems to project to the A laminae of the LGN and not to the SC, the W cell to the SC and not to the LGN, while the Y cell appears to project to both. The arrows point to field potentials generated by nearby Y- or X-cells. [Reproduced from the *Journal of Neurophysiology* with kind permission of the American Physiological Society.]

Sefton, 1979), and the retinal recipient zone of the pulvinar (Leventhal *et al.*, 1980). From the dLGN sites, W cell activity is relayed to areas 17, 18, and 19, but principally to area 19 (Holländer and Vanegas, 1977; Dreher *et al.*, 1978, 1980; Kimura *et al.*, 1980). Among W cells, central projections differ partially but significantly between suggested that as many as 10% of X cells project to the SC, the more direct evidence of Leventhal, Rodieck, and Dreher (personal communication) suggests that X cells reach the pretectal area, but not the SC itself. The receptive field properties of neurons in area 17 and the topography of area 17 reflect the strong representation of X cells in the geniculate input to area 17 (Stone and Dreher, 1973; Dreher *et al.*, 1980). *Y cells* project both to the SC and to laminae A, A1, and C and the MIN component of the dLGN (Cleland *et al.*, 1971; Hoffmann *et al.*, 1972; Dreher and Sefton, 1975, 1979; Mason, 1975; Sherman *et al.*, 1976; Dreher *et al.*, 1978). From the laminated dLGN, the relay cells with Y input project to areas 17 and 18 (Stone and Dreher, 1973) and from the MIN to areas 17 and 18 and the lateral suprasylvian area. The principal projection of *W cells* is to the SC, and they form approximately 90% of the retinal input to the SC (Hoffman, 1973; see Chapter 7). They also project to the pretectum (Leventhal, Rodieck, and Dreher, personal communication) and to several forebrain sites: the vLGN (Spear *et al.*, 1977), the MIN and parvocellular C-lamina components of the dLGN (Wilson and Stone, 1975; Cleland *et al.*, 1975, 1976; Wilson *et al.*, 1976; Dreher and small- and medium-soma γ cells. Most of the γ cells projecting to midbrain sites have small somas, and most projecting to forebrain sites have medium-sized somas. Two forebrain sites reached by W cells, the retinal-recipient zone of the pulvinar and the MIN

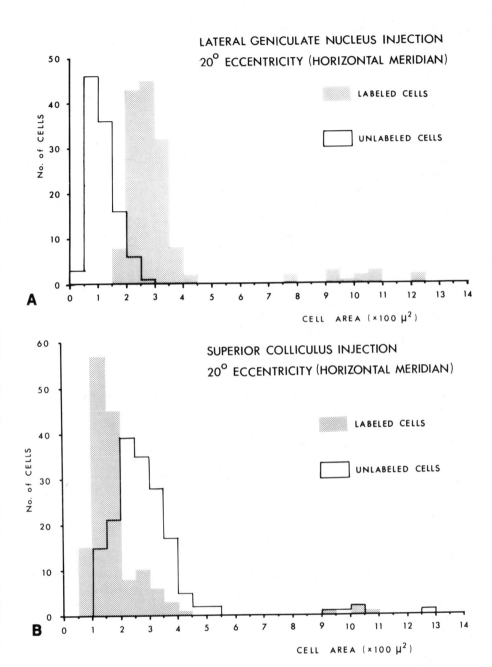

Figure 2.20. Central projections of different ganglion cell classes. Evidence (from Kelly and Gilbert, 1975) of different central projections of small (presumably γ/W cells), medium (presumably β/X cells), and large (presumably α/Y cells). (A) Following injection of HRP into the A laminae of the LGN, reaction product was found only in medium-sized and large cells (hatching). (B) Following injection of HRP into the SC, labeling was most widespread in small cells, with more limited labeling of medium-sized cells. Approximately 50% of the large cells were labeled.

Thus, there is a strong trend for small cells to project to the SC, medium cells to the A laminae of the LGN, and large cells to both areas. Perhaps half the large cells may project to both areas, by a branching axon. [Reproduced with kind permission of the *Journal of Comparative Neurology*.]

subdivision of the dLGN, receive projections from only the medium-soma γ cells (Rowe and Dreher, 1979, 1982*b*; Leventhal, Rodieck, and Dreher, personal communication).

The projections of Y, X, and W cells described above were strongly corroborated by studies employing the retrograde transport of HRP, (Kelly and Gilbert, 1975; Magalhaes-Castro *et al.,* 1975; LeVay and Ferster, 1977; Holländer and Vanegas, 1977; Ferster and Levay, 1978; Rowe and Dreher, 1979, 1982*b*; Leventhal, 1979; Wässle and Illing, 1980; Leventhal, Rodieck, and Dreher, personal communication; Stone and Keens 1978, 1980).

Systematic variation: There is limited evidence that Y-cell projections to area 17 and 18 differ systematically with eccentricity. Leventhal and Keens (1978) noted that the Y-cell projection to peripheral parts of areas 17 and 18 comes largely from the MIN, while the Y-cell projections to the area centralis regions of these two areas come principally via the laminated dLGN.

Residual variation: Very little is known concerning the degree of variation present in central projections of the major groups of ganglion cells. The studies of central projections just discussed were, by and large, studies of populations of cells. Because it is difficult to provide evidence that a particular cell does not project a given area, present techniques provide little evidence of the variation between otherwise similar cells in their central projections.

2.4. A TWO-GROUP (XY/W) CLASSIFICATION OF CAT RETINAL GANGLION CELLS

I have already imposed on the reader at some length concerning the value of developing multiple taxonomic levels in a classification of nerve cells, as of other biological entities; in this section I would pursue the argument one further step. In practice, the first step in developing such a classification is often quite mundane, a judgment that a particular variation in cell properties (such as the ON-center/OFF-center difference among X cells) is more fruitfully considered as evidence of two subgroups of the broader group X cells, than of two entirely different cell groups. The implications of that judgment are wide, however, for it assumes that classifications can comprise higher- and lower-order groups and that a particular cell can, depending on the context, be usefully regarded as a member of higher- or lower-order groups (just as, depending on context, we would describe a cat as a domesticated feline, a carnivore, a mammal, or a vertebrate).

Briefly stated, the advantages of multilevel classifications are (at least) three, when compared with the common alternative, the single-level classification in which all groups exist at the same taxonomic level:

1. Multiple-level classifications allow the multiple sources of variation in cell properties to be expressed, and allow a more complex and subtle interpretation of that variation. This point was discussed in Section 2.3.2.
2. Multiple-level classifications allow the variety of cell properties to be expressed without exaggerating the differences between them; and conversely, they allow

broad similarities to be traced without obscuring the differences. The lower-level groups express the variety of properties encountered, while the higher-level groups express the properties common to the subgroups.

3. When coupled with a hypothetico-deductive methodology (rather than the reliance on definitions that characterizes many single-level classifications), a multiple-level classification is heuristically effective; i.e., it both generates experimental questions and is responsive to the accumulation of new observations.

The main proposal of this section illustrates these three points. The evidence summarized in the preceding Section 2.3.3 seems insistently to lead to the conclusion that X and Y cells have so many common properties, when compared to the W-cell group, that they can fruitfully be regarded as subgroups of a higher-order group, which can then be viewed as existing at the same taxonomic level as W cells. The proposal, then, is that, in view of the evidence (point 3 above), a higher-order group of ganglion cells should be recognized that comprises X and Y cells; the name suggested for the group is *XY* cells (pronounced *zwi* cells). That recognition in no way diminishes the differences between X and Y cells; it does express their common properties (point 2 above). The classification of ganglion cells then proposed is then (hopefully) a more complete expression of the variation of ganglion cell properties, and of their functional significance (point 1 above). The argument for recognizing the XY grouping of cat retinal ganglion cells is as follows.

One of the features of the multiple interpretation of the properties of cat ganglion cells presented above is the prominence of systematic variation within the X- and Y-cell groupings, particularly in relation to retinal eccentricity, and of residual variation in the W-cell grouping. For example, among both X and Y cells, there is a tendency for the degree of tonicity and sensitivity to defocusing of the retinal image to increase between peripheral retina and the area centralis, and for receptive field size, axonal conduction velocity, and responsiveness to fast stimulus movements to decrease. These trends occur in properties that distinguish X from Y cells, and they occur in such a way that a central Y cell closely resembles a peripherally located X cell, although at any particular retinal location the X/Y distinction remains sharp. Similarly, the morphology of both α and β cells shifts systematically with eccentricity, again in such a way that the α/β difference remains strong at a particular eccentricity, while the difference between a peripherally located β cell and a central α cell is small. Boycott and Wässle (1974) note specifically that knowledge of eccentricity is needed to distinguish a central α from a peripheral β cell. The effect of this parallel, systematic variation in both X and Y cells is that the differences between them are maintained at all eccentricities. The parallel variation in the two cell systems suggests some common factor to their function; the maintenance of a relatively constant difference between them suggests that the differences are important aspects of the functional roles of, and interaction between, X and Y cells. The nature of this interaction is not well established. One possibility is that in animals with highly mobile eyes and restricted retinal areas specialized for high-resolution vision, e.g., the cat and many primates, a good deal of coordination is required between two distinct mechanisms, one (presumably the X cells) involved in high-resolution pattern vision, the other (presumably the Y cells) in the initiation and control of eye movements, so that moving or stationary targets can be detected and fixated. The saccadic and smooth pursuit eye movements of the cat are similar to those of primates, although somewhat less developed (Evinger and Fuchs, 1978; *see also* Dreher and Zernicki, 1969). In rabbits, on the other hand, in which the area centralis is weakly developed (Provis, 1979), both the small saccades and smooth pursuit eye movements observed in cats and primates are apparently absent (Collewijn, 1977); and Caldwell and Daw (1978) noted that the distinction between X and Y cells in rabbit retina is less clear than in the cat. In the monkey, by contrast, in which both the area of the retina specialized for fixation and the X-cell system seem highly developed, the differentiation between X and Y cells is reported to be very marked (Dreher *et al.,* 1976; Sherman *et al.,* 1976). Thus, the relationship between the presence of a central specialization of the retina and the occurrence of certain eye movements (Collewijn, 1977; see also Steinman, 1975) suggests a functional interaction between X- and Y-cell systems, and may be related to the maintenance in all retinal areas of sharp and relatively constant differences in the properties of the two cell groups. The pronounced sensitivity of Y cells to deprivation of patterned visual input during early postnatal life (Chapter 10, Section 10.2.1) further suggests that these cells play some role in pattern vision; the recent work of Lehmkuhle *et al.* (1980*a,b*) has considerably developed this idea.

By contrast, W cells and their morphological counterparts, γ cells, form a population that seems distinct from X and Y cells at all eccentricities; and, so far, they have been shown to vary with eccentricity in only one property, receptive field size (Section 2.3.3.1). As a group, they show far more residual variation in parameters such as receptive field size, axonal conduction velocity, and morphology than either X or Y cells. If a higher XY grouping is recognized, X and Y cells then would exist at the same taxonomic level as the major subgroups of W cells, i.e., W1 and W2 cells. This rearrangement of the relative taxonomic levels of W, X, and Y cells is an important corollary of the proposal of an XY-cell group.

What might be the functional categories for each of the taxa (groups) of ganglion cells being distinguished? The suggestion in Table 2.2 follows (as does much of this section) the analyses set out in Rowe and Stone (1980*b*), Stone *et al.* (1979), Section 2.2.1 and Section 11.2. Briefly, it is suggested that X and Y cells subserve what Trevarthen (1968) called *focal vision* and Stone *et al.* (1979) and Rowe and Stone (1980*b*) termed *foveal vision;* and that within that function, X cells subserve high-resolution pattern vision and Y cells movement detection, their combined function being to detect and direct the fixation of objects of interest whether stationary or moving, and to analyze the features of the fixated object. W cells, it is suggested, subserve what Trevarthen and several later workers have termed *ambient vision,* a form of vision that is not centered around any particular part of the visual field but which, Trevarthen suggested, forms a spatial framework for focal vision. Further discussion of the empirical basis and value of the focal/ambient distinction is set out in Chapter 11 (Section 11.2). That basis is far from compelling; a close reading of the evidence shows that its use in the present context requires extrapolation from human to animal contexts, and the assumption that the residual visual ability of animals with lesions to striate cortex is usefully regarded as a single subcomponent of normal vision. However, the focal/ambient distinction seems the clearest psychophysical correlate available for the XY/W classification of ganglion cells.

This difficulty in specifying compelling functional categories for all the groups of ganglion cells that it seems natural to form on the basis of their physical properties is of course disappointing; for it stems from inadequacies in our understanding of visual psychophysics, or of the visual pathways, or of both. Taking a positive view, however, the inadequacies of the classification in Table 2.2 suggest areas of work, in both the psychophysics and neurobiology of vision, that might strengthen both the classification itself and our understanding of the relation between visual performance and visual neurons. That work may require radical recasting of the classification of ganglion cells suggested above; but then the value of that classification is not its perfection, but its ability to generate experimental questions and respond to the results.

2.5. TWO NOTES ON THE CLASSIFICATION OF NERVE CELLS

Without anticipating too closely the substance of Chapters 4 and 5, the following two notes on the methodology of classification are offered to illustrate how inseparable are the problems of classification and of scientific methodology.

2.5.1. Incommensurable Classifications

In his influential 1962 book, T. S. Kuhn argued that periods of debate ("crisis," "revolution") in a scientific field are marked by the currency of alternative, competing "paradigms" for the understanding of that field. The rivalry between two "paradigms" (or theories, or hypotheses, etc.) is ideally settled by experimental test, but proponents of rival views may hold fiercely to their ideas and may not agree on what constitutes an appropriate experimental test; and, further, they may interpret the same observations quite differently. so that the choice between the paradigms may prove unresolvable by experiment. Because of the psychology of investigators, therefore, competing paradigms may, Kuhn suggested, prove "incommensurable."

To take a topical example, a number of groups of workers interpret linearity and nonlinearity of summation as defining characteristics of X and Y cells, and refer to Enroth-Cugell and Robson (1966) in support. On the other hand, Rowe and Stone, seeking on the basis of a different methodology of classification to delineate functional groups of cells by description of many of their properties (rather than by definition in terms of any one), interpret the same report of the linearity and nonlinearity of two groups of ganglion cells as a description of variation in one of their properties, rather than as a definition of their distinct types or essences. Some workers have gone one way on this point, some the other. Whatever the reason for their choice, its influence on classification and terminology is fundamental, leading to the distinct parametric and feature extraction classifications of visual neurons discussed in Chapter 1 and Chapter 2 above.

A comparable choice of methodology separates the parametric $\alpha/\beta/\gamma$ classification from the 23-group classification proposed by Kolb (1981). This latter grouping, in which many groups are defined by a single morphological feature, seems typological or essentialist in approach and vulnerable to the criticisms of that approach formulated by (among others) Tyner (1975) and Rowe and Stone (1977, 1979, 1980b), and in Chapter

5. However, the relative advantages of these approaches are not the present issue. Kuhn's point (though he expressed it differently) is that the choice of interpretation is not, itself, experimentally testable. That choice depends on the assumptions of the classifier as to what methodology he should follow, and perhaps on his or her "psychology," and there seems to be no way of experimentally proving or disproving such assumptions. In this sense, at least, different classifications are often "incommensurable."

Kuhn's concept of the "incommensurability of paradigms" has been criticized (e.g., Lakatos, 1970) because it seems to imply that scientists are purely irrational both in their choice and in their changes of view. Even granting this criticism, however, the concept does express the mutual exclusiveness of different sets of presuppositions, which tends to divide scientists into two (or more) very distinct schools of analysis, such as the "parametric" and "feature extraction" schools discussed in this and the previous chapter.

2.5.2. Mixed Classifications: The Best of Both Approaches?

Interestingly, some scientists mix presuppositions that others find incommensurable, and their readiness to do so seems to be an important exception to Kuhn's concept. Rodieck (1979), for example, proposed a classification of ganglion cells that mixes the parametric Y/X groupings with a typological definition of other groups and seems to me to mix the underlying presuppositions. Thus, in the context of a detailed analysis of the evidence, Rodieck proposes retaining the X- and Y-cell groupings and labels. Since, however, the W-cell grouping seems more heterogeneous than the X- or Y-cell groups, he proposes abandoning it and replacing it with five new groups, each less heterogeneous and each named and defined by certain physical features (color-coding, local edge detector, direction-selective, phasic, tonic, suppressed-by-contrast). Arguably, this approach overcomes the "problem" of the heterogeneity of W cell receptive field properties but it also involves the disadvantages of the definitional approach. Those disadvantages are (at least) two.

In the long term, there is the disadvantage that a classification established by definitions will, for reasons discussed in Rowe and Stone (1977, 1979), be untestable, a codification of old ideas instead of an avenue to new ideas. In the short term, the defining of groups by particular characteristics has the effect, as Mayr (1969) has commented for animal taxonomy, of exaggerating "the constancy of taxa and the sharpness of the gaps separating them." Two examples may be sufficient to make the latter point.

First, although Rodieck's reclassification of W cells into five separate groups seems to seek the physical homogeneity of cell groups, the groupings he proposes instead still contain substantial variation. Among both direction-selective and suppressed-by-contrast cells, for example, considerable variation in properties has been described and discussed (Stone and Fukuda, 1974a; Cleland and Levick, 1974b; and Rowe and Stone, 1976a). The defining of groups by single properties implies that the cell groupings distinguished are homogeneous, at least in the "important" properties of the cells. The effect is to obscure variation within groups; in Mayr's words, "to exaggerate the constancy of taxa."

Conversely, in Rodieck's classification, local edge detectors are given separate taxonomic status from phasic W cells, despite the evidence of a continuity of properties between them (Stone and Fukuda, 1974a; Rowe and Stone, 1977). The definitional approach thus tends to exaggerate the sharpness of the gaps between taxa.

However, the fundamental problem in the "mixed" approach to cell classification seems (to me) to be not the mixing of alphanumeric with descriptive terminologies, or of description with definition, or even the exaggerations just mentioned, but the mixing of methodologies. When considering the situation in the cat, Rodieck (1979) keeps the X- and Y-cell groups of previous workers, with their noncommittal terminology and nondefinitional basis, but takes a definitional, typological approach with W cells. Again, Dreher *et al.* (1976), in their study of the visual pathway of the monkey, propose X- and Y-like cell groups among relay cells of the LGN and, by implication, for the retina. Their proposal is supported by what seems to me a sound analysis, an articulate example of the parametric approach to classification discussed above. However, it is fair and relevant to comment that the X-like grouping they proposed for relay cells of the monkey LGN (and by implication for retinal ganglion cells) included both color-coding and broad-band cells; whereas in the cat the color-coding property of a minority of cells was taken by Rodieck as the single definiting characteristic of a separate cell group.

Perhaps this issue of methodology is really less important than it seems to me to be; perhaps scientists should be free to be inconsistent in these matters (consistency being the "last refuge of the unimaginative"). My own judgment (and, in the final analysis, it is a matter of judgment) is different; it is that classification is fundamental to any body of empirical knowledge and that the creative resolution of the problems of nerve cell classification depends on the degree of understanding we have of the presuppositions of the neuroscientists working in the area; in the present case, of our own presuppositions.

Ganglion Cell Classification   3
in Other Species

Classifications of retinal ganglion cells have been proposed for several species other than the cat, including both mammals and nonmammals. The "parametric" and "feature extraction" approaches to cell classification distinguished in studies of cat ganglion cells (Chapters 1 and 2) are also apparent in the work discussed here; and, as in the cat, some of the variety in the way cells have been classified and named can be usefully understood in those terms. Perhaps the most important aspect of a comparative survey of ganglion cell groupings, however, is the insight such comparisons provide into the phylogenetic history of ganglion cells, and their groupings.

3.1. IN THE MONKEY

3.1.1. Conduction Velocity Groupings

Two studies have reported specifically on conduction velocity groupings in monkey optic nerve; both relied on field potential recordings from the optic nerve or tract. Ogden and Miller (1966) concluded that two major groupings are present in the nerve of the rhesus macaque monkey *(Macaca mulatta)*, with conduction velocities of 8 and 4 m/ sec. Griffin and Burke (1974) reported two clear groups in the optic nerve of the cynamolgus monkey or crab-eating macaque *(M. irus)*, with much higher velocities, 22 and 11 m/sec. Further, they traced both groups to the LGN, and suggested that the faster-conducting axons terminate in the magnocellular laminae of that nucleus, and the slower fibers in the parvocellular laminae. The difference in velocities reported by the two studies could, of course, be a species difference, but I suggest below that Griffin and Burke's estimate is probably the more accurate, for the rhesus as well as the crab-eating macaques.

Estimates of axonal conduction velocities of monkey ganglion cells have also been made in single-unit studies (Gouras, 1969; DeMonasterio *et al.,* 1976; Schiller and Malpeli, 1977*a*). Because of the sampling limitations inherent in them, such studies do not provide a strong guide to the major velocity groupings present. On the other hand, they have, as in the cat, detected the presence of ganglion cells with very-slow-conducting axons, whose activity was not apparent in field potential recordings. Gouras (1969) reported that the two groups of ganglion cells he distinguished in the rhesus macaque retina by their receptive field properties had largely distinct axonal conduction velocities (mean velocity for phasic cells was 3.8 m/sec, for tonic cells 1.8 m/sec, with some overlap in the ranges; Fig. 2.4). These values are much lower than Ogden and Miller had reported for the same species, probably for technical reasons. Gouras' estimates were based on the latency of a cell's antidromic spike response, recorded at its soma, after stimulation of its axon in the optic tract. The action spike traversed both the myelinated, extraretinal part of the axon (along which Ogden and Miller had measured velocity) and the much-slower-conducting, unmyelinated intraretinal segment of the axon, so that Gouras' estimates are a mean of extra- and intraretinal values. Since the length of the intraretinal segment varied from cell to cell, this problem may also have contributed to some of the overlap of conduction velocity that Gouras reported between tonic and phasic cells. DeMonasterio *et al.* (1976) reported similarly slow conduction velocities, for apparently the same reason.

Working with the rhesus macaque, Schiller and Malpeli (1977*a*) measured the velocities of intra- and extraretinal segments of the axons separately, by recording the antidromic spike responses evoked in individual cells by stimulation of the SC, LGN, and optic chiasm. They reported that the extraretinal portions of the axons of one of the major ganglion cell groups they distinguished (broad-band cells) had a mean conduction velocity of 22 m/sec; the corresponding value for the axons of the second major group (color-opponent cells) was 12.9 m/sec. These values are close to those reported by Griffin and Burke (1974) in *M. irus.* For the same groups of axons, Schiller and Malpeli estimated intraretinal velocities of 1.3 and 0.9 m/sec, respectively, suggesting that the extraretinal segment of an axon conducts over 10 times faster than its intraretinal segment.* Schiller and Malpeli also recorded antidromic responses from a small number of ganglion cells that they termed *rarely encountered.* These have receptive fields remarkably similar to those of cat W cells and, like W cells, project to the SC. They further noted that "the rarely-encountered neurons tend to have long latencies, especially those that were demonstrated to be retinotectal." Schiller and Malpeli's Fig. 17 suggests that the rarely encountered cells resemble cat W cells in having axons more slowly conducting than the two major conduction velocity groups of the optic nerve. Estimates of their conduction velocity were not made, however.

In summary, the optic nerve of the monkey resembles that of the cat in comprising two prominent conduction velocity groups (which correspond to the axons of X- and Y-like cells) and a group of very-slow-conducting axons whose activity does not form a substantial field potential, and which project to the SC. Differences between the two species in the conduction velocity groupings of their axons certainly exist; in absolute

---

*The corresponding ratio in the cat was estimated at 6.3–9.5 by Stone and Freeman (1971).

values, for example, the cat groups conduct at nearly twice the velocity of the monkey groups. The general pattern of groupings is remarkably similar, however.

3.1.2. Physiological Classifications: Parametric and Feature Extraction

Classifications of monkey retinal ganglion cells based on physiological observations have been proposed by Gouras (1969), DeMonasterio and Gouras (1975), DeMonasterio *et al.* (1976), Schiller and Malpeli (1977*a*), and DeMonasterio (1978*a,b*); and physiologically based classifications of relay cells of monkey dLGN, which directly imply classifications of retinal ganglion cells, have been proposed by Marrocco and Brown (1975), Dreher *et al.* (1976), and Sherman *et al.* (1976). The retinal studies are largely (but not entirely) typological in approach, while the geniculate studies of Dreher *et al.* (1976) and Sherman *et al.* (1976) took a parametric approach. Despite differences in terminology as well as approach, however, the ganglion cell groupings suggested by these studies correspond remarkably closely to the Y-, X-, and W-cell groupings of cat retina.

The first physiological classification of monkey ganglion cells that allows comparison with the classifications developed for the cat was proposed by Gouras (1969), who reported that ganglion cells recorded in the foveal region of the rhesus macaque *(M. mulatta)* retina could be characterized as either "tonic" or "phasic," by several properties. First, "tonic" cells responded to a flashing spot stimulus with a tonic pattern of firing, i.e., a burst of spikes that was sustained as long as the spot was presented. "Phasic" cells, by contrast, gave only short, transient bursts of spikes, which died away before the stimulus ended. Second, the tonic cells concentrated near the fovea, while phasic cells were relatively more common in peripheral retina. Third, the tonic cells had distinctly slower-conducting axons than phasic cells (Fig. 2.4). Further, Gouras inferred from their axonal conduction velocities (correctly, as subsequent work has shown) that tonic cells are smaller than phasic cells. In these four properties, tonic cells differ from phasic cells in the same ways as cat X cells differ from Y cells. In addition, Gouras showed that many of the tonic cells were "color-coding" (i.e., that their responses vary with the wavelength of the stimulus light), while phasic cells did not show this specificity.

Gouras' descriptions have been confirmed and expanded in subsequent studies. DeMonasterio and Gouras (1975) described three groups of monkey ganglion cells, one group being "concentric" and "color-coding," a second group being "concentric" and "broad-band," and the third being "nonconcentric." That is, they followed Cleland and Levick (1974*a,b*)* and Levick (1975)* in taking "concentricity" of receptive field orga-

---

*These authors considered a receptive field "concentric" if it could be shown to comprise an ON-center and a concentric OFF-surround, or, conversely, an OFF-center and an ON-surround. Other concentric arrangements [e.g., an ON-OFF-center and a concentric, inhibitory surround or the concentric but coextensive receptive field mechanisms present in Wiesel and Hubel's (1966) type II cell] are considered "nonconcentric," along with fields with clear nonconcentric organization, such as direction- and orientation-selectivity. Effectively "concentric" receptive fields resemble the patterns that were described by Kuffler (1953) and were for many years considered to be the only patterns present in the cat retina; "nonconcentric" fields have more recently recognized patterns.

nization as a key property on which a dichotomy could be based, and selected another property, color specificity, as the basis of a second dichotomy. The "broad-band/concentric" cells generally resembled the phasic cells described by Gouras (1969) and cat Y cells. Their responses seemed independent of stimulus color, and they were relatively more common in peripheral retina. They gave phasic responses to standing contrast stimuli, and had larger receptive fields (Fig. 3.1). The "color-opponent" cells, on the other hand, resembled Gouras' tonic cells and cat X cells in being more numerous (63% of their sample as against 24% for phasic cells), in giving tonic responses to standing contrast stimuli, in having small receptive fields, and in being particularly frequent at the fovea. The "nonconcentric" cells were a newly recognized group, and comprised cells with two patterns of receptive field organization, ON-OFF-center and "movement-sensitive." They had generally large receptive fields (Fig. 3.1), comprised only 9% of the sample, and did not seem to concentrate either at the area centralis or in peripheral retina. In several ways they resemble cat W cells, extending the similarity between cat and monkey groupings.

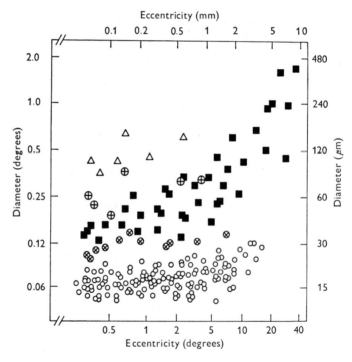

Figure 3.1. Physiological groupings of monkey ganglion cells (from DeMonasterio and Gouras, 1975). Plot of the size of the receptive field center regions of monkey ganglion cells as a function of the distance of the cell from the center of the fovea (eccentricity). Data are shown for concentric color-opponent cells (○), for color-opponent cells receiving input from two cone mechanisms to the center (⊗), for broad-band cells (both varieties) (■), for nonconcentric color-opponent cells (⊕), and for nonconcentric cells with phasic responses (△). [Reproduced with kind permission of the *Journal of Physiology (London)*.]

Schiller and Malpeli (1977*a*) also proposed three broad groupings of monkey ganglion cells. Like DeMonasterio and Gouras, they relied on color sensitivity to classify most of their sample of monkey ganglion cells into two groups, "broad-band" or "color-coding" (Fig. 3.2), which, these authors noted, closely resemble the Y and X cells of the cat, respectively. The broad-band cells gave more transient responses to standing contrast stimuli (Fig. 3.3) and had lower spontaneous firing rates than color-opponent cells; when stimulated with slowly drifting grating stimuli, they showed a nonlinear increase

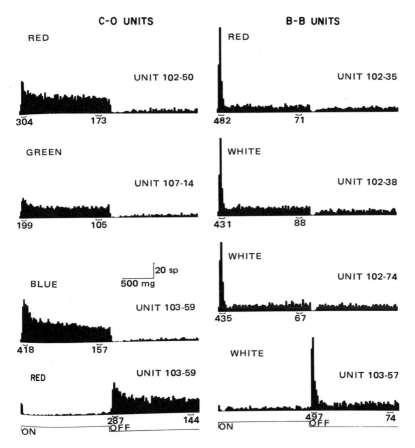

Figure 3.2. The color-opponent/broad-band grouping proposed for monkey ganglion cells by Schiller and Malpeli (1977*a*). The responses of monkey ganglion cells are shown as averaged response histograms. Those in the left column were obtained from color-opponent cells, those in the right column from broad-band cells. In each case, the stimuli were small, stationary discs of light of the color indicated, centered on the receptive field. The stimuli were turned on and off at the times shown at the bottom. The color-opponent cells gave tonic responses, i.e., the response continued as long as the stimulus remained; the responses of the broadband units were more phasic. The bottom two histograms in the left column were obtained from the same cell and demonstrate its color-opponency (it gave an ON-response to a blue spot and an OFF-response to a red spot). [Reproduced from the *Journal of Neurophysiology* with kind permission of the American Physiological Society.]

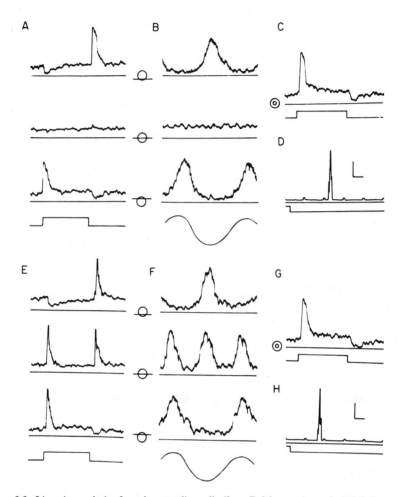

Figure 3.3. Linearity analysis of monkey ganglion cells (from DeMonasterio *et al.,* 1976). Responses of linear (A–D) and nonlinear (E–H) ganglion cells of monkey retina to a range of stimuli. The three histograms in A show the response of a linear cell to square-wave modulation of a bipartite field stimulus; i.e., the stimulus comprised darker and lighter panels separated by a sharp, horizontal border. The contrast between the two panels was alternated (right half bright then left half bright), the changes occurring in sudden step fashion. When the border was centered on the receptive field center (middle histogram), the cell did not respond; it apparently summed the influences of the two halves of its receptive field linearly. When the border was a little above or below the center of the field, strong responses were elicited. The histograms in B show a comparable set of responses when the contrast of the stimulus was modulated sinusoidally; again, a "null" response was obtained when the border was centered on the receptive field. The histogram in C shows the response of the cell to a small centered spot of light turned on and off at times indicated by the steps in the lower trace. The histogram in D shows the antidromic response of the cell to a brief electrical stimulus delivered to the optic tract at the time indicated by the step in the lower trace. For A–C, the time calibration is given by the length of the step in the bottom trace of column A, which represents 100 msec. For D, the small periodic marks in the histogram mark intervals of 2 msec.

Histograms E–H show a matched set of responses for a nonlinear cell. Note the strong, double-frequency responses obtained in the middle histograms in E and F; these indicate the cell's nonlinearity. This cell's antidromic response to stimulation of the optic tract is shown in H; note that it occurs earlier than the response to this stimulus of the linear cell (D). The nonlinear cell has a faster-conducting axon. [Reproduced with kind permission of *Vision Research.*]

in mean firing rate; and they had, on the average, faster-conducting axons than color-opponent cells (mean axonal velocity of 22 m/sec as against 12.9 m/sec). Further, they comprised a relatively small percentage of the population of cells and projected to the SC (which the more numerous color-opponent cells do not). However, Schiller and Malpeli noted considerably more overlap between "broad-band" and "color-opponent" groups in their axonal conduction velocities and in the time courses of their responses to standing contrast stimuli than has been reported for cat X and Y cells. A third group of cells was also distinguished, which resembled DeMonasterio and Gouras' "nonconcentric" cells and cat W cells. Their receptive field patterns were varied, some having ON-OFF-center regions, others being suppressed by stimuli of either contrast, very few showing any color specificity. Further, like W cells, they had large receptive fields, low spontaneous activity, slow-conducting axons, and projected to the SC [also shown by Marrocco and Li (1977) and Marrocco (1978)], apparently comprising the major component of the retinal input to the SC. Despite their common properties, Schiller and Malpeli considered their receptive fields too heterogeneous for these cells to form a distinct functional group and the cells were therefore given a negative name: *rarely encountered*. Subsequently, however, Schiller *et al.* (1979) termed the same cells *W-like*, thus to some extent recognizing their common properties and moving toward a parametrically based grouping of them.

In summary, then, both DeMonasterio and Gouras (1975) and Schiller and Malpeli (1977a) distinguished "broad-band" and "color-opponent" cells that generally resemble Gouras' phasic and tonic cells, and cat Y cells and X cells, with rather greater overlap of properties than in the cat; both also distinguished another group of cells that resemble cat W cells in several properties, but which were given negative names, viz., *nonconcentric* and *rarely encountered*.

DeMonasterio *et al.* (1976) extended the analysis of monkey ganglion cells to include (Fig. 3.3) the linear/nonlinear spatial summation properties described as a distinguishing feature of cat X and Y cells by Enroth-Cugell and Robson (1966), and axonal conduction velocity. Their results indicate that broad-band cells further resemble Y cells in the nonlinearity of their spatial summation, while color-opponent cells resemble X cells in being linear; and they confirmed Schiller and Malpeli's observation that the axons of the broad-band cells conduct at about twice the velocity of the axons of color-opponent cells. They too noted considerable overlap in the conduction velocity ranges of broad-band and color-opponent cells, despite the mean difference. Marrocco's (1978) report of the properties of relay cells in the dLGN also described a lack of correlation between tonicity of response and color-specificity or conduction velocity, confirming the impression that any groups that might be distinguished among monkey ganglion cells are less distinct than in the cat, with considerable overlapping of properties.

DeMonasterio's later (1978a,b) reports on the properties of monkey retinal ganglion cells employed the X/Y terminology developed in the cat, yet they also added to the impression that Y, X, and W groupings are not as clearly distinguishable as in the cat. It was not that Y- and X-like cells could not be distinguished; DeMonasterio (1978a) applied a single linearity test to define these groupings. However, when distinguished by this single test, X- and Y-cell groupings did not show the sharp differences in axonal conduction velocity characteristic of cat X and Y cells. Nevertheless, the X cells so defined had generally slower axons than Y cells and more tonic responses to standing contrast stimuli. Furthermore, X cells tended to concentrate at the fovea centralis, as they do at the cat area centralis, and were shown to project exclusively to the dLGN, and not to the SC, again resembling cat X cells.

DeMonasterio (1978*b*) reported on the properties of monkey ganglion cells with "atypical" receptive fields; "atypical" here seems to mean the same as "nonconcentric," as discussed in the last footnote. Unfortunately, for a comparison with cat W cells, DeMonasterio included as atypical many cells that had the class 2 pattern of color-sensitive mechanisms described by Wiesel and Hubel (1966). These were considered atypical because they lack a surround region to their receptive field; they were grouped separately from X cells despite the fact that, like X cells, their spatial summation properties were linear, they were color-opponent, tended to concentrate at the fovea centralis, and projected to the dLGN and not to the SC. However, among the other atypical cells were cells that had ON-OFF receptive fields and projected to the SC, and a third group that resembled cat suppressed-by-contrast W cells. Broadly, I would argue, these and the earlier studies provide striking evidence of Y-, X-, and W-like groupings among monkey ganglion cells, despite the variety of approaches and terminologies used; and despite the different phylogenetic histories of the two species. It is true that all the above studies reported more overlap of properties between groups than reported in the cat; but it is also possible that this overlap stems not from the indistinctness of the groups, but the typological emphasis of the above studies, which all tended to define cell groups in terms of particular physical properties (color-opponency, tonicity, concentricity, linearity).

It is noteworthy, therefore, that evidence of much sharper groupings of monkey ganglion cells emerged in Dreher and co-workers' (1976) classification of relay cells in the macaque dLGN into Y-like and X-like cells, and in Sherman and co-workers' (1976) classification of relay cells in owl monkey, dLGN into X and Y cells. Both groups of workers took a parametric approach, noting that relay cells in the parvocellular laminae of the LGN differ from magnocellular lamina cells in many of the ways in which cat X cells differ from Y cells, but not defining their groups in terms of any one property. Thus, Dreher and co-workers noted that cells in the parvocellular laminae generally gave tonic (as against phasic) responses to stationary contrast stimuli, were capable of higher spatial resolution, were less responsive to fast-moving stimuli, and received retinal input from a slower-conducting group of optic tract axons (Fig. 3.4). There were, of course, species differences between X cells in the cat and X-like cells in the monkey and between Y cells and Y-like cells; for example, the monkey cells had generally smaller receptive fields, and the X-like group contained many color-opponent cells. However, the differences between parvo- and magnocellular layer cells in the monkey closely resembled the Y/X difference seen in the cat. Moreover, the differences seemed better developed in the primate. Dreher and co-workers commented:

> The X/Y distinction, described in cats, is even more striking in monkeys. Every lateral geniculate cell we studied could be correctly classed as either X-like or Y-like on the basis of any of three tests. . . .

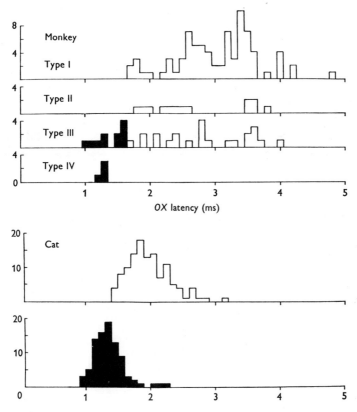

Figure 3.4. Conduction velocity analysis of optic tract afferents to relay cells in monkey LGN (from Dreher *et al.,* 1976). Frequency/latency histograms are shown from a study of relay cells of monkey LGN. The latency measure is the time from a brief electrical stimulus delivered to the optic chiasm (OX) to the cell's earliest action spike response; it is a measure of the conduction velocity of the ganglion cell axon providing the principal drive to the cell.

The types I–IV follow the groupings of Hubel and Wiesel (1968); type I and II cells are color-opponent (though in distinct ways), and types III and IV are nonopponent. The filled and open segments of the histograms represent cells with Y-like (closed) and X-like (open) responses. Thus, the Y-like cells have fast-conducting afferents and are spectrally nonopponent, while the X-like group has slower afferents and includes all the spectrally opponent cells.

The two histograms at the bottom show reference data for the cat; the authors note that the overlap of latencies of Y-like and X-like cells is less in the monkey than in the cat. [Reproduced with kind permission of the *Journal of Physiology (London)*.]

These were tonicity, responsiveness to fast-moving stimuli, and afferent conduction velocity (Fig. 3.4).

Sherman and co-workers noted similarly that "the difference between X- and Y-cells based on the tonic–phasic distinction was particularly dramatic in the owl monkey" (Fig. 3.5). They too observed the segregation of X cells to the parvocellular and Y cells

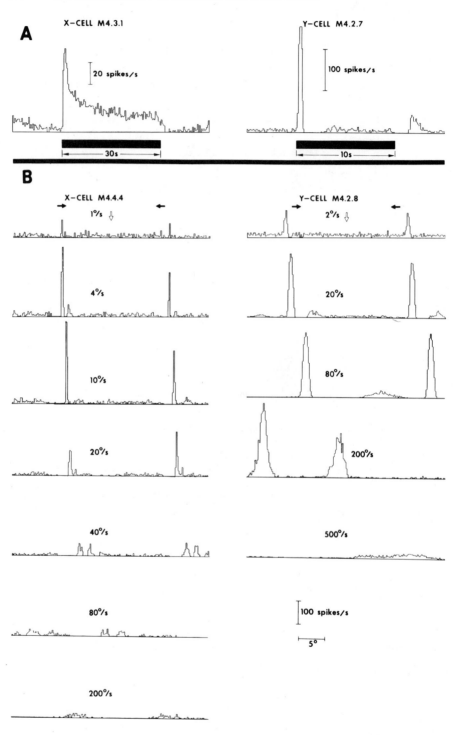

to the magnocellular layers, and added the valuable observation that, as in cats, the conduction velocity of a cell's axon was closely correlated with the velocity of its afferent. Y cells, with fast-conducting afferents, had fast-conducting axons, while for an X cell both its afferent and axon were, on the average, slower-conducting.

One feature of Dreher and co-workers' study that seems important in relating the geniculate studies to the earlier retinal work is that, while many cells in their X-like group were color-opponent (and all color-opponent cells were X-like in their other properties), a significant minority of X-like cells (32%) were not color-opponent. In DeMonasterio and Gouras' (1975) and Schiller and Malpeli's (1977a) classifications, these were presumably grouped with Y-like cells as "broad-band"; hence, the overlap these workers reported in many of the properties of broad-band and color-opponent cells. One point in support of the parametrically based X/Y grouping used in the geniculate studies is that it matches well the morphology of monkey dLGN (X-like relay cells being found in the parvocellular laminae, Y-like cells in the magnocellular laminae) and of geniculate terminations in the visual cortex (Chapter 8, Section 8.2.1). Schiller and Malpeli's (1978) report of the functional organization of monkey dLGN confirmed these patterns as well as adding new evidence of a segregation of ON-center from OFF-center relay cells.

Both Sherman and co-workers' and Dreher and co-workers' studies concerned geniculate relay cells, but it is a reasonable assumption that the properties of these cells indicate the existence of corresponding groups of retinal ganglion cells. These parametric studies lead to the conclusion that X-, Y-, and (arguably) W-cell groupings exist among monkey ganglion cells that are at least as striking and sharp as those in the cat.

3.1.3. The W-like System of Ganglion Cells

The suggestion that a group of ganglion cells can be identified in the monkey that corresponds closely to the W-cell group of cat retina is to a considerable extent my interpretation of others' studies, but it is in good agreement with Schiller and co-workers' (1979) suggestion in the Introduction to their study of corticotectal projections and with Malpeli and co-workers' (1981) summary of ganglion cell classes. The evidence for the suggestion may be summarized as follows. Ganglion cells have been recognized in monkey retina that resemble cat W cells

1. In having receptive field organizations similar to those found in cat W cells, such as ON-OFF-centers and suppressed-by-contrast properties.

Figure 3.5. X/Y analysis of owl monkey LGN. The histograms show the averaged responses of cells in owl monkey LGN to stationary and moving stimuli (from Sherman *et al.*, 1976). The left column shows responses of an X cell, the right column of a Y cell. The two histograms in (A) show the responses of the cells to a small, stationary spot of light positioned in the receptive field center and flashed on and off; the response is sustained for the X cell, transient for the Y cell. The histograms in (B) show responses to a slit of light moved across the receptive field at the speeds indicated. Note that the X cell ceases to respond clearly at speeds higher than 40°/sec, while the Y cell still responds clearly at 200°/sec. [Reproduced from *Science*, Vol. 192, with kind permission of the American Association for the Advancement of Science; copyright 1976 by the American Association for the Advancement of Science.]

2. In having slow-conducting axons (Schiller and Malpeli, 1977*a*).
3. In projecting to the SC and perhaps being predominant in that projection (Schiller and Malpeli, 1977*a*).
4. In having relatively large receptive fields.

Moreover, as set out in the following section,

5. Morphological studies show the presence of γ-like ganglion cells in monkey retina. Such cells were apparent in Polyak's drawings (Fig. 1.16), and their occurrence has recently been confirmed by Leventhal *et al.* (1981). Their properties match those of the W-like cells seen physiologically and they resemble cat γ and W cells in soma size, axonal caliber, and central projections.

3.1.4. Morphological Classifications

Three principal groups of workers have described the classification of monkey ganglion cells according to their morphology, viz., Polyak (1941, 1957), Boycott and Dowling (1969), and Leventhal *et al.* (1981). In discussing them, an interesting parallel emerges with the morphological classification of cat ganglion cells, namely this: The early classifications tended to be typological, with considerable numbers of groupings defined by certain physical features. It was when (1) the cells were seen in whole mounts, in which their full dendritic trees could be seen, and (2) descriptions of physiological classifications were available, that simpler, parametrically based classifications came to be proposed.

Cajal's (1893) classifications of vertebrate ganglion cells, which did not extend to primates, relied almost entirely on the level of the inner plexiform layer at which a ganglion cell's dendrites branched (see Fig. 1.15). Polyak adopted this approach, but relied also on the shape of the dendritic field as seen in sections cut across the thickness of the retina, distinguishing, for example, "parasol" and "shrub" types (Fig. 1.16). His classification was typological and has proved difficult to relate to subsequent physiological studies. The exception has been Polyak's "midget" ganglion cell, which shows a characteristic morphology (Fig. 1.16), but was also characterized by another two parameters, its topography (it is most numerous and clearly differentiated at the fovea), and connections (Polyak concluding that it receives input from a single cone bipolar and hence provides a "private line" for a single cone into the brain). Perhaps (as already commented in Chapter 1) it is because of this multiparameter basis that the midget grouping has proved particularly useful.

Boycott and Dowling (1969) confirmed many of Polyak's observations on monkey ganglion cells, their descriptions largely following Cajal's emphasis on the sublamina of the inner plexiform layer at which the cell's dendrites branched. The groups they distinguished were: midget (following Polyak), diffuse (dendrites spreading in all sublaminae), stratified diffuse (dendrites spreading in two or three adjacent sublaminae), unistratified (dendrites spreading in only one sublamina), and displaced (cell body in the inner plexiform layer). These classes too are strongly typological, and the groupings are not exclusive. Midget cells are all unistratified, for example, and displaced ganglion cells must also be classifiable into one of the other groups according to their dendritic formations. With the exception again of the more widely based midget group, those classifications have not proved useful in subsequent advances in the understanding of these cells. The problem is not that the criteria used in these early classifications are not of importance. The sublamina at which a cell's dendrites spread has been shown, at least in the cat, to relate to the ON-/OFF-polarity of its center region (Chapter 2, Section 2.1.4); and in birds (Reiner *et al.*, 1979), though not in rabbits (Oyster *et al.*, 1980), displaced ganglion cells have been shown to have a particular central projection (to the nucleus of the basal optic root, perhaps a homolog of the mammalian accessory optic tract nuclei). The problem is that, as argued previously (Rowe and Stone, 1977, 1979) and in Chapters 4 and 5, definition of cell groups by physical features places undue emphasis on those features, and distracts attention from others. As in the cat (Boycott and Wässle, 1974), a more parametric approach (combined with retinal whole mounts) has led to a simpler, yet more broadly based and heuristically effective classification of monkey retinal ganglion cells.

Leventhal *et al.* (1981) examined the morphology of monkey ganglion cells as shown by the retrograde transport of HRP from various central injection sites; they used a cobalt intensification technique to gain a better demonstration of dendritic trees. They propose a four-group classification of monkey ganglion cells based on soma size, dendritic morphology, axonal caliber, and central termination of the cell's axon. The classes were named A, B, C, and E because of the close correspondence between them and respectively the α, β, γ, and ϵ groups the same authors observed among cat ganglion cells (Figs. 3.6 and 3.7).

A cells are large cells (Fig. 3.6). They have large somas and wide-spreading dendritic fields, and the stoutest axons of all four types. They are the only cell type to project to the magnocellular laminae of the dLGN, and do not project to the parvocellular laminae. A minority of them project to the SC. In all these features, they seem closely to match the properties of monkey Y-like cells (which have fast axons, large receptive fields, projections to the magnocellular dLGN and to the SC).

Conversely, *B cells* (Fig. 3.6) have small cell bodies, and compact dendritic trees, much smaller than those of A cells. They are the only cell type to project to the parvocellular laminae of the dLGN and do not project to the magnocellular laminae, or to the midbrain. Leventhal and co-workers note that B cells correspond closely to the midget group described by Polyak. Their properties match closely the properties of X-like cells.

C cells closely resemble the γ cells described by Boycott and Wässle (1974) in the cat and the small diffuse ganglion cells described in monkey retina by Polyak (see Fig. 1.16). They have small cell bodies, and loosely branched, wide-spreading dendritic fields and fine axons. They project to the SC in considerable numbers, and do not appear to project to the dLGN. Their properties and projections match closely those of W-like ganglion cells.

E cells have medium to large cell bodies, fine- to medium-caliber axons, and large dendritic fields and were shown to project to the pretectum and not, so far, to any other site. In several respects (having medium-sized somas and thin axons, and wide-spreading dendritic fields), they resemble cat ϵ cells [the "medium-soma γ cells" of Stone and

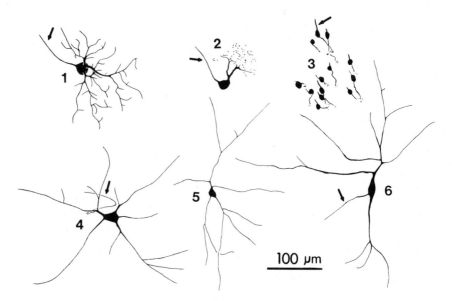

Figure 3.6. Morphological groups of ganglion cells distinguished in monkey. Leventhal *et al.* (1981) distinguished several groups comparable to those distinguished in the cat. For example, the cells labeled 2 and 3 resemble cat β cells and were termed B cells; these seem identical with Polyak's midget cells. Like β cells, they are relatively small in size at the fovea (area) centralis; compare the group of cells labeled 3, which were found near the fovea, with cell 2, which was more peripherally located. These workers also observed cells that seemed to correspond to α cells (cell 1, termed *A cells*), to γ cells (cell 5, termed *C cells*), and to ε cells (cell 4, termed *E cells*). Cell 6 was "unclassified". The arrows point to axons. [Reproduced from *Science,* Vol. 213, with kind permission of the American Association for the Advancement of Science; copyright 1981 by the American Association for the Advancement of Science.]

Clarke (1980)]. A physiological grouping clearly corresponding to them has not yet been described, as is the case too with ε cells.

These descriptions do not amount to a comprehensive classification, as the authors themselves make explicit, but they do provide the first parametrically based classification of monkey ganglion cells. The similarity between monkey and cat that emerges from these descriptions is, it seems to me, compelling, and is enhanced by two further points.

First, some issue was made in Chapter 2 (Section 2.3.3.6) of evidence that among cat ganglion cells, systematic variation in properties with retinal eccentricity is strong among cat α and β cells, and much less marked among γ cells. Correspondingly, Leventhal and co-workers note that eccentricity-related changes in morphology are as clear in A and B cells as in cat α and β cells and less pronounced in C and E cells. Second, in other animals, particularly the rat and the rabbit (see Sections 3.2 and 3.3 below), the same approach to cell classification suggests substantial differences between those species and the cat in the major groupings of ganglion cells present. The similarity argued above between the monkey and the cat in the properties of their retinal ganglion cells is not an artifact of experimental approach.

3.1.5. Summary

The similarities between cat and monkey in the properties and groupings of their ganglion cells indicate what features of this system may derive from a common ancestor, or perhaps by parallel evolution. It is equally valuable to note the differences:

1. X cells seem more numerous, and W cells less numerous, in the monkey than in the cat.
2. A strongly differentiated color-opponency is present within the X-cell group of the monkey; the rare color-opponent ganglion cells in cat retina seem more aptly considered part of the W-cell class.
3. In monkey dLGN, X and Y cells are segregated to different components of the dLGN, whereas they are intermingled in the same laminae of cat dLGN.
4. The full range of receptive field properties found among cat W cells has yet to be observed among monkey ganglion cells.
5. As yet, no evidence is available that the W-cell system of the monkey projects to the forebrain, as it does in the cat.
6. The difference in properties between X and Y cells has been reported to be more strongly developed in the monkey than in the cat.

Taken together, these differences suggest that the W-cell system is less prominent in the monkey than in the cat and, conversely, that the X-cell system is more highly developed and differentiated. The strong development of the X-cell system is presumably at least part of the basis for the higher spatial and chromatic acuity of primates.

Figure 3.7. Quantitative comparison between cat α and β cells and monkey A and B cells (from Leventhal *et al.*, 1981). The graph plots variation in soma and dendritic field size in two classes of cat ganglion cells (α and β cells) and the two corresponding classes of monkey ganglion cells (A and B cells). In both species, the soma and dendritic field size are closely correlated, and the gap between the two groups in each species is very similar; the four groups even seem to lie along a single linear relationship. The data add weight to the suggestion that the groups correspond between the two species. [Reproduced from *Science*, Vol. 213, with kind permission of the American Association for the Advancement of Science; copyright 1981 by the American Association for the Advancement of Science.]

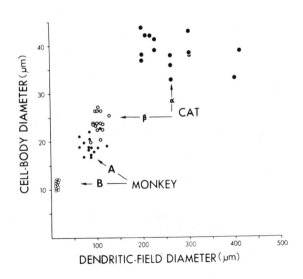

3.2. IN THE RAT

3.2.1. Conduction Velocity Groupings

Evidence of conduction velocity groupings among axons of rat retinal ganglion cells was provided by the work of Sefton and Swinburn (1964), who distinguished three groups (with mean velocities of 13.5, 5.5, and 3.0m/sec). Their estimates were based on recordings of the antidromic field potential evoked in the nerve by stimulation of the optic tract. Estimates have also come from the single-unit studies of Noda and Iwama (1967), Sumitomo *et al.* (1969*a*), and Fukuda (1973, 1977), who distinguished at first two, and subsequently three groups among the afferents to the relay cells of the LGN; and from the study of Sumitomo *et al.* (1969*b*), who also distinguished three velocity groups among ganglion cell axons, all projecting to both the SC and LGN.

Fukuda's (1977) study provided the first, and so far the only, recordings from the somas of individual ganglion cells in the rat, and he was able therefore to measure conduction latencies over longer distances than earlier workers. His recordings, supplemented by measurements of ganglion cell somas, provide perhaps the clearest evidence of the presence of three conduction velocity groupings among the axons of rat retinal ganglion cells, and the values he arrived at for their characteristic velocities (Fig. 3.8) were close to those of Sefton and Swinburn (1964). Fukuda also confirmed that all three axon groups project to the LGN and probably also to the SC.

3.2.2. Receptive Field Correlates: Is There an X-like Group?

Although he recorded from the somas of single rat ganglion cells in the intact eye, Fukuda (1977) did not characterize their receptive field properties, so that the only reports so far available of the receptive fields of rat ganglion cells are the studies of Brown and Rojas (1965) and Partridge and Brown (1970). Recording from axons in the optic tract, they reported two patterns of receptive field organization. One pattern was common to other vertebrate retinas, the receptive fields comprising concentric center and surround components, either ON-center and OFF-surround, or vice versa. In the other pattern, the receptive field lacks any surround mechanism, comprising only a circular ON- or OFF-region. Brown (1965) suggested that these two patterns may be related to the "loose" and "tight" patterns of dendritic field formations he distinguished morphologically (Section 3.2.3). The different receptive field patterns have not been related to conduction velocity groupings among ganglion cell axons.

Some evidence of the receptive field/conduction velocity correlates present in rat retina may, however, be found in the studies of LGN relay cells reported by Hale *et al.* (1976, 1979) and Fukuda *et al.* (1979). Assuming that, as in the cat and monkey (Chapter 6), the receptive fields of LGN cells resemble those of retinal ganglion cells, these reports suggest that ganglion cells with the fastest axons have a number of Y-like properties. That is, the LGN cells receiving input from the fastest axons of the optic nerve have large receptive fields, give phasic responses to flashing spot stimuli, and are particularly responsive to fast-moving stimuli. Geniculate cells receiving input from slower-conducting ganglion cells generally resembled cat W cells. Some had ON-OFF receptive

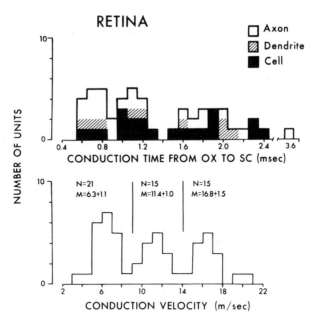

Figure 3.8. Conduction velocity groupings in rat optic nerve (from Fukuda, 1977). Fukuda's analysis was based on recordings from the somas of single retinal ganglion cells. He activated each cell antidromically from both the optic chiasm (OX) and the SC. For each cell, the difference between those latencies represents the conduction time of the axon between those two sites; the values obtained are shown in the upper histogram. Measurements of the distance between the two sites then allowed conversion of the latency values to conduction velocity values, as in the lower histogram. The lower histogram confirms previous estimates based on field potential recordings of three conduction velocity groups. [Reproduced with kind permission of Elsevier/North-Holland Biomedical Press.]

fields (Fig. 3.9), others responded only to slow stimulus movement, had large ON- or OFF-center receptive fields, and gave phasic responses to stationary contrast stimuli. Relay cells with X-like properties (small receptive fields, medium-velocity axons, tonic responses) could not be clearly distinguished. Both groups of workers suggest that the rat lacks a well-developed X-like system of ganglion cells.* Hale and co-workers relate this lack to the poor development of central vision in this species, i.e., to the poor development of the area centralis of the retina and of fixational eye movements.

3.2.3. Morphological Classifications

Studies of the morphology of rat ganglion cells tend to confirm Hale and co-workers' (1979) suggestion that, while cell groups can be recognized that correspond to the

---

*Note that this suggestion depends on a parametric approach to the delineation of an X-like grouping. If X cells were considered defined by a single feature, such as linearity of spatial summation, then it is highly likely that an X-like (linear) group of cells could be distinguished.

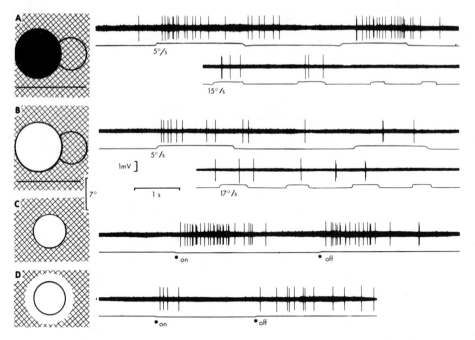

Figure 3.9. W-like relay cell in rat LGN (from Hale *et al.*, 1979). The oscilloscope traces show the actual responses of the cell to stimulus situations presented at the left. The trace below each spike trace shows the modulation of luminance at the center of the receptive field (an upward deflection indicates a decrease in luminance). (C) The cell responded to a stationary centered spot of light, both when it flashed on and when it turned off, with a burst of action spikes, i.e., it gave an ON-OFF-response. (D) With a larger flashing spot, the cell's response was weaker, indicating that the region around the ON-OFF-center was inhibitory to the cell. (A, B) The cell responded to both dark and white spots moved across the receptive field, provided the velocity of movement was less than about 20°/sec. [Reproduced with kind permission of Springer-Verlag.]

Y (α) and W (γ) cells of the cat, a clear X-(β) like group is not present. However, paralleling the experience in the cat (Chapter 2, Section 2.1.4) and monkey (Section 3.1.4), present understanding of morphological groupings has required the use of retinal whole mounts and a parametric approach to cell classification. Thus, Fukuda (1977), working with Nissl-stained retinal whole mounts, suggested a grouping of ganglion cells according to soma size (Fig. 3.10), and noticed similarities between large-, medium-, and small-soma groups and, respectively, the Y-, X-, and W-cell systems of the cat. For example, the cells with the largest somas (L cells, 15- to 21-μm diameter) form only a small fraction of the population and are most frequent relatively (7.5%) in peripheral retina, resembling cat α (Y) cells. The medium-sized cells (M cells, 9- to 15-μm diameter) formed 25% of the population [as against 49% of the cat population formed by β (X) cells], but like X cells they appear to be most numerous relatively in central retina. The smallest cells (S cells, < 10-μm diameter) formed 70% of the population as against 50–55% for cat W cells.

Conversely, Brown (1965), working with retinal whole mounts stained intravitally with methylene blue, proposed a grouping based on the density of dendritic branching (Fig. 3.11). He distinguished two groups; the dendrites of cells of one group ("tight" cells) branched more frequently, so that in whole mounts the dendritic field appeared dense (cells A, C, E, and G in Fig. 3.11), while the dendritic fields of cells of the other group appeared "loose" by comparison. Brown noted further that the dendrites of "loose" cells tend to spread more superficially in the inner plexiform layer (i.e., nearer the ganglion cells) than those of "tight" cells, and he suggested that tight and loose cells correspond to the two receptive field types described in an accompanying paper by Brown and Rojas (1965), respectively, to those lacking and those with a demonstrable surround mechanism. A third approach was followed by Bunt (1976), working with Golgi-stained sections of the retina (Fig. 3.12). She proposed a grouping based (follow-

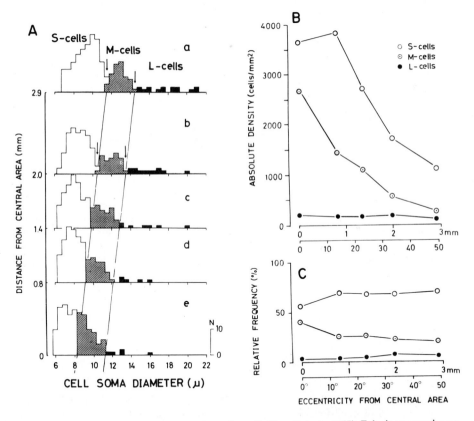

Figure 3.10. Soma-size analysis of rat retinal ganglion cells (from Fukuda, 1977). Fukuda measured soma size at the area centralis [bottom histogram in (A)] and in regions of the retina at successive distances into peripheral retina (e through a). He suggested that each histogram be regarded as comprising three soma-size components, large, medium, and small, as indicated. The graphs in (B) and (C) show that the small-soma cells are the most frequent in all parts of the retina, but that medium-soma cells reach their peak, in both absolute and relative terms, at the area centralis. [Reproduced with kind permission of Elsevier/North-Holland Biomedical Press.]

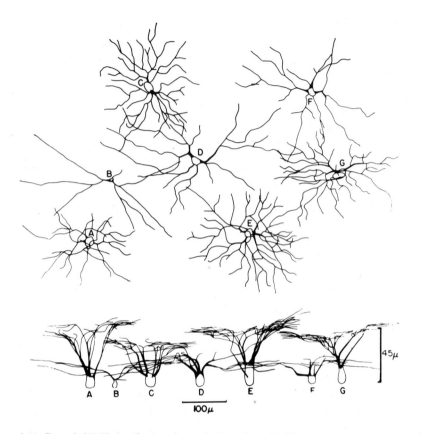

Figure 3.11. Brown's (1965) classification of rat retinal ganglion cells. The grouping was based on whole-mount material stained intravitally with methylene blue. Brown distinguished "loose" cells (B, D, and F) from "tight" cells (A, C, E, and G) on the basis of the density of branching of their dendrites. He also reconstructed how the cells would appear in sections, and noted a tendency for the dendrites of the "loose" cells to branch more superficially in the inner plexiform layer (i.e., nearer the layer of ganglion cell somas). [Reproduced from the *Journal of Neurophysiology* with kind permission of the American Physiological Society.]

ing Cajal, see Chapter 1, Section 1.3, and Section 3.1.4) on the sublamina of the inner plexiform layer at which the dendrites of the cell spread, distinguishing "diffuse" (Fig. 3.12), "giant," "unistratified," and "bistratified" groups of cells. She recognized the "tight"/"loose" distinction but considered it one aspect of the groupings according to laminar spread of dendrites. These groupings have still to be related to physiological groupings of ganglion cells, although it seems likely that, as in the cat (Chapter 2, Section 2.1.4), the stratification of ganglion cell dendrites will be found to be related to the ON/OFF difference in responses to stimulus contrast.

Working with Golgi-impregnated retinal whole mounts, and with prior work in the cat as context, Perry (1979) proposed a classification that was based on a consideration of several criteria rather than one; it was a parametric classification in the sense discussed in Chapters 1 and 2. On the basis of soma size, axon size, and the size and branching density of dendritic fields, he distinguished three classes of ganglion cells, which he termed types I, II, and III. The three classes seemed to correspond, respectively, to the α, δ, and γ classes of the cat.

Thus, type I cells had the largest somas (mean diameter 20.4 μm), fairly large dendritic fields (mean diameter 312 μm), and the thickest axons of the three classes (Fig. 3.13). Their dendritic fields resembled the "loose" pattern described by Brown (1965), but Perry noted that the dendrites of type I cells seemed to terminate in the outer part of the inner plexiform layer, rather than in the inner part where Brown had reported the spreading of "loose" dendritic fields. Perry comments that the similarity between type I cells and cat α cells is striking.

Perry's type II cells have intermediate-size cell bodies (mean diameter 13.5 μm) and the smallest dendritic fields (mean diameter 150 μm). One feature that seemed characteristic of them was the large number of short branches coming off each of the main dendrites. This gave the tree a dense appearance, resembling Brown's description of "tight" dendritic fields. Perry does not explicitly discuss why he regards these as analogs of cat δ cells, rather than of β cells. In parametric terms, type II cells resemble β cells in having medium-sized somas and the smallest dendritic fields of any type; on the other hand, type II cells more closely resemble δ cells in the pattern of dendritic branching, and in the ratio of dendritic field size to soma size. This ratio is low in β cells, and gives them a characteristic "small-field" appearance. Perry's conclusion (that there are few if any β-like cells in the rat) also matches well the independent physiological work [especially that of Hale *et al.* (1979)] discussed above, which indicates the absence of an X-like group of ganglion cells from rat retina.

Perry's type III cells have the smallest somas of the three types (mean diameter 10.1 μm), the thinnest axons, and their dendritic fields are the most loosely branched. The branching pattern of their dendritic fields seemed more variable than in type I or II cells, and the diameter of their fields varied more widely, generally being large (range 133–694 μm, mean 339 μm). Perry considered them "very similar" to cat γ cells.

Perry described a fourth class of cell in the ganglion cell layer, but considered them probable amacrine cells. In subsequent papers based on Golgi-stained whole mounts, Perry and Walker (1980a,b) have described the variety of amacrine cells present in rat retina and the morphogenesis of neurons present in the ganglion cell layer.

3.2.4. Summary

The rat provides a valuable reference point in the comparative analysis of the visual pathways. The species belongs to an order (Rodentia) long distinct from either carnivores or primates; its retina shows far weaker central specializations (Fukuda, 1977) than either cat or monkey, and its visual environment and behavior are very different. Nevertheless, it shows clear velocity groupings among the axons of the optic nerve. Moreover, considering a wide range of their properties, groups of ganglion cells can be recognized that seem to correspond clearly to the Y and W groupings of cat and monkey.

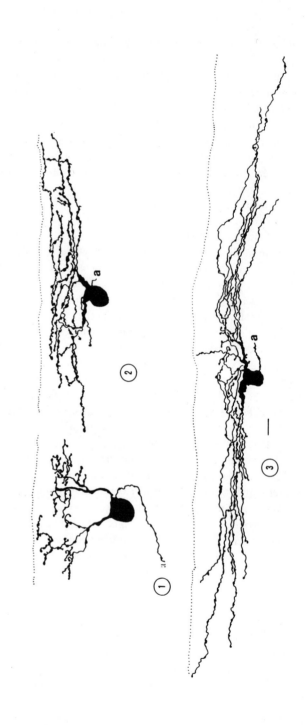

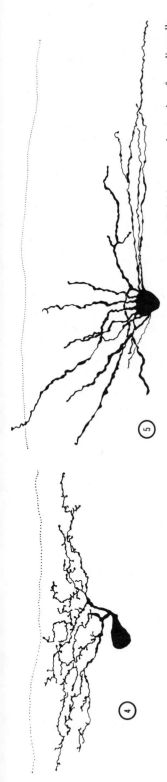

Figure 3.12. Rat ganglion cells seen in radial sections. Camera lucida drawings, from Bunt (1976), of the appearance in Golgi-impregnated rat retina of ganglion cells with "diffuse" dendritic fields, i.e., the dendrites branch throughout the thickness of the inner plexiform layer. The scale indicates 10 μm. [Reproduced with kind permission of Elsevier/North-Holland Biomedical Press.]

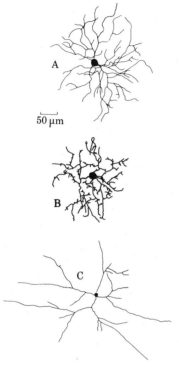

50 μm

Figure 3.13. Classes of ganglion cells in rat retina (from Perry, 1979). (A) Example of Perry's type I cells. They have relatively large cell bodies, three to six primary dendrites with smooth branches and thick axons. Perry suggested that these cells correspond to the α cells of cat retina. (B) Type II cells have somas of intermediate size, with dendritic trees characterized by many short branches. They have somewhat smaller dendritic fields than the other types. Perry considered them to correspond to the δ cells of cat retina. (C) Type III cells closely resemble cat γ cells. They have small somas and widely branched dendritic fields, with a relatively loose pattern of branching. Perry suggested that these and type I cells may correspond to Brown's (1965) "loose" cells (Fig. 3.11). [Reproduced with kind permission of the Royal Society (London).]

These correspondences suggest (but of course do not prove) that these features of ganglion cell differentiation may be part of the common phylogenetic inheritance of these mammalian orders. Conversely, the X-cell system, which is so strongly developed in monkeys, seems poorly differentiated in the rat; this may, as Hale *et al.* (1979) suggest, reflect the weak central specialization of the retina and poor central vision in this species. Judging from these three species, the X-cell group seems more labile in its phylogenetic development than the other two cell classes.

3.3. IN THE RABBIT

3.3.1. Conduction Velocity Groupings

The rabbit was the first mammalian species in which the velocity groupings of optic nerve axons were studied. Bishop (1933) suggested that three conduction velocity groups are present (Fig. 1.1); for the fastest group he estimated velocity of conduction at 20–50 m/sec, for the intermediate group at about 6–14 m/sec, and for the slow group at 4 m/sec.

Granit and Marg (1958), however, suggested the presence of five conduction velocity groups with maximum velocities of 56, 35, 23, 16, and 10 m/sec; the fastest three of these groups are within the range embraced by the single fast group proposed by Bishop. Subsequently, Lederman and Noell (1968) investigated properties of the fastest-conducting ganglion cells, but used photic stimulation to elicit responses in them. Because of the considerable latency of ganglion cell responses to photic stimulation, they could not estimate conduction velocity. Caldwell and Daw (1978), Vaney *et al.* (1978), and Reuter and Hoffmann (1980) all measured the latency of ganglion cell responses to antidromic stimulation of the optic nerve, but did not convert those latencies into conduction velocity values or comment on velocity groupings suggested by the latency values.

Two more recent studies have, however, provided confirmation of the groupings described by Bishop (1933). Semm (1978) observed early and late components to the field potential recorded at the optic disc following stimulation of the optic chiasm or SC; and he also recorded single units with considerably longer latencies, indicating the presence of cells with markedly slower-conducting axons. As with cat W cells, the velocities of these very-slow-conducting axons seemed relatively scattered, so that their antidromic responses to even a brief electrical stimulus would be too asynchronous to generate a field potential. Semm estimated the conduction velocity of the axons in the fastest-conducting group at 22–33 m/sec, and of the slower axons at 13–19 m/sec; he did not estimate a conduction velocity for the very-slow-conducting axons. Molotchnikoff *et al.* (1979) also recorded the field potential evoked at the optic disc by stimulation of the optic pathway (specifically of the optic tract). They described three successive components to the potentials, suggesting three groups of axons with conduction velocities of 21–23, 15–17, and 10–12 m/sec.

Overall, there is good agreement between these latter studies and Bishop's original estimates. There is also a considerable similarity between rabbit and other mammals. In both cat and monkey (Chapter 1, Section 1.1, and Section 3.1.1), for example, two clear groups have been described in the field potential recorded from the optic nerve following stimulation of the optic tract, and single-unit studies indicate the presence of a third group of axons with lower and more scattered velocities.

3.3.2. The Feature Extraction Classification of Rabbit Ganglion Cells

Papers contributing to the classification of rabbit retinal ganglion cells include those of Barlow *et al.* (1964), Levick (1967), Caldwell and Daw (1978), Vaney *et al.* (1978), Semm (1978), and Reuter and Hoffmann (1980). The classification developed by Barlow *et al.* (1964), and extended by Levick (1967) and Vaney *et al.* (1978), follows a feature extraction approach. The classes distinguished by Barlow and co-workers were:

Receptive field type	Most effective stimulus
ON-OFF direction-selective	Movement of small object in particular direction
ON-center direction-selective	Movement of small object in particular direction
ON-center concentric	Local brightening
OFF-center concentric	Local dimming
Large-field	Fast-moving objects

Their analysis of an ON-OFF, directionally selective cell is shown in Fig. 1.14. In discussion, Barlow and co-workers argue that

> The important step for the coding problem [is] to discover what part of the information provided by the animal's normal environment each class of unit transmits: what normally triggers its response? . . . one's guide to the trigger feature is the type of stimulus which is most effective in eliciting a response.

Levick (1967) reexamined the rabbit retina, in particular the visual streak region, where the ganglion cells are densely packed (Chapter 9, Section 9.4.1), and described three new classes of ganglion cell:

Receptive field type	Most effective stimulus
Orientation-selective	Elongated stimulus with long axis either horizontal or vertical
Uniformity detector	Absence of any contrast in receptive field; cell inhibited by any stimulus
Local edge detector	Small object with many edges (such as a grating)

Levick's analysis of the last of these three types is shown in Fig. 3.14. Two features of these descriptions were of particular interest. First, the considerable variety of receptive fields described in the rabbit had not previously been observed in a mammal. A comparable variety was first described in the frog (Chapter 1, Section 1.2, and Section 3.5.1), and several comparable receptive field patterns had been reported in the pigeon (Maturana and Frenk, 1963; Section 3.5.2); moreover, a comparable variety has subsequently been described in the cat (Chapter 2, Section 2.3.3) and the monkey (Section 3.1.3). These studies in the rabbit, however, provided the first descriptions of the range of receptive field patterns now known to be common in mammals. Second, Levick's (1967) observation that several receptive field patterns were particularly frequent in the visual streak of rabbit retina was the first evidence of a regional specialization of the retina in terms of the classes of ganglion cells present. Subsequently, evidence has been presented that in the cat, W cells concentrate in the visual streak (Rowe and Stone, 1976a; Chapter 2, Section 2.3.3.5) and that X cells are particularly frequent at the area centralis (Section 2.3.3) and X-like cells at the monkey fovea (Section 3.1.2).

In a brief report, Vaney *et al.* (1978) have extended the above classification to take in the brisk/sluggish and transient/sustained dichotomies proposed by Levick (1975) for cat ganglion cells. Instead of dividing the "concentric" class of cells into ON-center and OFF-center, as did Barlow *et al.* (1964), they divided concentric cells into the brisk or sluggish and sustained or transient types. They included the "large-field" cells of Barlow and co-workers in the concentric group, and they also described a receptive field property not previously described in the rabbit, color-opponency; a small proportion of cells had receptive fields with a blue OFF-center and a green ON-surround. Vaney and co-workers also sought to correlate the receptive field groups with axonal conduction velocity, as had been done in the cat. They comment that:

> In marked contrast to the cat, retino-chiasmal conduction latency is not a clear indicator of functional class. The latency distributions of brisk-transient and brisk-sustained units overlap

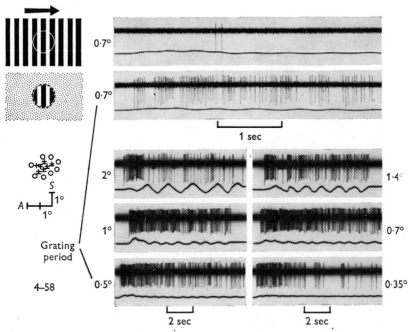

Square wave gratings, 90% contrast, mean luminance 17 cd/m²

Figure 3.14. Levick's (1967) analysis of the "local edge detector" type of rabbit ganglion cell. The receptive field plotted with flashing stimuli is shown at left middle. The ± signs represent locations at which an ON-OFF response was obtained. The o signs indicate locations at which no response was obtained. The field comprised an ON-OFF-center and an inhibitory surround. Top trace: The cell was unresponsive to a large grating pattern moved across its receptive field. Second trace: However, when the inhibitory surround of the field was masked and the grating appeared only in the center region of the field, the cell responded strongly. Remaining traces: When successively finer gratings were used (period 2° reducing to 0.35°), the cell's response increased down to the 0.7°-period grating. This increase was the basis for the name *local edge detector;* it was argued that the cell is specialized to detect lots of edges occurring within the receptive field center. [Reproduced with kind permission of the *Journal of Physiology (London)*.]

each other almost completely. They also extensively overlap the distribution of the sluggish concentric units. The on-off direction-selective units also provide a sharp contrast with the cat. Their latency distribution is aligned with that of the brisk concentric classes.

In summary then, although the same dichotomies of receptive field types could be distinguished in the rabbit and the cat, the groups so distinguished in the rabbit were not clearly correlated (as they are in the cat) with conduction velocity groupings of ganglion cell axons.

3.3.3. More Parametric Analyses

Caldwell and Daw (1978) proposed a classification of rabbit ganglion cells that mixes parametric and feature extraction approaches, and proposed the groupings listed in Fig. 3.15. They recognized all the receptive field classes described by Barlow *et al.*

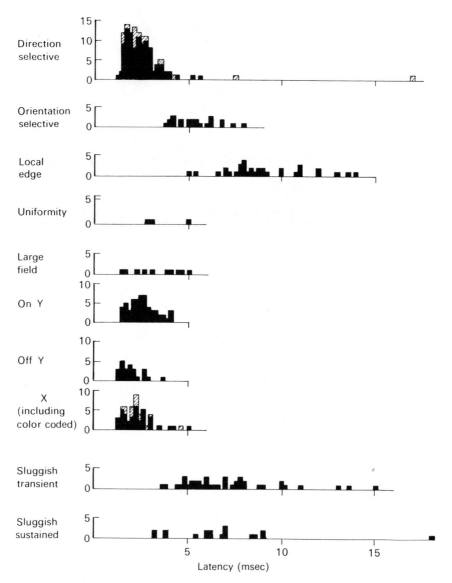

Figure 3.15. Physiological classification of rabbit ganglion cells (from Caldwell and Daw, 1978). Frequency/axonal-conduction latency histograms are shown for rabbit retinal ganglion cells. The cells are separated into 10 groups, named at left; the latencies are the latencies of the cells' antidromic responses to stimulation of the optic chiasm and indicate the conduction velocity of the cells' axons (short latencies mean fast-conducting axons, and vice versa). Several features of these data resemble the situation in the cat; for example, the local edge and sluggish cells have slow-conducting axons, as they do in the cat. On the other hand, there is little difference in latency between X and Y cells, and direction-selective cells tend to have fast- rather than slow-conducting axons. [Reproduced with kind permission of the *Journal of Physiology (London)*.]

(1964), Levick (1967), and Vaney *et al.* (1978). They grouped cells with "complex" receptive fields (direction-selective, orientation-selective, local edge detector, etc.) in separate groups, although they considered the color-opponent ganglion cells of rabbit retina (which they observed independently) to be part of the X group; and then subdivided the concentric (presumably "simple") receptive field cells into sluggish cells on the one hand, and X and Y (presumably brisk) cells on the other. Unlike Vaney and co-workers, they kept "large-field" cells separate from other brisk cells, and they comment that they could not find a reliable sustained/transient difference among brisk cells. They therefore distinguished (brisk) X from (brisk) Y cells with a test of linearity of spatial summation.

Overall, Caldwell and Daw rely fairly heavily on single physical features to define, and in some cases to name, their cell groups and, as with feature extraction classifications proposed in other species, their classification rests on a series of apparently unrelated dichotomies (sustained/transient, concentric/nonconcentric, linear/nonlinear). As a consequence, the groupings in their classification bear only limited resemblance to the Y/X/W groupings in the cat. In addition, the direction-selective cells in the rabbit have a wide range of axonal conduction velocities, some cells falling in all the major conduction velocity groupings; whereas in the cat, direction-selective ganglion cells all have very-slow-conducting axons characteristic of W cells. On the other hand, Caldwell and Daw do comment in parametric terms on some of the broader differences between the cell groups they distinguished and thereby provide some basis for comparison with the Y/X/W grouping. In particular, Caldwell and Daw remark in discussion that the brisk/sluggish distinction was easier to make in the rabbit than the distinction between X and Y cells. For example, there was no clear conduction velocity correlate of the X/Y distinction (Fig. 3.15), nor a difference in receptive field size. Caldwell and Daw suggest that the brisk/sluggish groupings may be much more important and fundamental to both cats and rabbits than the X/Y grouping among brisk cells. In one way, the idea seems implausible for no suggestion has ever been made as to the functional significance of the difference in peak firing rates that characterizes the brisk/sluggish distinction. But Caldwell and Daw presumably meant that the *overall* functional differences between "sluggish" (presumably W-like) cells on the one hand and "brisk" (X and Y) cells on the other might prove more fundamental than X/Y differences. Their comment then clearly foreshadows the two-group (W vs. XY) classification of ganglion cells proposed in Rowe and Stone (1980*b*) (Chapter 2, Section 2.4).

Semm's (1978) report provides further evidence that a parametric approach would lead to a classification of rabbit ganglion cells comparable to (though far from identical with) the Y/X/W groupings of the cat. He noted that cells with fast-conducting axons also resemble cat Y cells in giving phasic or transient responses to stationary contrast stimuli, while cells with intermediate-velocity axons further resemble cat X cells in giving tonic responses. Moreover, cells with very-slow-conducting axons further resembled cat W cells in having receptive fields with ON-OFF-centers and suppressed-by-contrast properties. On the other hand, direction-selectivity seems to have developed in all the major functional groupings of ganglion cells, reflecting its prominence and importance in the visual system of the rabbit. Semm's report is brief, however, and requires further experimental support and testing.

More recently, Reuter and Hoffmann (1980) reexamined the conduction velocity correlates of receptive field properties in rabbit ganglion cells. They reported on only two classes, X and Y cells, and noted a small but statistically significant difference in their conduction velocities. Unfortunately for a comparison with Caldwell and Daw's result, they distinguished X from Y cells with a number of tests (receptive field size, sustained/transient differences) that they found as useful in the rabbit as in the cat (Hoffmann *et al.,* 1972). Theirs was a parametric approach, whereas Caldwell and Daw relied on a single test (linearity); the groups of cells called X and Y might differ considerably between the two studies. Even so, Reuter and Hoffmann confirm the conclusion of earlier studies that the conduction velocity correlates of the X/Y difference are less marked in the rabbit than in the cat.

Finally, Rapaport *et al.* (1981*a*) have reported briefly a parametrically based morphological classification of rabbit ganglion cells. They suggest three major groupings, on the basis of soma size, pattern of dendritic branching, and axon caliber. Their class 1 cells resemble cat α cells, having large somas, thick axons, and wide-spreading, densely branched dendritic fields. Their class 2 cells, with loosely branched, wide-spreading but thin dendrites, thin axons, and relatively small somas, would seem comparable to cat γ cells. On the other hand, their class 3 cells, with medium somas, but wide-spreading dendritic fields with thin, wavy branches, seem unlike any class recognized in the cat or monkey. A β-like group of cells was not apparent in rabbit retina, supporting the view that the X/Y distinction is poorly developed. The class 1 (α-like) cells presumably are the somas of Y-like ganglion cells recognized physiologically; however, more extensive correlates between morphological and physiological studies have yet to be established.

3.3.4. Summary

It is difficult to provide a cogent summary of groupings of rabbit ganglion cells, principally (it seems to me) because most studies have chosen to define cell groupings by single physical features rather than to describe them parametrically. In addition, morphological classifications of rabbit ganglion cells are just beginning to be developed, and related to physiological groupings; this was an important step in the classification of ganglion cells in both cat and monkey. Nevertheless, two points of interest deserve mention. First, most workers agree that many rabbit ganglion cells resemble cat W cells in their receptive field patterns and axonal conduction velocities; and Caldwell and Daw stress the clarity with which W-like ganglion cells can be distinguished from X and Y cells. Second, several reports stress the lack of correlation between receptive field properties and axonal conduction velocity, at least in comparison with the cat. Either the correlations are weaker or, as Semm's report suggests, the question of correlates has yet to be correctly put.

3.4. IN OTHER MAMMALS: TREE SHREW, GOAT, AND GROUND SQUIRREL

3.4.1. Tree Shrew

Three studies have presented classifications of the retinal ganglion cells or geniculate relay cells of the tree shrew. Van Dongen *et al.*(1976) followed a typological, feature extraction approach; they recorded from axons in the optic tract and classified the ganglion cells into nine groups according to certain of their receptive field properties. In their sample of 93 cells studied, they distinguished the following groups:

Group	Number
Sustained	29
Transient	29
ON-OFF	7
Suppressed-by-contrast	2
Direction-selective	16
Orientation-selective	6
Opponent-color	1
Edge-inhibitory-off-center	1
Not classified	2

Van Dongen and co-workers' analysis of the responses of these cell types to stationary contrast stimuli is shown in Fig. 3.16. ter Laak and Thijssen (1978) have subsequently investigated other parameters of sustained and transient cells, showing strong similarities with the X- and Y-cell classes of cat ganglion cells. Sustained cells, for example, had smaller receptive fields than transient cells and tended to be more numerous in central retina; consideration of these extra parameters suggests a considerable similarity between tree shrew and cat in at least some of the major groupings of ganglion cells. If extended, this more parametric approach might establish groupings of tree shrew ganglion cells that can be more readily related to the Y/X/W groupings described in the cat.

For example, Sherman *et al.* (1975*a;* and see Chapter 6, Section 6.3.2) took a parametric approach in their study of the relay cells of tree shrew LGN and reported that the cells could be readily classified as X-like and Y-like. Assuming that, as in the cat, the receptive fields of LGN relay cells are very similar to those of the retinal ganglion cells that provide their input, their report suggests the presence of ganglion cells that resemble cat Y cells in having large receptive fields, fast-conducting axons, and in giving phasic responses to flashing stimuli and being particularly responsive to fast-moving stimuli; and of a class of cells that resemble cat X cells in having smaller receptive fields, slower axons, tonic responses, and lesser responsiveness to fast stimuli.

Sherman and co-workers' study apparently did not sample the full range of receptive field properties among tree shrew ganglion cells, as indicated by van Dongen and co-workers' study, perhaps because all types do not project to the LGN. Groups of cells clearly corresponding to cat X and Y cells appear to be present, however, and van Dongen and co-workers raise the possibility that "the great proportion of units other than sustained and transient ("W-cells"?) coheres with a very large superior colliculus in the tree shrew." The full range of correlations between receptive field/axon velocity destination, and between the tree shrew and other species, has still to be explored.

3.4.2. Goat and Ground Squirrel

In the goat, Hughes and Whitteridge (1973) followed a feature extraction approach to ganglion cell classification, describing two major classes of ganglion cell, *concentric*

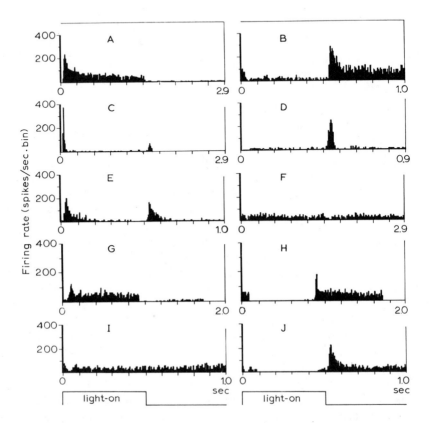

Figure 3.16. Classes of ganglion cells in tree shrew (from van Dongen *et al.,* 1976). Responses of tree shrew ganglion cells to stationary flashing spots of light. Histograms A and B show the responses of "sustained" cells with ON- and OFF-centers, respectively. Histograms B and C show the responses of ON- and OFF-center "transient" cells. E shows the response of an ON-OFF cell and F the response of a suppressed-by-contrast cell. The histograms in G and H show the responses of a color-opponent cell to a 440-nm and a 576-nm stimulus, respectively. I and J show the responses of an "edge-inhibitory OFF-center" cell to a 0.25° and a 1° spot, respectively. [Reproduced with kind permission of Springer-Verlag.]

including sustained and transient types, and *directional*. The sustained and transient cells both had receptive fields with ON- and OFF-centers and antagonistic surrounds. They resembled X and Y cells of the cat (respectively) in the sustained/transient difference, and also in the tendency for sustained cells to be relatively more frequent in central retina. A small proportion of the cells Hughes and Whitteridge encountered had ON-OFF receptive fields and were optimally responsive to a particular direction of stimulus motion.

In his pioneering studies of the ground squirrel, Michael (1969a,b) distinguished three classes of cells on the basis of particular receptive field properties: contrast-sensitive units, with ON- or OFF-center receptive fields and antagonistic surrounds; directionally selective cells optimally responsive to a particular direction of stimulus movement; and color-coded cells whose responses were dependent on the wavelength of the stimulus light. In this species too, it seems necessary to study parameters of ganglion cells other than receptive field properties, particularly their morphology, central connections, and retinal distribution, to gain a fuller understanding of the groupings present, and of their relation to groupings present in other species.

3.5. IN NONMAMMALS: FROG, TOAD, PIGEON, EEL, AND MUDPUPPY

3.5.1. Frog and Toad

Much of the pioneering work on the organization of the visual pathways has been done on these two species of amphibians, and some of that work has concerned the classification of ganglion cells. Hartline (1938, 1940a,b) and Barlow (1953) were the first to describe the receptive field properties of frog ganglion cells, distinguishing cells with ON-center, OFF-center, and ON-OFF-center receptive fields (Chapter 1, Section 1.2). A more formally developed classification was proposed by Maturana et al. (1960) and Lettvin et al. (1961) who followed a feature extraction approach in distinguishing five ganglion cell types. (Because of the significance of this work in the history of ganglion cell classification, it is discussed in some detail in Chapter 1, Section 1.2.2.) They noted several correlates of receptive field properties, providing evidence that cells with different receptive fields had distinct axonal destinations and axonal velocities and, it was argued (Lettvin et al., 1961), different morphologies. Although describing these correlates, however, these workers laid particular emphasis on those receptive field properties that seemed to indicate the "distinct and invariant" operation performed by that cell on the visual image (e.g., edge detection, dimming detection).

The adequacy of certain of these characterizations of the functions performed by the different cells has been questioned by later workers who followed a more parametric approach. For example, Grusser-Cornehls et al. (1963) found distinct "movement-detecting" ganglion cells not easily accounted for by Maturana and co-workers' grouping; and Gaze and Jacobson (1963) and Keating and Gaze (1970) argued that the "convex edge detector" and "sustained edge detector" groupings are really not distinct, and suggested a class 1, class 2, etc. terminology. Nevertheless, much of the detail of Maturana and co-workers' observations has been confirmed. For example, Witpaard and ter

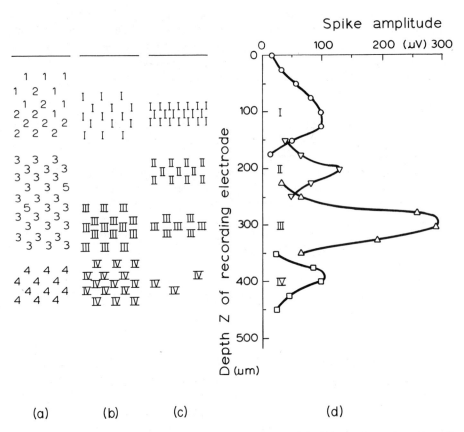

(a)   (b)   (c)   (d)

Figure 3.17. Witpaard and ter Keurs' (1975) summary diagram of the different classes of ganglion cells distinguished in the frog, and of the level of their termination in the optic tectum. The photograph at the right shows the fiber architecture of the optic tectum of the frog, and the diagram in (e) shows the distribution of degenerating terminals found in the tectum after enucleation of the contralateral eye; four levels of termination are distinguished (dark broken lines). The graphs in (d) represent the amplitudes of the action

Keurs (1975) proposed a four-group classification of frog retinal ganglion cells based on a parametric approach to classification. Their groupings are different from those of Maturana and co-workers but are easily related (Fig. 3.17). Thus, Witpaard and ter Keurs followed Gaze and Keating (1969) in including two types of edge detector in their type I, but distinguish two types of ON-OFF receptive fields that Maturana and co-workers had grouped together. Pomeranz and Chung (1970) described the dendritic morphology and receptive field properties of retinal ganglion cells in the tadpole, suggesting that one receptive field type and a corresponding morphological group of cells develop only in the frog. Their report supports Lettvin and co-workers' earlier (1961) suggestion of a correlation between the five receptive field types of ganglion cells distinguished in frog retina by Ramon y Cajal (1911). Reuter and Virtanen (1972) followed Maturana and co-workers' general scheme in investigating the spectral sensitivity of frog ganglion

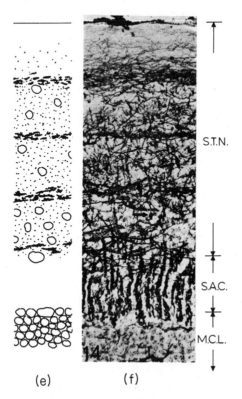

(e)   (f)

potentials of the four types of units (I–IV) that these workers distinguished; one peak appears to correlate with each layer of terminals. The diagrams in (a) to (c) show the distribution of ganglion cell types as proposed by Maturana *et al.* (1960), by Gaze and Keating (1969), and by Witpaard and ter Keurs themselves. Note that there is good general agreement between the three groups, but some variation in the naming of cells in the outer two layers. [Reproduced with kind permission of *Vision Research.*]

cells; and Backstrom *et al.* (1978) followed the same general scheme in their report of direction-selectivity among frog retinal ganglion cells.

In the toad, Ewert and Hock (1972) approached the classification of ganglion cells in the context of a "Gestalt" analysis of the shapes of objects that seem important in eliciting predatory and escape behavior in the free-moving animal. For example (Ewert, 1974), toads respond to small objects with prey-catching reactions, but if the object has long vertical extensions, the animal is more likely to seek to escape from it. In seeking a basis for the coding of the apparently important stimulus parameters (the size, contrast, and velocity of stimulus objects), Ewert and Hock described classes of ganglion cells termed (a), (b), and (c). The classes were not defined by a single physical feature, but were described in terms of a number of receptive field parameters. For example, type (a) cells had small receptive field centers (4-deg diameter) and an inhibitory surround. Accordingly, they were very responsive to small stimulus objects moved through the field center, but were unresponsive to changes in ambient illumination. They showed marked adaptation when a stimulus was repeated at short intervals. Type (b) cells had large center regions (8-deg diameter) and showed little adaptation; while type (c) cells had large receptive fields and were responsive to large dark objects. Ewert and Hock suggest that these three types correspond to types II, III, and IV described among frog ganglion cells by Maturana *et al.* (1960) (they did not observe a group corresponding to type I) and discuss their coding capabilities in considerable detail.

Reuter and Virtanen (1972) and Guiloff (1980) have provided evidence that a class of ganglion cells is present in toad retina that corresponds to type I of Maturana *et al.* (1960). That is, it responds with a sustained burst of firing when a contrast border is introduced into the center region of the receptive field. Further, the terminals of the (now) four classes of ganglion cells terminate at different levels in the superficial layers of the optic tectum. Not surprisingly perhaps, there appear to be strong similarities between frog and toad in several parameters of the ganglion cell population and its projection to the optic tectum. On the other hand, many of the parameters on which the Y/X/W classification of mammalian ganglion cells is based (retinal distribution, projections to forebrain as against midbrain, conduction velocity groupings, morphological correlates) remain unexplored, or only partially explored, in the amphibians. The degree of correspondence that can be traced between amphibians and mammals in the functional groupings of their ganglion cells is still undetermined.

3.5.2. Pigeon

In the pigeon, Maturana and Frenk (1963) reported the presence of six classes of ganglion cells. They took a feature extraction approach, classifying and naming the cells by their trigger features. In this report, they described only two classes, viz., directional movement detectors (responsive to a particular direction of movement) and horizontal edge detectors (maximally responsive to a horizontal edge stimulus moved up and down). Holden (1969, 1977*a,b*) has confirmed these descriptions and also recognized ON- and OFF-center receptive fields. Other parameters of pigeon ganglion cells (numbers, retinal distribution, morphology, central projections) have still to be explored.

3.5.3. Eel and Mudpuppy

These two species are discussed together because their ganglion cell groupings have been studied largely in terms of a single property, "linearity." Thus, in the eel, Shapley and Gordon (1978) distinguished two classes of ganglion cells, according to the linearity of their spatial summation. They did not examine other parameters of the receptive fields, so that the classification is typological, the groupings being defined by a single physical property. One group of cells was linear, i.e., each cell appeared to sum linearly the influences reaching it from the different subregions of its receptive field. This is the property described by Enroth-Cugell and Robson (1966) as characteristic of X cells, and Shapley and Gordon named the linear ganglion cells of the eel *X cells*. The nonlinear cells of the eel, on the other hand, seemed distinct from the nonlinear Y cells of the cat, and were termed $\bar{X}$ *cells.*

In the mudpuppy, Tuttle and Scott (1979) took a similar approach, distinguishing linear and nonlinear types of ganglion cells, and terming them X and Y *cells,* respectively, by analogy with the cat. However, they made two additional observations. First, they noted that the receptive fields they encountered had either ON-, OFF-, or ON-OFF-center regions; the ON-center and OFF-center cells could be further subdivided into sustained and transient types, by the duration of their responses to standing contrast stimuli. The nonlinear Y cells resembled cat Y cells in being transient, while the linear X cells resembled cat X cells in being sustained. Second, Tuttle and Scott note that the ON-OFF cells, initially called Y cells because of their nonlinear behavior, "seem to be more analogous to the ON-OFF W cells of the cat retina due to their on and off responses, their large suppressive surround, and their other receptive field properties." Thus, these workers seem to move from a strictly typological toward a parametric analysis, seeking correspondences with the Y/X/W classification of the cat. Clearly, however, much more must be learnt about these cells before the validity of these correspondences can be assessed.

On the Methodology of Classification

II

This book arose out of a passage in Borges, out of the laughter that shattered, as I read the passage, all the familiar landmarks of my thought. . . . The passage quotes a 'certain Chinese encyclopaedia' in which it is written that 'animals are divided into: (a) belonging to the Emperor, (b) embalmed, (c) tame, (d) sucking pigs, (e) sirens, (f) fabulous, (g) stray dogs, (h) included in the present classification, (i) frenzied, (j) innumerable, (k) drawn with a very fine camelhair brush, (l) et cetera, (m) having just broken the water pitcher, (n) that from a long way off look like flies'. In the wonderment of this taxonomy, the thing we apprehend . . . is the limitation of our own [system of thought], the stark impossibility of thinking that.

M. Foucault in the Preface to his *The Order of Things*, Tavistock Publications,
London, 1970.

Toward Certainty, Objectivity, or Testability?

4

Two Notes on Alternative Methodologies of Classification

It would seem a step designed to try the reader's patience for me now to comment that the classification of retinal ganglion cells is not a straightforward, objective task, but is as much a product of the observer's presuppositions as any other scientific proposition; and is likely, therefore, to provide its share of grist for the philosopher's mill. Yet it is a fair and relevant criticism of neurobiologists (including myself) that we have been largely unaware of the problems of methodology inherent in classification, and of the substantial literature that exists on those problems. There has been little awareness, either, of the central part played by classification (whether of nerve cells, plants, animals, aphasias, or rocks) in the organization of bodies of knowledge.* Perhaps as a consequence, much of the variety and inconsistency found among neuronal classifications stem from differences between scientists in our presuppositions; differences that, though less exotic than the gulf that intrigued Foucault, are nevertheless substantial and highly influential on our work.

Tyner (1975) was one of the first to draw the attention of neurobiologists to the relevance to our work of the literature on animal taxonomy, and to the importance of methodology in classification. His article was particularly concerned with classifications of motor system neurons; it prompted M. H. Rowe and myself to attempt a critique of ganglion cell classifications (Rowe and Stone, 1977, 1979, 1980a,b), parts of which are

---

*The same point was made by Pratt (1977) in his article on the French philosopher Foucault's contribution to the theory of classification:

> We shouldn't really have needed Foucault, of course, to tell us that classification is a central activity of living things ... the fact that taxonomy [has] dominated natural history ... shouts it at us; ... without classifying we have only individuals and therefore no possibility of scientific study. The mode of classification adopted gives any study its basic categories and thus structures it at the most fundamental level.

in Chapters 1 and 2 above and in Chapter 5 below. In this chapter, two notes on classification are presented that are preliminary to the case, presented in the next chapter, for a hypothetico-deductive approach to the classification of nerve cells.

4.1. ALTERNATIVE METHODOLOGIES OF CLASSIFICATION: THEIR BASIS IN THE CONCERNS OF CLASSIFIERS

In each of the three previous chapters, two alternative approaches have been described or referred to in the study and classification of ganglion cells. In Chapter 1, the terms *feature extraction* and *parametric* were used to distinguish alternative approaches in work on the cat up to 1966. In Chapter 2, the same approaches were traced in work on the cat in the years since 1966. In Chapter 3, the same distinction was extended to classifications of ganglion cells described for other species.

These distinctions and arguments have required interpretation on my part, as the scientists whose work is discussed did not characterize their methodology in these or comparable terms, perhaps because they were unconcerned with methodology. Even in such circumstances, however, classifiers do communicate something about their methodologies that, although implicit, can be detected in what they seem concerned to achieve by their classifications. An effective starting point to a discussion of alternative methodologies is, therefore, to distinguish the concerns of different classifiers.

4.1.1. Nominalism and Realism

Classifiers could take the position that their classifications are no more than convenient descriptions. The groupings in their classifications, they might argue, do not represent real groupings among nerve cells but are merely useful artifacts of their minds. However powerful and self-consistent such a classification might be, the classifier would claim no more for it than convenience. This approach to classification has been termed *nominalist* (e.g., Mayr, 1969). The same approach to the formulation of scientific theories generally has been termed *instrumentalist* (see Chapter 5, Section 5.3).

Very few classifiers of nerve cells, if any, have taken a nominalist position, however. Perhaps early in their studies many classifiers make initial groupings of cells that they regard as no more than convenient. As more data accumulate, however, classifiers of nerve cells have consistently taken the view that their developed categories reflect real, functionally meaningful, "natural" groupings among the cells studied, i.e., they adopt a "realist" position. The nominalist position may, of course, serve as a useful refuge when methodological problems emerge in a classification; but it has not been a common approach in work on the classification of neurons.

4.1.2 Different Realist Approaches: Toward Certainty, Objectivity, or Testability

Most of the variety we find in approaches to cell classification seems to stem from the distinct starting positions (assumptions, presuppositions, concerns) of different workers. Though all realists, classifiers of nerve cells have taken one of three general approaches, which seem best characterized by their different concerns.

A concern with *certainty* is characteristic of an "essentialist"* or "typological" approach to classification, and seems to underlie the feature extraction classification of retinal ganglion cells. Central to the descriptions of feature-detector groupings among ganglion cells, for example, is the assumption (or presupposition) that it has been possible, on the basis of one or a few "key features," to identify the essential operation performed by each cell type. A classification of cells can then readily be set up, each grouping comprising a cell type performing a distinct operation. The groupings of such classifications are defined in terms of properties of the cells, with the result that the classifications are not testable [for reasons set out previously (Rowe and Stone, 1977, 1979) and in Chapter 5], and are therefore immune· to disproof. These classifications thus acquire a certain certainty. Conversely, the criticism expressed by the essentialist classifier of more tentative, parametric classifications is their uncertain, constantly changing bases (see, for example, Hughes, 1979). The problem of the objectivity or fallibility of the observer in identifying the essential natures of different types of ganglion cells is generally not discussed; the identification is considered compelling and self-evident.

Objectivity is, however, a principal concern of the numerical taxonomist. For example, Sneath and Sokal (1973, p. 11) describe the chief advantages of numerical taxonomy as repeatability and objectivity. Animals (or neurons) are grouped according to the degree of similarity between them as assessed by measuring their properties. Objectivity is sought by including many, ideally all, features of a cell or animal in the data base for classification (rather than a few key features) and, ideally, giving each feature equal weight. This avoids any subjective influence on the resulting classification, such as must operate in the essentialist's choice of key features. Moreover, standard, quantitative (and presumably therefore objective) procedures are used to detect phenetic similarity between different animals. Mayr (1969) argues that numerical taxonomy is a form of nominalism in taxonomy, i.e., that the groupings in such classifications are mere artifacts of the classifier's mind, and he expresses several criticisms of the approach. Numerical taxonomists themselves do not see their position as nominalist, however. The approach has not yet been used in the classification of cells in the visual system (or, as far as I am aware, in other brain systems), although it is close to the approach recommended by Tyner (1975). For further discussion, the reader is referred to Sneath and Sokal (1973), Mayr (1969), and to Tyner's article.

The critique of numerical taxonomy proposed by Pratt (1972) seems to me compelling, however. Pratt argues the impossibility of knowing what constitutes a "feature," to be measured and weighed equally with other features. If a ganglion cell has an ON-·center and OFF-surround receptive field, for example, does that constitute one feature or two or four? If the number of features attributable to a cell is indeterminate, how can the number of features common to, or distinct between, two (or more) cells be assessed?

A concern with *testability* leads to the setting up of classifications as hypotheses,

---

*Aristotle argued (in *De Partibus Animalium*, Bk. 1, Ch. 1, 641/20–25) that it is possible, indeed that it is the task of the naturalist, to discover the true essence of animals (of Aristotle's view more below), and this discovery of the essential natures of different animals has been the aim of many taxonomists since. Popper (1962, p. 34) argues that essentialist philosophers/classifiers seek the name proper to the essences of the things being classified and regard those names as definitions of the essences.

for example that cells of different physical types fill different functional roles within the nervous system. The hypotheses that comprise the classifications are then tested by deduction and experiment. The only explicit discussion of this methodology in the visual literature is (as far as I know) that presented by Rowe and myself (1977, 1979) in support of the Y/X/W classification of ganglion cells. Our discussion followed Bock's (1974) argument for a hypothetico-deductive approach to animal classification. Bock notes that although few taxonomists discuss or advocate this methodology, it has been of central importance in modern, phylogenetic taxonomy, and that the beginnings of its use can be seen in Darwin's writings on taxonomy (see also Ghiselin, 1969). In setting up his classification as a testable hypothesis, the classifier attempts to absorb the problems of certainty and objectivity. Certainty (it is argued) may be unreachable, but truth can be approached by testing hypotheses, and rejecting those shown to be false. Objectivity may also be unattainable, but if the influence of the classifier on his classification cannot be eliminated by logical rigor, it presumably can, sooner or later, be eliminated (insofar as it is wrong) by falsification.

An important advantage of this last methodology is that the testing of hypotheses encourages the collection of more observations about the cells under study. When a hypothesis is confirmed, it encourages the formulation and testing of more adventurous hypotheses; where disproved, it encourages the formulation and testing of substitute hypotheses. This questioning property of the hypothetico-deductive approach makes it the natural epistemology for the parametric approach to ganglion cell classification (discussed in Chapters 1 and 2). My own judgment is that the hypothetico-deductive approach is the most useful and defensible of any of which I am aware, and a formal exposition of a hypothetico-deductive approach to classification is set out at the end of Chapter 5 (Section 5.6). In the present context of a comparison with the essentialist and numerical taxonomic approaches, the following features of the hypothetico-deductive approach deserve mention.

1. It is "polythetic" (Tyner's term) or "parametric" (my term), i.e., the classification is based on many parameters of cells rather than one or a few (it has this feature in common with numerical taxonomy).
2. Although the classifier bases his groupings on the physical properties of cells, the groupings are not defined in terms of any property or properties. The groupings are described, not defined.
3. The classification is presented as a set of testable hypotheses to the effect that the groups of cells being delineated subserve distinct functional roles. To preserve this feature of testability, the groups of cells distinguished are not named after either their properties or the functional role they are hypothesized to fulfill; generally, alphanumeric designations are used. In addition, the functional roles are defined in terms independent of properties of the nerve cells.

4.1.3. Summary

Different starting points could have been chosen for the present discussion. Instead of analyzing the concerns of classifiers, I could for example have examined whether they based their classifications on the physiological or morphological features of neurons, or on their supposed functional roles, or on their phylogenetic histories. All classifiers, however, are seeking to establish some order in their observations, and the nature of the order they seek seems to determine the way they relate their classifications to their data. It thus seemed more fundamental to consider what different workers were concerned to achieve with their classifications. This approach is not novel; Ghiselin (1969), for example, made the same point when he commented:

> . . . a scientist's attitude towards classification systems is profoundly affected by the structure of his language, often in a manner of which he is quite unaware.

Ghiselin's context makes it clear that by "language" he means what might also be termed "presuppositions"; for example, whether a classifier presumes that abstractions, such as species or phylum, are real or artificial. Clearly, such presuppositions will have a strong influence on the sort of classification a taxonomist produces.

And Popper (1972, p. 159) made the same point in a philosophical context with his comment on his own understanding of the Greek philosopher Heraclitus.

> My general approach to Heraclitus may be put in the words of Karl Reinhardt: the history of philosophy is the history of its problems. If you want to explain Heraclitus, tell us first what his problem was.

The same may be said, it seems to me, for classifiers.

The analysis advanced in the preceding section could be viewed as an attempt to classify classifications, in which case, of course, I should follow the methodology of classification proposed previously (Rowe and Stone, 1977) and in Section 5.6. In the terms set out there, "certainty," "objectivity," and "testability" are categories of the classification, and various classes of proposed classifications are the taxa. For example, the various "feature extraction" classifications might be considered one taxon (group of classifications), directed to establish a certain body of knowledge about ganglion cells; numerical taxonomic classifications would be another group directed to establish objectively true knowledge about ganglion cells; and hypothetico-deductive classifications another group directed to establish testable propositions about the functional groupings of ganglion cells. But my major point is less pedantic than this; it is to argue that an effective way of understanding various approaches to classifications is to consider what the classifiers were trying to achieve. In the classification of nerve cells, three distinct thrusts can be recognized in the writings of different workers: toward certainty, toward objectivity, and toward testability.

4.2. THREE STAGES IN THE TAXONOMY OF ANIMALS

It is convenient, though no doubt simplistic, to distinguish three stages in the evolution of animal taxonomy that bear relevance to the classification of neurons; these I term *metaphysical essentialism, typology,* and *post-Darwinian taxonomy.* The first two of these stages are commonly grouped together as *essentialist* or *typological* by contemporary writers (e.g., Mayr *et al.,* 1953), because in both stages, animals were classified in terms of their "essences" or "types." Aristotle, however, proposed that animals must be understood (hence classified) in terms of their *metaphysical* essence or "cause," whereas more modern typologists classify animals according to their *physical* essence or "type." Nevertheless, typology is commonly considered to have begun with Aristotle, presumably because he was one of the first to argue that animals have to be understood in terms of their "essences." The theory of evolution has had a dramatic effect on animal taxonomy, making possible several distinctly different classifications of animals (see, for example Bock, 1974), each based on phylogenetic relationships between animals.

4.2.1. Aristotle's Taxonomy: Metaphysical Essentialism

Although the typological approach to animal taxonomy, which was a principal casualty of the "Darwinian revolution," is commonly traced to the writings of Aristotle, these writings themselves (e.g., *De Partibus Animalium*) show that Aristotle was aware of the difficulties that ensue when animals are classified according to their physical features.* Rather, he stressed the importance for understanding the nature of animals of a concern with their heredity and function, and sought a basis for the classification of animals in their "final cause" or metaphysical "essence." Further, in *De Partibus Animalium* [translated by W. Ogle in *The Basic Works of Aristotle* (R. McKeon, ed.), 1941], Aristotle shows a clear awareness of the difficulties that result if consideration is limited to just physical features.

According to McKeon (p. 19 Introduction), Aristotle sought to combine the "lofty contemplation" of Plato with the empirical emphasis placed on physical features by Democritus and the atomists. He sought to give greater reality to Plato's forms and definitions, as they might apply to animals, by relating them causally to the physical features of animals. In the context of animal taxonomy, Aristotle argues (*De Partibus Animalium*, Bk. 1, Ch. 1, 640/10–15) that a major initial question in describing animals

> . . . is whether the proper subject of our exposition is . . . the process of formation of each animal; or whether it is not rather . . . the characters of a given creature when formed.

Aristotle recommends a concern with both, seeking to relate one to the other causally:

> The best course appears to be that . . . we should begin with the phenomena presented by each group of animals and . . . proceed afterwards to state the causes of those phenomena, and to deal with their evolution.

This is a functionally oriented, polythetic beginning to classification, very much in keeping with phylogenetic taxonomy. Aristotle put the case against physical particularism clearly (*ibid.*, 640/15–20):

> Empedocles then was in error when he said that many of the characters presented by animals were merely the results of incidental occurrences during their development. . . . In so saying

---

*This rather long essay is partly in penitence for a most superficial earlier treatment of Aristotle (in Rowe and Stone, 1977), but mostly in recognition of the modern relevance of several of Aristotle's insights, particularly his arguments against classification by physical features.

he overlooked the fact that propagation implies a creative seed endowed with certain forma-
tive properties . . . [so that] it is the possession of certain characters by the parent that determine the development of like characters in the child.

He went on to argue that the attempt to characterize animals by some physical
feature not only neglects their heredity but is inadequate conceptually (*ibid.*, 641/0–10):

Democritus says . . . that it is evident to everyone what form it is that makes the man, seeing
that he is recognisable by his shape and colour. And yet a dead body has exactly the same
configuration as a living one; but for all that is not a man. So also no hand of bronze or wood
. . . can possibly be a hand in more than name.

By what forces then is the shape of a hand determined?

The woodcarver will perhaps say, by the axe or auger; the physiologist, by air and by earth.
Of these two answers the artificer's is the better, but it is nevertheless insufficient.

Aristotle's criticism of the woodcarver's mechanistic explanation of his sculpture of
a hand is that it omits any account of "the reasons why he struck his block in such a
way as to effect [a carved hand]." Thus, he sees the carver's *intent* as a basic cause
underlying the formation of the wooden hand. In the same sense, he argues, the ultimate
explanation of an animal's form must be sought in terms of some underlying nonphysical
cause, "for the process of evolution is for the sake of the evolved, and not this for the
sake of the process."

This is the point at which the modern, polythetic, phylogenetically oriented taxonomist would part company with Aristotle. For the modern theory of evolution has
provided a compelling mechanistic framework for understanding the formation of animals, which requires no underlying plan, "final cause," or metaphysical "form." The
early physiologists' explanation of the formation of animals in terms of the action of the
"air and earth" can be replaced by reference to species, orders, and families of animals,
to genetic mutations, natural selection, and environmental pressures, and to the molec-
ular basis of heredity. Without these terms of reference, Aristotle sought a nonmechan-
istic solution to the question of what comprises the nature of an animal, and proposed
that the essence of an animal is its soul, a metaphysical entity that is the underlying,
basic cause of its formation. Nevertheless, he did not seek by this to take the essence of
animals out of the ambit of investigation (*ibid.*, 641/20–25).

. . . if now this something that constitutes the form of a living being be the soul . . . then it
will come within the province of the natural philosopher to inform himself concerning the
soul, to treat of it in its entirety or . . . of that part of it which constitutes the essential character of an animal; and it will be his duty to say what this . . . is.

Aristotle believed that the essential nature of an animal can be perceived, and
should be the basis of understanding the animal and of defining its relation to other
animals; in modern terms, of classifying it. Yet he believed that essence to be metaphysical rather than physical. No doubt he was influenced by Plato's metaphysics, as well
as by the inadequacy in his time of any mechanistic account of the properties of animals,
such as the theory of evolution now provides. But it seems clear that he was also influenced by a sense of the inadequacy of any attempt to comprehend the nature of an
animal in terms of its physical features. Some of the modern criticisms of essentialism

are, of course, applicable to Aristotle, particularly the objections raised by Popper as well as by taxonomists (Simpson, 1961; Mayr, 1969) to the arbitrary and authoritarian nature of any assertion a taxonomist might make that he has identified the essence or "type" of an animal. But the following passages suggest that responsibility for the problems that result from classification by physical features should not be Aristotle's.

Aristotle was an insistent critic of the "method of Dichotomy" proposed by Plato. His critique is presented in a little detail here because dichotomous classifications (linear/nonlinear, sustained/transient, brisk/sluggish, simple/complex) are still commonly used in cell classification in the visual system. Aristotle writes (*De Partibus Animalium,* Bk. 1, Ch. 2, 642/5–10):

> Some writers propose to reach the definitions of the ultimate forms of animal life by bipartite division. . . . If . . . natural groups are not to be broken up, the method of Dichotomy cannot be employed, for it necessarily involves . . . breaking up and dislocation.

A principal problem Aristotle saw is that dichotomies are either nonexclusive, e.g., feathered vs. footed, in which case animals may be encountered with both features; or they are exclusive, say feathered and featherless, the second term being "privative."* Aristotle notes (*ibid.,* Ch. 3, 642/20–25) that:

> . . . privative terms admit of no subdivision . . . yet a generic differentia must be subdivisible, for otherwise what is there that makes it generic rather than specific?

What Aristotle means by his proposition that "privative terms admit of no subdivision" has been analyzed by Balme (1975), who notes Aristotle's insistence that, when animals are divided and then subdivided into groups, the differentiations must be successive. That is, each new criterion for division must imply its predecessors (as *biped* implies *footed*). It would be an error, then, to divide the footed into the warm-blooded and the cold-blooded. Aristotle's reason is that the warm-/cold-blooded division may be appropriate to a wider genus than the footed animals. "Privative" classes, Aristotle seems to argue, cannot be subdivided because no differentia can be found that implies a privative predecessor. Aristotle concludes (*ibid.,* 643/10–15) that

> . . . the method of dichotomies is either impossible (for it would put a single group under different divisions or contrary groups under the same division), or it only furnishes a single ultimate differentia for each species.

In visual physiology, this approach has been pursued furthest by Levick (1975) who described three "dichotomizing subdivisions" in the properties of cat retinal ganglion cells. This was an outcome, perhaps a logical outcome, of Cleland and Levick's (1974a,b) choice of certain physical features of these cells as constituting their physical

---

*J. S. Mill was similarly critical of dichotomous classifications with groups named after their distinguishing feature:

> Take any attribute whatever, and if some things have it, and others have not, we may ground on the attribute a division of all things into two classes; and we actually do so the moment we create a name which connotes the attribute. The number of possible classes, therefore, is boundless; and there are as many classes . . . as there are general names. [M:ll (1843), Bk. I, Ch. 1, Sect. 4]

type or essence. Because Aristotle's essentialism was metaphysical, he could avoid the problems raised by a reliance on physical features.*†

Aristotle's metaphysical essentialism was, of course, quite impractical; he makes no claim either to have identified the essence of even a single species, or to have devised a method for identifying metaphysical essences and his is, like any form of essentialism, at least partly authoritarian. Nevertheless, Aristotle saw more clearly than many before him the difficulties of classifying by physical features, and two more passages indicate the depth of his insight.

First (*De Partibus Animalium*, Bk. 1, Ch. 3, 643/10–15):

> The method then that we must adopt is to attempt to recognize the natural groups, following the indications afforded by the instincts of mankind, which led them for instance to form the class of Birds and the class of Fishes, each of which groups combines a multitude of differentiae, and is not defined by a single one, as in Dichotomy.‡

Second (*Analytica Priora*, Bk. 1, Ch. 31, 46/30–35):

> It is easy to see that division [by Dichotomy] into classes . . . is . . . a weak syllogism; for what it ought to prove, it begs, and it always establishes something more general than the attribute in question.

Obviously, it cannot be concluded from these brief passages that Aristotle was proposing a hypothetico-deductive approach to classification. He was not, although the first passage commends a central role for intuition in the formulation of classifications, comparable to the role a falsificationist would suggest for intuition in the formulation of

---

*None of this is to deny either the logical rigor of dichotomies or their occasional usefulness. A remark attributed to Mae West shows the usefulness of a dichotomy based on physical attributes, when ambiguity is the aim of the distinction: "I go for two kinds of men, the kind with muscles and the kind without." And a judge of the Australian High Court (Windeyer, J., 1959, *Commonwealth Law Reports* **107**:272) has commented on the logical validity, yet overall uselessness, of some dichotomies: "Counsel . . . started with the proposition that disputes are either industrial or not industrial. . . . That is logically incontestable. . . . Like Sinclair's well-known division of sleep into two sorts, namely sleeping with or sleeping without a night cap, it would seem to exhaust the subject."

†The present view seems in close agreement with D. A. Balme's analysis of Aristotle's zoological classification (Chapter 12 in Barnes *et al.*, 1975). Balme argues that Aristotle made three major reforms to the method of logical division used by Plato and others. First, Aristotle sought a classification that would reveal the nature (essence) of the objects classified, and not just serve to put like things together, and keep unlike apart. Second, Aristotle argued for "successive differentiation," i.e., for a logical rather than arbitrary sequence to the selection of distinguishing features. The third reform, Balme argues, was Aristotle's rule "to divide the genus by a plurality of features, not by one at a time." In *De Partibus Animalium* (Bk. 1, Ch. 3), Aristotle argues that division by a single differentia "must always fail." The last point is, of course, central to the above discussion.

‡A similar recommendation is made by Mill in the following passage from *A System of Logic* (1843, Bk. IV, Ch. VII, Para. 2):

> The phrase Natural Classification seems . . . appropriate to such arrangements as correspond . . . to the spontaneous tendencies of the mind, by placing together the objects most similar in their general aspect; in opposition to those technical systems which, arranging things according to . . . some circumstance arbitrarily selected, often throw into the same group objects which in . . . general . . . present no resemblance, and into different . . . groups others which have the closest similarity.

hypotheses; and the latter passage implies the need to prove (test, hence confirm or falsify) a classification. But it seems clear that he understood these problems of "typology" that have led many modern taxonomists to the hypothetico-deductive position. Aristotle's solution to these problems was a metaphysical essentialism.

One final strand may be drawn out of this argument. Aristotle and modern phylogenetic taxonomists seem to be in substantial agreement on this: that an animal's physical properties, and its relation to other animals, are to be understood in terms of its function and heredity. Neither would rely for classification solely on physical features. Both would argue that physical features are expressions of heredity and function, and are to be understood in those terms. It is argued in Chapter 5 that, analogously, the physical features of neurons are best employed in classifying if they are viewed as expressions of the function and also (ideally) of the phylogenetic history of those neurons.

4.2.2. Physical Typology

Typology is the classification of animals according to their physical type, based upon consideration of their physical properties. It is clearly *not* an approach proposed by Aristotle, but presumably developed in reaction to the practical failure of his metaphysical essentialism. Typology was the dominant methodology of classification between Aristotle and Darwin, and reached its highest conceptual development in the *Systema Naturae* of Linnaeus (1735), in the 10th edition of which the trinomial system of nomenclature, still widely used, was first applied. Linnaeus' *Systema* covered plant and mineral "kingdoms" as well as the animal, and its basis in observable phenomena was consonant with the growing empiricist philosophies of Linnaeus' time. The common belief in the fixity of species would have discouraged any development of Aristotle's suggestion that an animal's heredity was an important part of its essence.

Typologists can of course be either nominalist or (the great majority) realists, in the senses of these terms used above (Section 4.1.1), and it is probably fair to suggest that the realists among them have been "essentialists" in the sense that they believed that the features of an animal by which they characterized its "type" were its "essence," or at least part of it. That is, the wings of birds and the vertebral column of vertebrates were considered part of the essential nature of these animals. In the final analysis, however, typology relies on arbitrary and stipulative approaches to the identification of the physical type or essence of an animal, and is beset by the logical problems that Aristotle foresaw. In its favor was its reliance on observable phenomena and, until the emergence of the theory of evolution, it was the only effective scheme of animal classification.

4.2.3. The Influence of the Theory of Evolution

The development of the theory of evolution in the mid-19th century had a revolutionary impact on the typological, fixed-species concepts of animal taxonomy that had preceded it. Many points of this impact can be distinguished. For example:

1. The theory of evolution thoroughly undermined the idea of the fixity of species; taxonomic theory henceforth allowed for dynamic changes in animals.

2. The theory seriously questioned the uniqueness of man among animal species, suggesting that man was quantitatively different (more intelligent, more complex in social interactions) from other species, but not qualitatively different.

3. The theory of evolution suggested that the scheme of animal development that it explained, and whose outlines had been understood before Darwin, was not a progression toward a species of unique perfection (man?), but occurred under the pressure of competition for survival, an influence acting "from behind." This was an implication that many found difficult to accept, even after they had accepted point 2 above.

4. Most importantly for the present context, the theory of evolution offered a solution to the problem that Aristotle had confronted. Aristotle had reacted to the logical weakness of physical typology by proposing the classification of animal according to a metaphysical "final cause" or essence. Because neither Aristotle nor any follower had developed a scheme for identifying that metaphysical essence, most subsequent taxonomists, less troubled than Aristotle by its logical weaknesses, reverted to typology as the only practicable approach to constructing usable classifications.

   The solution offered by the theory of evolution is that animal classification need involve neither the arbitrary selection of physical types, nor the uncharted search for metaphysical essences. Classification would become rather a series of hypotheses about the phylogenetic relationships between animals. That is, evolution provided a "functional basis" (Rowe and Stone, 1977) for animal classification. This solution was seen by Darwin who wrote (1872, p. 569) that "expressions such as that famous one by Linnaeus . . . that the characters do not make the genus but the genus makes the characters, seem to imply that some deeper bond is included in our classifications than mere resemblance."

   Darwin (1872, p. 576) went on to suggest

   > . . . that community of descent is the hidden bond which naturalists have been seeking and . . . not the mere putting together and separating objects more or less alike. . . .

5. The theory of evolution led to a wide acceptance of Aristotle's criticism of typological classifications. Mayr (1959), for example, lists as one of the three major contributions made by Darwin that "he replaced typological by population thinking." Mayr was commenting in a zoological context; the theory of evolution has led, he argues, to the view that animals can fruitfully be regarded as forming populations, with considerable variation of properties within populations, rather than as forming qualitatively distinct types, with only minor variation within types. The recognition of variation within biological populations helped overcome one of the difficulties of the theory of evolution, the problem of how one "type" of animal can transform into a qualitatively different "type"; the difficulty is greatly reduced when intragroup variation is recognized and the differences between groups are seen as quantitative rather than qualitative. Oldroyd (1980) makes much the same point:

   > Darwin shifted attention away from the search for single essential characters which might serve as marks of whole groups. Rather, he said, one should consider clusters or

aggregates of characters when attempting to identify natural kinds, and taxonomic distinctions should be made on the basis of genealogical differences.

The lesson seems equally relevant to the classification of nerve cells. These are components of animals and evolving populations in themselves. For biological as well as epistemological reasons, it seems necessary to view groups of nerve cells as populations containing significant variation within them, and to move away from classifying them as qualitatively distinct "types," toward parametrically based, functionally oriented classifications.

This rejection of typology as classification by "mere resemblance" is now widely, though far from universally, accepted in the taxonomy of animals; I here urge the same rejection in the taxonomy of nerve cells. The aim of cell classification should be to describe and discover the organization of the nervous system; just as, for many zoologists, the aim of animal taxonomy has become the elucidation of the phylogenetic relationship between animals, in Huxley's (1940, p. 2) words "the discovery of evolution at work." Classification could become part of the investigative procedure of the neurobiologist as well as of the zoologist. Consequent to this point, it is argued in Chapter 5 that for the classification of nerve cells to be heuristically effective, it must be as open to testing, and corroboration or falsification, as any other proposition that might be made about such cells. Classifications, it is argued there, should be set up as hypotheses about functional groupings among nerve cells, not as definitions of cell types. Classification by definitions based on physical features, however attractive the sense of certainty initially achieved, is likely to introduce into the literature the difficulties of arbitrariness and inconsistency that Aristotle foresaw.

Epistemological Background   5

Inductivism, Essentialism, Instrumentalism, Falsificationism, and Paradigms

> *I have tried too in my time to be a philosopher; but, I don't know how,*
> *cheerfulness was always breaking in.*
> Oliver Edwards, in Boswell's *Johnson*, April 17, 1778.

There are compelling reasons for an experimental scientist to avoid philosophical issues, yet equally compelling matters that demand his attention to them. Many neurobiologists already feel so distracted from their research by the duties imposed by their institutions and the technical management of their laboratories that they are impatient with the further distraction of philosophical debate. In any case, some of the problems raised by philosophers (can we really know anything?, can we prove anything?, can we disprove anything?) seem either meaningless or unanswerable and certainly, as Dr. Johnson's acquaintance commented, cheerless. There seems little to be gained by spending time on them; better get on with our experiments.

On the other hand, issues of methodology and epistemology do have to be considered if one seeks an understanding of how different scientists operate, and how their bodies of work interact; if one seeks to understand, in other words, the dynamics of one's field of research. Some neurobiologists have recognized, moreover, the impact of theories of knowledge on their day-to-day work. Eccles (1975) has commented, for example,

Through my association with Popper I experienced a great liberation in escaping from the rigid conventions that are generally held with respect to scientific research. Until 1944 I held the following conventional ideas about the nature of research: First, that hypotheses grow out of the careful and methodical collection of experimental data. (This is the inductive idea of science that we attribute to Bacon and Mill.) Second, that the excellence of a scientist can be judged by the reliability of his developed hypotheses which, it is hoped, stand as a firm and secure foundation for further conceptual development. Finally, and this is the important point, that it is in the highest degree regrettable and a sign of failure if a scientist espouses an hypothesis that is falsified by new data so that it has to be scrapped altogether. When one is liberated from these restrictive dogmas, scientific investigation becomes an exciting adventure opening up new visions; and this attitude has, I think, been reflected in my own scientific life since that time.

Sympathetic or unsympathetic to philosophical debate, the reader will have noted my own concern with methodology. This chapter is a brief attempt to distinguish some of the methodologies and philosophies of science that seem relevant to an understanding of modern work on the visual pathways. I have tried to keep my own adoption of the falsificationist methodology separate from the discussion, or at least to make it explicit. In that discussion I employ terms such as *falsificationism, inductivism, essentialism,* and *instrumentalism* that come from the current literature of the philosophy of science. I do this, first, because the various philosophies and assumptions discussed have all been recognized and labeled by philosophers of science and, second, because the present analysis does not go beyond that found in recent accounts of contemporary philosophy of science, such as Chalmers (1976) provides.

5.1. INDUCTIVISM

Many visual neurobiologists might, if asked to give an account of progress in their field, describe that progress as the gradual accumulation of facts established. "So-and-so demonstrated that phenomenon A occurs in the retina"; "Somebody-else showed that phenomenon B occurs in the lateral geniculate nucleus, but only in lamina K"; "Somebody-else's off-sider came up with the entirely new finding that nucleus C projects to the visual cortex"; and so on. On the basis of observations (facts) such as these, it might be argued, a general account of how the visual system works is being put together. It would be accepted in such an account that So-and-so, Somebody-else, and his Off-sider were reliable and thorough investigators who, by following appropriate scientific methods of testing and proof, had ensured that their demonstrations were clear and unequivocal, so that what they had reported could be accepted as objectively true. On the basis of many such established facts, theories of visual function can be soundly formulated, and their implications can be followed out deductively. Each careful study in the field might then be seen to contribute a brick to the structure of visual science and, while individual scientists might feel that their particular brick was especially well conceived and made, science is nevertheless seen as the cumulative effort of a community of scientists.

This is the inductivist view of scientific progress; it gains this name because, in this view, the fundamental process of scientific discovery is the induction of a general theory from empirically established facts. The inductivist assumes that objectivity is possible, that facts can be established beyond equivocation by empirical means, and that theories can be based on these facts with logical rigor and necessity. It is considered possible, therefore, to distinguish facts from theories and to attain certainty; and it is also considered that there is a formal logical methodology for the induction of theories, to which the scientist must adhere, if his theory is to be sound.

This model of science is (according to many accounts; see, for example, Chapter 1 of Chalmers, 1976) one of the most popular formulations of how science works. It allows a place for individuality (many workers with different ideas), and accounts for controversies (which bricks are unsound), for progress (the reliance of scientists on what has been done before), and for the obvious success of science. Historically, it can be traced back to Bacon's celebrated *Novum Organum* and, as Eccles' comment quoted above makes clear, it has been a commonly held view in the neurosciences. Eccles drew attention to what he felt were the restrictive effects of this view. Because a theory which is disproved must (according to this view) have been induced by an "unsound" methodology or, at best, have been premature, an obligation is placed on scientists to be not just earnest and honest, but right. Because their reputation is at stake with their theory, they cannot unemotionally enter subsequent debate about it; too often there is resort to authority and assertion. On epistemological grounds, moreover, many modern philosophers of science would argue that inductivism is a deeply flawed model of how science really operates. Chalmers, for example, terms this account of science *naive inductivism*, and points to two fundamental weaknesses in it as a theory of knowledge. First, there is no logic to induction, i.e., no valid way to draw a generalization (e.g., a theory) from observations, however repeatable and numerous. Hume was among the first to formulate the "problem of induction," arguing that there is no logical basis for the assumption "that those instances, of which we have had no experience, resemble those of which we have had experience" (*A Treatise on Human Nature*, Bk. 1, Pt. III, Sect. VI and XII). There have been counterarguments, notably by Mill (*A System of Logic*, especially Bk. III, Ch. VIII), who proposed five rules for the performance of inductive procedures, but few modern commentators accept the inductivist position. There is simply no way of knowing when sufficient "hard" observations have been made, or when one's assumptions have been sufficiently tested and realized, to establish a fact.* Second, the distinction between observations and theories cannot be sustained. That is, all observations are made in the context of a theoretical framework and cannot, therefore, become established as proven independent of the validity of any theory. Observations are not theory-independent; rather theory is deeply intertwined with every observation. Even if there were a valid logic to induction, i.e., to the formulation of a theoretical generalization on the basis of "hard", theory-free observations, we would not be able to find any such observations with which to start.

For example, a visual physiologist might do an experiment to determine whether the preferred orientations of cells in the cat's visual cortex are evenly distributed over

---

*There are many examples of inductive generalizations long held, but eventually proved false. These include the fixity of species, the propositions that all swans are white, that the earth is flat, and that the sun goes around the earth, and the belief of the inductivist turkey, that all turkeys are fed at 9.00 a.m. daily, a belief the turkey held until shortly before one Christmas.

the 0- to 360°-range possible. The question seems amenable to a clear answer; either the distribution is uniform or it is not. Our investigator might determine the preferred orientation of several hundred cells from half-a-dozen cats and conclude after appropriate statistical analysis that the preferred orientations are evenly distributed, or at least that there was no evidence of certain orientations being more commonly preferred. The work might be repeated on a sample of several hundred (or several hundred thousand) more cells, by another laboratory (or another hundred laboratories), and the result clearly confirmed. Can the result be considered proven and, therefore, beyond further question? Can it, and similar observations, be used as secure building blocks for a general theory of cortical function?

The answer appears to be in the negative, for there was an important hidden assumption in our investigator's experimental design, namely that in the context of his question, cortical cells are appropriately considered to be a single population. Leventhal and Hirsch (1977, 1978) have reported that if visual cortical cells are separated into two groups, which they termed *C* and *S cells* and felt were natural groupings among cortical cells, then a statistically significant grouping of preferred orientations was apparent for S cells (about the vertical or horizontal), but not for C cells. If, as Leventhal and Hirsch suggest, C and S cells are functionally distinct groups of cortical cells (perhaps related to the Y-cell/X-cell grouping at subcortical levels), then the answer to our question (are the preferred orientations of cortical cells evenly distributed?) becomes "yes" for C cells and "no" for S cells. With a different theoretical construct (two groups of cortical cells instead of one), different conclusions are obtained. It is not that one conclusion is wrong, and the other right. The point is that the conclusions reached depend on the observer's presuppositions.

It is arguable that with a simpler observation, say determining the preferred orientation of a single cortical cell, there would be no theoretical complications to the observation. There have, however, been substantial arguments about the measurement of orientation-selectivity. The arguments have concerned the distinction between orientation-, direction-, and axis-selectivity (see, for example, Henry *et al.,* 1974, and Pettigrew, 1974), the range of responses of a cell that must be studied to show orientation-selectivity, and the procedures that might be used to specify the preferred orientation of a cell from any particular set of measurements. Even the measurement of the single response of a cell, a more basic problem, requires some assumptions to be made. For example, should one measure the total number of action spikes evoked by a stimulus?, or the maximum firing rate reached during the response? (And if the latter, over what unit of time should rate be assessed?) How should the stimulus be chosen for a particular cell, given that cortical cells are differently sensitive to the size, velocity, or contrast of a stimulus? How carefully need the level of anesthesia be monitored?

Despite these fundamental problems, inductivism was one of the major formulations of scientific method that the 19th century bequeathed to the 20th. Because the problems of inductivism are intractable, the inductivist could overcome them (and remain an inductivist) only by resort to authority and arbitrary definition. It was largely in response to this authoritarian strand of argument that 20th century philosophers, notably Popper, sought a different, more creative solution to the "problem of induction" raised by Hume.

5.2. ESSENTIALISM (TYPOLOGY)

In the broadest sense, essentialism is an attempt to understand the world in terms of the essential natures of things. In the taxonomy of animals, essentialism has meant (see Chapter 4, Section 4.2) either the metaphysical essentialism advocated by Aristotle or the physical essentialism of many of his followers. That is, classifiers sought to discover the essential natures of animals, and to characterize the differences between them, in terms of either their metaphysical "cause" or their physical "type" (hence typology). In the classification of nerve cells, typology has been and still is a common approach; it seems to be the methodology underlying the feature extraction approach to the classification of nerve cells.

Essentialism has a good deal in common with inductivism as a model and methodology of science. In particular, it shares the assumption that careful observations can provide the basis for sound theories, but it is a somewhat bolder approach. The essentialist/typologist believes that he can elucidate the essential nature of the phenomenon in which he is interested in a single, inductive step (or at most a few steps) based on key observations. This contrasts with the almost open-ended observation-by-observation progress envisaged by the inductivist. Typology, therefore, shares the vulnerability of inductivism to the criticism that it is nonlogical (yet claims certainty) and that it assumes that observations can be made that are free of theoretical assumptions. In addition, however, the very boldness that makes it so attractive introduces substantial problems. For example (see also the discussion of these problems in Simpson, 1961; Mayr, 1969; Tyner, 1975; and Rowe and Stone, 1977, 1979):

1. There is no logic to the determination of the properties of a cell that comprise its essential nature. Typology depends, therefore, on an assertion by the observer that he has discovered the true nature of the cell. There is, consequently, rich ground for unresolvable disagreement between authorities as to what the essential nature might be. Typology thus allows an authoritarian strand into classification.
2. Physical typology always involves the choice of a particular property, or group of properties, as "key" features of a cell's type, other features being less important, even accidental. That choice throws great weight on features that may prove to be of little import, and discourages investigation of other, initially uninteresting features, which may prove to be of fundamental importance.
3. The emphasis on "key properties" means that objects (say nerve cells) that differ in quite fundamental ways may nevertheless be grouped together because they share a single "key" property. Thus, important distinctions between cells may be blurred.
4. Conversely, objects (or cells) that differ in the key feature may share many properties of fundamental importance, yet be put into different groups. In this way, substantial similarities between the objects may be obscured.
5. Most importantly, the use of key features to define categories of a classification reduces, indeed often eliminates, the capacity of that classification for self-correction. The groupings, being established by definition, are not readily modified;

the classification becomes a codification of particular ideas, instead of an avenue to new ones.

5.3. INSTRUMENTALISM

Instrumentalism is a much more pessimistic analysis of scientific knowledge and its growth than either inductivism or its near relative, essentialism. Popper, in his defense of a realist view of scientific knowledge ("Three Views concerning Human Knowledge" in his 1972), traces the origins of instrumentalism to the confrontation between Galileo and the Church over Galileo's dramatic evidence in support of the Copernican view that the sun, rather than the earth, is the center of the solar system. The Church, Popper notes, did not forbid Galileo to teach his new system, provided he prefaced it with the statement that, whatever its usefulness and power, no claim was being made as to its truth. Popper quotes a letter from Cardinal Bellarmino:

> Galileo will act prudently . . . if he speaks hypothetically. . . . To say that we give a better account of the appearances by supposing the earth to be moving and the sun at rest, than we could if we used eccentrics and epicycles, is to speak properly; there is no danger in that and it is all that the mathematician requires.

Provided Galileo presented his analysis as a mere tool or "instrument" of use in describing and relating various facts and measurements about the solar system, the Church saw no problem. The crisis between the parties arose when Galileo persisted in claiming the reality of his system.

The instrumentalist view of science, then, is that any of the abstractions the scientist makes on the basis of observations are the creations of his or her mind, have no necessary correspondence to reality, and are to be valued for their usefulness and analytical power, not for any insight they might provide into the real nature of the world.

Further discussion of instrumentalism is probably not warranted here, for it has not been a popular position among neurobiologists. Critiques of the view are given by Popper (1972) and Chalmers (Chapter 10 in his 1976). Popper in particular seems to regard it as an unduly cautious and defensive posture, which protects the modern instrumentalist from challenge to the reality of his theory, just as it might have protected Galileo from the insecurity of the Church. With the exception of Hughes (1979; and even he does not sustain the position), the visual neurobiologists of whom I am aware seem confident that their descriptions of the properties and functional groupings of nerve cells reflect something of the real organization of the visual system. The prevalence of the instrumentalist view among physicists, to which Popper gives attention, may be the result of the many controversies that science has seen, the physical smallness of the particles with which it deals, and the greater degree of abstraction required in its theories.

5.4. FALSIFICATIONISM

Falsificationism is a major strand of 20th century philosophy of science. It was formulated, by Popper [*The Logic of Scientific Discovery* (1959), being a translation of

Logik der Forschung (1935)], both as a description of how science progresses and as a prescription of how scientists can best formulate their ideas and experiments. As Popper himself has noted (quoted in Section 4.1.3), the history of philosophy is the history of its problems; and two or perhaps three problems seem to have stimulated the development of Popper's theme. First, Popper's thesis was a response to the problem of induction, that is, to the lack of any logically rigorous way by which a theory could be based on observations. Second, it sought to delineate the boundary between scientific knowledge and other forms of knowledge; Popper traces his own analysis to "the autumn of 1919, when I first began to grapple with the problem, 'when should a theory be ranked as scientific?'" (from "Science: Conjectures and Refutations" in Popper, 1972). Third, Popper saw strong analogies between the positivist epistemologies of the early 20th century and the philosophical bases of political absolutism, which he argues in his two volumes of *The Open Society and its Enemies*. Openness both in the political life of nations and in the scientist's pursuit of knowledge seemed to require a constant questioning of assumptions, and a persistent denial that any proposition was beyond challenge.

Popper argued that what makes scientific theories distinct from nonscientific propositions is that they are testable, refutable, or falsifiable by recourse to observation; this contrasts with the prior, widely held view that scientific propositions are distinguished by their firm basis in observation. Popper's formulation deals with the problem of induction by agreeing with Hume that induction is not logically compelling. Indeed, the falsificationist sees a central place in science for the drawing of imaginative, nonlogical generalizations from more particular observations/theories, i.e., for a nonlogical form of induction. Such generalizations are, however, regarded as hypotheses (conjectures) to be tested, not as proven. Empirical observations then serve to test the nonlogical hypotheses. For a falsificationist, observations are thus the source of criticism and falsification; whereas for an inductivist the role of observation is to provide the sure basis of sound theory. Falsificationism also provides a solution to the problem of the theory-dependence of observations by accepting that all observations are made in the context of some theoretical framework; and then insisting that any scientific proposition (whether it is called "observation" or "theory") must be testable. Further, it provides a basis for arguing that scientific analysis can be realist, i.e., can provide insight into reality. The constant testing of propositions means that, where they rest on unwarranted assumptions or erroneous theories, these errors can be eliminated by testing and falsification. Science progresses, and truth can be approached, by the nonlogical formulation of hypotheses, by the testing of those hypotheses and the rejection of those that prove false.

The implications of this view are many and by now familiar. It implies, for example, that we can never arrive at established truth, but only approach it by falsifying wrong ideas; that scientists should welcome and seek, instead of resisting, a disproof of their ideas, because such disproof is scientific progress; that scientists should make bold (but falsifiable) conjectures and not fear being wrong (this is Eccles' point, quoted above); that hypotheses developed out of research work are only as valuable as they are testable, indeed are more valuable the more risky and testable they are. The view accepts that both certainty and objectivity are unattainable (thus avoiding the criticisms to which inductivism and essentialism seem vulnerable); and it provides for openness in scientific discourse both by insisting on the need for challenge, and removing from individual scientists the stigma that would, in the inductivist/essentialist view, have attached to honest error.

Falsificationism is, itself, surrounded by the arguments of its critics and the responses of its supporters, as, for example, in the work of (among others) Kuhn (1962, 1970, 1974), Lakatos (1970), Musgrave (1976), and Feyerabend (1975). One criticism is that it ignores how scientists really operate, particularly their tendency to work in communities that do not, at least for long periods, question many of their assumptions; this is part of the thrust of Kuhn in his "The Structure of Scientific Revolutions" (see next section). Perhaps a more fundamental criticism, however, is that summarized by Lakatos (1970) in his development of "sophisticated" or "methodological" falsificationism. Early formulations of falsificationism assumed an asymmetry between proof and disproof. Proof is impossible, disproof possible; we must progress, therefore, by disproof. Against this it is argued by many that disproof may be no more attainable than proof; that is, the disproof of a hypothesis may itself be disproved. Thus, the attempt to progress by falsification seems as ill-fated as the search for sure proof. Some critics (notably Feyerabend) then argue that, since neither proof nor disproof is available, science is in fact anarchistic; the scientist is free to believe what he wants, and indeed has always done so. Lakatos' own response was more restrained. Without certainty in either proof or disproof, the scientist should, he argues, look for systems of ideas that seem to be more productive or fruitful, tractable questions than their alternatives; and he proposes a distinction between "progressive" and "degenerating" research programs. Degenerating programs are marked by the defensive use of multiple, ad hoc hypotheses to cope with anomalies arising between predictions and observations; progressive programs can deal more easily with known observations, and are generating useful hypotheses about new areas of investigation. Musgrave, on the other hand, argues that falsificationism provides a considerably more robust alternative to anarchism than Lakatos had allowed. For much more scholarly and learned discussion of these problems, the reader is referred to Lakatos' article, and to Feyerabend (1975), Chalmers (1976), and Musgrave (1976).

5.5. PARADIGMS AND REVOLUTIONS

Kuhn ("The Structure of Scientific Revolutions," 1962) argued that whatever methodologies scientists think they follow or ought to follow, science is best understood as conducted by communities of scientists in various fields. For long periods of "normal science," those scientists collect data and make careful but unadventurous experiments in the context of a generally held view of the scientific issues in their field (i.e., of a scientific paradigm). These periods of development of an established paradigm are separated by periods of crisis and revolution, which are precipitated when anomalies in the reigning paradigm accumulate and alternative paradigms are proposed, debated, and perhaps accepted.

Kuhn's view has attracted a good deal of criticism (many of the arguments are included in Lakatos and Musgrave, 1970), partly on the grounds that his concepts of scientific communities and of paradigms were loosely defined, and partly because of the implications for scientific methodology of two of his concepts, the "normal science" of interrevolutionary periods and the "incommensurability of paradigms." By "normal science" Kuhn meant periods in which workers in a particular field accept certain basic propositions; in visual science, for example, the correlation between rod function and scotopic vision, the photoreceptor function of outer segments, or the neuron theory. If his results seemed in conflict with these basic propositions, which comprise the current paradigm of the field, an investigator would assume his experiments to be at fault, not the paradigm. Like someone working on a crossword puzzle, he would feel himself to be on his mettle. The criticism of this concept from the falsificationist's point of view is that "normal science" is bad science because it hesitates to question its assumptions (Popper, 1970). Whether or not it is what scientists do, it is not what they should do. On this point, the falsificationist position has perhaps yielded a little; Lakatos (1970) argues, for example, that most scientists do not, and need not, question all their assumptions all the time, and he builds a distinction between the "hard core" of a research program (which is not under test), and a "protective belt" of hypotheses under test, into his scheme of methodological falsificationism. A second criticism of the "normal science" concept is that crises and revolutions are a lot commoner than Kuhn had originally seemed to assume, and in some fields might be going on most of the time, making the distinction between normal and revolutionary science meaningless. Kuhn (1970) has subsequently accepted this point, while still arguing for the usefulness of the concept of "normal" science.

In his concept of the "incommensurability of paradigms," Kuhn argued that there was a psychological aspect to the conflicts that characterize periods of revolutionary science. Scientists who have worked in an established paradigm, for perhaps the whole of their scientific life, simply cannot grasp radical new ideas; while, conversely, the young turks are so carried away with their new vision that they have neither understanding of nor patience for the earlier framework. A scientist's change of allegiance from the old to the new (if it ever occurs) contains an element akin to religious conversion, so that nothing seems the same afterwards as before. Scientists, Kuhn argued, would cling to old paradigms long after the contrary evidence was overwhelming, at least until a viable alternative was available, and often to their deathbeds. To many critics, however, the "incommensurability of paradigms" seemed to argue that scientific progress was an irrational process, a matter of "mob psychology" (Lakatos, 1970), rather than a process of rational debate about the issues. In any case, Kuhn's analysis seems to underestimate the ability of an able intellect to entertain two ideas at once, and he has subsequently (1970) modified his position on this point too.

Nevertheless, there was much in Kuhn's thesis that attracted attention from practicing scientists. It seems to me, for example, that the simple/complex paradigm of the organization of visual cortex is still widely held despite the accumulation of a great deal of contrary evidence (see Chapter 8), and may well survive until a viable alternative paradigm is seen to have been developed, and perhaps a good deal longer than that. Contrary evidence, Kuhn argued, is not sufficient to destroy a paradigm; scientists must have an alternative immediately. We seem, like nature, to abhor a vacuum.

Of course, some visual scientists rejected the hierarchical paradigm some years ago, many others continue to accept and use it, and others are waiting to hear more about the alternatives. Kuhn seems to assume that scientists within a community are more homogeneous than they really are in their response to scientific controversies, an assumption that may be necessary for the normal/revolutionary distinction. Nevertheless, he may have been fundamentally correct in drawing attention to the tendency of scientists to avoid questioning long-established patterns of thinking, even when others challenge them.

5.6. A FALSIFICATIONIST APPROACH TO THE CLASSIFICATION OF NEURONS

It was doubtless an impertinence for me to write on philosophy of science; the account given is oversimplified, inaccurate, too general, and yet too limited in scope. My defense must be that, however poor the above discussion, it is one of the few available in the literature of visual science. Of the several methodologies or models of science discussed, my clear preference is for the falsificationist approach, particularly as a way of designing experiments around ideas. Rationality seems more central to falsificationism than to the other approaches, for there is no place in it for argument by authority, and it seems best suited to lead to the continuous formulation of tractable yet fruitful questions. This is not to ignore the problems that seem to persist with falsificationism as a methodology, but rather to argue that, compared to the alternatives, it seems the best approach (perhaps in the same way that democracy, to Winston Churchill, seemed "the worst conceivable form of government, except for all the others"). The following paragraphs are based closely on, but also go beyond, the last section of the article by Rowe and Stone (1977). We sought there to formulate a hypothetico-deductive or falsificationist methodology for the classification of neurons.

Rowe and Stone (1977) set out two propositions on the methodology of classification, as a step toward a theoretical basis for cell classification. This approach is developed to some degree here, though not changed in its fundamentals. The two propositions were: (1) that any classification should be viewed not as a definition of cell groupings, but as a hypothesis that the cell groupings being delineated fulfill distinct functional roles; and (2) that the functional roles referred to by the classification should be defined in terms quite independent of the properties of the cells being classified. I will not argue further the case for organizing classifications as hypotheses. The logical structure proposed for hypothetico-deductive classification deserves closer comment, however.

1. Classification is an inductive process, a set of general propositions being proposed on the basis of many specific propositions or observations (Mayr, 1969). For reasons discussed above, it is nonlogical and derives its scientific validity from its hypothetical form and falsifiability. Irrefutable classifications are heuristically valueless.
2. A classification involves two sets of hypotheses, one set concerning categories and the other set concerning taxa. Categories are simply the conceptual divisions that the classifier builds into the classification. In the phylogenetic taxonomy of animals, for example, categories are species, genus, family, order, and phylum. One set of hypotheses in the classification proposes that these categories are meaningful definitions of the phylogenetic relationships possible between animals.

Taxa, on the other hand, are groups of the animals being classified, in our example groups of animals such as *Felis domesticus*. The second set of hypotheses proposes that certain taxa fit into particular categories.

In the classification of nerve cells in the visual pathways (I argue), appropriate categories are the functional roles performed by visual cells; some cells might subserve high-resolution vision for example, others peripheral movement vision, others color vision. These categories would be a modifiable set of functional roles, hypothesized to be meaningful and distinct components of visual function. The taxa would be groups of cells (such as X, Y, or W cells) hypothesized to subserve the functional roles set out as categories. Ongoing work in cell classification should provide tests both of the adequacy with which the functional roles have been defined, and of the validity of the division of cells into taxa.

3. To preserve the falsifiability of the classification, the categories are defined in terms independent of the properties of the taxa. For example, it could be proposed that the taxon X cells subserves the category high-resolution vision. Note that X cells are identified by their physical properties; high-resolution vision is quite independently defined as the ability to discriminate fine patterns.

4. The two sets of hypotheses have to be set up largely simultaneously to form the classification, since the properties of the cells may influence the choice of categories.

5. Once the classification has been set up, a particular cell may be identified as belonging to a certain taxon by a logical, deductive process. For example, certain properties of retinal ganglion cells are good identifiers of which taxon they belong to (a strong periphery effect is a good, but not perfect, identifier of Y cells, a small receptive field center of X cells, slow axonal conduction velocity of W cells). The certainty of the identification can be increased by considering other identifying characters, but the logical, deductive process of identification remains distinct from and subsequent to the nonlogical, inductive formulation of the categories and taxa that comprise classification.

Heuristic Value of Classification by Hypothesis

The value of a classification, it might be argued, can be seen in its ability to generate testable predictions. In Lakatos' (1970) terms, a classification of nerve cells may comprise the "protective belt" of a research program. That is, a classification of nerve cells should ideally generate a wide range of fruitful and tractable questions about the organization of the brain; indeed, this book was written because the Y/X/W classification has done just that. Enroth-Cugell and Shapley (1973) commented, for example, on the "renaissance of interest" in ganglion cell properties that had occurred since the X/Y formulation in 1966 (by Enroth-Cugell and Robson); that interest has continued unabated. The Y/X/W classification of ganglion cells has led to the formulation of a great many fruitful hypotheses about and observations on the visual centers of the brain; these are reviewed in Part III. It is a measure of the impact of ganglion cell classification on the understanding of the visual pathways that, of the three parts of this book, Part III is much the largest.

The Impact of Ganglion Cell Classification   III

On the Understanding of $\quad$ 6
Visual Processing in the
Diencephalon

The axons of retinal ganglion cells terminate in at least three nuclei of the diencephalon, the lateral geniculate nucleus (LGN) of the thalamus, the pulvinar nucleus of the thalamus, and the suprachiasmatic nucleus of the hypothalamus. Of these, the terminations in the LGN are by far the heaviest and best analyzed; the hypothalamic and pulvinar terminations are much less substantial and even their existence was debated until relatively recently.

The function of the LGN, or at least of its large dorsal subdivision (the dLGN), seems well understood; it is the thalamic nucleus specialized to relay retinal activity to the visual cortex. Because retinal activity must cross a synapse in this nucleus to reach the visual cortex, the LGN is a site at which nonretinal factors, such as the activity of brain-stem centers controlling level of arousal, can affect visual processing. The function of the retinohypothalamic projection is also known; it receives and relays a visual input for the control of circadian rhythms. The functional importance of the retinal-recipient zone of the pulvinar is not known however; perhaps it should be regarded as part of the LGN. A good deal is known about how the dLGN is organized to process the activities of Y-, X-, and W-class ganglion cells, and recent studies of its small ventral subdivision, of the suprachiasmatic nucleus, and of the pulvinar have provided relevant data for these nuclei as well.

6.1. THE LGN OF THE CAT

6.1.1. Evidence of Parallel Processing in the LGN

The LGN is a concentration of neurons found, in the cat, at the lateral posterior margin of the thalamus. Axons from a large proportion of retinal ganglion cells (at least

50% in the cat) terminate within the LGN, and most of its neurons are relay cells, i.e., they receive input directly from the axons of retinal ganglion cells and send their axons to a distant structure. In the case of the dorsal component of the LGN (the dLGN), the target of the relay is the cerebral cortex (indeed, the area of the cortex to which the dLGN projects is often termed the "primary" visual cortex), while the relay cells of the ventral component of the LGN (the vLGN) project to several subcortical centers. A minority of neurons of the dLGN are considered to be interneurons, i.e., their axons are thought to terminate within the LGN. There is a substantial debate, however, about the numbers and even the existence of such interneurons. These cells are thought to produce the marked inhibitory phenomena described in the dLGN.

This section concerns the organization of the LGN of the cat in terms of the Y-, X-, and W-cell components of its retinal input. The pattern is emerging that different neuron groups within the dLGN receive input from different functional groups of ganglion cells and relay the activities of the different groups to the cortex, separately and in parallel. This parallel organization was anticipated in several pioneering studies of the LGN, for example by Walls (1953):

> The LGN is a station on the route which visuophysiological information travels from the retina to the cortex. This information travels through the LGN in neuron channels. Each of these channels is, then, a diageniculate (dia = through) path. . . .
>
> In a given kind of animal there will be as many . . . diageniculate path types . . . as the number of behaviorally different opticus-fiber types that can be demonstrated, say with microelectrode techniques and with manipulation of intensive, extensive, chromatic and temperal aspects of stimuli. Intraretinally, what must this mean? That the animal's visual system incorporates certain standardized or modal patterns of relation of visual-cell assortments to bipolar etc. assortments, and thence to individual ganglion cells which, themselves, are of various types.

Walls' insight was remarkable, but his was not the first suggestion that different groups of geniculate relay cells process different aspects of the retinal image, in parallel. Le Gros Clark (1940) and Chang (1951) had argued that the three subcomponents of color vision are relayed through different components of the LGN; neither suggestion has been confirmed, however. Walls' idea of "diageniculate paths" determined by functional groupings of ganglion cells is particularly striking for its accuracy, but also for the neglect it has suffered in the literature. I would here make amends for my part in that neglect, and also note that Polyak (1957, p. 333) proposed an interpretation of geniculate lamination that involved a similar insight.

One of the studies that began the experimental confirmation and elaboration of Walls' idea of "neuron channels" within the LGN related to different classes of retinal ganglion cells was reported by Bishop and MacLeod (1954). They showed (Fig. 6.1) that the fast- (t_1) and slow-conducting (t_2) axons of cat optic tract, when activated by an electrical shock, generate separate postsynaptic field potentials in the dLGN (labeled r_1 and r_2 in Fig. 6.1), suggesting that t_1 and t_2 axons terminate on different groups of relay cells. Similarly, Bishop and Clare (1955) concluded that t_1 (fast-conducting) axons of the optic nerve and tract terminate on large relay cells, whose axons are also fast-conducting, while t_2 axons terminate on smaller relay cells with slower-conducting axons. This relationship was confirmed and extended by the single-cell study of Noda and

Figure 6.1. Evidence of functionally separate populations of cells in the dLGN. The traces show responses recorded in the dLGN of the cat following stimulation of the contralateral (g) and ipsilateral (h) optic nerves (from Bishop and MacLeod, 1954). The time intervals (j) represent 0.2 msec. The labels t_1 and t_2 mark potentials considered to be generated by the

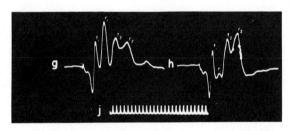

afferent volleys arriving respectively via fast- and slow-conducting fibers in the optic nerve and optic tract (see also Fig. 1.3). The labels r_1 and r_2 mark potentials considered to be generated by the synaptic action of t_1 and t_2 axons, respectively. These potentials are temporally distinct, suggesting that separate populations of nerve cells may be involved. [Reproduced from the *Journal of Neurophysiology* with kind permission of the American Physiological Society.]

and Iwama (1967) in the dLGN of the rat (Fig. 6.2) and has since been confirmed in single-cell studies for the rat (Fukuda, 1973, 1977; Hale *et al.*, 1976, 1979), cat (Stone and Hoffman, 1971; Cleland *et al.*, 1971; Fukada and Saito, 1972; Hoffmann *et al.*, 1972), monkey (Marrocco and Li, 1977; Sherman *et al.*, 1976; Schiller and Malpeli, 1978), and tree shrew (Sherman *et al.*, 1975a).

The "organizing principle" suggested for the dLGN by these studies seems clear, and has two parts. First, fast- and slow-conducting retinal axons terminate on and excite separate subgroups of geniculate relay cells; the visual information carried by fast- and slow-conducting retinal axons can, therefore, reach the cortex separately. Second, the target relay cells approximately match the retinal cells that provide their input in axonal velocity and, probably, in soma size. The later studies included investigations of receptive field properties as well and showed that in the cat, monkey, and tree shrew, the fast-

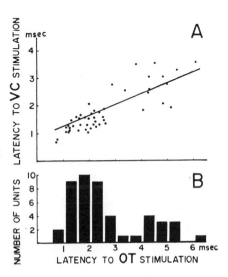

Figure 6.2. Single-cell evidence of conduction velocity grouping among dLGN relay cells (from Noda and Iwama, 1967). (A) Relationship between the orthodromic spike latencies of 48 neurons in the dLGN of the rat to stimulation of the optic tract, (OT) and their antidromic spike latencies to stimulation of the visual cortex, (VC). A grouping into short- and long-latency groups is clear to inspection. The correlation coefficient was 0.80. (B) Frequency/OT latency histograms for the same cells. Two modes are apparent. [Reproduced with kind permission of *Vision Research*.]

axon/slow-axon difference is closely correlated with a Y/X difference in receptive field properties. That is, individual relay cells were shown to relay to the cortex the activity of Y- or of X-class ganglion cells, and were therefore termed *Y- or X-class relay cells.* In the cat, for example, Cleland *et al.* (1971) showed that the receptive field properties of Y- and X-class geniculate relay cells resemble those of the Y- and X-class ganglion cells that provide their input; for example, in their linearity of summation and velocity-selectivity, and in the "transience" or "sustainedness" of their responses to stationary contrast stimuli. Cleland and co-workers also reported that, as in the rat (Fig. 6.2), the conduction velocity of a relay cell's axon was positively correlated with the velocity of the retinal afferent reaching it. That is, they showed that each Y-class relay cell has a fast afferent (or afferents), a fast axon, and many of the receptive field properties char-acteristic of retinal Y cells; while each X cell has a slower afferent (or afferents), a slower axon, and many of the receptive field characteristics of an X cell. Hoffmann *et al.* (1972) confirmed these findings (Fig. 6.3), and traced out changes in the receptive field sizes and relative proportions of Y and X cells related to eccentricity (Fig. 6.4). In 1972, without having read Walls' monograph (quoted above), I attempted to review what was known of the X/Y organization of cat dLGN and suggested that:

> ... the activity of X- and Y-cells of the retina is relayed by different LGN cells to the visual cortex, through functionally separate, parallel neuronal channels, one (the Y-channel) fast-conducting, the other (the X-channel) slow-conducting.

Subsequent work has both extended and qualified the "parallel channel" concept of geniculate organization in significant ways, reviewed in the following sections. The concept has been extended to embrace the third class of ganglion cells recognized in cat retina (W cells), the subdivisions of the LGN, the electron microscopy of the retinogen-iculate synapse, the morphology of relay cells, inhibitory phenomena within and retic-

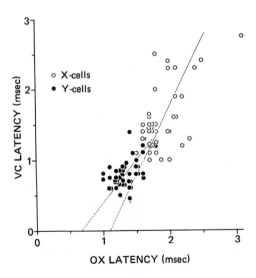

Figure 6.3. Evidence of X/Y correlates of conduction velocity in cat dLGN (from Hoffmann *et al.,* 1972). The graph shows the relationship between orthodromic (OX) and antidromic (VC) latencies for 85 relay cells. It is thus closely comparable with Noda and Iwama's (1967) data shown in the previous figure. Orthogonal regression lines are shown separately for Y cells (closed circles) and X cells (open circles), and their extrapolations to the abscissa are shown. Y cells generally have short latencies (and hence faster-conducting afferents and axons) than X cells. [Reproduced from the *Journal of Neurophysiology* with kind permission of the American Physiological Society.]

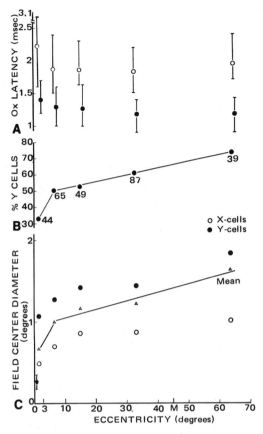

Figure 6.4. Gradients in the properties of dLGN relay cells, related to the eccentricity of their receptive field (from Hoffmann *et al.*, (1972). (A) The orthodromic (OX) latencies of X (open circles) cells are consistently longer than those for Y cells (closed circles). Latencies are fairly independent of eccentricity except that they seem distinctly longer for the area centralis cells (i.e., those with eccentricities near zero). (B) The proportion of Y cells increased with eccentricity, especially just outside the area centralis. (C) The receptive field sizes of both X and Y cells increased with eccentricity, Y cells having distinctly larger fields at all eccentricities. The triangles represent average diameter. [Reproduced from the *Journal of Neurophysiology* with kind permission of the American Physiological Society.]

ular influences on the LGN, the pattern of termination of geniculocortical axons, and a variety of species. The principal qualification to the concept comes from evidence of convergence of X- and Y-cell activity onto single geniculate relay cells.

6.1.2. The W-Cell Relay in the dLGN

The "parallel channel," X/Y model of the geniculate relay just discussed was established by studies of the prominent A laminae of cat dLGN. It has been known since Rioch's (1929) study that a less prominent, ventrally located lamina (or set of laminae) was (were) present in the dLGN, and comprised smaller neurons than those of the A laminae; Rioch termed this region the *lamina parvocellularis*. It corresponds to the pars dorsalis B of Thuma's (1928) description (Fig. 6.5), and to the C laminae distinguished by Guillery (1970) and Hickey and Guillery (1974); specifically to the ventral part of lamina C plus laminae C1 and C2. Pearlman and Daw (1970) and Daw and Pearlman (1970) described the presence in these laminae of small numbers of relay cells whose responses to visual stimuli were color-sensitive. Comparable color-sensitivity was not detected among retinal ganglion cells until Cleland and Levick (1974*b*) described similar properties in a small minority of ganglion cells that arguably (Rowe and Stone, 1977; see Chapter 2, Section 2.3.3.3) form a subgroup of the W-cell class. Subsequently, Cleland *et al.* (1975, 1976), Wilson and Stone (1975), and Wilson *et al.*

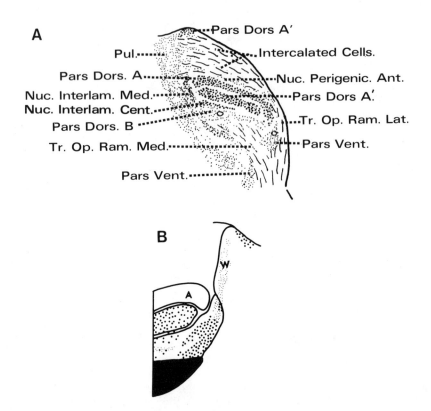

Figure 6.5. (A) Lamination of cat LGN described by Thuma (1928) from Nissl-stained coronal sections. The drawing represents a coronal section through the middle of the LGN. Nuc. Interlam. Cent., central interlaminar nucleus; Nuc. Interlam. Med., medial interlaminar nucleus; Nuc. Perigenic. Ant., perigeniculate nucleus; Pars Dors. A, A′, B, laminae A, A1, and B of dLGN; Pars Vent., vLGN; Tr. Op. Ram. Med., medial ramus of optic tract; Tr. Op. Ram. Lat., lateral ramus of optic tract. Redrawn from Thuma (1982). (B) The termination of fine and coarse axons in cat dLGN (from Guillery *et al.* 1980). The diagram represents a coronal section of the medial side of the LGN, at an anteroposterior level similar to that illustrated by Thuma in (A). Here the fine and coarse dots represent the terminals of fine and coarse axons, as demonstrated by degeneration techniques. The black segment represents the optic tract at the ventral margin of the nucleus. Note the fine and coarse terminals in the MIN, and the fine terminals in the RRZ (W) (the patch of coarse terminals at the top of the RRZ is the upturned "tail" of the nucleus). The fine terminals are considered the terminations of W-cell axons, the coarse terminals most likely originating from X- and Y-cell axons (mostly Y cells in the MIN). Note that the label W in this diagram is used because the authors term the RRZ the "wing" of the LGN. [Reproduced with kind permission of the *Journal of Comparative Neurology.*]

(1976) demonstrated that the activity of W cells (including cells with several receptive field patterns) is relayed to the visual cortex via the small relay cells present in these laminae (Fig. 6.6A). Mitzdorf and Singer (1977) used a technique of current source density analysis to detect the long latency activity of W cells in the ventralmost laminae of the dLGN (Fig. 6.7). The distribution of W-, X-, and Y-class relay cells through the laminae of cat dLGN, as indicated by these studies, is shown schematically in Fig. 6.8. Cleland and co-workers' evidence of a small number of W-class relay cells in the A laminae was not confirmed by Wilson *et al.* (1976) or by Mitzdorf and Singer (1977), although their presence could not be entirely excluded. The antidromic latencies of relay cells to an electrical shock to the visual cortex were generally longer for W cells than for X cells, and longer too for X cells than for Y cells (Fig. 6.6). That is, W cells have slower axons on the average than cells in the A laminae, as well as slower afferents and smaller cell bodies. More recently, Dreher and Sefton (1975, 1979) and Rowe and Dreher (1979, 1982*b*) have provided evidence that W-class relay cells are also present in the medial interlaminar component of the dLGN, as described in the following section.

6.1.3. The Medial Interlaminar Nucleus

The medial interlaminar nucleus (MIN) was first described by Rioch (1929) as a collection of large neurons at the medial side of the dLGN. Hayhow (1958) demonstrated that it receives direct retinal projections from both eyes and suggested that it functions as "a relatively independent pars dorsalis" of the LGN. The reports of Dreher and Sefton (1975, 1979), Mason (1975), and Kratz *et al.* (1978) provided independent evidence that the MIN is particularly involved in the relay of Y-cell activity to the visual cortex, 80–90% of the neurons encountered physiologically being clear-cut Y-class relay cells (Fig. 6.9). Bowling and Michael (1980) have recently provided a beautiful morphological demonstration (Fig. 7.1B) of Y-cell afferents to the MIN: they are branches of optic tract axons that also branch to reach laminae A and C, or A1, of the dLGN and the SC.

In all the physiological studies of the MIN just discussed, the identification of Y cells was unqualified. Some differences were noted between Y cells in the MIN and those of the main part of the dLGN, but they strengthened rather than weakened the identification of the cells as Y cells; MIN Y cells have considerably larger receptive fields, for example, and on the average slightly shorter orthodromic latencies. Dreher and Sefton comment that this could result either from MIN Y cells receiving input from a particular subclass of retinal Y cells (with large fields and fast axons) or (which seems to me more likely) from a greater degree of excitatory convergence of retinal Y cells onto MIN cells. Both Dreher and Sefton (1975, 1979) and Mason (1975) reported the presence of small numbers of X cells at the lateral margin of the MIN (i.e., near its border with the main part of the dLGN), and Dreher and Sefton (1979) also reported a small proportion of W-type relay cells within the MIN. Of the eight W cells they identified in the MIN, two had ON-OFF receptive fields; the remainder had ON- or OFF-centers with antagonistic surrounds, three being phasic and three tonic in their responses to

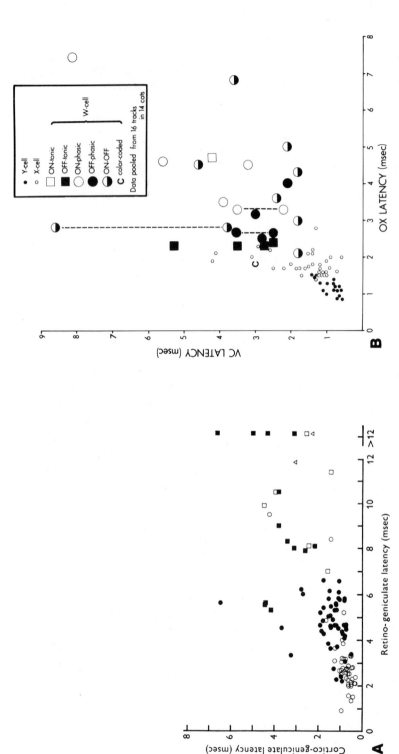

Figure 6.6. Conduction velocity relationship among relay cells in cat dLGN, now including W cells. (A) Relationship between antidromic and orthodromic latencies of geniculate relay cells ((from Cleland *et al.*, 1976). A cell's antidromic latency is the time taken by an action potential to travel backwards along its axon, from the end of the axon in the visual cortex to the body of the cell in the dLGN; the latency is inversely proportional to the conduction velocity and caliber of the axon. A cell's orthodromic latency is the time taken by an action potential to travel from the retina along the ganglion cell axon(s) innervating the cell and also cross the synapse to activate the geniculate cell. Except for the brief synaptic delay, this latency is inversely proportional to the conduction velocity of the ganglion cell axon(s) involved. Open circles represent "brisk transient" (Y) cells, closed circles "brisk sustained" (X) cells, open squares "sluggish transient" (W) cells, closed squares "sluggish sustained" (W) cells, and triangles "local edge detector" (W) cells. [Reproduced with kind permission of the *Journal of Physiology (London)*.] (B) Analogous data from Wilson *et al.* (1976). Note the different symbols used to represent Y, X, and the various subclasses of W cells. For three W cells, two VC latencies are shown (joined by dotted lines). They were obtained from different sites of cortical stimulation. [Reproduced from the *Journal of Neurophysiology* with kind permission of the American Physiological Society.]

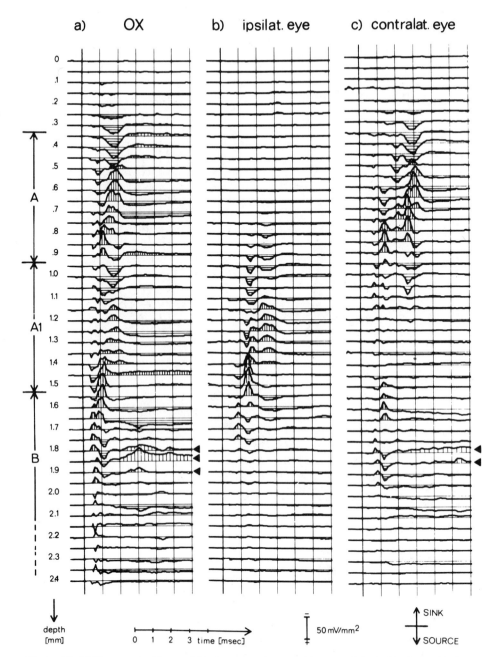

Figure 6.7. Field potential evidence of W-cell input to the ventral laminae of the dLGN (from Mitzdorf and Singer, 1977). Field potentials were recorded in the dLGN at 100-μm steps along a vertical electrode penetration of the dLGN. The levels of laminae A, A1, and B (B comprises laminae C, C1, and C2) are shown at the left. The traces show estimates, based on a "current source density" analysis, of the sinks (upward deflections) and sources (downwards) generated by a stimulus applied (a) to the optic chiasm (OX), (b) to the ipsilateral and (c) to the contralateral optic nerve. The arrowheads indicate traces that show the occurrence of long-latency sinks in lamina B (i.e., in the C laminae), following stimulation of the optic chiasm and contralateral optic nerve. The absence of a sink in the C laminae (specifically in lamina C1) following stimulation of the ipsilateral nerve is not readily explicable. [Reproduced from the *Journal of Neurophysiology* with kind permission of the American Physiological Society.]

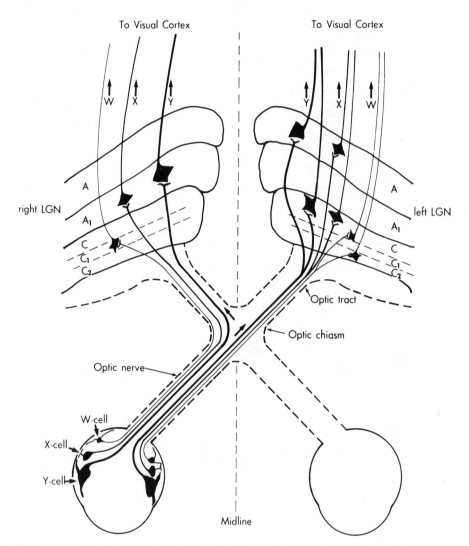

Figure 6.8. Schematic diagram of the relay of Y-, X-, and W-cell activity through cat dLGN (from Wilson *et al.,* 1976). X- and Y-class relay cells are intermingled in laminae A and A1 and in the dorsal part of lamina C. W-Class relay cells are segregated to the ventral part of lamina C, and to laminae C1 and C2. Note that each class of retinal cell relays through one lamina (A1 or C1) of the ipsilateral dLGN, and through two laminae of the contralateral dLGN (X and Y cells via laminae A and C, W cells via laminae C and C2). [Reproduced from the *Journal of Neurophysiology* with kind permission of the American Physiological Society.]

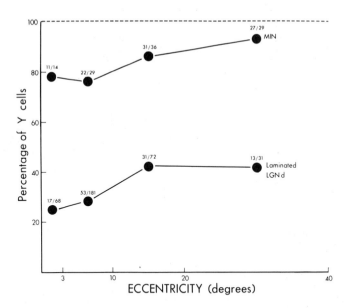

Figure 6.9. Comparison of proportions of Y cells encountered physiologically in the MIN and the main part of the dLGN as a function of eccentricity (from Dreher and Sefton, 1979). At all eccentricities, Y cells form a much higher proportion of MIN cells than of cells in the main part of the dLGN. [Reproduced with kind permission of the *Journal of Comparative Neurology.*]

standing contrast. Mason (1981) has confirmed the occurrence in the MIN of cells with W-like receptive fields; in his results too, W cells formed only a minority (7/79) of cells encountered with extracellular recording techniques.

Thus, when assessed by physiological single-cell sampling techniques, the MIN appears to be predominantly a Y-cell relay nucleus. It is also distinct from the main part of the dLGN in its cortical projections (see Section 6.1.7) and contains a separate representation of the visual field (Kinston *et al.*, 1969; Sanderson, 1971; Guillery *et al.*, 1980; Rowe and Dreher, 1982*b*), all these factors supporting Hayhow's suggestion that it serves a distinct functional role.

There is, however, some discrepancy between anatomical and physiological assessments of the relay cells of the MIN. Although Thuma (1928) noted the large size of the neurons of the MIN, and their relatively large size has been confirmed by detailed measurement (e.g., in Kratz *et al.*, 1978), Kratz and co-workers also note that there are many small neurons in the nucleus, and Leventhal (1979) showed that the small as well as the large cells of the MIN send axons to the visual cortex, i.e., that some small cells at least are relay cells. Furthermore, Rowe and Dreher's (1979, 1982*b*) HRP study of the retinal input to the MIN indicates that Y cells may constitute only about 50% of the relay cells of the MIN. They observed that when small amounts of HRP are injected into the MIN, only about half of the retinal ganglion cells subsequently found carrying HRP label (and whose axons presumably project to the MIN) had the large soma size characteristic of retinal Y cells. The other labeled cells had medium-sized somas and were shown to be medium-soma γ cells (see Chapter 2, Section 2.1.4). Correspondingly, Mason and Robson (1979) reported that fine as well as coarse axons reach the MIN from the optic tract. They note that "these fine axons appear to be an extension of the fiber population in laminae C1 and C2," i.e., of the W-cell axons reaching these layers. Guillery *et al.* (1980) confirmed that the MIN receives coarse and fine afferents (Fig. 6.5B), and argue for some degree of segregation of fine and coarse terminals within the nucleus. These, of course, are assessments of the retinal input to the MIN, and do not entirely preclude that a relatively high proportion of MIN cells are relaying Y- rather than W-cell activity. The high proportion of Y cells encountered physiologically could result, however, from a sampling bias of microelectrodes for large cells (e.g., Stone, 1973), and it seems likely that as much as 50% of the retinal input to the MIN is derived from retinal W cells. Rowe and Dreher (1979, 1982b) observed no β-class ganglion cells labeled from the MIN, suggesting that the X cells reported in the MIN near its medial border may, in fact, have been in the adjacent main section of the dLGN.

6.1.4. The vLGN

The vLGN of the cat is a group of relatively small cells found at the ventrolateral margin of the dLGN. It is considered to have a distinct embryological origin from the ventral region of the diencephalon, which also gives rise to the hypothalamus and subthalamus. Its cells relay retinal activity to subcortical centers, including the SC, pretectum, subthalamus and pontine nuclei, and the suprachiasmatic nucleus of the hypothalamus (Edwards *et al.*, 1974; Swanson *et al.*, 1974). The vLGN has long been recognized as comprising small neurons (Thuma, 1928; Rioch, 1929), and Jordan and Holländer (1974) have described several cytoarchitectonic subdivisions within it. O'Leary (1940) demonstrated that in the cat, retinal afferents to the vLGN arise as collaterals of optic tract axons (Fig. 6.10B), and provided a valuable early clue to the ganglion cell input to this subnucleus when he noted that, although both coarse and fine optic tract fibers sweep around or through the vLGN to reach the dLGN,

> No collaterals have been observed to depart from the large fibres . . . but small fibres issue right angled collaterals which enter the nucleus and arborize within it.

In retrospect, O'Leary's observation suggests a strong W-cell input to the vLGN, but this point was not specifically demonstrated for nearly 30 years. In the meantime, direct retinal projections to the vLGN were extensively confirmed by several groups using more modern techniques (e.g., Hayhow, 1958; Laties and Sprague, 1966; Hollander and Sanides, 1976), and both Hayhow and Holländer and Sanides confirmed that the retinal axons reaching the vLGN are of small caliber. Specific evidence that the vLGN receives W-cell input came from the physiological study of Spear *et al.* (1977). They noted that the axons reaching the vLGN are much slower-conducting (mean velocity 5.3 m/sec) than the X- and Y-cell axons reaching the A laminae of the dLGN (Fig. 6.10A) and that the receptive field properties of vLGN cells resemble those of retinal W cells. Morphological support for the view that the vLGN receives principally W-cell input comes from Leventhal, Rodieck, and Dreher's (personal communication)

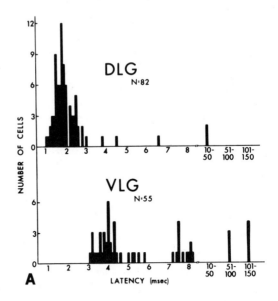

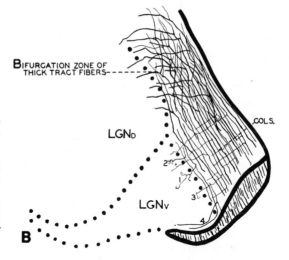

Figure 6.10. Evidence of slow-conducting, fine-caliber input to cat vLGN. (A) Frequency/latency histograms for samples of neurons encountered in the dLGN (upper) and vLGN (lower). The latency plotted is the latency of each cell's orthodromic response to stimulation of the optic chiasm. The latencies are markedly longer for the vLGN cells, indicating that the afferents to them are slow-conducting. [From Spear *et al.* (1977). Reproduced from the *Journal of Neurophysiology* with kind permission of the American Physiological Society.] (B) Diagram from O'Leary (1940) showing (1, 2, 3, 4) collaterals arising from fine-caliber axons of the optic tract and passing into the vLGN. The drawings were made from Golgi material. [Reproduced with kind permission of the *Journal of Comparative Neurology.*]

observation that injection of HRP into the vLGN causes the retrograde labeling of only γ-class (both small-soma and medium-soma or ε) retinal ganglion cells.

Spear and co-workers noted a very high proportion of ON-center receptive fields among vLGN cells, and discussed the possible involvement of the vLGN in the pupil reflex and in the control of eye movements. They demonstrated a separate representation of the visual field in the vLGN, supporting the idea that it functions independently of the dLGN; in this case, the function is mediated by W-class retinal ganglion cells. The study of Hughes and Ater (1977) of the receptive fields of cells of the vLGN provides some confirmation and expansion of these conclusions showing, for example, color-coded responses in a small proportion of vLGN cells, presumably derived from color-coded ganglion cells in the W class.

In summary, the vLGN appears to function as a relay of W-cell activity to a number of subcortical sites. As with the W-cell relay through the C laminae of the main part of the dLGN, the neurons forming the relay are relatively small in soma diameter. Several of the sites to which the vLGN projects, also receive direct retinal input (e.g., the SC, pretectum, suprachiasmatic nucleus), suggesting an "integrative" function for the vLGN (Hughes and Ater, 1977). A more specific indication of its function does not seem available.

6.1.5. The Lamination of the dLGN

The dLGN of the cat is "laminated"; i.e., its cells are spatially segregated sheets or laminae, often separated by narrow, relatively cell-free interlaminar zones. In the cat, the laminae lie largely on top of each other, so that the pattern of lamination is best seen in sagittal or coronal (Fig. 6.5) sections. Demonstration of the full pattern of lamination requires a range of techniques. Much of the pattern is readily apparent in Nissl-stained sections and was described from such sections in the early studies of Winkler and Potter (1914), Thuma (1928), and Rioch (1929). Thuma, for example, described three principal laminae (Fig. 6.5), pars dorsalis A, A1, and B. The later studies of Hayhow (1958), Guillery (1970), and Hickey and Guillery (1974) showed that one functional correlate of lamination is eye-dominance, different laminae receiving input from different eyes. Thuma's pars dorsalis A [termed *lamina A* by Hayhow (1958)], for example, receives input from the contralateral eye, and pars dorsalis A1 *(lamina A1)* from the ipsilateral eye. Thuma's pars dorsalis B *(lamina B)*, on the other hand, comprises several laminae when analyzed in terms of the inputs it receives from the two eyes. These laminae, termed *C, C1,* and *C2* by Guillery (1970) and Hickey and Guillery (1974), are not apparent in Nissl-stained sections; they were distinguished only when axon-tracing techniques were used to map the projection of one eye to the LGN. Layers C and C2 receive input from the contralateral eye, and layer C1 from the ipsilateral eye, as shown schematically in Fig. 6.8. The MIN and vLGN also receive input from both eyes, with a less complete separation of the inputs.

The analysis of geniculate organization in terms of W, X, and Y cells provides a second correlate of lamination, as illustrated in Fig. 6.8. This pattern was worked out in the same studies as discussed in Chapter 2 (Section 2.3.3.7), concerning the central projections of retinal ganglion cells. Laminae A and A1 contain a mixture of Y- and X-class relay cells, with the latter predominating. The dorsal part of lamina C also contains a mixture of Y and X cells, with the former predominating. The ventral part of lamina C, together with laminae C1 and C2, contain W-class relay cells. The MIN contains both Y- and W-class relay cells, while the vLGN contains W-class relay cells. Thus, the cytoarchitectural subdivisions of the LGN can be related on the one hand to the laterality of retinal input and, on the other, to the class of retinal ganglion cell providing their input.

6.1.6. Morphology of Relay Cells

Guillery (1966) described three classes of relay cells in the cat LGN, as seen in Golgi-impregnated material. He termed them *class 1, class 2,* and *class 4 cells;* class 3 cells were considered to be short-axon interneurons (Fig. 6.11). His descriptions were confirmed by LeVay and Ferster (1977) (Fig. 6.12). Wilson *et al.* (1976), LeVay and Gilbert (1976), and LeVay and Ferster (1977) have proposed that class 1 cells correspond to Y-class relay cells, class 2 cells to X cells, and class 4 cells to W cells. If this suggestion proves correct, then the correlation between physiological and morphological properties of Y, X, and W cells is as striking among geniculate relay cells as among retinal ganglion cells.

The evidence in favor of this suggestion is as follows. Class 1 and 2 cells are found intermingled in laminae A, A1, and the dorsal part of lamina C, which all contain a mixture of X and Y cells. Class 2 cells are of medium size, with thick primary dendrites characterized by the presence of "grapelike appendages"; these are clusters of protrusions that concentrate around the first branch-point of the dendrites. Class 1 cells are bigger in soma size, with regularly branching dendrites. Correspondingly, it was long suspected that Y cells would, because of their faster-conducting and therefore stouter axons, have larger somas than X cells, an expectation confirmed by the morphological studies of Ferster and LeVay (1978) and Leventhal (1979). Again, class 1 cells are fewer in number than class 2 cells within the A laminae (where Y cells are in the minority) but are more numerous than class 2 cells in the MIN and the dorsal part of lamina C (Szenthagothai, 1973), areas in which Y cells are found in high proportions in physiological sampling experiments. The two cell classes thus meet the expectation that the somas of Y cells are bigger than those of X cells, and that their relative numbers in different components of the dLGN should match the known distribution of X and Y cells. LeVay and Ferster also presented evidence that class 1 cells become an increasingly high proportion of the relay cell population, going from the medial to the lateral edge of the A laminae, matching the increasing proportion of Y cells reported along this parameter by Hoffmann *et al.* (1972).

LeVay and Ferster (1977) reported evidence that the striking but little-understood cytoplasmic organelles observed in the dLGN of the cat by and described as laminated "cytoplasmic bodies" by Morales *et al.* (1964; Fig. 6.13) are found only in class 2 (X) cells. They suggested that these bodies can be used as identifiers of X cells. Kalil and Worden (1978) confirmed LeVay and Ferster's finding that these "cytoplasmic laminated bodies" (CLBs) are found chiefly in medium-sized cells and are most common toward the medial side of laminae A and A1 (Fig. 6.14). However, they observed CLBs in a much smaller proportion of dLGN cells than did LeVay and Ferster (20% as against 50%) and questioned the tight correlation between CLBs and class 2 cells that the latter workers had proposed. Geisert's (1980) report also provides limited support of LeVay and Ferster's view, noting that CLBs are found only in medium- and small-soma cells and that "virtually all" CLB-containing cells project to area 17 (the cortical target of X cells). On the other hand, Geisert confirmed Holländer (1978) in observing CLBs in cells of the MIN, which include few if any X cells. Schmidt and Hirsch (1980), observing the CLBs with a different technique, were even less hopeful of a correlation

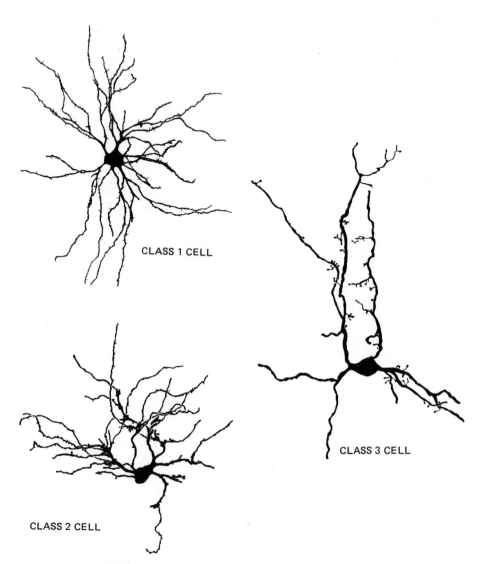

Figure 6.11. Neuron types in the dLGN, described by Guillery (1966). Class 1 cells (example shown at 270×, magnification reduced 40% for reproduction) found in lamina A are relatively large cells with spiny dendrites. Class 2 cells (example drawn at 430×, magnification reduced 40% for reproduction) are relatively small cells whose stem dendrites bear clusters of grapelike appendages close to their initial branching points. Class 3 cells are also relatively small (example shown at 260×, magnification reduced 40% for reproduction) with slender stalked appendages to their dendrites. [Reproduced with kind permission of the *Journal of Comparative Neurology.*]

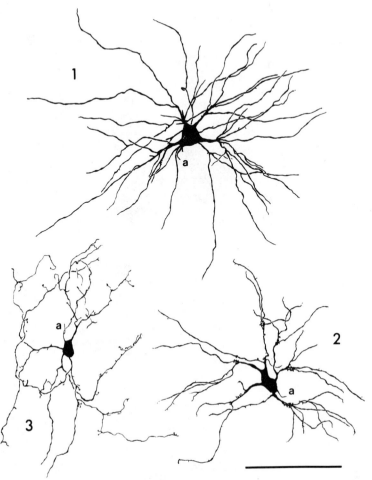

Figure 6.12. Confirmation of neuron types described by Guillery (1966), reported by LeVay and Ferster (1977). The three cell classes illustrated match the class 1, 2, and 3 cells of Guillery (1966) illustrated in the preceding figure. The scale represents 100 μm. [Reproduced with kind permission of the *Journal of Comparative Neurology*.]

between CLBs and a particular class of relay cells. They could not find evidence in favor of a mediolateral gradient in the frequency of CLB occurrence, and observed significant numbers of CLBs in regions where class 2 and X cells are considered not to occur, viz., in the MIN, in the vLGN, and in the vicinity of the optic tract. Schmidt and Hirsch concluded that there is no clear correlation between CLB-containing and X or class 2 cells.

The suggestion that class 4 relay cells are the morphological correlates of W-class relay cells is relatively easy to sustain. Both cell types have been observed only in the parvocellular C laminae and are the only class of relay cell described there. Guillery

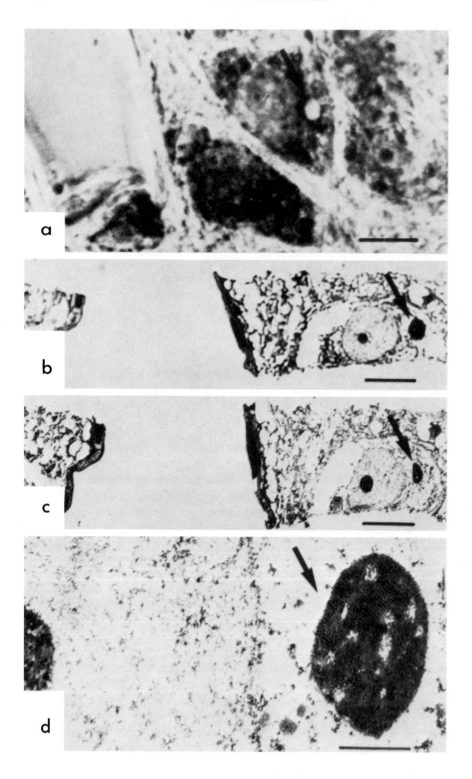

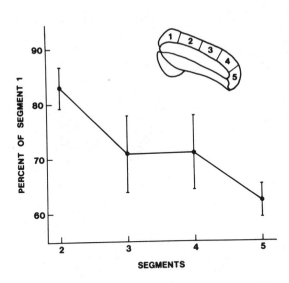

Figure 6.14. Distribution of CLBs in cat dLGN (from Kalil and Worden, 1978). CLB densities were measured as CLBs/mm^2 of section. Measurements were made in five segments of lamina A (as shown in the diagram at the top right) extending from medial (segment 1) to lateral (segment 5). The graph shows CLB densities expressed as a percentage of the density found in segment 1. The area centralis is represented in segment 1 and increasingly peripheral areas of the retina in segments 2–5. CLB density decreases steadily with eccentricity. This confirms LeVay and Ferster (1977), and can be correlated with the decrease in the proportion of X cells reported in physiological sampling experiments (see Fig. 6.4B). [Reproduced with kind permission of the *Journal of Comparative Neurology.*]

(1966) described class 4 cells as comprising a medium-sized soma, often fusiform in shape, with a simple pattern of dendritic branching oriented along the breadth of the lamina. The soma sizes of these cells were measured by Holländer and Vanegas (1977), who identified them as the somas in the parvocellular C laminae that were retrogradely labeled by HRP injected into the visual cortex. Their data (for example, their Table 2) suggest that these cells are distinctly smaller (mean soma area 131 μm^2) than relay cells in the A laminae (240–260 μm^2, including both class 1 and 2 cells). The histograms in their Fig. 2 further suggest that the C-lamina cells are smaller than the most common cell class in the A laminae, presumably (see above) class 2 cells. This supports Rioch's (1929) choice of the name *lamina parvocellularis* to refer to what are now called the *deep C laminae* (i.e., the ventral part of lamina C plus laminae C1 and C2).

A number of features of the ultrastructure of the dLGN can be related to the Y/X/W and class 1/2/4 classifications just discussed. Most particularly, descriptions of the synaptic organization of the dLGN emphasize a structure termed the *synaptic glomerulus* (Szenthagothai, 1973) or *encapsulated synaptic zone* (Guillery, 1969b). This is a glial-bounded set of synapsing processes (Fig. 6.15), in which retinal ganglion cell

---

← **Figure 6.13.** Holländer's (1978) demonstration of the identification of cytoplasmic laminated bodies (CLBs) in cat LGN by phase-contrast microscopy. If the refractive index of the mounting medium is relatively high (> 1.56), CLBs can be seen as bright spots under phase-contrast conditions. (a) Phase-contrast view of a 15-μm-thick celloidin-embedded section of cat LGN. The arrow indicates a CLB. (b) Same CLB after reembedding in araldite and cut at 3 μm. (c) Same CLB seen under electron microscopy, in an ultrathin section. (d) Same CLB seen at higher power. Note the "fingerprint" appearance of its membrane structure, first noted by Morales *et al.* (1964). Calibration marks represent 10 μm for (a), 5 μm for (b) and (c), and 1 μm for (d). [Reproduced from *Microscopica Acta,* with kind permission of Hirzel-Verlag.]

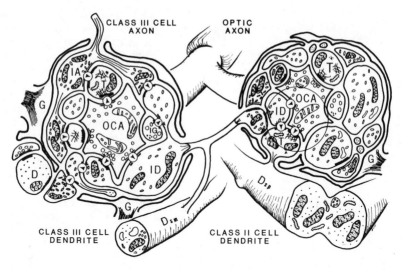

Figure 6.15. Famiglietti and Peters' (1972) interpretation of the synaptic glomerulus of cat dLGN. In this interpretation, the optic axon breaks up into terminal branches (OCA) that form synapses onto processes labeled T. These are synaptic thorns of a dendrite (Dsp) of a geniculate relay cell. The "thorn" is attached to the dendrite by a fairly narrow, short stalk. The optic axon also synapses on processes labeled ID, which are the characteristic appendages of the dendrites of class 3 cells, probably interneurons. The ID terminals arise from the class 3 cell dendrites by long, slender stalks.

   The region of synapse is bounded by processes of glial cells (G), forming the synaptic glomerulus. Within the glomerulus, synapses are formed by optic axons onto both ID and T processes, and by ID onto T processes. Also present are processes (IA) of the axons of class 3 cells that synapse onto both T and ID processes. No synapses are formed onto OCA processes. The circle marks a "triadic" synaptic relationship between OCA, ID, and T processes. [Reproduced with kind permission of the *Journal of Comparative Neurology.*]

axons terminate on dendritic appendages of relay cells, with processes presumed to derive from class 3 cells (interneurons) closely apposed. Some of these latter processes have properties of both dendrites (they receive retinal synapses) and axons (they form synapses onto relay cells) (Famiglietti, 1970), and are believed to be inhibitory in function; for example, they have the flattened synaptic vesicles generally associated with an inhibitory synaptic action. These synaptic complexes may serve, therefore, as particular sites of inhibitory action on geniculate relay cells.

   There is some reason to expect that this inhibitory action may be relatively specific to class 2 (X) cells. The processes of relay cells that enter the synaptic complexes are considered to be the "grapelike dendritic appendages" characteristic of class 2 relay cells (Fig. 6.11, 6.12). The complexes have been described in laminae A, A1, and, less commonly lamina C, matching the distribution of X and class 2 cells. Conversely, the synaptic zones are rare, and relatively simple, in areas receiving Y- or W-cell input, such as the MIN and the deep C laminae (Guillery and Scott, 1971). Further, Fukuda and Stone (1976) reported physiological evidence that inhibition is more marked in X- than in Y-class relay cells, a finding that received some support from the study of Rodieck and Dreher (1979).

One recent series of studies has challenged in a fundamental way the correlation between morphological and physiological cell classes just discussed. Friedlander *et al.* (1979, 1981) recorded intracellularly with HRP-filled microelectrodes from individual relay cells in the cat dLGN, identified them by physiological criteria as X or Y class, showed by antidromic activation from the visual cortex that many were relay cells, and then passed electrical current through the electrode to "fill" the cell with HRP. They found that many Y cells have class 1 morphology (as suggested by earlier workers), but that a significant proportion had class 2 morphology. Further, most X cells in their sample had either class 2 morphology, or resembled the class 3 cells previously considered to be interneurons. Comparing the morphology of X and Y cells demonstrated in this way, Friedlander and co-workers conclude that Y cells

1. Differ from X cells in that their dendrites cross laminar boundaries.
2. Have larger somas and thicker dendrites than X cells.
3. Have thicker axons than X cells.
4. Have radially symmetric dendritic fields, whereas the dendritic fields of X cells tend to be elongated along projections lines.
5. Have simpler appendages on their dendrites.

In many ways these observations are consistent with earlier views of the morphology of X- and Y-class relay cells. If confirmed, however, and Friedlander and co-workers' evidence is arguably more direct than that provided in any other reports, these studies seem to require a fundamental reassessment of not only the morphological basis of the Y/X grouping of LGN relay cells, but also:

1. The number of interneurons within the LGN: The observation that many class 3 cells in laminae A and A1 are relay cells suggests that few cells found within the dLGN can be interneurons.
2. The structural basis of inhibition in the LGN: The axon terminals previously thought, on the basis of the pleiomorphy of their vesicles and the symmetry between membranes on both sides of the synapse, to be inhibitory in function, were believed to be formed by interneurons with class 3 morphology. But if class 3 cells are relay cells, and presumably therefore excitatory in their synaptic action, which cells form the inhibitory terminals?
3. The numbers of Y cells present: If many of the cells others considered to be interneurons are in fact X cells, then the proportion of Y cells relative to X cells is lower (about 30%) than previously estimated [50% (LeVay and Ferster, 1977)].

Fortunately, another study in this series (Stanford *et al.*, 1981) confirmed that W-class relay cells have the class 4 morphology Guillery described for relay cells in the C laminae.

6.1.7. Cortical Projections of Y-, X-, and W-Class Relay Cells

Anatomical evidence has been available since the work of Wilson and Cragg (1967), Glickstein *et al.* (1967), Garey and Powell (1967), Niimi and Sprague (1970), and Rossignol and Colonnier (1971) that the LGN of the cat sends axons directly not only to area 17 of the cerebral cortex, but also to areas 18 and 19 and to the lateral part of the suprasylvian gyrus. Physiological evidence of these multiple projections goes back to the work of Talbot (1942), and includes the subsequent studies of Doty (1958), Toyama and Matsunami (1968), and Ohno *et al.* (1970). More recent evidence has amply confirmed these reports and provided evidence that the multiple cortical projections from the dLGN are not repetitive of each other. Rather, each projection to a cortical area involves a different component of the Y-, X-, and W-cell activity being relayed through the dLGN. Put another way, there are significant differences in the cortical projections of Y-, X-, and W-class relay cells, differences that contribute substantially to the distinct functions of the cortical areas concerned. The pattern of projections is shown schematically in Fig. 6.16.

X-Cell Projections

X-Class relay cells of the dLGN project to area 17; indeed, area 17 is the only known target of these relay cells. X cells form a major component of the geniculate input to area 17 (it also receives substantial Y- and W-cell projections), and presumably subserve the one aspect of the cat's visual behavior that is known to be affected by the destruction of area 17, namely, the ability to resolve high spatial frequencies (Chapter 11, Section 11.2.3). The evidence for this conclusion is follows (note that several lines of evidence rely on the assumption that X cells have smaller somas than Y cells):

First, Garey and Powell (1967) noted that many smaller neurons of the dLGN degenerate after destruction of area 17 (presumably because their axons are severed there), whereas the larger neurons do not degenerate unless both areas 17 and 18 are destroyed. They suggested that the smaller cells (presumably including the X cells of layers A and A1) degenerate because they project only to area 17, while the larger cells (presumably the Y cells) survive destruction of area 17 because they are "sustained" by an undamaged branch of their axon, which reaches area 18 (Fig. 6.17)

Second, Stone and Dreher (1973) advanced physiological evidence that X cells project to area 17 but not area 18, and provide the geniculate input to a majority of area 17 cells (Fig. 6.18). Their evidence is discussed in more detail in Chapter 8, Section 8.1.1. It comprised analyses both of the latencies of neurons in areas 17 and 18 to afferent volleys in the visual pathways, and of the receptive field properties of the same cells. Stone and Dreher drew attention to similarity between the X- and Y-cell projections they observed and the pattern reported by Garey and Powell (1967) for medium and large cells, respectively. Singer *et al.* (1975) confirmed Stone and Dreher's conclusion that X cells form the predominant input to area 17.

Third, Maciewicz (1975), Gilbert and Kelly (1975), Holländer and Vanegas (1977; Fig. 6.19), LeVay and Ferster (1977), Garey and Blakemore (1977), Ferster and LeVay (1978), Leventhal (1979), and Geisert (1976, 1980) all used the retrograde axonal transport of HRP to locate and measure the somas of geniculate relay cells projecting to the visual cortex. All these reports gave evidence that medium-sized relay cells in the A laminae project in large numbers to area 17. LeVay and Ferster considered these medium-sized cells likely candidates to be X cells, the larger cells to be Y cells,

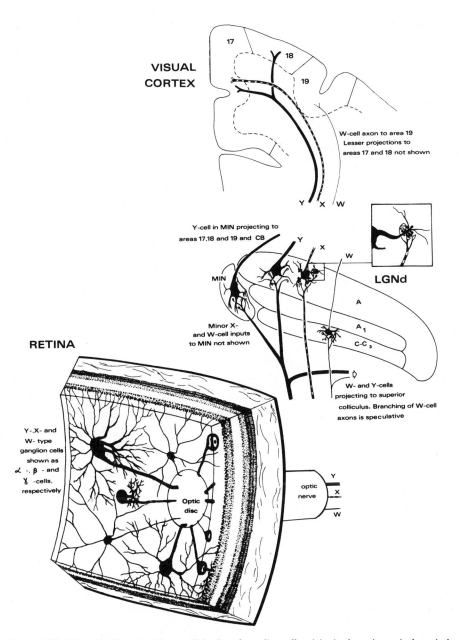

Figure 6.16. Schematic diagram of the parallel relay of ganglion cell activity in the retinogeniculocortical pathway (from Stone *et al.*, 1979). Lower: A segment of the retina is shown with different functional classes of ganglion cells represented as their postulated morphological counterparts (α, β, and γ cells). Their axons reach the dLGN, represented by a schematic coronal section of the nucleus (middle). There they contact different classes of relay cells, also represented as their postulated morphological counterparts, the class 1, 2, and 4 cells of Guillery (1966). Class 1 cells are also shown in the MIN. The inset shows the postulated manner of termination of ganglion cell axons on the "grapelike dendritic appendages" characteristic of class 2 cells. The axons of class 1, 2, and 4 cells project differently to areas 17, 18, and 19 and the Clare-Bishop area (CB) of the visual cortex, which is represented at the top. [Reproduced with kind permission of Elsevier/North-Holland Biomedical Press.]

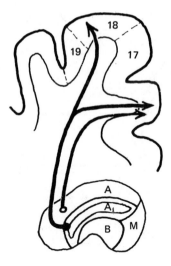

Figure 6.17. Different patterns of geniculocortical projection for large and small relay cells of cat dLGN (from Garey and Powell, 1967). Large cells project to both areas 17 and 18 (closed circle) and small cells project only to area 17 (open circle). [Reproduced with kind permission of the Royal Society (London).]

and the smaller cells interneurons; and Ferster and LeVay traced medium-caliber axons, which they argued were derived from medium-soma relay cells, to their terminations in layers 4C and 6 of area 17. Small numbers of medium-sized cells in laminae A1 and C were shown to project to area 18, and small numbers of medium-sized cells in lamina C to project to area 19 [see Figs. 3 and 4 of Holländer and Vanegas (1977)]; these could represent X cells projecting to areas 18 and 19. Taken together, the anatomical and physiological evidence support the view that X cells project at least principally to area 17, and comprise a major, and perhaps the principal, geniculate input to area 17.

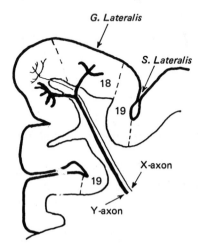

Figure 6.18. Different projections of X- and Y-class relay cells to areas 17 and 18 in the cat (from Stone and Dreher, 1973). From physiological evidence, we suggested that Y cells project to both areas 17 and 18, many by a branching axon, thus resembling the large cells of Garey and Powell's model (Fig. 6. 17). X cells appeared to project to area 17, but not area 18, resembling the small cells of Garey and Powell's model. [Reproduced from the *Journal of Neurophysiology* with kind permission of the American Physiological Society.]

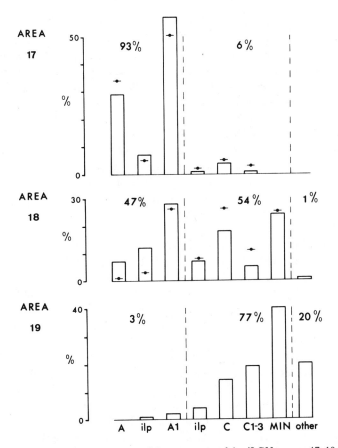

Figure 6.19. Evidence of different projections of the components of the dLGN to areas 17, 18, and 19 (from Holländer and Vanegas, 1977). The histograms show the percentage distribution among different components of the dLGN of the cells labeled with HRP following HRP injections into area 17, 18, or 19. Note that the distribution of labeled cells shifted from the A laminae to the C laminae and MIN as the injection site was changed from area 17 (top histogram) to area 18 to area 19 (bottom histogram). "Other" refers to labeled cells found outside the LGN, "ilp" to labelled cells found in interlaminar plexuses. [Reproduced with kind permission of the *Journal of Comparative Neurology.*]

Y-Cell Projections

Y-Class relay cells are found in laminae A, A1, and C, and the MIN. Current evidence indicates that (1) Y cells project from the MIN strongly to area 18, probably also to area 17 and to the suprasylvian gyrus, and perhaps to area 19; and (2) Y cells project from laminae A, A1, and C to areas 17 and 18.

The evidence concerning Y-cell projections from the MIN comes from anatomical studies using retrograde degeneration (Garey and Powell, 1967), anterograde degeneration (Burrows and Hayhow, 1971; Wilson and Cragg, 1967; Niimi and Sprague,

1970), orthograde axonal transport (Rosenquist *et al.*, 1974), and the retrograde axonal transport of HRP [Maciewicz, 1975; Holländer and Vanegas, 1977 (this paper includes a valuable analysis of prior literature); Leventhal and Keens, 1978; Leventhal, 1979]; and from physiological studies that suggest that the MIN contains a high proportion of Y-class relay cells (Dreher and Sefton, 1975, 1979; Mason, 1975; Kratz *et al.*, 1978). Taken together, these studies indicate clearly that many of the large-soma (presumably Y-class) relay cells of the MIN project to area 18. There is some disagreement concerning a projection from the MIN to area 17, but recently Leventhal (1979), using highly localized injections of HRP in area 17, reported that small numbers of MIN cells do terminate in area 17. Some of the cells had large somas characteristic of Y cells, and their axons terminated, as do the axons of Y cells of the main part of the dLGN, in layer 4AB of area 17. There is also disagreement as to whether MIN cells project to area 19, Gilbert and Kelly (1975) reporting negative results on this point. Leventhal *et al.* (1980) have reported an observation that may resolve this particular issue. They noted that many MIN cells were labeled with HRP following an injection of the enzyme into area 19, but most were small in soma diameter. It may be the case that the W cells of the MIN do project to area 19, while the Y cells do not.

Anatomical evidence that Y cells project from laminae A, A1, and C to areas 17 and 18 comes from studies employing retrograde degeneration (Garey and Powell, 1967) and anterograde degeneration (Rossignol and Colonnier, 1971) techniques and, more recently, from HRP studies (Gilbert and Kelly, 1975; Geisert, 1976, 1980; LeVay and Ferster, 1977; Holländer and Vanegas, 1977; Leventhal, 1979; Fig. 6.19). Taken together, these studies show that large relay cells in laminae A, A1, and C send large-caliber axons to areas 17 and 18. Physiological analyses of these projections came from studies of the afferent latencies and receptive field properties of single neurons of areas 17 and 18 [Watanabe *et al.*, 1966; Toyama and Matsunami, 1968; Ohno *et al.*, 1970; Stone and Dreher, 1973 (Fig. 6.18); Singer *et al.*, 1975; Ferster and LeVay, 1978; Leventhal, 1979; Dreher *et al.*, 1980; Kimura *et al.*, 1980]. These provided evidence that many cells in both areas receive a monosynaptic input from the fast-conducting axons of geniculate Y cells.

The same papers indicate that Y cells project differently between areas 17 and 18. They form a distinct input to area 17, terminating in layers 4AB and 6, but may be less numerous than the X-cell afferents. By contrast, Y cells form at least the predominant geniculate input to area 18. The evidence concerning Y-cell projections to area 19 is problematical. From HRP tracing material, Holländer and Vanegas (1977) and Geisert (1980) described a number of medium- and large-soma cells in lamina C projecting to area 19; and Ohno *et al.* (1970) and Kimura *et al.* (1980) reported physiological evidence of area 19 cells directly activated by fast-conducting, presumably Y-cell afferents. On the other hand, Dreher *et al.* (1980), in a physiological study, searched for but found very little evidence of a direct Y-cell input to area 19.

One other discrepancy among these reports must be noted. Garey and Powell (1967) and Stone and Dreher (1973) argued that the axons of many large, Y-class relay cells of the LGN branch to both areas 17 and 18 (see Figs. 6.17 and 6.18). LeVay and Ferster (1977), on the other hand, found so few cells in the A laminae labeled with

HRP after injections into area 18 that they concluded that very few, if any, Y cells can have such branching axons; rather, they argued, separate populations of Y cells project to the two areas. The cell size measurements of Holländer and Vanegas (1977) provide clear support for this argument. On the other hand, Geisert (1976, 1980) has provided evidence using a double-HRP technique that large relay cells in the cat dLGN send axons to both areas, indeed that as many as 90% of geniculate cells that project to area 18 also send an axon to area 17. In addition, Holländer and Vanegas (1977) describe a branching of geniculocortical axons in the white matter just under areas 17 and 18. It seems likely that the axons of many Y cells do branch in this way; the proportion they form of Y-class relay cells is for the moment at issue.

W-Cell Projections

W-class relay cells have been described in the C laminae of the dLGN and in the MIN. The relay cells of the ventral part of lamina C and of laminae C1 and C2 are relatively small in soma size and are known to be the somas of W-class relay cells. These cells have been shown by HRP studies to send axons to areas 17, 18, and 19 but principally to area 19 [Maciewicz, 1975; Gilbert and Kelly, 1975; Holländer and Vanegas, 1977 (Fig. 6.19); LeVay and Ferster, 1977; Ferster and LeVay, 1978; Leventhal and Keens, 1978; Leventhal, 1979]. In area 17, it is now recognized that W cells form a distinct component of this area's direct input from the dLGN, terminating in a characteristic way in layer 1, and to a lesser extent in layers 3 and 5. The W-cell axons are finer and harder to detect than X- or Y-cell axons, and probably numerically a minority, but their contribution to the physiology of area 17 is beginning to be recognized (Leventhal and Hirsch, 1980; Dreher *et al.*, 1980; Chapter 8, Section 8.1.2).

W cells may form the predominant geniculate input to area 19; but, as with area 17, the significance of this input has been recognized only gradually. Maciewicz (1975) reported evidence that some cells in the C laminae project to area 19. Subsequently, Holländer and Vanegas (1977) noted that area 19 appears to be the main cortical target of relay cells in the C laminae, including small-, medium-, and large-soma cells. Leventhal and Keens (1978) confirmed this general finding but noted that the cells projecting to area 19 were "usually small." More recently, the physiological studies of Kimura *et al.* (1980) and Dreher *et al.* (1978, 1980) have indicated that W cells provide the major geniculate input to area 19; indeed, Dreher and co-workers concluded that the direct geniculate input to area 19 is derived entirely from W cells. This strong W-cell projection provides a clue to understanding the functional significance of both area 19 and the W-cell system. The smaller neurons of the MIN are also presumably W cells (Section 6.1.3) and have been shown to project to areas 17, 18, and 19 (Leventhal *et al.*, 1980).

In summary, W cells appear to project from the C laminae and the MIN to areas 17, 18, and 19 and the lateral suprasylvian area. Of these target areas, area 19 seems to receive the greatest part of the W-cell projection. Physiological evidence of direct W-cell input is now available for areas 17 and 19, but not yet for area 18 or the lateral suprasylvian area.

6.1.8. Corticogeniculate Projections

Updyke's (1975) study of corticogeniculate projections to the main part of the dLGN of the cat shows a striking reciprocity of connections between the various laminae of the dLGN and areas 17, 18, and 19. Area 17 was found to project heavily and uniformly to the full thickness of the main part of the dLGN, including laminae A, A1, and C1–3, and the interlaminar zones betweeen A and A1 and between A1 and C. Area 18 projects to the same laminae but generally less strongly; however the projection from area 18 to the interlaminar zones and to lamina C is particularly heavy. Thus, the area 17 projection reciprocates the input it receives from all major cell types in all laminae of the dLGN. The relatively weak projection of area 18 to laminae A and A1 may reflect the fact that its input is derived from the Y cells of these laminae, and not from the X cells. Conversely, the strong projection of area 18 to lamina C seems to reciprocate the presence in the dorsal part of this lamina of a high proportion of Y-class relay cells (Wilson *et al.,* 1976), and the strong projection of lamina C to area 18 (Holländer and Vanegas, 1977). Within the main part of the dLGN, area 19 projects only to the deep C laminae, reciprocating in a striking way the strong input that this area receives from W cells in these laminae, largely or entirely to the exclusion of X and Y cells (Dreher *et al.,* 1978, 1980; Kimura *et al.,* 1980).

The "parallel-wiring" concept of the organization of geniculocortical projection seems to carry over strongly into corticogeniculate projections. It should be noted, however, that Updyke's subsequent (1977) description of the projections of areas 17, 18, and 19 to other parts of the LGN (the vLGN and the MIN), and to other thalamic nuclei, shows a considerable convergence of the projections of the three areas to these sites. The "parallelism" just stressed appears limited to the projections to the main part of the dLGN.

6.2. THE LGN OF PRIMATES

The appearance of superficial large-celled laminae (one or two) is a characteristic feature of the lateral geniculate body of Primates, and so far as is known it is conspicuously present in every Primate, with the curious exception of Tarsius. [LeGros Clark, 1932]

One may suspect that the larger phasic [ganglion] cells, which are relatively more common away from the fovea, synapse with cells in the [magnocellular] two layers (of the LGN) and the smaller, tonic cells, which are both absolutely and relatively more common near the fovea, synapse with cells in the [parvocellular] four layers. [Gouras, 1969]

The dLGN of primates shows a sharp and characteristic differentiation of large-cell from small-cell laminae, and later physiological studies have established Gouras' suggestion that small-and large-cell laminae process the activities of different functional groupings of ganglion cells. Considering a range of primates, as in Fig. 6.20, the large-cell laminae are placed either lateral or ventral to the small-cell laminae, but in all cases nearer the outer (pial) surface of the thalamus. Clarke (1932) suggested that the ventral location of the magnocellular laminae in "higher" primates (monkeys, apes, and man) allows for the strong development of the parvocellular laminae in these species. Confirming this idea, Rakic (1977) has described how the LGN of the rhesus macaque monkey rotates during gestation, in essentially the pattern that Clarke had anticipated.

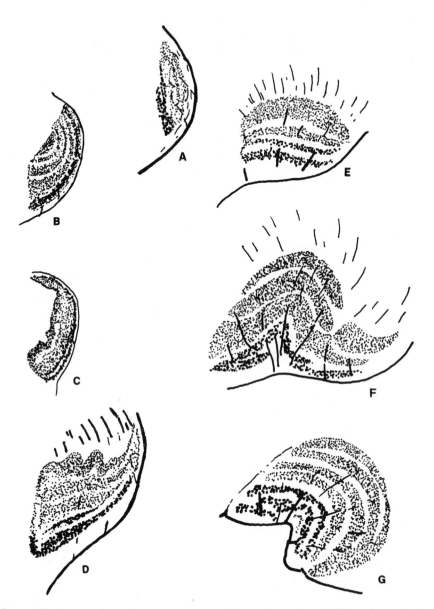

Figure 6.20. Evidence of a consistent magno–parvocellular division of the dLGN in primates (Le Gros Clark, 1932). The diagrams show coronal sections of the dLGN from several primates. The lines show the direction of entering blood vessels. The lateral side of the LGN is toward the right. Magno- and parvocellular divisions of the dLGN are represented for (A) tree shrew, (B) mouse lemur, (C) Coquerell's dwarf lemur, (D) *Lemur catta,* (E) orangutan, (F) human, and (G) *Cercopithecus* (Old-World monkey). [Reproduced with kind permission of the Editor of the *British Journal of Ophthalmology.*]

The cells that form the magnocellular layers are born earliest from the ventricular zone of the thalamus, migrate outwards, and collect at the pial surface of the thalamus. The cells that will form the parvocellular layers are born later and, as they migrate outwards, they accumulate inside the future magnocellular layers. Eventually, after the LGN has rotated ventrolaterally, they are found in their adult location, dorsal to the magnocellular layers.

6.2.1. X/Y Analysis of Parvo- and Magnocellular Laminae

In the species so far investigated (the Old-World macaque monkeys and the New-World owl monkey), the parvo–magnocellular division of the LGN has been shown, confirming Gouras' suggestion quoted above, to correlate strongly with a Y/X division of function. That is, relay cells in the parvocellular laminae relay that activity of X-like ganglion cells to the visual cortex, while neurons in the magnocellular laminae relay Y-cell activity. Thus, relay cells of parvo- and magnocellular laminae differ in the same properties, and in the same way, as cat X cells differ from Y cells. Moreover, the X/Y differentiation is at least as strong as in the cat, perhaps stronger. It is tempting to suggest that the parvo–magnocellular differentiation of the dLGN found in a wide range of primates, including man, indicates that a strong Y/X differentiation is a general feature of the retinogeniculocortical pathways of primates.

The evidence for this functional difference between parvo- and magnocellular layers comes from the studies of Dreher *et al.* (1976), Sherman *et al.* (1976), and Schiller and Malpeli (1978). Dreher *et al.* (1976) and Sherman *et al.* (1976) presented evidence that in both Old-World macaque monkeys (*M. irus* and *M. nemestrina*) and the New-World owl monkey (*Aotus trivirgatus*), the relay cells of the dLGN can be classified into X-like and Y-like groups. The X-like cells of the monkey resemble cat X cells in having (on the average) smaller receptive fields, giving sustained or tonic responses to standing contrast stimuli, and in having slow afferents from the retina and slow axons projecting to the visual cortex. The Y-like cells resemble cat Y cells in having (on the average) larger receptive fields, in giving phasic responses to standing contrast stimuli, in having faster-conducting afferents and axons, and in being particularly responsive to high stimulus velocities. The X-like cells of the macaque differ from cat X cells in that many of them show color-opponent properties not found in cat X cells, and some also lack an antagonistic surround (see Chapter 3, Section 3.1.2). Both Dreher and co-workers and Sherman and co-workers comment, however, that the X/Y distinction was as clear-cut in the monkey as in the cat, and in some respects even more striking.

Both Dreher and co-workers and Sherman and co-workers noted a strong tendency for X cells to be located in the dorsal, parvocellular laminae of the dLGN, while Y cells are restricted to the more ventral, magnocellular laminae, with apparently no intermingling. There appears, therefore, to be a sharp anatomical segregation of X- and Y-class relay cells in the LGN of both New- and Old-World monkeys. Schiller and Malpeli (1978) extended these findings to the rhesus macaque (Fig. 6.21), and also reported a tendency for ON-center X cells to concentrate in the most dorsal two parvocellular laminae, while OFF-center X cells concentrate in the more ventral two parvocellular lami-

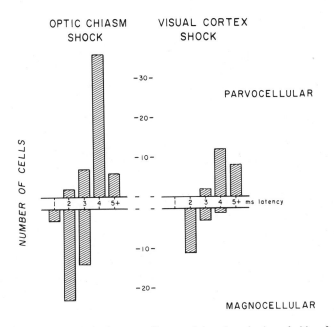

Figure 6.21. Evidence of a correlation between afferent and axonal conduction velocities of relay cells of monkey dLGN (from Schiller and Malpeli, 1978). In the parvocellular laminae (upgoing histograms), both the orthodromic responses of cells to stimulation of the optic chiasm and their antidromic responses to stimulation of the visual cortex are relatively long (modal value 4 msec). For the magnocellular laminae (downgoing histograms), both latencies are considerably shorter (modal value 2 msec). These differences indicate that the afferents to relay cells in the parvocellular laminae are slower-conducting than those to cells in the magnocellular laminae, and that the parvocellular cells have slower axons. [Reproduced from the *Journal of Neurophysiology* with kind permission of the American Physiological Society.]

nae. These results suggest that, despite the distinct phylogenetic histories of the cat and of Old- and New-World monkeys, the "paralled channel" model of geniculate organization is applicable in all three classes of animal.

Two recent receptive field studies have suggested different interpretations of the magno-/parvocellular difference in the monkey LGN. Shapley *et al.* (1981), for example, identified parvocellular cells as X cells (as did the studies just discussed) but considered that magnocellular cells include both X and Y types. They therefore suggested that the magnocellular layers may be homologous to the dLGN of the cat, the parvocellular layers representing "a visual neural pathway which is not present in cats." Blakemore and Vital-Durand (1981) make a similar suggestion. My own view is that these latter studies, by relying on a single criterion (spatial summation) to identify cells as X or Y type, provide a poorer basis for judging of homologies than the earlier studies, which distinguished Y-like from X-like cells by several parameters, (axonal conduction velocity, receptive field size, and velocity-selectivity). Moreover, Leventhal *et al.* (1981) have demonstrated the morphologies of ganglion cells projecting to the parvo- and magnocellular laminae and their data support Dreher, Fukada and Rodieck's interpretation of parvocellular cells as X-like and magnocellular cells as Y-like (Chapter 3, Section

3.1.4). Discussion of the advantages of parametric over single-criterion classifications can be found in Chapter 2, Section 2.3.

Reports of the morphology of geniculate relay cells in the monkey indicate a partial analogy with the cat. Szenthagothai (1973) described distinct types of relay cells in the magno- and parvocellular laminae of monkey dLGN. The relay cells characteristic of the magnocellular laminae are large and in dendritic branching patterns resemble class 1 relay cells of cat LGN. However, their main dendrites also bear some appendages similar to, but more irregular than, the grapelike appendages of class 2 cells of the cat. Szenthagothai considers them "a more specialized kind of neuron" than cat class 1 cells, although very similar. The characteristic relay cell of the parvocellular laminae is a smaller neuron that "corresponds undoubtedly to the class 2 cell of the cat." In qualification, however, Szenthagothai reports "midget" and "radiate" relay cells, which lie outside this grouping. Moreover, Wong-Riley (1972), in a study of squirrel monkey LGN (not yet analyzed physiologically), reported clear analogies of the class 1 and class 2 cells of cat LGN, but reports both types in both magno- and parvocellular laminae. Saini and Garey (1981) also noted that of the neuron types they distinguished in Golgi studies of the LGN of both New- and Old-World monkeys, none seemed segregated to the parvo- or magnocellular laminae. However, "multipolar" cells comprised the bulk of the neurons in the parvocellular (X) laminae, while "bipolar" cells were most common in the magnocellular laminae, suggesting an approximate correlation between "multipolar" and X-class relay cells, and between "bipolar" and Y-class cells. Guillery and Colonnier (1970), on the other hand, observed no differences in synaptic structure between magno- and parvocellular laminae of the macaque dLGN. Overall, the evidence supports only an approximate correlation between physiologically and morphologically distinguished groups of relay cells in monkey LGN.

6.2.2. Other Components of the LGN

Recent studies have provided increasingly compelling evidence (Wong-Riley, 1976; Hendrickson *et al.,* 1978; Yukie *et al.,* 1979; Yoshida and Benevento, 1981; Benevento and Yoshida, 1981; Fries, 1981) that a minority of cells in the dLGN of the macaque monkey project to the prestriate cortex. In all these studies, the principal evidence consists of labeling of neurons of the dLGN following injection of HRP into the prestriate cortex. It seems unlikely that the labeling of geniculate cells following HRP injections into the prestriate cortex resulted from spread of HRP into area 17. Yukie and coworkers, for example, noted several characteristic differences between geniculate cells projecting to the prestriate cortex and those projecting to area 17. First, the prestriate-projecting cells were distributed differently within the nucleus, being found in interlaminar zones as well as in the main laminae. Second, although the prestriate-projecting cells were found in both parvo- and magnocellular parts of the dLGN, their soma size did not vary significantly between the two parts, being on the average larger than parvocellular cells and smaller than magnocellular cells. Third, the prestriate-projecting cells formed only a minority of LGN cells and were more sparsely distributed through the nucleus. Similarly, in Yoshida and Benevento's report, the cells labeled from the prestriate cortex were few in number and sparsely distributed and were distributed through the nucleus quite differently from 17-projecting cells; specifically, they were found in interlaminar zones. These workers corroborated their HRP findings with autoradiographic tracing, which indicated that the cells in question terminate principally in layers 4 and 5 of area 19. Fries (1981) and Benevento and Yoshida (1981) both reported that the LGN–prestriate projection is retinotopically organized, but less precisely than the LGN–striate pathway. So far, only the outlines of this pathway have been described; further investigation should reveal its relationship to the X-like and Y-like components of the geniculate projection to area 17, and give some indication of its functional significance. It is still not clear, for example, whether the LGN cells concerned receive direct retinal input; Benevento and Yoshida (1981) note, for example, that the cells concentrate in interlaminar regions of the LGN where they are contacted by afferents from the SC.

A vLGN has been described as a consistent feature of primate LGN. In addition, two components of the dLGN have been distinguished that seem distinct from the prominent magno- and parvocellular laminae: the ventral (superficial) S lamina, to which Campos-Ortega and Hayhow (1970) drew attention (Fig. 6.22), and the intermediate cell group of Minkowski (1919) that Campos-Ortega and Hayhow (1971) suggested

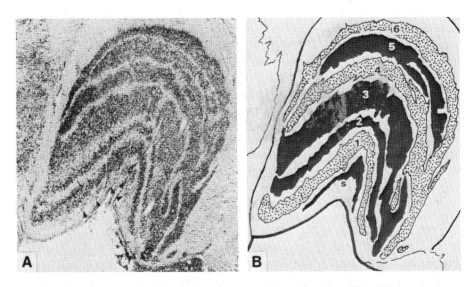

Figure 6.22. Evidence of a lamina lying superficial (ventral) to the magnocellular laminae of primate dLGN (from Campos-Ortega and Hayhow, 1970). The LGN of the bush baby *Galago crassicaudatus* is shown in coronal section; (A) is a photomicrograph, (B) is a schematic drawing. Laminae 2, 3, and 5 show evidence of degenerating axons, 4 days after enucleation of the ipsilateral eye, while laminae 1, 4, and 6 are degeneration-free. Hence, laminae 2, 3, and 5 receive input from the ipsilateral eye, laminae 1, 4, and 6 from the contralateral. Ventral to laminae 1 and 2 (the magnocellular laminae) and near the ventral surface of the dLGN (hence "superficial" or "s") is a lamina of cells in which degenerating terminals are apparent, the S lamina. This lamina has been shown subsequently (Kaas *et al.*, 1978) to comprise separate sublaminae receiving contralateral as well as ipsilateral input, to be present in several primates, and to project to area 17 of the cerebral cortex. It contributes to the thalamic relay of ganglion cell activity to the visual cortex. [Reproduced with kind permission of Elsevier/North-Holland Biomedical Press.]

may prove to be homologous to the MIN component of cat dLGN. None has yet been analyzed in terms of the Y/X/W paradigm.

The S lamina has been shown (Kaas *et al.,* 1978) to receive input from both eyes, and to project to area 17. One intriguing, still-untested possibility is that, like the most ventral laminae of the dLGN of the cat, its cells relay W-cell activity to the visual cortex.

6.2.3. Corticofugal Projections

The report of Lund *et al.* (1975) suggests that, as in the cat, some "parallelism" is apparent in corticogeniculate projections, reciprocating the X/Y channels in the geniculocortical projections just discussed. The cells of area 17 that project back to the dLGN are located, apparently exclusively, in cortical layer 6. Lund and co-workers note that different populations of cells project to parvo- (X) and magnocellular (Y) parts of the dLGN: those located in the upper part of layer 6 projecting to the parvocellular laminae, those in the lower part of layer 6 projecting to the magnocellular laminae. As discussed in more detail in Chapter 8, Section 8.2.1, Lund and co-workers note further that the dendrites of the parvo-projecting cells may, following Lund and Boothe's (1975) descriptions, spread close to the regions of termination of axons originating in the parvocellular laminae; and, conversely, that the dendrites of the magno-projecting cortical cells spread in, or close to, the lamina of termination of axons originating in the magnocellular laminae. Thus, the anatomical basis appears to exist for a parallel, reciprocal interconnection of X-channel and Y-channel neurons, between the dLGN and the visual cortex.

6.2.4. Summary

In several parameters, the dLGN of primates seems strongly oriented to the parallel relaying of the activities of different functional classes of retinal ganglion cells. Moreover, the parallel organization seems to persist into the projections from the visual cortex back to the dLGN. Several groups of workers have commented that the X/Y differentiation in monkeys seems even better developed in the cat; one index of this better differentiation is the segregation of X- and Y-like relay cells into distinct (parvo- and magnocellular) laminae of the nucleus. An interesting trend is that the primate retinogeniculocortical pathway seems to show, by comparison with the cat, a much stronger development of the X-cell system, and a weaker development of the W-cell system. The strong development of the X-cell system (as for example in the prominence and separateness of the parvocellular components of the dLGN) may reflect the strong development in many primates of high-resolution, binocular vision.

6.3. OTHER SPECIES

6.3.1. The LGN of the Rat

Some of the early observations fundamental to the parallel processing model of the LGN were made in the rat, and it is possible to trace strong analogies of geniculate organization between the rat, cat, and monkey in several major parameters (such as the organization of the geniculate relay according to conduction velocity, the morphology of relay cells, and geniculocortical projections), and partial analogies in receptive field properties.

Sefton and Swinburn (1964) obtained field potential evidence for the presence of three conduction velocity groupings in the optic nerve of the rat, with velocities of 13.5, 5.5, and 3.0 m/sec, and found evidence that all three project to the LGN (Fig. 6.23). The presence of three conduction velocity groups among optic nerve fibers, and the termination of all three in the LGN, were corroborated by the work of Sumitomo *et al.* (1969*b*), Fukuda (1977), and Fukuda *et al.* (1979).

Noda and Iwama (1967) recorded from single relay cells in rat LGN, and showed a strong correlation between the latencies of the cells' responses to orthodromic and antidromic stimulation (Fig. 6.2). This is the same relationship suggested for the cat by Bishop and Clare (1955) and confirmed in several studies of single-cell latencies in the cat and monkey (Sections 6.1.1 and 6.2.1). It suggests that the information carried by fast and slow axons does not converge on geniculate relay cells, but is relayed to the cortex by different subgroups of relay cells.

Noda and Iwama's report provided the first single-cell data on this point for any species. Their observation has been confirmed and extended by Fukuda (1973, 1977), who concluded that three groups of relay cells could be distinguished and related to each of the three conduction velocity groupings of the optic nerve. These groups are analogous in conduction velocity, but not necessarily in their receptive field properties, to the Y-, X-, and W-cell groups among cat relay cells.

Further, the Golgi study of Grossman *et al.* (1973) of relay cells of rat dLGN provides some evidence of a morphological distinction among them in some way analogous to the class 1/2/4 among cat relay cells. The most common relay cell type they observed was a medium-sized (15- to 20-μm soma diameter) neuron whose dendrites often carried "clusters of dendritic appendages" similar to, but "never as prominent as" those seen on class 2 cells of the cat. These cells seemed most common in the posterior parts of the nucleus, where Lund and Cunningham (1972) described the presence of "complex encapsulated synaptic zones" that may be related, as has been suggested in the cat, to the grapelike dendritic appendages of relay cells. Grossman and co-workers described another type of relay cell "perhaps sufficiently distinctive to comprise a sep-

Figure 6.23. Evidence of three conduction velocity groups in rat optic nerve (from Sefton and Swinburn, 1964). The potential recorded from the optic nerve following stimulation of the contralateral optic tracts shows three components (t_1, t_2, t_3). The traces a, b, and c were obtained using stimuli of increasing strength, showing that successively later components have increasingly high thresholds. The conduction velocities estimated for the three groups were: t_1, 13.5 m/sec; t_2, 5.5 m/sec; t_3, 3.0 m/sec. [Reproduced with kind permission of *Vision Research*]

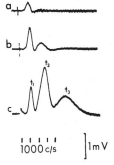

arate subclass," whose dendrites were free of appendages and which was common in the anteromedial portion of the nucleus. This may correspond to the anteromedioventral segment of the dLGN in which Lund and Cunningham (1972) found simple nonencapsulated synaptic zones to be most common, together with relatively simple encapsulated zones. Lund and Cunningham suggested that "two major ganglion cell types each end on different principal cells . . . in the lateral geniculate nucleus and, further, that the synaptic arrangement involved may be different for each."

Further morphological evidence of the parallel organization of rat dLGN comes from Brauer and co-workers' (1979) description of two morphologically distinct forms of terminals formed in the dLGN by optic tract axons. They distinguished (Fig. 6.24) a coarser, loosely branched terminal, from a finer axon that branched more profusely and formed smaller terminal boutons. Further, there was some degree of segregation in their occurrence in the nucleus, the coarser axons being more frequent anteriorly, the finer more posteriorly and laterally, suggesting some correlation with the different relay cell classes and synaptic formations just discussed. The correlation between these various parameters so far appears only partial, however. No description is yet available, for example, of a third class of relay cell or optic tract terminal.

Fukuda and Sugitani (1974) sought to test the analogy between the dLGN of the rat and cat by determining the cortical projections of relay cells. They distinguished two classes of relay cells, fast- and slow-axon, and presented evidence that the slow-axon cells project only to area 17 of the visual cortex, while the fast-axon cells project to both area 17 and the adjacent area 18; and they pointed out that these projections are closely analogous to those of cat X and Y cells. Subsequently, however, Fukuda (1977) concluded that the fast-axon group of relay cells distinguished earlier (Fukuda, 1973; Fukuda and Sugitani, 1974) may comprise two groups, with fast- and intermediate-velocity axons. If these groups of relay cells (fast-, intermediate-, and slow-axon) are in fact analogous to cat Y, X and W cells, then the analogy between cat and rat in their cortical projections is lessened. Moreover, H. C. Hughes (1977) reported that, assessed by autoradiography following the injection of tritiated amino acids, the dLGN of the rat projects to areas 17 and 18A, but not to area 18.

A further partial analogy between the dLGN of rat and cat emerges from the receptive field/latency studies of the rat dLGN reported by Hale *et al.* (1976, 1979) and Fukuda (1977). They confirmed the projection to the dLGN of axons of all three conduction velocity groups of the optic nerve and the finding that individual relay cells receive input from one velocity axon or another, with little intermingling. Moreover, many of the cells receiving fast-axon input had several receptive field properties similar to those of cat Y cells (phasic responses, responsiveness to fast-moving stimuli), while cells with slow-axon input had receptive field properties similar to the "phasic" W cells of Stone and Fukuda (1974a). However, a group of cells with X-like receptive field properties was not clearly distinguishable, and the cells whose receptive fields resembled those of cat W cells did not form a distinct, slow-axon group. The analogy between cat and rat in the physiological organization of the dLGN is substantial but incomplete. Hale and co-workers suggested that the absence of a clear X-cell group in the rat may reflect the poor central vision in this species.

One study is available of the physiological organization of the vLGN of the rat

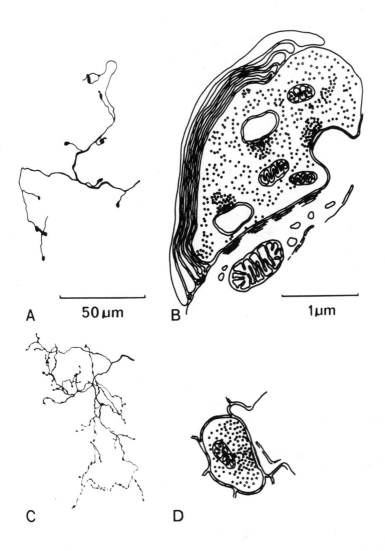

Figure 6.24. Evidence of two classes of retinal terminals in rat LGN (from Brauer *et al.,* 1979). These workers distinguished a relatively coarse terminal tree (A) from the finer, more densely branched terminal shown in (C). They suggested that these different terminals could be recognized in the electron microscope as forming the large and small terminal boutons shown in (B) and (D). [Reproduced with kind permission of Springer Verlag.]

(Hale and Sefton, 1978). By contrast with the comparison between the dLGN of the cat and rat (which seem similar in neuronal connectivity, but not in the differentiation of receptive field properties), there are clear analogies between the vLGN of the two species in their receptive field properties, but not in neuronal connectivity. Thus, Hale and Sefton noted that in the rat, the vLGN appears to receive input from all three conduction velocity groupings, whereas in the cat, it receives an exclusively slow-axon W-cell input (Spear *et al.,* 1977; Section 6.1.4). Conversely, vLGN cells in cat and rat share two receptive field features in common. First, a very high proportion of LGN cells in both species have ON-center receptive fields, whereas OFF-center receptive fields are equally frequent among retinal ganglion cells and dLGN relay cells. Second, in both species many vLGN relay cells are present whose maintained activity varies monotonically with ambient illumination, suggesting that they are coding information about ambient illumination, perhaps for control of the pupil reflex. The behavioral study of Legg and Cowey (1977) provides support for this suggestion, showing that the vLGN is particularly important for the rat's ability to discriminate light intensity.

In summary then, the idea that the LGN is organized to relay the activities of different functional groups of ganglion cells in parallel to the visual cortex (and other projection targets of this nucleus) seems useful in understanding this nucleus in the rat. Differences are apparent between rat, cat, and monkey that presumably are related to the different visual capabilities of the species. In the rat, the functional groupings seem rather less distinct than in the other two species, the X-cell group being hard to detect, a trend that can be related to the poorer development in this species of the area centralis of the retina and of fixational vision. Much remains, of course, to be learnt of many parameters of geniculate organization in this species.

6.3.2. The LGN of the Tree Shrew

Sherman *et al.*(1975*a*) have provided evidence of a functional subgrouping of relay cells in tree shrew dLGN closely analogous to the X/Y grouping of the cat. Indeed, they termed the two groups of relay cells they distinguished *X* and *Y cells*. X-class relay cells had smaller receptive field centers than Y cells, and responded tonically to standing contrast stimuli, while Y cells responded phasically. Moreover, X cells were less responsive to fast-moving visual stimuli than Y cells, were driven by slower-conducting optic nerve axons, and had slower-conducting axons than Y cells (Fig. 6.25). Of their sample of 52 identified relay cells, three had "mixed" X/Y properties, but generally the X/Y dichotomy seemed very distinct, and closely analogous to that of the cat. Sherman and co-workers did not comment on the laminar distribution of the two cell classes within the LGN, and little evidence is available concerning the morphology of relay cells, or of the retinogeniculate synapse.

6.3.3. The LGN of the Mink

Guillery and Oberdorfer (1977) studied the pattern of axon degeneration produced in the LGN of the mink by enucleation of one eye. They reported evidence of a close analogy with the cat, in the following ways.

First, ventrally located parvocellular laminae can be identified in mink dLGN that seem closely analogous to the C laminae of cat LGN. Indeed, Guillery and Oberdorfer use the same terminology to name them. These parvocellular laminae receive terminals

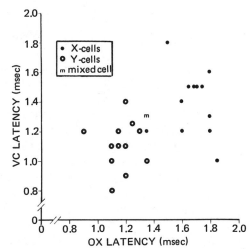

Figure 6.25. Evidence of a conduction velocity correlate of an X/Y classification of relay cells in tree shrew dLGN (from Sherman *et al.,* 1975*a*). As in the cat and monkey, X-class geniculate cells have longer latencies to both OX and VC stimulation than do Y-class relay cells, Hence, it is likely that X cells have slower afferents and slower axons than do Y cells. [Reproduced with kind permission of Elsevier/North-Holland Biomedical Press.]

from fine-caliber retinal axons, as they do in the cat. Moreover, the dorsal, magnocellular part of laminae C and the A laminae of the dLGN receive axons of coarser caliber, as in the cat. Second, the MIN in the mink recieves distinct fine- and coarse-caliber axons, analogous to the predominant Y-cell and minor X- and W-cell projections to cat MIN.

On the basis of their analysis of mink LGN, Guillery and Oberdorfer propose a strong similarity between the pathways of the cat and mink, suggesting that the W-cell system is relatively strongly represented in the mink.

6.3.4. Summary

It is one of the charms of comparative neuroanatomy that something is learnt whether a comparison shows a close similarity of properties between species, or a total contrast. The present discussion of the "parallel" organization of the LGN in different mammals (which has attempted to deal with all species for which there are significant data) indicates that the LGN is in each case organized to relay the activities of different functional groups of ganglion cells to the visual cortex (or other sites) separately and in parallel. In none of the species examined is there substantial mixing of those activities within the LGN. On the other hand, the degree of differentiation of the resulting functional groups of relay cells clearly varies, from the very obvious segregation apparent in primate LGN to the more diffuse situation in the rat. That variation can be related to the development of certain aspects of visual behavior in the different species, so that the degree of development of parallel processing in the LGN may prove a significant index of the development of visual behavior.

6.4. QUALIFICATIONS TO THE PARALLEL PROCESSING MODEL OF THE LGN

The parallel processing model of geniculate function proposes that geniculate relay cells comprise a number of groups, each of which is relaying the activity of a different class of retinal ganglion cells. Central to the model is the proposition that there is little excitatory convergence at the retinogeniculate synapse of the activity of different functional classes of retinal ganglion cells, e.g., that (in the cat) the activities of Y-, X-, and W-class ganglion cells project to different groups of relay cells.

The evidence supporting this proposition was traced out at the beginning of this chapter. A number of authors have drawn attention to evidence of "mixing," particularly of X- and Y-cell activity in cat LGN. Cleland *et al.* (1971) mention a minority of relay cells that appeared by their double-recording technique to be driven by both X- and Y-class ganglion cells. Hoffmann *et al.* (1972) mention that only 5 out of 184 relay cells were not "predominantly X- or Y-class" and encountered two cells with apparently two orthodromic latencies for spike discharge after stimulation of the optic chiasm, one latency characteristic of Y-cell input, one of X-cell input. Stone and Dreher (1973) noted a small number of relay cells with mixed X- and Y-class properties. Cleland *et al.* (1976) present further evidence of synaptic convergence of X and Y activity of single relay cells. Wilson *et al.* (1976) reported that 6% of the cells they encountered in laminae A, A1, and C had mixed X and Y properties. Sherman *et al.*(1975a) also reported that a minority (3/52 or 6%) of relay cells in the dLGN of the tree shrew had mixed X and Y properties. More recently, Bullier and Norton (1977) have described a rather larger proportion of relay cells in cat dLGN (15/68 or 22%) that they termed *IM cells* because they were intermediate in latency, receptive field size, and visual response properties to X and Y cells. They comment that "the homogeneity of the response characteristics of IM cells lead us to believe that they may constitute a separate group" of relay cells. In their subsequent (1979a,b) comparison of the receptive fields of geniculate relay cells with those of retinal ganglion cells, however, they concluded, after a detailed parametric analysis, that the "intermediate" cells were probably X cells; indeed, their study provided valuable evidence of variation of properties within the X-cell group.

In monkey dLGN, Sherman *et al.* (1976) and Dreher *et al.* (1976) reported very few mixed-property cells. The cells encountered in the parvocellular laminae were unambiguously X-like, those in the magnocellular Y-like. Sherman and co-workers mentioned that 3 out of a sample of 59 cells were "unclassified" but these may have been encountered in interlaminar regions. [Schiller and Malpeli (1978) reported less distinctive differences between parvo- and magnocellular cells. I argued in Chapter 3, Section 3.1.2, that this indistinctness is a result of the typological approach to cell classification that these authors followed.] Even in the rat, the difficulty experienced by Hale *et al.* (1979; see Section 6.3) in delineating strong functional groups of relay cells in rat LGN may not stem from excitatory convergence at the retinogeniculate synapse. Indeed, clear evidence of the parallel (noncovergent) organization of rat dLGN was obtained in 1967 (by Noda and Iwama, 1967), and subsequent evidence makes the wiring of rat dLGN seem as specific as in the cat. It is possible, as Hale and co-workers suggest, that the lack of distinct receptive field classes among rat dLGN relay cells results from a lack of specialization among retinal ganglion cells rather than excitatory convergence of different cell classes in the dLGN.

To summarize, it seems clear that a certain amount of excitatory convergence of different ganglion cell classes does occur at the retinogeniculate synapse. It seems to occur to a lesser degree in the monkey, where X- and Y-class relay cells are in different laminae, than in the cat, where the two cell classes intermingle in the same laminae. Even in the cat it seems to occur only in a minority of relay cells.

6.5. THE HYPOTHALAMUS

The use of autoradiographic techniques by Moore and Lenn (1972) in the rat, by Hendrickson *et al.* (1972) in the rat, guinea pig, rabbit, cat, and monkey, by Moore (1973) in the American marsupial opossum and in the hedgehog, tree shrew, cat, and several primates, and by a number of subsequent workers in these and other species such as the Australian marsupial possum (Pearson *et al.*, 1977), has established a direct retinohypothalamic projection as a feature general among mammmals. The retinohypothalamic axons appear to arise as collaterals of optic nerve axons and terminate in the suprachiasmatic nucleus of the hypothalamus. Their function is refreshingly easy to describe. Ablation studies (e.g., Moore and Eichler, 1972; Moore and Klein, 1974) have shown that the retinal input to the hypothalamus provides a strong visual input to the control of circadian rhythms.

Mason *et al.* (1977), using a cobalt precipitation technique, provided direct evidence that in the rat, the retinohypothalamic axons are collaterals of optic nerve axons. Interestingly, they observed that the collaterals appeared to arise from fine-caliber axons, suggesting that a W-like system of ganglion cells might be involved. Another indication that the axons involved might be of fine caliber was the circumstance that the projection could not be clearly demonstrated with fiber-degeneration techniques (see Hendrickson *et al.*, 1972, for a brief review) but was readily apparent with autoradiographic techniques based on the axonal transport of radioactive amino acids. It was the latter technique that Hickey and Guillery (1974) used to demonstrate details of the W-cell projections to the C laminae of cat dLGN. It seems better able to show the projections of fine-caliber axons than techniques based on axonal degeneration. It may be the case, however, that while the collaterals that reach the suprachiasmatic nucleus are themselves fine, some may arise from quite coarse axons. Millhouse (1977) reported, from a study of Golgi-impregnated sections of rat brain, that the collaterals of optic tract axons that reach the suprachiasmatic nucleus arise from coarse as well as fine optic tract axons (Fig. 6.26). Subsequently, Pickard (1980) has reported that the ganglion cells that appear, by retrograde HRP transport techniques, to project to the hypothalamus in the golden hamster are large in soma diameter, and presumably give rise to coarse axons. At this stage, it is not clear whether a particular functional group of ganglion cells provides the retinal input to the control of circadian rhythms by way of the hypothalamus.

Very recently, Riley *et al.* (1981) have provided evidence of a second retinohypo-

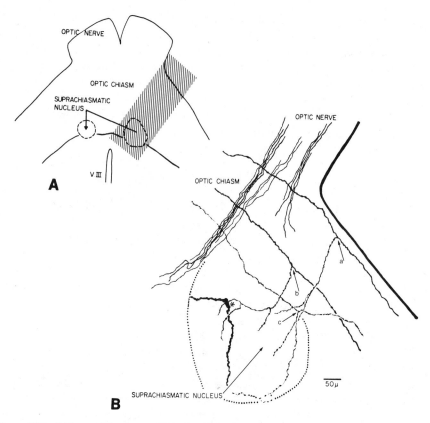

Figure 6.26. Evidence of the axon caliber of retinal afferents to the suprachiasmatic nucleus of the rat (Millhouse, 1977). (A) The relationship between the suprachiasmatic nuclei and the optic chiasm as seen in a 150 μm thick horizontal section. The right side of the section is more dorsal than the left, accounting for the difference in size between the two nuclei. The shaded area indicates the region illustrated in (B). (B) Axons of varying caliber in the chiasm and tract give off collaterals (a, b, c) that terminate in the suprachiasmatic nucleus. The asterisk denotes the soma of a suprachiasmatic neuron. From a 25-day-old rat. [Reproduced with kind permission of Elsevier/North-Holland Biomedical Press.]

thalamic projection in the rat. Cells in the lateral hypothalamic area extend dendrites into the optic tract, and those dendrites receive synapses from optic tract axons. The functional significance of this pathway is not understood.

6.6. THE PULVINAR: EVIDENCE FOR AN EXTRAGENICULATE W-CELL RELAY

Campos-Ortega *et al.* (1970) reported evidence of a weak but direct retinal input to the inferior pulvinar of two primates, the rhesus monkey and baboon, but not in the squirrel monkey or bush baby. Hassler (1966) had earlier raised the possibility of a similar projection in man, after noting the small size of the inferior pulvinar in cases of congenital anophthalmia. Because of the well-established projections of the pulvinar to the peristriate cortex, both reports raised the possibility that retinal activity is relayed to the visual cortex through the pulvinar quite directly.

Although these reports in primates have yet to be confirmed with more recent techniques, evidence of a retinal input to the pulvinar in other species has been provided by several subsequent studies. For example, the electrophysiological study of Ohno *et al.* (1975) provided evidence of monosynaptic input to neurons in the pulvinar of the tree shrew, via slow-conducting (1.5–2.3 m/sec) optic nerve axons. The autoradiographic study of Hubel (1975) has shown radioactive labeling in tree shrew pulvinar, following injections of labeled acids into the eyeball; and the EM–degeneration study of Somogyi *et al.* (1981) provides evidence of retinal afferents synapsing onto cells of tree shrew pulvinar. The autoradiographic studies of Berman and Jones (1977), Berson and Graybiel (1978), and Leventhal *et al.* (1979, 1980) have shown an accumulation of radioactivity over the most lateral part of the pulvinar of the cat following injections of radioactive amino acids into the eye, indicating a termination there of optic nerve axons. Berman and Jones describe the retinopulvinar projection as bilateral, forming a dorsoventrally oriented sheet interrupted by bundles of axons emerging from the internal capsule. Hedreen (1969) had previously noted evidence of a retinopulvinar projection in the cat, and Rockel *et al.* (1972) reported comparable observations in the marsupial brush-tailed possum. Itoh *et al.* (1979) used the anterograde transport of HRP from the eye to confirm the bilateral termination of retinal afferents in the pulvinar regions of the cat thalamus.

None of these results provided evidence of whether retinopulvinar axons are of large or fine caliber, although the infrequency with which they were observed with degeneration techniques was very suggestive. The recent studies of Itoh *et al.* (1979), Kawamura *et al.* (1979), Guillery *et al.* (1980), Leventhal *et al.* (1980; see Fig. 6.27), Mason (1981), Leventhal, Rodieck, and Dreher (personal communication) and Lee *et al.* (1982) have made it possible to relate the retinopulvinar projection in the cat to the W/X/Y organization of the visual pathway.

Taken together, these studies indicate:

1. The projection to the pulvinar may be (as earlier workers had suggested) part of a direct retino-pulvinar-cortical relay, since the cells in the retinal-recipient zone (RRZ) of the pulvinar send axons to area 19, the lateral suprasylvian area and area 21a.
2. The projection from the retina to the RRZ of the pulvinar is topographically organized, and the cells of the RRZ are relatively small, about the size of W-class relay cells in the parvocellular C laminae of the dLGN.
3. The ganglion cells that project to the RRZ have fine axons and appear to be medium-soma cells (termed ϵ *cells* by Leventhal and co-workers. They have medium-sized somas, but differ from β-class ganglion cells in the following ways:

   a. They are widely scattered over the retina and do not tend to concentrate at the area centralis.

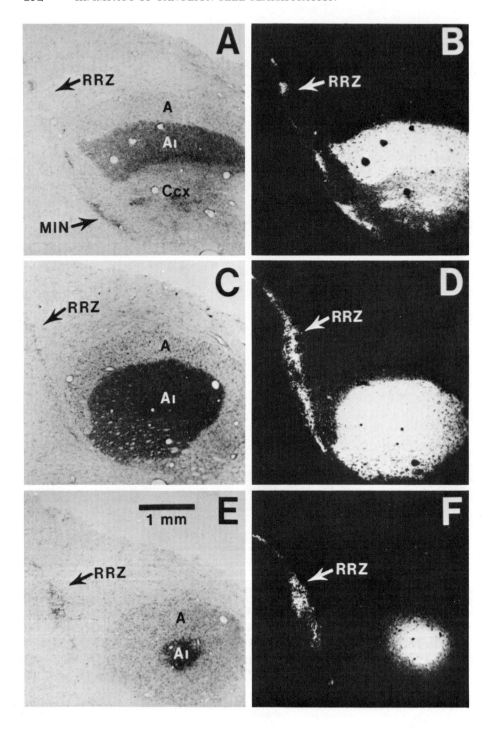

    b. They show no centroperipheral gradient in soma size.

    c. They show no nasotemporal difference in soma size.

    d. Their dendrites are wide-spreading and loosely branched, very distinct from those of β cells, and they have finer-caliber axons than β cells.

    e. They resemble in morphology many of the ganglion cells that project to the C laminae of the dLGN, a known target of retinal W cells.

4. Some cells in the RRZ have receptive field properties characteristic of retinal W cells (Mason, 1981).

The RRZ of cat pulvinar may thus be a site of an extrageniculate relay of W-cell activity to the visual cortex. Indeed, given its close apposition to the LGN and its possible relay function, the RRZ may come to be viewed as part of the LGN (in which case this discussion of the RRZ should be part of Section 6.1). Mason (1979) and Guillery *et al.* (1980) both argue in this direction, from their work in the cat. The generality of these findings to other species has yet to be tested.

Figure 6.27. Evidence of retinal projection to cat pulvinar (from Leventhal *et al.,* 1980). A and B, C and D, and E and F are pairs of photomicrographs of three successively more anteriorly located coronal sections of the LGN and neighboring thalamic nuclei of the cat. A, C, and E are light-field views; B, D, and F are corresponding dark-field views. Autoradiographic label appears dark in A, C, and E and bright in B, D, and F. Tritiated proline was injected into the eye ipsilateral to this nucleus 2 days before death, producing label in layer A1, layer C1, in the MIN, and in the RRZ of the pulvinar nucleus. [Reproduced with kind permission of the *Journal of Comparative Neurology.*]

On the Understanding of the Visual Centers of the Midbrain

7

7.1. THE MIDBRAIN/FOREBRAIN DIVISION OF THE VISUAL PATHWAYS: BY BRANCHING OR GROUPING OF GANGLION CELLS?

The axons of retinal ganglion cells reach several midbrain centers, in addition to the diencephalic centers discussed in Chapter 6; and there has long been great interest in the relationship between the retinomesencephalic and the retinodiencephalic pathways. These two pathways are formed by a major division of the optic tract, a division that raises several fundamental issues: Are the two pathways formed by branches of the same axons (and hence by the same ganglion cells) or are different groups of axons (and hence of ganglion cells) involved? If different, in what ways? My own reading of the literature [see also Giolli and Towns' (1980) review of axonal branching in the visual system] suggests no single answer to these questions. In many species, for example the rat, ganglion cells of all major functional groups project to both mid- and forebrain centers by means of branching axons; while in other species, such as the monkey, there is a substantial degree of grouping, whereby different functional groups of ganglion cells project separately to mid- or forebrain. In short, there is considerable variation between species in the relative prominence of branching and grouping of ganglion cells in forming this major division of the visual pathway.

    The occurrence and relative prominence of these two mechanisms of dividing the visual pathway have been debated for many decades, and interpretations have varied with technique as well as species. Von Gudden (1886) was one of the first to notice the different calibers of axons in different branches of the visual pathways. and to suggest differences in the pattern of connections of thick and thin fibers. He concluded from

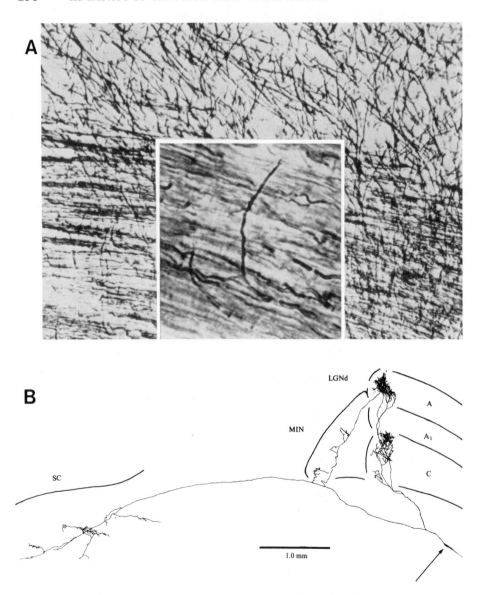

Figure 7.1. (A) Pyridine–silver impregnation of axons in cat optic tract branching to reach the LGN as well as, presumably, the midbrain. Magnification is 60×. The inset (magnification 1100×) shows a single collateral branching from an optic tract fiber. [From Barris *et al.* (1935). Reproduced with kind permission of the *Journal of Comparative Neurology*.] (B) Bowling and Michael's (1980) reconstruction of the axon of a Y-class ganglion cell, traced with intracellularly injected HRP. Note that the axon branches to reach both the SC and the LGN, and also branches to reach a number of subcomponents of the LGN, the MIN and layers A and C. The thinner axons of X- and W-class cells have not yet been demonstrated in this way. [Reprinted by permission from *Nature (London)*, Vol. 286; copyright 1980 by Macmillan Journals Limited.]

ablation experiments that in the rabbit. the small fibers of the optic tract pass to the superior colliculus (SC), while the larger fibers pass more anteriorly and subserve the pupil reflex. Later studies in other species suggested different conclusions, however. Barris *et al.* (1935), for example, concluded from pyridine–silver preparations (Fig. 7.1A) that in the cat, both small and large fibers branch to reach both fore- and midbrain (Fig.

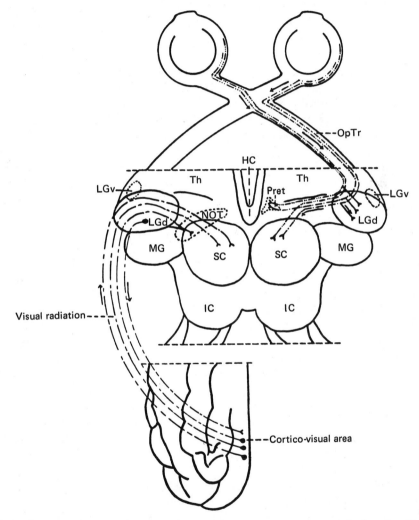

Figure 7.2. Diagrammatic representation by Barris *et al.* (1935) of retinocollicular and corticocollicular pathways in the cat. On the left are shown projections from the visual cortex to the midbrain and on the right, retinocollicular pathways. Note the branching shown in the retinocollicular pathways. These authors suggested that fibers of all calibers branch. [Reproduced with kind permission of the *Journal of Comparative Neurology.*]

7.2), suggesting no difference in the classes of ganglion cells involved. Bishop and O'Leary (1938), on the other hand, concluded from physiological experiments, also in the cat, that fast-conducting (and therefore thick) and slow-conducting (thin) axons distribute very differently between fore- and midbrain, the thicker axons passing to the LGN, the thinner axons to the SC. More recent techniques have made easier the compelling demonstration of axonal branching; for example, Bowling and Michael (1980) traced the multiple branching of single optic tract axons, by filling them with the tracer enzyme HRP (Fig. 7.1B), and Geisert (1976, 1980) and Illing (1980) have used dual retrograde tracing techniques to show that individual cells project to more than one target nucleus.

The thrust of the present chapter is to argue that present understanding of these and other issues concerning the organization of the SC is best expressed in terms of the functional groupings of ganglion cells and their patterns of projections to fore- and midbrain centers. The retinal projection to the SC (the most prominent visual center of the midbrain) is analyzed in these terms, for the species for which data are available, in Sections 7.2–7.4. Because of the variation apparent between species, the analysis is presented for particular species separately. The smaller target nuclei, including the pretectal nuclei, the nucleus of the optic tract, and the terminal nuclei of the accessory optic tract, are discussed in Section 7.5.

7.2. THE SUPERIOR COLLICULUS OF THE CAT

The SC of the cat receives strong input from two major classes of retinal ganglion cells, the Y and W cells (Fig. 7.3). It receives little if any input from X-class ganglion cells. Present evidence suggests that the Y-cell afferents to the SC are branches of axons also projecting to the LGN, while the W-cell input comes from a subgroup of W cells that does not branch substantially, at least between the LGN and the SC. Thus, in the cat, the midbrain/forebrain division of the visual pathways involves both branching, particularly of Y-cell axons, and grouping, a major subclass of W cells projecting predominantly to the midbrain, X cells predominantly to the forebrain.

Within the SC, Y- and W-cell afferents terminate in different laminae of the superficial gray stratum of the SC, and appear to activate different subgroups of SC cells; thus, as in the LGN, different groups of neurons appear to process the activities of different classes of ganglion cells in parallel. Moreover, the effect of visual deprivation on the SC can also be interpreted in terms of its effect on a particular subgroup of retinal ganglion cells (Y cells). On the other hand, the SC also receives a strong cortical input, part of which originates from retinal Y cells, and is relayed through area 17; but the cortical input originates from all of areas 17, 18, and 19, and appears to converge on individual neurons of the SC. Overall, therefore, the cortical input to the SC does not seem organized to process activities of different ganglion cell classes in parallel. So far, the Y/X/W analysis of collicular circuitry has not been extended to the deeper layers or onward projections of the SC.

Some of the above ideas are summarized in Fig. 7.3; their emergence is traced in the following paragraphs.

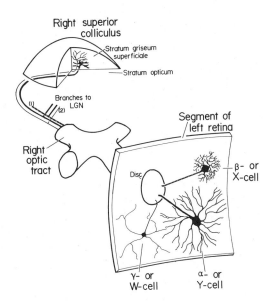

Figure 7.3. Diagram of the projections of cat Y-, X-, and W-type ganglion cells to the SC. The three cell types are represented as their morphological types, α, β, and small-soma γ cells. The α and γ cells are known to project the superficial gray stratum to the SC with the γ cells terminating more superficially. A minority of β cells are believed to send a branch to the midbrain (1), but their site of termination is not yet established. Some small-soma γ cells do project to the LGN, perhaps by a branching of their axon (2).

The diagram is incomplete in several ways, for example in not showing the relatively small projection from temporal retina to the ipsilateral SC, or the relatively small number of medium-soma γ cells known to project to the SC.

7.2.1. Early Evidence: Conduction Velocity Analysis of the Retinocollicular Projections

Early evidence of "parallel processing" in the retinal projection to the SC of the cat came from the reports of Bishop and O'Leary (1938, 1940, 1942), discussed in Chapter 1 (Section 1.1). Bishop and O'Leary (1940) concluded that slow-conducting (in modern terms, W) axons project strongly to the SC but, somewhat surprisingly in retrospect, failed to note evidence of fast-fiber (Y) input. Bishop and O'Leary (1942) and Bishop and Clare (1955) expanded these observations, noting the presence in the optic tract of a medium-velocity (X) group of axons that projects to the LGN, and confirming the strong projection of slow-conducting axons to the SC. These reports also provided the first evidence that the medium-sized (t_2, X-cell) axon group does not project to the SC. Perhaps the earliest indication that the large-axon (t_1, Y-cell) group of ganglion cells does project to the SC was Bishop and Clare's (1955) tentative conclusion that "a few" large fibers reach the SC, and synapse there. Soon after, Chang (1956) showed (see Fig. 1.4) that t_1 axons can be readily excited antidromically from the SC, indicating that they project there in considerable numbers; while t_2 (X) axons could not be similarly excited.

Since Chang's report, it has been widely accepted that t_2 (X) axons do not project to the SC in substantial numbers [but see discussion below of Wässle and Illing's (1980) study]. Surprisingly, however, there was persisting uncertainty about both the fast-axon (Y) and the slow-axon (W) components of the retinal input to cat SC. Altman and Malis (1962), for example, provided strong physiological evidence of a slow-axon (about 5 m/sec) input to the SC from the optic nerve, but did not detect the fast-axon input. On the other hand, Marchiafava and Pepeu (1966) observed only the fast-conducting group of afferents, and Hayashi *et al.* (1967) also found evidence of the fast- but not the slow-axon input. Hayashi and co-workers reported evidence of an intermediate-velocity group of afferents to the SC; the potentials attributed to this group were relatively weak and inconstant, however, and the authors suggested that relatively few such axons are involved. In an important study, however, which seems in retrospect to have laid the basis for present understanding, Bishop *et al.* (1969) confirmed the projection to the SC of both fast- (t_1) and slow-conducting axons (which they term T_3 *axons*) (Fig. 7.4).

Bishop and co-workers made two new and significant points about T_3 axons: that they are very numerous (up to 60% of the population) and that their principal projection is to the SC. At first, these conclusions made little impact on the work of other laboratories, principally because, even as late as 1969, the presence of slower-than-t_2 axons in cat optic nerve was not widely accepted, and the idea that they comprise a substantial proportion of the nerve seemed quite radical. Both points have been substantially confirmed, of course, but during the 1950s and 1960s, the emphasis of physiological studies of vision shifted to the study of the receptive fields of single cells recorded with microelectrodes; and those microelectrodes had a strong sampling bias against small somas and thin axons. The reality of the fine-fiber component of cat optic nerve was accepted only when their distinct receptive field properties were described; and, more generally, the value of conduction velocity in the analysis of the circuitry of the SC became widely accepted only when receptive field correlates were established for the conduction velocity groups in the retinocollicular pathway. These correlations were established by Hoffmann (1972, 1973).

7.2.2. Receptive Field Correlates: Hoffmann's Three-Channel Model of the Retinocollicular Projection

The description of functional subgroupings among cat retinal ganglion cells (Chapter 2) stimulated the analysis of collicular receptive fields in terms of those groupings. Hoffmann (1972, 1973) presented evidence that two groups of ganglion cells, the Y cells and the newly described W group cells project strongly to the SC where they activate

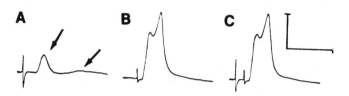

Figure 7.4. Field potentials recorded from the optic nerve following stimulation of: (A) SC; (B) optic tract (OT); (C) OT and SC, timed so that the T_1 inflections are superimposed. Calibrations represent 1 mV and 3 msec. The arrows in (A) indicate T_1 (left) and the weaker T_3 (right) deflections. The T_3 deflection in (A) was the basis for the authors' (Bishop *et al.,* 1969) conclusion that as many as 60% of optic nerve fibers have very slow-conducting axons. It seemed unlikely, but has proved correct. For comparison, the two large, early peaks in (B) represent activity of T_1 and T_2 axons. (B) and (C) show that the T_1 potentials in (A) and (B) do not summate when both stimuli are given. [Reproduced with kind permission of Academic Press.]

different subgroups of collicular cells. His report thus confirmed earlier evidence (presented above) that the fastest (t_1) and slowest (T_3) axons of the optic nerve project to the SC, while the medium-velocity (t_2) axons do not. The receptive fields of Y- and W-input SC cells reflected their retinal input, most notably in their large receptive fields, and in the responsiveness of Y-input cells to fast stimulus motion. Hoffmann concluded (Fig. 7.5) that 89% of the cells of the strata griseum superficiale and opticum of the SC that received direct retinal input were driven by retinal W cells; i.e., that about 90% of the retinal input to the SC was found to be derived from this newly recognized and numerous class of small-soma, slow-axon ganglion cells. Hoffmann noted evidence of a third strong retinal input to the colliculus, which he called *fast indirect;* "fast" because it was mediated by the fast-axon Y cells, "indirect" because it involved a relay through the dLGN to the visual cortex, and thence to the SC.

To summarize, Hoffmann argued for three projections from the retina to the SC (Fig. 7.6): *fast direct,* involving the direct projection of retinal Y cells; *slow direct,* involving the direct projection of retinal W cells; and *fast indirect,* involving the relay of Y-cell activity to the SC via the dLGN and visual cortex. Other pathways by which retinal activity might reach the SC were not ruled out, however.

Many aspects of Hoffmann's model have been well corroborated. For example, the following studies provide support for his suggestion of a "fast-indirect" pathway from the retina to the SC, via the visual cortex: Palmer and Rosenquist (1974) confirmed Hayashi's (1969; Fig. 7.7) description of cortical cells that could be driven antidromically from the SC, i.e., that projected to the SC. Palmer and Rosenquist noted, moreover, that the cells are located in layer V, and presumably correspond to the corticofugal cells located in layer V by Toyama *et al.* (1969; Fig. 7.8) and to the corticocollicular cells located in layer V by morphological analysis (Holländer, 1974; Fig. 7.9). These cells have receptive field properties (large size, responsiveness to fast-moving stimuli) that indicate that they receive input from geniculate Y cells. It still has not been demonstrated that these cells are monosynaptically activated from the visual radiation, but their dendrites extend into laminae (e.g., 4 and 6) where geniculate axons terminate. They could well correspond to the Y-input "complex" cells described by Stone and Hoffmann (1971), and hence provide a basis for Hoffmann's "fast-indirect" pathway from the retina to the SC. McIlwain and Fields (1971) and McIlwain (1973, 1977a) showed a powerful, excitatory influence of the visual cortex on cells in the SC.

Similarly, the following studies corroborate Hoffmann's description of direct fast (Y) and slow (W) retinocollicular projections: First, in physiological studies of cat ganglion cells, Fukuda and Stone (1974) and Cleland and Levick (1974b) confirmed by antidromic activation techniques that Y- and W-class ganglion cells project in numbers to the SC. Second, in anatomical studies, Kelly and Gilbert (1975) used the retrograde transport of HRP to show that the SC receives input from two distinct classes of retinal ganglion cells; the majority have the small somas characteristic of W cells, while a minority have large somas characteristic of Y cells. Only a small proportion of ganglion cells with medium-sized somas were shown to project to the SC. This pattern was described independently by Magalhaes-Castro *et al.* (1975) and has been confirmed in several later studies, such as Stone *et al.* (1980), Stone and Keens (1980), Wässle and

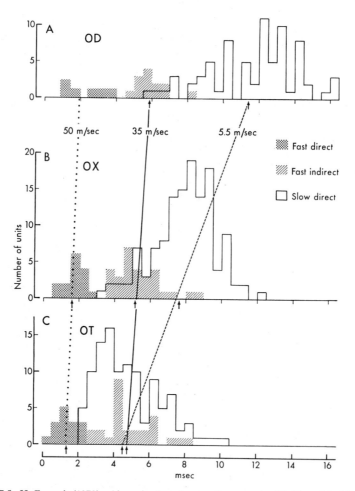

Figure 7.5. Hoffmann's (1973) evidence for three input pathways to the SC. The original legend reads:
"Diagram showing the latency-frequency distribution of units in the superior colliculus after stimuli applied to optic disc, chiasma, and tract. *A:* latencies after optic disc stimuli (OD); *B:* latencies after optic chiasma stimuli (OX); *C:* latencies after optic tract stimuli (OT). Diagrams spaced according to the conduction distances between OD, OX, and OT. Vertical axes: number of units; horizontal axes: latency in milliseconds (msec). Units are classified according to their conduction velocity as fast direct (CV > 35 m/sec), fast indirect (CV > 35 m/sec but OX latency longer than 3.0 msec), and slow direct (CV < 15 m/sec). Dark bars: fast direct = Y-fiber input; striped bars left to right: fast indirect = Y-axons in the optic tract and optic radiation, complex cell axons from visual cortex to superior colliculus. Open bars: slow direct = W-fiber input. Arrows under horizontal axes indicate the mean latencies for the three groups of units from the three different stimulation sites. The latency shifts for the different retinofugal fiber groups involved when stimulation sites are changed are indicated by the dotted line for the Y-axons in the fast direct pathway, by the solid line for the Y-axons in the fast indirect pathway, and by the broken line for the W-axons in the slow direct pathway." [Reproduced from the *Journal of Neurophysiology* with kind permission of the American Physiological Society.]

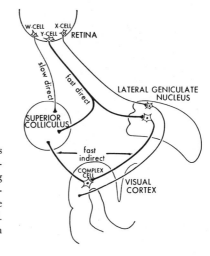

Figure 7.6. Hoffmann's (1973) model of three pathways from the retina to the SC in the cat. W cells with slow-conducting axons project directly to the SC; fast-conducting Y-cell axons branch to both the LGN and the SC; and Y-class relay cells in the LGN activate "complex" cells in the visual cortex, which send their axons to the SC. [Reproduced from the *Journal of Neurophysiology* with kind permission of the American Physiological Society.]

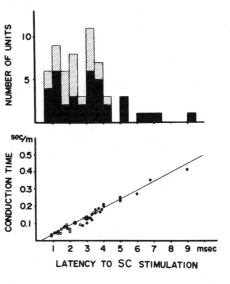

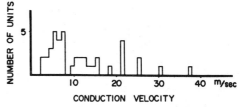

Figure 7.7. Frequency/latency histograms for cells in cat visual cortex that project to the SC. The latency is the delay before the cortical cell responded with an antidromic action spike, following a brief electrical stimulus delivered to the SC. Latencies vary from 1 to 9 m/sec. Short latencies indicate fast-conducting (up to 40 m/sec) axons, long latencies slow-conducting axons (down to 4 m/sec). Stippled bars indicate normal cats; black bars, cats with the optic nerve sectioned. [From Hayashi (1969). Reproduced with kind permission of *Vision Research*.]

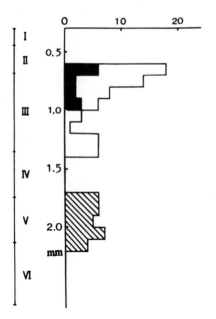

Figure 7.8. Frequency/depth histograms for cells recorded intracellularly in area 18 of cat visual cortex. The Roman numerals indicate the cortical layers I–VI. Commissural cells (filled bars) were found in layer II, association cells (open bars) in layers II and III, and corticofugal cells (most of which projected to the SC; hatched bars) were found in layer V. [From Toyama *et al.* (1969*a*). Reproduced with kind permission of Elsevier/ North-Holland Biomedical Press.]

Illing (1980), and Leventhal, Rodieck, and Dreher (personal communication). Indeed, Leventhal and co-workers' ability to observe the dendritic trees of many cells enabled them to support Hoffmann's conclusion that X cells do not project to the SC. They observed that, although a minority of medium-soma ganglion cells do project to the SC, their dendritic trees did not resemble those of β (X) cells.

Kelly and Gilbert's (1975) observations allowed them to comment on the suggestion of Hayashi *et al.* (1967), Hoffmann (1973), Singer and Bedworth (1973) and Fukuda and Stone (1974) that individual retinal Y cells project to both the SC and the dLGN by means of a branching axon. Because virtually all large (Y-class) ganglion cells appeared to project to the dLGN, and about 50% to the SC as well, Kelly and Gilbert concluded that the axons of about half the Y-cell population must branch to reach both the SC and the dLGN. Subsequently, however, two studies have provided evidence that the proportion of Y cells branching to reach both the SC and the dLGN may be nearly 100%. First, Bowling and Michael (1980) injected HRP into nine optic tract axons that they also identified physiologically as the axons of retinal Y cells Fig. 7.1B). They were thick axons (2- to 5-μm diameter, including myelin), as expected, and all nine branched to reach both the SC and the dLGN. Second, Wässle and Illing (1980) reported a much higher proportion (over 90%) of large ganglion cells labeled following HRP injection into the SC.

Little direct evidence is available as to whether the thin (W-cell) axons bifurcate to reach both the SC and the LGN. However, recent evidence (Chapter 2, Section

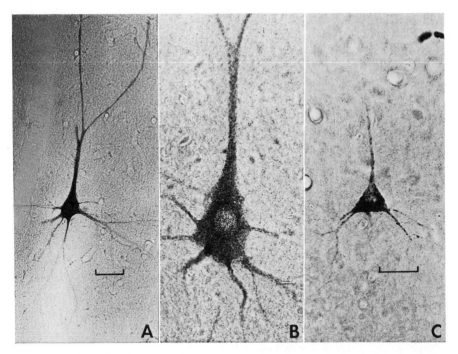

Figure 7.9. Large pyramidal cells in cat cerebral cortex labeled with HRP injected into the SC. All are in layer V, the cell in (A) in the lateral bank of the suprasylvian sulcus, the cell in (B) in the medial bank of the same sulcus, the cell in (C) in striate cortex, area 17. The scales represent 25 μm; the scale for (B) is the same as for (C). [From Hollander (1974). Reproduced with kind permission of Springer-Verlag.]

2.3.3.7) that the W cells projecting to the SC and LGN differ considerably in soma size (the former having small somas, the latter medium-sized somas) suggests fairly clearly that most W-cell axons do not branch to reach both mid- and forebrain (or at least to reach both the SC and the LGN).

Concerning the termination of Y- and W-cell axons in the SC, Hoffmann (1973) noted that the target cells of Y and W axons are distinctly segregated within the stratum griseum superficiale (SGS). Specifically, the W afferents terminate, and the W-input cells are located, superficially (i.e., near the surface of the SC), while the Y afferents terminate and the Y-input cells are found more deeply, near the stratum opticum. McIlwain and Lufkin (1976) confirmed this depth segregation (shown diagrammatically in Fig. 7.3) and presented field potential evidence that W-input cells are found in the SGS and Y-input cells at the junction of the SGS and the adjacent stratum opticum. They also noted that Y-input cells occur least frequently near the anterior pole of the SC, where the area centralis is represented, and increase in relative numbers more posteriorly, where peripheral retina is represented. This trend in Y-cell representation matches observed variations with eccentricity in the proportion of Y cells among retinal ganglion cells (Chapter 2, Section 2.3.3.5). More recently, Itoh *et al.* (1981) have confirmed this dorsoventral segregation or Y- and W-cell terminations by making layer-specific injections of HRP into the SGS. Superficial injections labeled a fairly homogeneous population of small ganglion cells, while injections into the deeper part of the SGS labeled only large ganglion cells (Fig. 7.10).

McIlwain (1978a) described an unusual type of action potential, a "juxtazonal potential," which is recorded in the most superficial layer of the SGS and is apparently an action potential in the dendritic processes of collicular cells receiving W-cell input (Fig. 7.11).

Finally, McIlwain (1978b) and Kawamura *et al.* (1980) used HRP-tracing techniques to locate the cells of the SC that project to thalamic nuclei. Of particular interest is Kawamura and co-workers' observation that cells that project to the tectal-recipient zone of the posterolateral thalamic nuclei are located in the deeper part of the SGS.

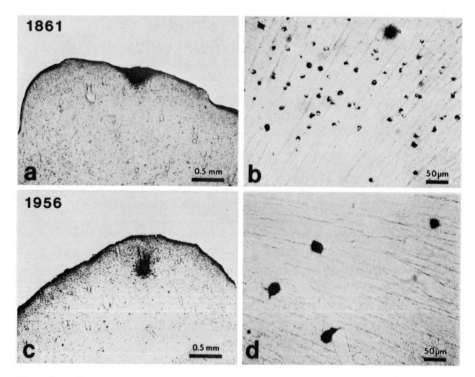

Figure 7.10. The photomicrographs in (a) and (c) show coronal sections through the SC of the cat. The dark patches represent the regions into which HRP was injected. In (a), the injection site is very superficial, and the ganglion cells found labeled with HRP are shown in (b); they are almost all relatively small, with a few medium-soma cells present. In (c). the injection site is in the deeper part of the SGS, and the labeled ganglion cells (d) are relatively large. These results (from Itoh *et al.*, 1981) confirm conclusions from physiological studies that γ (W) cells terminate in the superficial part of the SGS, and α (Y) cells in the deeper part. [Reproduced with kind permission of Elsevier/North-Holland Biomedical Press.]

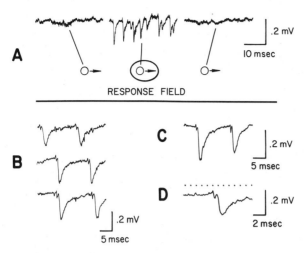

Figure 7.11. Evidence of an unusual form of postsynaptic potential associated with the W-cell input to cat SC. The potentials were recorded in the superficial part of the SGS, just deep to the stratum zonale, in the area of termination of W-cell axons, and were termed *juxtazonal potentials* (JZPs) by McIlwain (1978*a*). These potentials show the all-or-none behavior typical of action spikes but, unlike spikes recorded extra-cellularly from the region of cell somas, are entirely negative in polarity and several milliseconds in duration. McIlwain suggested that JZPs are postsynaptic events generated by W-cell afferents to the SC, i.e., that they occur in the processes (presumably dendrites) of collicular cells that receive W-cell input. [Reproduced from the *Journal of Neurophysiology* with kind permission of the American Physiological Society.]

while cells that project to the LGN are located in the upper parts of the SGS. The result implies that cells with distinct retinal input have distinct axonal projections, confirming the idea that different groups of SC cells are processing different components of retinal activity separately, and in parallel. Kawamura and co-workers note that the LGN-projecting cells lie in the region of termination of W-cell afferents to the SC, and project quite discretely to the ventral laminae of the LGN, where the W-input cells of this nucleus are located. Kawamura and co-workers comment on the remarkably specific connection between two centers that both receive strong W-cell input.

In summary, therefore, considerable evidence is available to support Hoffmann's (1972, 1973) division of retinal input to the SC into fast-direct (Y), slow-direct (W), and fast-indirect (Y, via visual cortex) components, and for the existence of groups of collicular cells processing these activities separately and in parallel. Further support for the model comes from its usefulness in understanding the influence of the visual cortex and of visual deprivation on the physiological properties of SC cells, as summarized in the following section.

7.2.3. The Influence of the Visual Cortex and of Visual Deprivation on the Superior Colliculus

Hoffmann and Sherman (1974, 1975) studied the influence on the organization of the SC of monocular and binocular visual deprivation (effected by eyelid suture). They concluded *inter alia* (see also Chapter 10, Section 10.4) that monocular deprivation has little effect on the Y-direct and W-direct pathways to the SC, but severely weakens the Y-indirect pathway. Binocular visual deprivation seemed to leave unaffected the direct-W component of the retinal input to the SC, but to weaken both the direct-Y and the indirect-Y components. Hoffmann and Sherman suggested that the weakening caused by deprivation in the Y-indirect pathway occurs in the geniculocortical part of the pathway, in which a specific effect of monocular deprivation on Y-cell activity had been previously described (Sherman *et al.*, 1972; Chapter 10, Section 10.2).

Hoffmann and Sherman also noted that two receptive field properties of normal SC cells, direction-selectivity and binocularity, are markedly reduced by monocular and binocular deprivation of visual experience, and suggested that the reduction may result from a loss or weakening of the Y-indirect component of the retinocollicular pathway. Support for Hoffmann and Sherman's analysis came from the observation that the corticocollicular part of the Y-indirect pathway most probably originates. at least partially, from a group of cells located in lamina V of the visual cortex (described in the preceding section). These cells were identifed physiologically by Palmer and Rosenquist (1974), who found that the visual properties of these cells (most were direction-selective and binocular) were "precisely those which are lost in the colliculus when the influence of the cortex is removed," whether by ablation of the cortex or by visual deprivation (see, for example, Wickelgren and Sterling, 1969*a,b;* Rizzolatti *et al.*, 1970; Rosenquist and Palmer, 1971).

7.2.4. Qualifications and Limitations

Despite the success of the Y-direct/W-direct/Y-indirect analysis of the retinocollicular input, several observations have been reported that are not readily accountable in terms of it. For example, the Y-indirect pathway that Hoffmann observed to reach the SC via the visual cortex may be only the fastest part of the corticocollicular projection. Hayashi (1969; Fig. 7.7) noted, for example, that antidromic latencies of corticotectal cells to electrical stimulation of the SC were commonly short (1–2 msec) and therefore appropriate for a "fast-indirect" pathway, but in a minority of cells were considerably longer (up to 9 msec). This range of latencies was confirmed by Palmer and Rosenquist (1974) and indicates the presence of a slow-conducting component in the corticocollicular pathway, whose influence on the colliculus has yet to be described.

Furthermore, the SC receives separate projections from areas 17, 18, and 19 (Kawamura *et al.*, 1974; Gilbert and Kelly, 1975; Updyke, 1977; McIlwain, 1977*a;* Kawamura and Konno, 1979), in each case from pyramidal cells in layer V (Holländer, 1974; see Chapter 8, Section 8.1.4). Areas 17 and 18 receive Y-cell input from the retina via the dLGN, which might be selectively routed to the SC. However, area 17 also receives a strong X-cell input and a distinct W-cell input from the LGN, and area 19 receives a predominantly W-cell input (Chapter 8, Section 8.1.1), and it seems inevitable that these different thalamic inputs will prove to be reflected in the physiological properties of their collicular projections as well. Similarly, McIlwain's (1977*a*) study indicated that the conduction times to the SC varied little between corticotectal cells in areas

17, 18, and 19, and his results also showed evidence of slow-conducting corticocollicular axons that may, because of their small caliber, be more numerous than physiological studies indicate. By this analysis too, it is possible that there are "X-indirect" and "W-indirect" pathways from the retina to the SC, via the visual cortex, as well as the Y-indirect pathway suggested by Hoffmann (1973). Some morphological evidence of diversity in the corticocollicular projection comes from Kawamura and Konno's (1979) observations on the variety of soma sizes of the cortical neurons whose axons form the projection.

An additional qualification comes from McIlwain's (1977a) evidence that the corticocollicular projections from areas 17, 18, and 19 converge onto individual collicular cells. If, as argued in Chapter 8, areas 17, 18, and 19 are processing separately the activities of different groups of ganglion cells (i.e., Y, X, or W), then the corticocollicular projection may be an important site of excitatory convergence of Y-, X-, and W-cell activity; it may be a component of the visual pathway in which parallel processing breaks down. Even in this projection, however, some evidence of parallel processing is apparent. Using anatomical techniques, Kawamura *et al.* (1974) and Updyke (1977) reported considerable similarities between the patterns of projection of areas 17, 18, and 19 to the SC. Axons of corticocollicular cells in all three areas terminate in laminae I, II2, and II3 of the SGS. This similarity of termination agrees well with McIlwain's (1977a) observation of physiological convergence of the projections of areas 17, 18, and 19 onto single collicular cells. However, both Kawamura *et al.* (1974) and McIlwain (1977a) also noted evidence that area 18 projects to a relatively deep level within the SGS. This matches, perhaps imperfectly, the dorsoventral segregation of W- and Y-cell terminals within the SGS, since in both retinocollicular and corticocollicular projections the Y-cell component terminates more deeply.

Another possible qualification to the three-channel model was raised by Cleland and Levick (1974a), who concluded from physiological experiments that small numbers of X cells project directly to the SC. Magalhaes-Castro *et al.* (1975a) also raised the possibility of a direct X-cell projection to the SC, because their HRP-tracing experiments indicated that small numbers of medium-sized ganglion cells project there. More recently, Wässle and Illing (1980) confirmed that medium-soma ganglion cells appear by HRP-tracing techniques to project to the SC, and suggested that as many as 10% of retinal X cells may contribute to the retinocollicular projection. My own judgment of the evidence is that a few if any X cells project to the SC. For example, Fukuda and Stone (1974) noted, in agreement with Cleland and Levick, that a small minority of X cells can be activated antidromically from electrodes placed in the SC; but we also noted that such activations of X cells always required strong stimuli (whereas many Y and W cells could be activated by much weaker pulses), and suggested that the axons involved may not enter the SC, but enter a nearby site, perhaps the pretectum. Conversely, Wässle and Illing's report was written before descriptions became available of medium-soma cells (Chapter 2, Section 2.1.4). Leventhal, Rodieck, and Dreher (personal communication) have subsequently observed that at least the large majority of medium-soma ganglion cells that project to the SC are γ-like (ϵ cells in their nomenclature) rather than β-like (presumably X cells). If X cells do project to the SC, they are very few in number.

7.3. THE SUPERIOR COLLICULUS OF THE MONKEY

As noted in Chapter 3 (Section 3.1), classes of ganglion cells have been described in monkey retina that seem clearly analogous to the Y/X/W classes of the cat. In several important ways, the analogy between the two species can be extended to include the retinal input to the SC.

Retinal projections to mid- and forebrain: In one substantial respect, however, the retinal input to primate SC seems distinct from that in other species investigated; viz., in the degree to which the ganglion cells that project to the SC are distinct from those projecting to the diencephalon. The first report on this issue in the monkey (Bunt *et al.*, 1975) suggested that most ganglion cells branch to reach both the LGN and the SC. This was one of the early studies to employ the retrograde transport of HRP in the primate visual system, however, and subsequent physiological studies and the HRP study of Leventhal *et al.* (1981) indicate rather that the majority of monkey ganglion cells project to the LGN or to the SC, but not to both. Thus, the X-like cells, whose morphological counterparts are the B cells of Leventhal and co-workers, appear to project to the LGN but not to the SC, and the W-like cells (presumably C and E cells) project to the SC and pretectum but not to the LGN. The exceptions may be the large-soma Y-like (A) cells. These form less than 10% of the total population; many, perhaps all, project to the magnocellular laminae of the LGN and at least some project to the pretectum and SC, perhaps by branching axons. Overall, branching of axons to reach both the SC and the LGN is more limited in the monkey than in any other species investigated.

Analysis of retinocollicular projections: In most other respects, however, the pattern of retinocollicular projections in the monkey is remarkably similar to that of the cat. The SC receives input from Y-like and W-like ganglion cells, and not from X-like cells (Schiller and Malpeli, 1977*a*; Leventhal *et al.*, 1981). Schiller and Malpeli suggest that the W-like cells (which they termed *rarely encountered*) form "a substantial proportion and probably a majority" of the ganglion cells that project to monkey SC. Confirming this, Marrocco (1978) reported that 9% of the retinal input to the SC comes from fast-axon ganglion cells and 91% from slow-conducting ($<$ 4 m/sec) afferents (Fig. 7.12). The analogy between this and Hoffmann's (1973) estimate of 11% Y-cell and 89% W-cell input to cat SC is striking.

As in the cat, available evidence suggests that Y- and W-like afferents terminate on different populations of collicular cells, laying a basis for the parallel processing of the activities of Y- and W-like ganglion cells within the SC. Thus, the receptive field properties of collicular cells are well accounted for in terms of individual cells receiving a W- or Y-cell input from the retina. For example, collicular cells are not color-sensitive and are usually phasic in their responses to flashing stimuli (Marrocco and Li, 1977; Schiller and Malpeli, 1977*a*), supporting the idea that X-like cells do not project to the SC. Among collicular cells, fast-afferent (Y-input) cells are responsive to faster velocities than slow-afferent (W-input) cells (Marrocco and Li, 1977), reflecting properties of the ganglion cells.

Finally, Finlay *et al.* (1976) have identified corticotectal cells in monkey area 17. As in the cat, they form a fairly homogeneous population of cells with "complex" properties, broadly-turned orientation specificities, and high degree of binocularity.

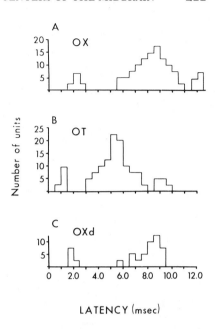

Figure 7.12. Marrocco's (1978) evidence of fast- and slow-axon inputs to the SC of the monkey. These are frequency/latency histograms of cells in monkey SC following electrical stimulation of the optic chiasm (OX) and optic tract (OT), and also of the optic chiasm of an animal with area 17 removed (OXd). (A) Short and long latency responses following OX stimulation. (B) Similar latency distribution but with latencies reduced because the stimulus site (OT) is nearer the SC. (C) Fast and slow groups in a decorticate animal. Some of the longer latency input to the SC apparent in (A) has been abolished, presumably because it traversed the visual cortex. Nevertheless, fast and slow groups remain. The fast group corresponds to the fast-axon (Y-like) group of the overall nerve, the slow group to the axons of W-like ganglion cells. [Reproduced with kind permission of Elsevier/North Holland Biomedical Press.]

However, the analogy between the cat and the monkey in the organization of the SC is not complete; in several respects, differences in organization are already apparent. For example, the influence of the visual cortex on the colliculus is, in the monkey, concentrated in its deeper layers (Schiller *et al.,* 1979) rather than on cells of the SGS, as in the cat. Again, Marrocco (1978) found no change in the afferent conduction velocities of collicular cells as a function of their depth within the SGS; whereas there is in the cat a dorsoventral separation of W- and Y-input cells in cat SC. Nevertheless, the degree of analogy between monkey and cat in the "parallel" Y-/W-cell circuitry of the SC seems considerable.

7.4. THE SUPERIOR COLLICULUS OF OTHER SPECIES

7.4.1. The Rat

Sefton (1969) investigated the conduction velocity groupings among optic tract axons projecting to the SC in the rat. She concluded (Fig. 7.13) that three conduction velocity groups are present among optic tract axons (with mean velocities of conduction of 18, 7.8, and 3.1 m/sec), and her evidence indicates that at least the majority of axons in each group bifurcate to reach both the LGN and the SC. Sumitomo *et al.* (1969*b*) and Fukuda (1977; Fig. 3.8) confirmed Sefton's observations of three conduction velocity groupings, using single-unit techniques. Similarly, the report of Fukuda *et al.* (1978), although arguing for a four-group classification of the collicular cells that receive retinal input, delineates three conduction velocity groupings among the retinocollicular axons themselves, confirming the prior studies on this point. Bunt *et al.* (1974), using an HRP-

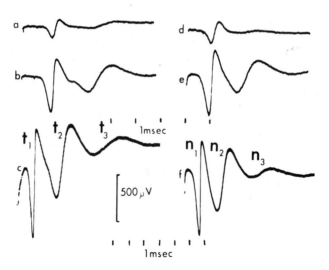

Figure 7.13. Sefton's (1969) evidence of three conduction velocity groups in rat optic nerve, each projecting to both the SC and the LGN. At the left is a trace recorded from the optic nerve following stimulation of the LGN, and showing three deflections, labeled t_1, t_2, and t_3; at the right the potential recorded following stimulation of the SC, and also showing three deflections (n_1, n_2, and n_3). [Reproduced with kind permission of *Vision Research*.]

tracing technique, have provided morphological support for the conclusion that the major conduction velocity groupings of rat optic nerve reach both the LGN and the SC, by means of axonal branching. They observed that the ganglion cells labeled from injections in the SC and LGN comprised similar spectra of soma sizes; there was no evidence, for example, of small-soma cells projecting preferentially to the SC. The more recent HRP study of Lund *et al.* (1980), however, provides evidence of a distinct pattern of central projections. They suggest that the largest of the three size groupings of ganglion cells [such as those distinguished by Fukuda (1977)] projects to both the LGN and the SC, by means of a branching axon; but that the medium-soma cells project predominantly to the LGN; while the small-soma cells project predominantly to the SC. It is probably fair to comment that these data are in need of confirmation; but they raise the possibility that the retinofugal pathways of the cat and rat are more alike than previously thought.

Fukuda and co-workers' (1978) report has provided some evidence of a laminar separation of the collicular cells that receive input from the different conduction velocity groups. They found that, as in the cat, the cells in the most superficial part of the SGS receive input from the slowest-conducting retinal axons (< 4 m/sec). The cells in the deeper part of the SGS receive input from an intermediate-velocity (4–8 m/sec) group, while cells in the stratum opticum receive input from fast-conducting (> 8 m/sec) axons. In addition, they noted that a number of cells in the stratum opticum receive slow-axon input. Corresponding cells have not been described in the cat; these cells comprise the fourth of Fukuda and co-workers' four groups of collicular cells.

Fukuda and Iwama (1978) investigated the receptive field correlates of the afferent conduction velocities and laminar position of cells in rat SC. Although they described several correlates of these parameters, they were not able to relate many of the receptive field properties of the collicular cells to those of the ganglion cells providing their input. This is partly because the receptive field properties of rat ganglion cells have yet to be related to their axonal velocities (Chapter 3, Section 3.2.2); and may partly result from the relatively poor differentiation of certain ganglion cell groupings in this species, an issue discussed by Hale *et al.* (1979) (see Chapter 3, Section 3.2.2).

By comparison with the cat, the organization of the retinocollicular pathway in the rat is distinguished by the branching of all ganglion cell groups to reach both the SC and the LGN; yet there is also a partial but considerable similarity between the rat and the cat in the laminar distributions of collicular cells within the SGS. The laminar segregation of collicular input according to conduction velocity suggests that, to some degree at least, the rat SC processes input from different ganglion cell classes in parallel.

7.4.2. The Rabbit

As for the rat, limited information is available concerning afferent conduction velocities and receptive field properties of cells in rabbit SC. Concerning conduction velocity groupings, Takahashi *et al.* (1977) described two groupings in the retinocollicular projection (mean velocities 15 and 6 m/sec), with the slower group numerically predominant, as in the cat and monkey. Molotchnikoff *et al.* (1979), on the other hand, suggested the presence of three conduction velocity groupings in rabbit optic nerve [thus confirming Bishop (1933)], with velocities of 21–23, 15–17, and 10–12 m/sec, all reaching the SC. Further, they suggested that the same groups of axons also reach the LGN, although they were unable to conclude whether individual axons branch to reach both targets. Recently, Vaney *et al.* (1981) concluded from an HRP transport study that virtually all ganglion cells in rabbit retina send an axon to the SC, supporting the view that projections to any other site must be achieved by axonal branching. Concerning receptive field properties, Masland *et al.* (1971) and Hughes (1971) described a considerable variety of receptive field types among collicular cells, most being accountable for in terms of the properties of retinal ganglion cells. However. correlations between afferent conduction velocity and receptive field properties of collicular cells have yet to be established.

7.4.3. The Hamster

Rhoades and Chalupa (1979) have provided a detailed report of properties of cells in hamster SC, concluding that in several ways the organization of hamster SC is distinct from that of cat SC. In particular, they found evidence that all three of the major conduction velocity groupings in hamster optic nerve and tract project to the SC, rather than two groups, as in the cat. Second, Rhoades and Chalupa found no analog for the "fast-indirect" pathway from the retina to the SC observed in the cat (Hoffmann, 1973). Third, no evidence was found that afferents of different conduction velocity terminate in distinct laminae of the SC, as they do in the cat. Fourth, there was no evidence of a correlation between retinal eccentricity and the properties of afferent fibers, as described for the cat (McIlwain and Lufkin, 1976). Finally, Rhoades and Chalupa found only one significant receptive field correlate of afferent conduction velocity. The proportion of cells showing direction-selectivity was higher (90%) among slow-afferent cells than among fast-afferent cells (41%). Rhoades and Chalupa suggest, however, that this property is determined by the influence of the visual cortex on the tectum, rather than by the direct retinotectal input.

Subsequently, Chalupa and Thompson (1980) employed the retrograde transport of HRP injected into hamster SC to test some of these conclusions. They found that ganglion cells of all sizes appear to project to the SC, confirming the idea that, as in the rat (another rodent) and rabbit, all major functional groups of ganglion cells project there.

We still have little evidence as to whether the activities of these different ganglion cell groupings are processed by separate populations of SC cells, providing a basis for a "parallel processing" model of hamster SC. It can be pointed out, however, that Rhoades and Chalupa did not report evidence of direct excitatory convergence of fast- and slow-conducting afferents on individual SC cells so that some correlations between retinal and tectal properties may one day be established.

7.4.4. The Opossum

Rapaport and Wilson (1983) used the HRP-tracing technique to determine the central projections of different classes of ganglion cells in the retina of the North American opossum *(Didelphis virginiana)*. Previously (Rapaport *et al.,* 1981*b;* Rowe *et al.,* 1981), they had distinguished four groups among opossum ganglion cells, basing the classification both on soma size and on axonal conduction velocity. Rapaport and Wilson provide evidence that all four of these groups project to the SC in considerable numbers; as in the rat, hamster, and rabbit, the SC in the opossum appears to process the activities of all the recognized ganglion cell classes. Rapaport and Wilson note further that cells in three of these four groups also project to the LGN; it is not yet clear whether the same cells reach both the LGN and the SC, by means of a branching axon. The group of cells with the smallest somas and slowest-conducting axons appears not to project to the LGN.

7.4.5. Summary

Unfortunately for the comparison attempted in the above paragraphs, our information on the groupings present among the ganglion cells of different species is limited; and understanding of their central projections is similarly incomplete. Two generalizations may perhaps be drawn from the above discussion. First, there is considerable interspecies variation in the groups of ganglion cells that project to the midbrain. Second, that variability particularly concerns ganglion cells with somas of medium size. In all species studied, the large-soma, fast-axon class of ganglion cells (such as cat α or Y cells) seems to project to the SC, and to the LGN as well; in some species at least, individual large cells reach both sites by means of a branching axon. Similarly, the small-soma cells project in all species to the SC. The medium-soma cells seem to vary more in their projections; in the hamster and opossum, for example, they reach the SC and LGN in broadly equal numbers; but in the monkey and cat, medium-soma cells project predominantly to the LGN. It is difficult to speculate on the significance of these differences and similarities between species, except to note that the medium-soma ganglion cell group "withdraws" from the SC in species (cat and monkey) in which the group is most specialized. That is, it is in the monkey and cat that the X-like group of ganglion cells is most distinct, numerous, and functionally important.

7.5. OTHER MIDBRAIN CENTERS

7.5.1. The Pretectal Nuclei and the Nucleus of the Optic Tract

It has been established for some time that the nuclei that comprise the "pretectal" area of cat midbrain [the pretectal nucleus and the nucleus of the optic tract (Garey and Powell, 1968)] receive direct input from the retina. Several lines of evidence suggest that this input is derived principally from W-class retinal ganglion cells. First, Hoffmann and Schoppmann (1975) recorded from single cells in the nucleus of the optic tract (NOT) of the cat, showing that many NOT cells receive direct excitation from axons with conduction velocities in the 5–10 m/sec range, which is characteristic of W cells. Further several receptive field properties of the NOT cells were also compatible with their receiving W-cell input. Hoffmann *et al.* (1976) showed subsequently that many NOT cells project to the inferior olive, providing a source of visual input to the cerebellum.) Second, Schoppmann and Hoffmann (1979) provided a similar analysis for the pretectal nucleus of the cat. Analyzing the responses of individual pretectal cells to electrical stimulation of the optic nerve and chiasm, they distinguished two broad groups of cells, those receiving fast- (Y-) axon input, and those receiving slow- (< 15 m/sec, W-) axon input. Moreover, certain receptive field properties correlated well with afferent type; for example, the cells responsive to high velocities of movement of a visual stimulus were those receiving Y-cell input. Overall, the W-input cells were several times more common than Y-input cells; as in the SC, the W component of retinal afferents seems numerically predominant. Third, Magoun and Ranson (1935) showed that the pathway of the pupilloconstrictor reflex traverses the pretectum. More recently, Hultborn *et al.* (1978) traced this pathway again, confirming that it traverses the pretectum, apparently synapsing there. They concluded that the afferent fibers to the pretectum are axons of retinal ganglion cells, with velocities less than 10 m/sec (i.e., in the W-cell range). Fourth, the retinal ganglion cells that seem most suited to provide the retinal input to the pupilloconstrictor reflex, the "luminance units" of Barlow and Levick (1969), were classified by Stone and Fukuda (1974*a*) as a variety of W cells; Cleland and Levick (1974*a*) also observed these cells and also noted that they had slow-conducting axons. Fifth, Ballas *et al.* (1981) demonstrated by retrograde transport of HRP from injections limited to the NOT, that the great majority of retinal ganglion cells projecting to the nucleus are small-soma γ cells; interestingly, these cluster in the visual streak.

Four studies provide evidence of a projection of X cells to the pretectum. First, Fukuda and Stone (1974) found evidence that as many as 20% of retinal X cells send an axon to the midbrain; but because none of the X cells in their samples were excited at low threshold by stimulation of the SC, these workers suggested that their axons might project to other midbrain centers, such as the pretectum. Second, Cleland and Levick (1974a) also traced X-cell projections to the midbrain by antidromic activation. They were able to excite rather more X cells from the pretectum than from the SC, suggesting that many terminate in the pretectum. Third, Schoppmann and Hoffmann (1979) noted that some pretectal cells responded to electrical stimulation of the optic nerve and chiasm at latencies indicating that they receive input from intermediate-velocity (X-cell) axons. However, such cells provided only a small fraction (4%) of their sample of pretectal cells, a much lower percentage than might be expected if 20% of retinal X cells terminate there. Fourth, Leventhal, Rodieck, and Dreher (personal communication) have observed that a "minority" of the ganglion cells labeled by an HRP injection into the pretectal nucleus of the cat were identifiable as β cells. It seems likely therefore that the pretectal nucleus is one midbrain region in which X-class ganglion cells terminate; but it also seems possible, even likely considering the substantial proportion of X cells that may be involved, that X cells terminate in other, as yet unidentified midbrain nuclei.

7.5.2. Nuclei of the Accessory Optic Tract and Nucleus Raphe Dorsalis

Some evidence is available concerning the caliber of axons entering the accessory optic tract and passing to its terminal nuclei, which are arranged around the basis pedunculi of the midbrain. In several species, such as the cat (Hayhow, 1959; Garey and Powell, 1968), rat (Hayhow et al., 1960), and possum (Hayhow, 1966), the fibers are of fine caliber, suggesting that they are the axons of W cells. Farmer and Rodieck (1983) have recently used a HRP tracing technique to demonstrate the morphology of ganglion cells projecting to the terminal nuclei of the accessory optic tract in the cat. Their results indicate that the cells concerned are γ-cells with small to medium sized somas and thin axons (although Farmer and Rodieck considered the cells they observed to be distinct from γ-cells).

In other species, it remains possible that the axons reaching the terminal nuclei are fine branches of rather coarser parent axons; the axons reaching the suprachiasmatic nucleus in the rat have been shown, for example (Chapter 6, Section 6.5), to be fine branches of distinctly coarser axons. Indeed, in the rabbit, Oyster et al. (1980) injected HRP into one of the terminal nuclei of the accessory optic tract, and found that the ganglion cells labeled in the retina were relatively large, among the largest 20% of ganglion cells. Oyster and co-workers argue that these cells have a particular receptive field "type" (ON-center direction-selective), and may (Winfield et al., 1978) provide a fairly direct visual input to the cerebellum, via the medial terminal nucleus of the accessory optic tract.

Foote et al. (1978) have advanced evidence of a direct projection of retinal Y cells to the nucleus raphe dorsalis of the midbrain. The possibility is intriguing, since this nucleus has not previously been associated with visual function, and deserves further investigation.

On the Understanding of   8
Visual Cortex

The description of functional groupings among retinal ganglion cells and geniculate relay cells has obvious implications for the understanding of the visual cortex; for the cortex, like other visual centers, is likely to bear a strong imprint of the retina. Many of the intriguing aspects of the physiology of cells in the visual cortex, however, such as their orientation-selectivity, binocularity, and much of their susceptibility to visual deprivation, appear to be cortical in origin, and are not related in any predictable way to the functional grouping of ganglion cells. These "cortical" properties of cells in the visual cortex were described, and concepts of their mechanisms were developed, before the description of ganglion cell groupings. Perhaps as a consequence, and despite the many studies of Y-, X-, and W-cell activity in the visual cortex now available, a synthesis of the Y/X/W analysis of the visual cortex with earlier concepts of its organization has been slow to develop.

     Some of the issues involved in attempting this synthesis are discussed in Stone and co-workers' (1979) review of this problem, and in the final two sections of this chapter. The principal concern of the chapter, however, is to summarize evidence of how the cortex processes the activities of different functional groups of retinal ganglion cells. It is argued that, at least at the initial geniculocortical synapse, and in some instances throughout the circuitry of the visual cortex, Y-, X-, and W-cell activities remain substantially separate. The parallel processing "model" of the visual pathways developed in the retina, LGN, and SC can, therefore, be extended to the visual cortex. I would stress, however, that only certain aspects of the organization of the visual cortex can presently be encompassed by this analysis; these aspects are the subject of this chapter.

8.1. CAT VISUAL CORTEX: PROCESSING OF GENICULATE INPUT

8.1.1. Parallel Pathways to Different Cortical Areas

Several distinct areas of the cerebral cortex of the cat receive projections from sub-cortical visual centers. Specifically, cortical areas 17, 18, and 19 (Fig. 8.1) and part of the suprasylvian gyrus all receive input from visual components of the thalamus, such as the dLGN and the RRZ of the pulvinar; these areas will be termed the *visual cortex* of the cat. (Recent evidence [Lee *et al.* (1982)] that the RRZ projects to area 21a may lead to this, and perhaps other areas, also being considered visual cortex in the present sense). Area 19 and the suprasylvian gyrus also receive input from the tectal-recipient zone of the pulvinar; this is part of the "second" visual pathway discussed in Section 8.3. It has gradually become apparent that much of the rather complex pattern of projections from the thalamus to the visual cortex is better understood when analyzed in terms of the functional groupings of ganglion cells (Fig. 8.2.).

Early studies: Evidence that the different areas of cat visual cortex are connected "in parallel" to thalamic nuclei emerged very early in the physiological analysis of the visual cortex. Talbot (1942) noted a second representation of the visual fields, immediately lateral to the representation found in area 17, and suggested that it corresponds to area 18 recognized anatomically. Moreover, he noted that this "lateral representation" (which he termed *V2*) survives cautery and narcosis of, and the application of convulsant drugs to area 17, suggesting that it is not simply an association visual area processing information relayed to it from area 17. Marshall *et al.* (1943) further noted that an area of the lateral bank of the suprasylvian gyrus was also responsive to photic stimuli, and to electrical stimulation of the optic nerve. The responsiveness survived removal of both striate cortices, and the separation of the "total tectal region . . . from the geniculothalamic region by a sagittal knife cut," suggesting that this area, too, receives an independent visual input from the thalamus. Subsequently, Doty (1958) confirmed the independence of the visual responsiveness of area 18, and provided evidence that the LGN projects to area 18, and Vastola (1961) provided physiological evidence that the lateral suprasylvian area receives a direct visual input from the thalamus.

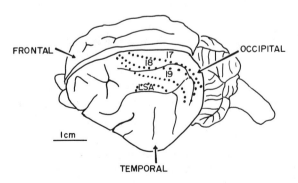

Figure 8.1. The locations of areas 17, 18, 19 and the lateral suprasylvian area (LSA) of cat visual cortex. Following Tusa *et al.* (1979) and Leventhal, Dreher, and Hale (personal communication). The brain is seen from a dorsolateral aspect: frontal, occipital, and temporal lobes of the hemispheres are labeled. Only portions of each area can be seen; a substantial part of each area forms the walls of sulci formed by folding of the cerebral cortex.

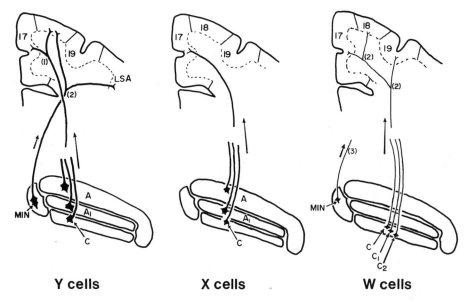

Figure 8.2. Projections of the LGN to areas 17, 18, and 19 in the cat. Separate diagrams show the projections of Y-, X-, and W-classes of relay cells of the LGN to areas 17, 18, and 19. Y cells project from laminae A, A1, and C to areas 17 and 18, and from the MIN to area 18 and the LSA; X cells project from laminae A, A1, and C to area 17; W cells project from laminae C, C1, and C2 to area 17, possibly to area 18, and to area 19.

Viewed conversely, area 17 receives all the X-cell projection plus substantial input from Y and W cells as well; the input to area 18 is derived predominantly from Y cells; and the input to area 19 is derived principally from W cells.

Notes: (1) The occurrence of this branching has been questioned (LeVay and Ferster, 1977) and supported (Geisert, 1976, 1980). It was proposed by Stone and Dreher (1973), following Garey and Powell (1967). (2) These branchings are speculative; it may be that different populations of cells project to these different areas. (3) The W cells in the MIN project to area 17 (Leventhal, 1979), area 19 (Leventhal *et al.*, 1980) and area 21a (Lee *et al.*, 1982).

Two important studies of the 1960s provided further information about the visual areas of cat cerebral cortex. Otsuka and Hassler (1962) described the cytoarchitecture and extent of areas 17, 18, and 19, and Hubel and Wiesel (1965) described, on the basis of single-cell recording techniques, the topography of areas 17, 18. and 19. These studies did not investigate the existence of parallel thalamocortical pathways; indeed, Hubel and Wiesel suggested that areas 18 and 19 receive their principal input from area 17 and function in series with it. However, several studies in the late 1960s established with considerable clarity and detail the projection of visual areas of the thalamus (principally the LGN but also certain neighboring nuclei of the posterior thalamus) to areas 17, 18, and 19 and to the lateral suprasylvian gyrus.

Degeneration studies of thalamocortical projections: Glickstein *et al.* (1967) used anterograde degeneration techniques to gain evidence that the LGN projects to areas 17 and 18 and the lateral wall of the suprasylvian sulcus (LSA), and also provided evidence that different components of the LGN project to different cortical areas, the medial interlaminar (MIN) component projecting particularly strongly to area 18. Wilson and Cragg (1967) used similar techniques and also concluded that the LGN projects to areas 17 and 18, and to the LSA. Using retrograde degeneration techniques, Garey and Powell (1967), Burrows and Hayhow (1971), and Niimi and Sprague (1970) reached similar conclusions. Garey and Powell (1967) suggested cell size correlates of three parts of the geniculocortical projection, the smaller cells of the A laminae projecting to area 17, the larger cells by a branching axon to both areas 17 and 18, (Chapter 6, Fig. 6.17) while the MIN and central interlaminar components of the LGN were considered to project to area 19. Burrows and Hayhow (1971) argued that the MIN projects only to area 18, while the adjacent posterior nucleus projects to area 19 and the LSA. Niimi and Sprague (1970) confirmed several of these patterns, showing also that area 19 receives input from the central interlaminar nucleus of the LGN and the medial part of the posterior nucleus [a nuclear group just medial to the LGN, distinguished by Rioch (1929)]. Niimi and Sprague concluded that because the LGN projects to a number of cortical areas, the designation of areas 17, 18, and 19 as V1, V2, and V3 (as had been suggested) was misleading, since these areas are connected in parallel to the thalamus.

Rossignol and Colonnier (1971) and Garey and Powell (1971) found further evidence that the LGN of the cat projects to areas 17, 18, and 19 and the LSA, and added significant new observations of the caliber of axons reaching different areas. They noted that both fine and coarse axons reach area 17 from the LGN, with some fine fibers extending up to layer 1 (these patterns anticipate the W-, X-, and Y-cell inputs to area 17 described in Section 8.1.3); the fibers to area 18 appeared predominantly coarse (anticipating subsequent evidence that area 18 receives Y-cell input); and the fibres to area 19 were predominantly fine in caliber (anticipating recent evidence of a substantial W-cell input to area 19). These results clearly match Garey and Powell's (1967) evidence mentioned above of a differential projection of medium and large LGN relay cells to areas 17 and 18.

In summary, therefore, these studies provided (1) strong and consistent evidence that the LGN of the cat sends parallel projections to several areas of the cerebral cortex and (2) early evidence that the parallel projections differ in morphology, thus raising the possibility of a functional differentiation among them.

Subsequent work on geniculocortical projections in the cat has concentrated on analyzing that differentiation and has followed two distinct lines, physiological and neuroanatomical; this work is summarized in the following two subsections.

Physiological studies: Early physiological evidence of differences in the geniculate projections to areas 17, 18, and 19 is apparent from a comparison of the intracellular recordings of Watanabe *et al.* (1966) in area 17, of Toyama and Matsunami (1968) in area 18, and of Ohno *et al.* (1970) in area 19. All three reports describe intracellular events following stimulation of the optic radiation, optic chiasm or tract. Watanabe and co-workers reported some range in the latencies of excitatory postsynaptic potentials (EPSPs) evoked in cortical cells by an afferent volley and a bimodality in the latency distribution of inhibitory postsynaptic potentials (IPSPs). They suggested that there are two latency groups among the cortical cell responses they observed, which may reflect conduction velocity groupings among the afferents to area 17. The suggestion has been substantially corroborated and expanded by evidence that Y-, X-, and W-type relay cells project to area 17 (see next section). Toyama and Matsunami (1968), by contrast, reported only a single latency grouping of EPSPs in area 18 cells, with uniformly short latencies (e.g., 0.94 msec following a stimulus to the optic radiation), and the geniculate input to area 18 has been shown subsequently to derive principally from geniculate Y cells, which have fast-conducting axons. Ohno *et al.* (1970) reported that area 19 neurons respond to electrical stimuli delivered to the afferent visual pathways with relatively long-latency EPSPs (e.g., 2 msec or greater following a stimulus to the optic chiasm). They suggested that the geniculate input to area 19 is mediated by relatively slow-conducting axons, shown subsequently to derive from geniculate W cells.

Several subsequent studies combined analyses of the response latencies of cortical cells, following electrical stimulation of the visual pathway, with receptive field analyses. Stone and Dreher (1973), for example, studied both areas 17 and 18 and reported that some area 17 cells seemed to be directly excited by fast-conducting afferents (Fig. 8.3) and, confirming Hoffmann and Stone (1971), that their receptive field properties differ characteristically (see next section). The fast-afferent cells resembled retinal Y cells in having large receptive fields and responding to fast-moving stimulus patterns, while slow-afferent cells resembled X cells in having smaller receptive fields and responding only to slower-moving stimuli. By contrast, area 18 cells seemed all to receive fast-conducting, Y-cell afferents. Consistent with that idea, Riva Sanseverino *et al.* (1973) reported a consistent trend for cells in area 18 to respond better to fast-moving visual

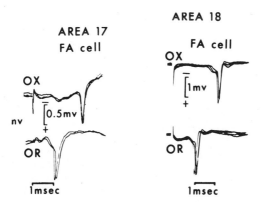

Figure 8.3. Evidence of Y-cell input to areas 17 and 18. Oscilloscope traces show the responses of cells in areas 17 and 18 to stimulation of the optic chiasm (OX) and optic radiation (OR). For the cells shown, the latencies to both stimuli are both short (approximately 1 msec for OR and 2 msec for OX), indicating that the cells are monosynaptically activated by fast-conducting, Y-cell afferents. The cells were therefore referred to as FA (fast-afferent) cells. [From Stone and Dreher (1973). Reproduced from the *Journal of Neurophysiology* with kind permission of the American Physiological Society.]

stimuli than cells in area 17. This conclusion was corroborated by Tretter *et al.* (1975) and Orban and Callens (1978), who examined the responsiveness of neurons in areas 17 and 18 of the cat, and concluded that some of the differences between neurons of the two areas resulted from their different geniculate inputs (X cells predominating in the input to area 17, Y cells to area 18). Again, Harvey (1980) also reported that the neurons he studied in area 18 received visual input from fast-conducting (Y-cell) axons. Recently, Kimura *et al.* (1980) and Dreher *et al.* (1980) have extended this analysis to area 19, both groups concluding that the main geniculate input to this area comes from W cells of the retina and LGN. Their studies followed Ohno and co-workers' (1970) evidence of a slow-conducting geniculate projection to area 19. Some of the latency observations from Dreher and co-workers' study are shown in Fig. 8.4.

Finally, four studies (Dreher and Cottee, 1975; Donaldson and Nash, 1975; Sherk, 1978; Kimura *et al.*, 1980) have tested the idea that areas 17, 18, and 19 receive parallel inputs from the thalamus by examining the effect on areas 18 and 19 of ablating area

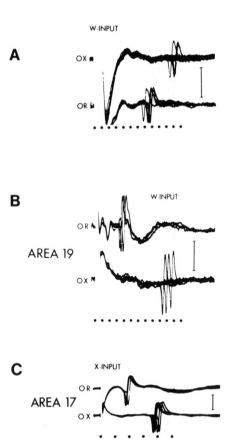

Figure 8.4. Evidence of X-cell input to area 17 and of W-cell input to areas 17 and 19. Action spikes of single cells in area 17 and 19 of cat visual cortex were observed following electrical stimulation of the optic chiasm (OX) and optic radiation (OR). The dots represent millisecond intervals, and the vertical bars represent 2 mV, with positive down. (A) Spike responses of a cell in area 17. The OX and OR latencies are relatively long (10 and 6.5 msec, respectively), indicating that the cell received its direct (monosynaptic) input from slow-conducting (W-cell) afferents. (B) Responses of a cell in area 19. The OR and OX latencies are also relatively long (approximately 3 and 9 msec, respectively), indicating that the cell received monosynaptic W-cell input. (C) Responses of a cell in area 17. The OR latency (about 1.8 msec and the OX latency (3.6 msec) suggest that the cell received direct (monosynaptic) X-cell input.

In all cases, the receptive field properties of the cell supported the interpretation of its input based on OX and OR latencies. [From Dreher *et al.* (1980). Reproduced from the *Journal of Neurophysiology* with kind permission of the American Physiological Society.]

17. The first three groups of workers studied the properties of area 18 cells after destruction or cooling of area 17. Dreher and Cottee noted that

> . . . there were no significant changes after ablation in the receptive field sizes of area 18 cells or in the proportion of orientation and directionally selective cells. Furthermore, the sharpness of orientation selectivity was not altered. We did not observe changes in proportion of cells with hypercomplex receptive field properties. Width and contrast sensitivity were also not significantly altered by the lesions.

Both Donaldson and Nash and Dreher and Cottee noted some increase in the proportion of unresponsive cells in area 18 after destruction of 17, but Dreher and Cottee noted that even this effect was eliminated when area 17 was destroyed by the less drastic technique of cautery of its superficial blood vessels. Sherk observed a drop in the peak responsiveness of area 18 cells after cooling of area 17, but little loss of stimulus specificities.

Donaldson and Nash concluded that area 17 may impose a degree of stimulus specificity on area 18 cells. This conclusion was not shared by Dreher and Cottee or by Sherk, except in the case of the selectivity of some cells in area 18 for slow stimulus motion. Such cells are normally rare in area 18; their numbers seem further reduced by destruction of area 17. However, all three groups of workers concluded that a majority of area 18 cells remain responsive to visual stimulation after destruction of area 17, and that a good deal of their stimulus specificity persists. Conversely, Dreher and Winterkorn (1974) reported that the response properties of area 17 cells are also little affected by destruction of area 18.

Kimura *et al.* (1980) studied a limited sample of area 19 cells following ablation of area 17, and concluded that their properties seem largely unaffected by ablation.

HRP and autoradiographic studies: Modern techniques, such as the autoradiographic detection of labeled amino acids transported along axons in an anterograde direction, or the histochemical localization of antero- or retrogradely transported proteins (in particular, the enzyme HRP), have confirmed and expanded the patterns of geniculocortical projection just described.

Rosenquist *et al.* (1974) and LeVay and Gilbert (1976) injected small amounts of tritiated amino acids into different components of cat LGN, and studied the subsequent distribution of radioactivity in the cerbral cortex. That distribution is considered to be an accurate localization of the terminal branches of axons of cells located in the injected portion of the LGN. Rosenquist and co-workers reported that the "laminar" part of the LGN (i.e., all the dorsal component of the LGN except the MIN) projects to areas 17 and 18 but not to area 19 while the MIN projects to areas 18, 19, and the LSA. LeVay and Gilbert confirmed that laminae A and A1 of the LGN project to areas 17 and 18, but not 19, but found evidence that the ventrally located C laminae project to areas 17, 18, and 19 and the LSA. These authors noted that the distinctive cortical projection of cells in the C laminae implies that "the differences in their properties are maintained at early stages of the processing of visual information in the cortex."

The HRP-tracing studies of Maciewicz (1975), Geisert (1976), LeVay and Ferster (1977), and Holländer and Vanegas (1977) have already been considered in some detail in Chapter 6 (Section 6.1.7). The distinct projections of Y-, X-, and W-type relay cells are described there and have been incorporated into Fig. 8.2.

Other thalamocortical projections and behavioral studies: Two other lines of evidence suggest that in the cat, the nonstriate areas of the visual cortex (i.e., areas 18, 19, and the LSA) function to a considerable degree independently of area 17. First, there is now substantial evidence (reviewed in Section 8.3.1) that the nonstriate areas receive input from the posterior thalamic complex, part of which may be involved in relaying to the cortex visual activity from the SC. Second, a growing body of studies is providing evidence of the behavioral capabilities of animals with only the nonstriate regions of the visual cortex intact; this is reviewed in Chapter 11.

Summary: Evidence from a wide range of techniques supports the idea that the different component areas of cat visual cortex are connected "in parallel" to different components of the visual thalamus, that they process different components of the output of the retina, and that they can function largely independently of each other. An important part of present understanding of these patterns stems from the functional classification of retinal ganglion cells.

8.1.2. Parallel Organization of Area 17: Correlations between Afferent Input and Receptive Field Properties

Many studies have provided evidence that the visual input to area 17 of the cat can usefully be regarded as comprising two, and more recently three, functional "channels," originating from different groups of retinal ganglion cells, and relayed by corresponding groups of LGN relay cells; and that the activities conveyed in these "channels" are, to a considerable degree, processed in parallel within area 17 by different groups of cortical neurons.

Perhaps the earliest physiological evidence of parallel processing within the visual cortex came from Watanabe and co-workers' (1966) intracellular study of the responses of cells in cat visual cortex (the region of recording was not more closely specified) to electrical stimulation of the optic chiasm or tract. EPSPs were evoked at latencies ranging from 2 msec to 4 msec for the optic tract stimulus, and IPSPs tended to occur about 1 msec later. Watanabe and co-workers noticed evidence of two groups among the IPSP latencies and suggested that there may be two conduction velocity groups among the afferents to visual cortex. Their results imply, moreover, that these different afferents activate or inhibit different groups of cortical cells.

Hoffmann and Stone (1971) sought a correlation between the receptive field properties of cortical cells and the conduction velocity of their afferent input. We presented evidence (Fig. 8.5) that many (but not necessarily all) of the "complex" cells distinguished by Hubel and Wiesel (1962) and Pettigrew *et al.* (1968) receive monosynaptic input from fast-conducting afferents, which several workers have shown to originate from Y-type relay cells of the LGN (Chapter 6, Section 6.1.1). Conversely, the afferents to the "simple" and "hypercomplex" cells in our sample, where they could be demonstrated, appeared to be slower-conducting. Reviewing this evidence subsequently (Stone, 1972), I argued that some of the properties of the slow- and fast-input cortical cells that

Figure 8.5. Evidence of X and Y input to different "types" of cortical cells. Data from Hoffmann and Stone (1971) indicate their interpretation of the input to cortical cells with different types of receptive fields. "Complex" cells were excited by fast-conducting afferents, responding to OX stimulation at 2.0–2.5 msec and to OR stimulation at about 1 msec, while "simple" and "hypercomplex" cells were excited at longer latencies, suggesting that they receive the slower-conducting X-cell input. The original legend reads:

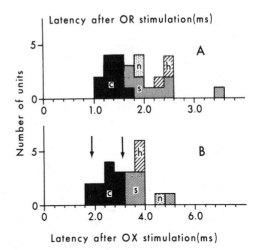

"Frequency/latency histograms for cortical units. A, shows the latency distribution of 24 units to OR stimulation; B, shows the distribution for the OX stimulus. Different fields are shaded separately: c for complex, s for simple, h for hypercomplex and n for non-oriented. Eleven of the 12 complex fields in our sample, one of the 14 simple fields, two of the 5 hypercomplex fields and one of the two non-oriented fields appear in both histograms. The total number of units was thus 33. The arrows in B represent the latency at the cortex after OX stimulation of the two most prominent conduction velocity groups of afferents." [Reproduced with kind permission of Elsevier/North-Holland Biomedical Press.]

Hoffmann and Stone described matched, respectively, the properties of X- and Y-class ganglion cells and geniculate relay cells. In particular, fast- (Y-) input cortical cells have larger receptive fields and are more responsive to fast-moving visual stimuli than slow- (X-) input cells. Further, it seemed possible that X afferents provide the input to "simple" cells and Y afferents to "complex" cells.

It is worth noting that this argument had two components: (a) that different groups of cortical cells process X- and Y-cell activity and (b) that those groups may correspond to "simple" and "complex" cells, respectively. The former suggestion, which seems to me to be central to the idea of parallel processing of visual information within area 17, has received wide and consistent support in subsequent work. On the latter point, however, evidence is conflicting; many studies support, and many deny, the view.

The following studies provide explicit support for both suggestions (a) and (b) above: Maffei and Fiorentini (1973) examined the spatial resolving power of individual geniculate relay cells and cortical cells. They reported that, as might be expected from their large receptive fields, Y-class relay cells have poorer spatial resolution than X cells and that "complex" cortical cells have poorer resolution than "simple" cells. They suggested, therefore, than X cells may provide the input to "simple" cells and Y cells to "complex" cells. Similarly, Movshon (1975) examined the velocity-selectivity of cells in area 17, and in a control sample of X- and Y-class geniculate relay cells. He concluded that, in terms of their velocity-selectivity (Fig. 8.6), many cortical cells seem to receive their input from X cells, many others from Y cells. Movshon also reported a strong correlation between receptive field "type" and speed-selectivity. The slow-selective (presumably X-input) cells were "simple" or "hypercomplex type I," while the fast-selective

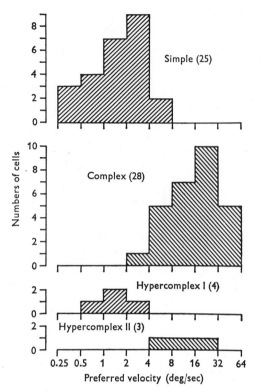

Figure 8.6. Velocity-selectivity of cortical cells. Frequency histograms, from Movshon (1975), showing the distribution of preferred velocities of various types of cortical cells. For "simple" cells and "hypercomplex type I" cells, preferred velocities were less than 4 deg/sec. For "complex" and "hypercomplex type II" cells, preferred velocities were greater than 4 deg/sec. These data support the view that "complex" cells cannot receive their input from "simple" cells, and that "simple" and "complex" cells might receive input from X- and Y-cell afferents, respectively. [Reproduced with kind permission of the *Journal of Physiology* (London).]

cells were "complex" or "hypercomplex type II." Subsequently, Movshon *et al.* (1978*a,b,c*) extended the comparison to another receptive field parameter, seeking a correlation between cortical receptive field type ("simple" vs. "complex") and the linearity of the cells' summation of influences reaching them from different parts of their receptive fields. Linearity of summation was one of the X/Y differences originally reported by Enroth-Cugell and Robson (1966; see Chapter 2); it is thus a property determined at least partly in the retina. Movshon and Tolhurst (1975) employed a drifting grating stimulus similar to that used by Enroth-Cugell and Robson (1966) and concluded that within area 17, most "simple" cells appear linear in their summation properties, thus resembling X cells, while most "complex" cells appeared grossly nonlinear, resembling Y cells. Wilson and Sherman (1976) noted a tendency for cortical "simple" cells to have X-like properties (small receptive fields, selective for slow stimulus speeds) and [confirming Stone and Dreher (1973)] to be more selective for stimulus orientation; and they showed further that such cells have more prominent inhibitory components to their receptive fields (a "cortical" property). They also noted that such cells resemble X-class relay cells of the LGN in being most frequent relatively at the area centralis (another "retinal" property). Conversely, they noted that "complex" cells have Y-like properties, are less orientation-selective, show fewer inhibitory components in their receptive fields, and are more frequent in regions of the cortex in which peripheral regions of the visual field are represented (Fig. 8.7). Riva Sanseverino *et al.* (1979) reported a partial correlation between the "simple"/"complex" classification of cells in cat areas 17 and 18 and their afferent input, which was assessed on the basis of the cells' receptive field properties as coming from either X or Y cells. Kulikowski *et al.* (1979) noted that the responses of "simple" cells in area 17 to standing contrast stimuli were relatively sustained, and therefore X-like, while those of "complex" cells were transient, and hence Y-like. Finally, Citron *et al.* (1981) have compared the responses of X- and Y-class relay cells in the LGN, and of "simple" and "complex" cells in areas 17 and 18, to computer-controlled patterns of visual stimulation designed to map the distribution of excitatory and inhibitory components of the cells' responses in both space and time. They emphasize an "extreme similarity" between the properties of X and "simple" cells, and between those of Y and "complex" cells, again corroborating the original suggestions.

However, an equally impressive string of studies can be quoted that question the correlation between X and "simple" cells, and between Y and "complex" cells. For example, Toyama *et al.* (1973) described two types of neurons recorded from the region of the area 17/18 border, which they also demonstrated to receive direct thalamic input. The two types differed in the layout of their receptive fields in a way that made them

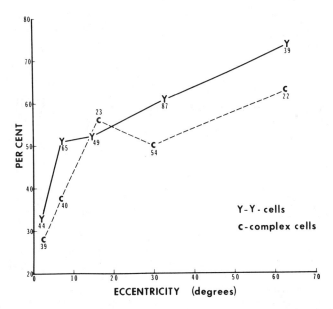

Figure 8.7. Variations with eccentricity in occurrence of cortical cell "types." Data from Wilson and Sherman (1976) show that the frequency of "complex" cells encountered in the visual cortex, relative to "simple" cells, increases with the eccentricity of receptive field position. The increase matches approximately a previously reported increase with eccentricity in the relative frequency of Y cells in the LGN. Wilson and Sherman suggest that their data support the view that "complex" cells in the visual cortex receive input from Y cells in the LGN. [Reproduced from the *Journal of Neurophysiology* with kind permission of the American Physiological Society.]

resemble "simple" and "complex" cells. Interestingly, cells of both types responded to afferent stimulation at short latencies (1.2–1.7 msec to stimulation of the LGN, 3.2–4.4 msec to stimulation of the retina; see Fig. 8.8), indicating that both receive direct activation by Y-cell afferents. Subsequently, Ikeda and Wright (1975) reported that they could not find any correlation between the sustainedness or transientness of the responses of cortical cells to standing contrast stimuli (which they took as an index of whether the cell received X- or Y-cell input) and their "simple"/"complex" typing. Both "simple" and "complex" cells, they concluded, could receive X- or Y-cell input (but not both). Similarly, Singer *et al.* (1975), in their studies of cells in the visual cortex, reported little evidence of correlates of X- or Y-cell input in terms of receptive field "types." In their analysis too, "simple" and "complex" cells could receive either X- or Y-cell input, either mono- or polysynaptically (see their Table 2). Lee *et al.* (1977) used a sophisticated and difficult double-recording technique to determine retinal input to cortical cells. They too found no correlation between X or Y input to a cell and its characterization as "simple"

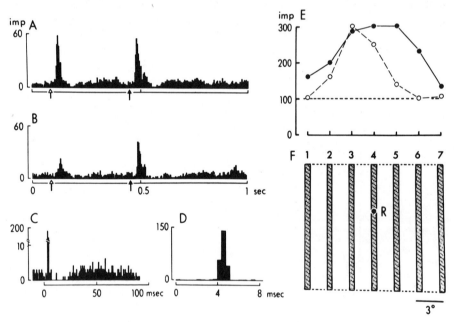

Figure 8.8. Evidence of direct geniculate input to "complex" cells. These data, from Toyama *et al.* (1973), provide evidence of fast-conducting (Y-cell) input to a "complex" cell of cat visual cortex. The cell gave ON-OFF-responses to an appropriately oriented flashing slit of light, as shown in histograms (A) and (B). (C) and (D) are histograms of the response of the cell to electrical stimulation of the retina, at time 0. Histogram (D) in particular shows that the latency of the cell's initial spike response to stimulation of the region of the retina where its receptive field was located was little more than 4 msec. This latency is consistent only with direct (one synapse in the LGN, one in the visual cortex) activation of the cell by fast-conducting afferents. (E) shows the distribution and magnitude of ON- (O) and OFF- (●) response areas across the receptive field; (F) shows the stimulus size and positions used for (A), (B), and (F). Note that ON- and OFF-regions of the field overlap, a widely used criterion for identifying "complex" cells. [Reproduced with kind permission of Elsevier/North-Holland Biomedical Press.]

or "complex." Similarly, Toyama *et al.* (1977*a*) distinguished three groups of cortical cells: "simple," "complex," and "cells with exclusively ON or OFF area." They activated cells of all three types by electrical stimulation of the appropriate region of the retina. All responded at about 4 msec, indicating that all three types receive fast-conducting input. Movshon *et al.* (1978*a,b,c*) repeated the "linearity" analysis of "simple" and "complex" cells, which had lead Movshon and Tolhurst (1975; see above) to support the X-"simple"/Y-"complex" correlation. Using different stimuli (for example, two flashing bars of variable relative position), they confirmed their earlier conclusion that "complex" cells are consistently more nonlinear than "simple" cells, but concluded that their nonlinearity results from the activities of linear "subunits" within their receptive fields. Thus, such cells could receive their input from geniculate X cells or (as Hubel and Wiesel suggested originally) from "simple" cells. [They note, however, that this analysis does not hold for "complex" cells described by others (such as Palmer and Rosenquist, 1974); they do not discuss the possibility that the subunits they propose might be retinal rather than cortical in origin, for example the subunits that Hochstein and Shapley (1976) suggested as components of Y-cell receptive fields; and they did not test their sample of cells for properties (such as velocity-selectivity and afferent conduction velocity) that would have contributed to identifying their input.] Bullier and Henry (1979*a,b,c*) and Henry *et al.* (1979) also reinvestigated the afferent input to cortical cells, relating it to their receptive field properties and laminar position. They use a distinctive scheme for the classification of cortical cells [the S, C, A, B, N scheme of Henry (1977), further discussed in Section 8.4] and reported only a partial correlation between these receptive field types and afferent input. For example, although virtually all C cells (approximately equivalent to "complex" cells) in their sample seemed to receive Y-cell input, some cells in their S ("simple") group received X input and others Y input. Similarly, Gilbert and Wiesel (1979) concluded that, among the "simple" cells they identified in layer 4 of area 17, some received input from X cells and others from Y cells. Indeed, "simple" cells receiving Y-cell input have been described in layer 4ab of area 17 (Gilbert and Wiesel, 1979), in layer 6 of area 17 (Leventhal and Hirsch, 1977), and in area 18 (Tretter *et al.,* 1975).

Clearly there is a substantial discrepancy between two sets of studies in their conclusions. What is its source? The problem has been, I believe, a tendency for investigators to rely on single physical features to identify cells as X or Y class, or as "simple" or "complex." For the X/Y distinction, some authors have relied entirely on the sustainedness or transientness of the cortical cells' responses to standing contrast stimuli; yet, as Kulikowski *et al.* (1979) note, "the sustained component [is] particularly sensitive to the state of the preparation and the level of anaesthesia." Other workers have relied entirely on tests of linearity. Among cortical cells, as discussed elsewhere (Stone, *et al.,* 1979), "simple" cells have been distinguished from "complex" by the layout of ON-, OFF-, and inhibitory areas within their receptive field, or by their velocity-selectivity. As a consequence, the identification of cortical cells as "simple" or "complex" and/or of their afferent input as from X or Y cells may have varied considerably between laboratories, despite the common terminologies. The problem has been the classification/identification of cell groups by single physical features; it is a problem of physical typology [see Tyner (1975), Henry (1977), Rowe and Stone (1977, 1979), Mann (1979), Stone *et al.*

(1979), and Chapters 2 and 4, for more detailed critiques of this approach to cell classification].

Because of these problems, several groups of workers have bypassed the question whether X- and Y-cell properties can be related to the "simple"/"complex" grouping and have asked instead: how can we recognize the cortical cells that receive and process the activities of Y, X, and (more recently) W cells? Are they separate groups of cells? or do the activities of different ganglion cell groups converge onto cortical cells? For example, Stone and Dreher's (1973) main emphasis was to compare geniculate inputs to areas 17 and 18, but we provided evidence of slow- and fast-afferent cells within area 17. In our sample of 43 area 17 cells responsive to electrical stimulation of the optic chiasm or radiation, a minority (10) had fast-conducting afferents, while the majority (30) appeared to have slow afferents. In their receptive field size and responsiveness to fast-moving stimuli, the fast-afferent cells resembled geniculate Y cells and the slow-afferent cells resembled X cells. Leventhal and Hirsch (1977) extended the correlation between "retinal" and "cortical" properties of cells in area 17 in two striking ways. First, they confirmed that cells with small receptive fields and low cutoff velocities ("retinal" properties) tend to be more highly orientation-selective (a "cortical" property) than cells with large receptive fields and high cutoff velocities. They then showed that, among populations of these two classes of cortical cells, the preferred orientations of the former tend to be grouped around the horizontal and vertical, while in the latter class the preferred orientations appear evenly distributed. Further, they noted that small-field, slow-velocity cells are more sensitive to binocular visual deprivation in the development of the property of binocularity, and less sensitive in the development of orientation-selectivity (see Chapter 10, Section 10.5).

Leventhal and Hirsch (1978) raised in discussion the possibility that "complex" cells in the superficial layers of area 17 might receive W-cell input, a suggestion that meshes well with evidence (discussed in Section 8.1.3) that the principal region of termination of W-cell afferents to area 17 is in layer 1. Dreher *et al.* (1980) and Leventhal and Hirsch (1980) provided the first physiological evidence that a proportion of cells in area 17 receive their principal excitatory drive from W cells. Dreher and co-workers' (1980) study concerned principally the W-cell input to area 19, but they also showed that in a significant proportion of area 17 cells, latencies to stimulation of both optic chiasm and optic radiation were long, suggestive of slow-conducting (i.e., W-cell) input. Further, the latency difference between the cells' responses to optic chiasm and optic radiation stimulation was also long (2.5 msec or greater), confirming that the conduction times along the different segments of the afferent pathway were all long, and similar to those reported for W-class relay cells of the LGN (Chapter 6, Section 6.1.2). They also confirmed that, at least in their responses to electrical stimulation of the optic chiasm or radiation, very few cells of areas 17, 18, or 19 (less than 2%) showed evidence of an excitatory convergence of Y-, X-, and/or W-cell afferents onto individual cortical cells. Again, W-, X-, and Y-cell afferents seem to impinge upon, and their activities seem to be processed by, different groups of cortical cells. Leventhal and Hirsch (1980) analyzed the receptive field properties of area 17 cells, much as they had in their 1978 study. They now distinguished a third group of cells (large-field cells, responsive to slow-moving visual stimuli) that, they suggest, receive their principal excitatory input from W cells. They found that the W-input cells concentrate in layers 2–4 or area 17, while X-input cells were found in layer 4 and Y-input cells in layers 5 and 6.

In short, the question of the organization of the visual cortex to process the activities of different groups of retinal ganglion cells is being successfully pursued without employing the "simple"/"complex" distinction among cortical cells. And this brings me to the substantive point of this section: Although many studies disagree quite explicitly on point (b) of the two-part question I discussed toward the beginning of this section, the same studies are in clear and consistent agreement on point (a): that different, recognizable groups of cortical cells process the activities of the major groupings of ganglion cells.

8.1.3. Parallel Organization of Area 17: Analyses of Its Lamination

The segregation of the cells of the cerebral cortex into layers according to the size and density of the neurons present (Fig. 8.9) has long provided an intriguing puzzle as to its significance. The answer to the puzzle is not simple or single, and in area 17 of the cat several correlates of lamination have now been described: for example, the morphology of cortical neurons differs between layers, as do the destinations of their axons and the sources of their afferents. The Y/X/W analysis of the visual pathway from the retina to the LGN to the cortex has added another parameter to this understanding. Three correlates of lamination are discussed here, viz., evidence that the W-, X-, and Y-cell afferents to the striate cortex terminate in different layers; evidence that the target cells of the different afferents show an associated distribution between layers; and evidence of different onward projections of the cells in different cortical layers.

Laminar terminations of W, X, and Y afferents: Perhaps the earliest evidence that different geniculate afferents terminate in distinct layers of area 17 was provided by Hubel and Wiesel's (1972) study of geniculate terminations in area 17 of the monkey, discussed in Section 8.2.1. In the cat, early evidence of distinctive patterns of cortical terminations of W-, X-, and Y-class relay cells came from autoradiographic studies of the geniculocortical pathway (Rosenquist *et al.*, 1974; LeVay and Gilbert, 1976), in which small injections of radioactive tracer were made into different components of the LGN. Rosenquist and Palmer, for example, showed that the MIN component of the LGN (which contains Y- and W-class relay cells) projects to areas 17, 18, and 19, while the laminated part of the LGN projects only to areas 17 and 18. LeVay and Gilbert showed that the A and C laminae of the LGN project differently to area 17. The A laminae (which contain X and Y cells) project to the full thickness of layer 4, to the deeper part of layer 3, and to layer 6; while the C laminae (now known to include W as well as X and Y cells) project to layer 1 as well as 4, but not to layer 6.

More detailed analyses have come from the work of Ferster and LeVay (1978) and Leventhal (1979), both studies having been undertaken since the recognition of a W-cell relay through the deeper parts of the C laminae of the LGN. Ferster and LeVay injected small amounts of HRP into the white matter under area 17 and traced the terminations of thick, thin, and medium-caliber axons. They identified the thick fibers as the axons of Y cells on the grounds that Y cells are known to have fast-conducting, and therefore presumably thick, axons. They further showed that axons of this caliber

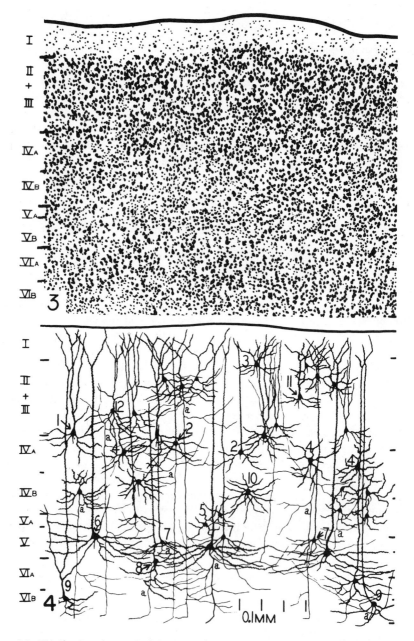

Figure 8.9. The layering of area 17 of the cat, as described by O'Leary (1941). The numbering system used by O'Leary is largely in use today, for example by Lund *et al.* (1979; Fig. 8.16). Others, for example Ferster and LeVay (1978), refer to O'Leary's IVA as layer IVab and to O'Leary's IVB as IVc (see Figs. 8.10, 8.12, and 8.15). O'Leary's original legends read:

"Fig. 3. A projection drawing of a typical locus of . . . the area striata so selected that comparison of its left and right halves demonstrates the variability in cellular distribution which can occur in a limited region."

"Fig. 4. A composite illustrating the makeup of the protoplasmic plexuses of the . . . area striata Numerical designations of cells and the significance of the stratification plan explained in the text."

[Reproduced with kind permission of the *Journal of Comparative Neurology*.]

arise from the class 1 LGN relay cells, which may be the morphological counterparts of Y cells (see Chapter 6, Section 6.1.6). On similar grounds, the medium-caliber axons were identified as the axons of X (class 2) cells, and the thin axons as the axons of W (class 4) cells. Leventhal injected small amounts of HRP into separate layers of area 17 and showed that the size and laminar location within the LGN of the cells projecting to the injection site varied characteristically with the layer injected. These two sets of results are in close agreement, and in several ways are complementary. Figures 8.10–8.12 summarize some of the findings. With respect to X and Y cells, these descriptions have been confirmed by Gilbert and Wiesel (1979) and by Bullier and Henry (1979c). The thick axons, from geniculate Y cells, terminate in the upper part of layer 4, specifically in layer 4ab, and extend into the lower part of layer 3. They also branch to provide some terminals to layer 6. The medium-caliber axons, from geniculate X cells, also branch to layer 6, but their principal termination is in the deeper part of layer 4, layer 4c, immediately deep to the main termination of Y cells. The thin axons from geniculate W cells terminate with a spectacular ramification in layer 1 and also give branches to layers 3 and 5.

Locations of target cells of X- and Y-cell afferents: Several anatomical studies have provided evidence of the layers of area 17 in which the cell bodies of the target cells of W-, X-, and Y-cell afferents are located. (Since geniculate afferents terminate principally on the dendrites of cortical cells, the somas may and in many cases do lie in layers other than those where their afferents terminate.) In particular, the Golgi analysis of Lund et al. (1979; Fig. 8.13) and the HRP-labeling study of Gilbert and Wiesel (1979) showed that the dendrites of layer 4 stellate cells are largely restricted either to the upper (Y-input) or to the lower (X-input) strata of the layer, indicating that X- and Y-cell afferents must activate different populations of layer 4 stellate cells. Moreover, Lund and co-workers' analysis and the Golgi–EM work of Peters et al. (1979) in the rat indicate that the terminals of geniculate axons in layers 1, 3, and 5 reach the dendrites of pyramidal cells in layers 2, 3, 5, and 6, and hence quite different cells from the layer 4 stellates. The W-cell input to area 17, for example, may exert its principal influence on pyramidal cells of layers 2, 3, 5, and 6, whose apical dendrites extend into layer 1. These studies have not, however, provided any indication of whether the W, X, and Y afferents reaching layers 5 and 6 converge on neurons of this layer, or reach separate groups of neurons.

Several physiological studies of the laminar distribution of target cells have been reported, though it is a common problem among them that little evidence is provided as to whether the electrode was recording from the cells' soma, axon, or dendrites. Palmer and Rosequist (1974) identified a group of "complex" cells in layer 5 that project to the SC and whose large receptive fields and responsiveness of fast-moving stimuli suggested that they receive Y-cell input. This suggestion was confirmed by Bullier and Henry's (1979c) report that a proportion of layer 5 cells, with C-type receptive fields, receive monosynaptic Y-cell input. Rosenquist and Palmer's findings were also expanded by Leventhal and Hirsch (1978), who noted that many cells in layers 5 and 6 show properties (high cutoff velocity, large receptive field size, high spontaneous activity and also high degrees of binocularity, weak orientation-selectivity, and high peak responses) that directly or indirectly suggest that they are innervated by Y cells. They also found some

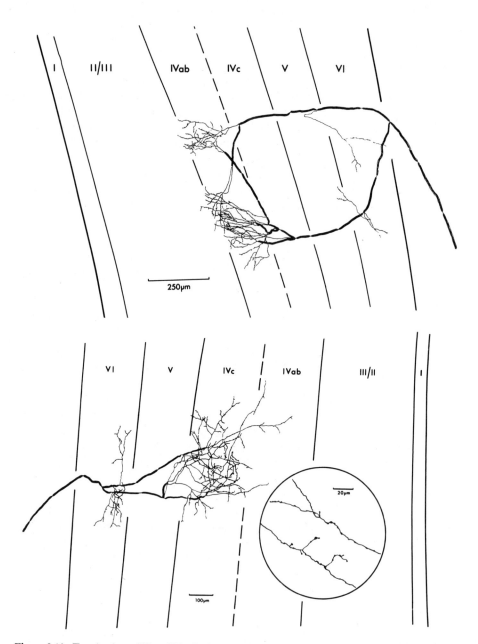

Figure 8.10. Terminations of X- and Y-cell afferents in area 17 of the cat (from Ferster and LeVay, 1978). The upper diagram shows the mode of termination in area 17 of an axon of large caliber (2-μm diameter). The axon spreads principally in layer IVab, forming two clusters that may correspond to ocular dominance columns. A few small branches extend into layer III, and the axon also gives collaterals to the upper part of layer VI. The lower diagram shows the mode of termination in area 17 of an axon of medium caliber (1.3 μm). The axon terminates principally in layer IVc, although a few small branches extend into layer IVab. This axon also gives collaterals to the upper part of layer VI. The lateral spread of its terminals is more restricted than is seen for the thicker axon at top.

Ferster and LeVay argue that large-caliber axons are the axons of Y-class geniculate relay cells, while medium-caliber axons are the axons of X cells. [Reproduced with kind permission of the *Journal of Comparative Neurology.*]

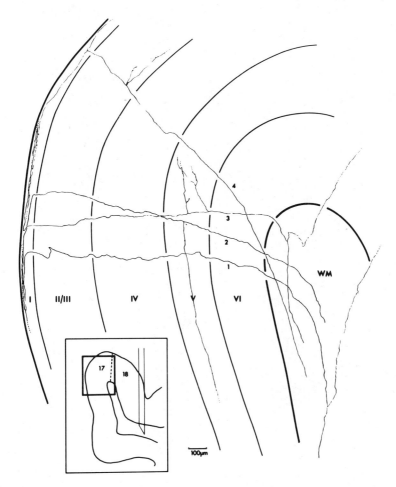

Figure 8.11. Termination in cat area 17 of putative W-cell axons (Ferster and LeVay, 1978). Axons of this group (four are shown, labeled 1–4) spread principally and widely in layer I and also give collaterals to the lower part of layer III and to layer V. They are thin (less than 1 μm in diameter), and Ferster and LeVay suggest that they are the axons of W-class relay cells located in the deep C laminae of the LGN. [Reproduced with kind permission of the *Journal of Comparative Neurology*.]

layer 5 and 6 cells with receptive field properties that suggested that the cells receive predominantly X-cell input, consistent with the anatomical evidence just described of X-cell terminations there. Conversely, they noted that many of the cells in layer 4c show properties (low cutoff velocities, small receptive fields, low spontaneous activity and also strong orientation-selectivity, relatively low degrees of binocularity, and low peak responses) that directly or indirectly indicate that they are innervated by geniculate X cells. They did not find cells in layer 4ab whose receptive field properties reflected the Y-cell input to this layer, but some of the cells that Leventhal and Hirsch localized to the deeper part of layer 3 (i.e., immediately above layer 4) showed several properties

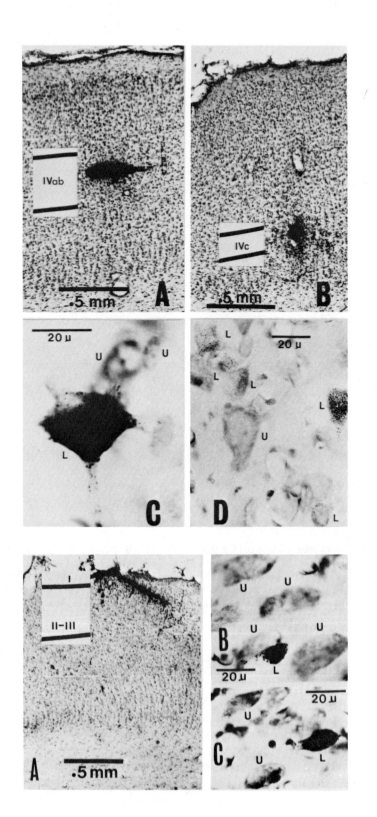

characteristic of Y input, which may have been determined by the Y-cell axons terminating principally in layer 4 (see Fig. 8.10).

Gilbert's (1977) laminar analysis of cat area 17 did not refer to possible Y/X/W
determinants of the properties of cortical cells, but many of his observations were confirmed by Leventhal and Hirsch (1978) and can be related to the different laminar
terminations of W, X, and Y cells. For example, Gilbert described a class of "special
complex" cells that may have received Y-cell input, since they had large receptive fields
and high cutoff velocities, showed high spontaneous activity, and were found in two
strata (layer 5 and the layer 3/4 border) that are in or close to layers of Y-afferent
termination.

Bullier and Henry (1979c) reported (Fig. 8.14) that, as assessed by electrical stimulation techniques, cells receiving monosynaptic X-cell input are located mainly in the
layers in which X-cell afferents terminate (4c and 6) but also in layer 5. Similarly, cells
receiving monosynaptic Y-cell input were found mainly in layers 4ab and 6, and at the
border of layers 3 and 4 (all regions where Y afferents terminate), and also in layer 5.
They did not detect or discuss the W-cell input to layers 1, 3, and 5. Within layer 4,
but not within layers 5 or 6, they recognized receptive field correlates of afferent input.
That is, cells receiving Y input had larger receptive fields than X-input cells and
responded to higher velocities of stimulus movement. Gilbert and Wiesel (1979) similarly found evidence that cells in layer 4ab (Y input) have properties distinct from cells
in layer 4c (X input).

It must be noted, however, that physiological studies have not found correlates for
all the components of cortical circuitry shown anatomically. For example, little physiological evidence has been reported of Y-cell input to layer 3, whose presence is suggested by evidence (Lund et al., 1979; Gilbert and Wiesel, 1979) that the stellate cells
of layer 4ab (which receive Y input) send substantial axon collaterals to layer 3 (Fig.
8.16). Similarly, receptive field correlates have yet to be described of the X input to layer
6.

So far, only tentative identification can be suggested for the target cells of W-cell
afferents to area 17. Camarda and Rizzolatti (1976) described a high proportion of
"hypercomplex" cells in layers 2 and 3, and Gilbert (1977) and Leventhal and Hirsch
(1978) noted that most "complex" cells of layers 2 and 3 differ from those of deeper

---

Figure 8.12. Evidence of different termination in cat area 17 of different classes of LGN relay cells (from
Leventhal, 1979). Upper: When HRP was injected electrophoretically into layer IVab (A), most labeled
(L) cells in the LGN were large in soma diameter [as in (C)] and were located in laminae A, A1, and C.
When HRP was injected into layer IVc [as shown in (B)], most labeled cells were still located in laminae
A, A1, and C, but they were smaller in soma diameter [as shown in (D)], while most large cells were
unlabeled. Lower: When HRP was injected into the superficial layers I and II [as in the case illustrated in
(A)], labeled cells were located in the deep C laminae [C, C1, and C2, example in part (C) of the montage]
or in the MIN [example in (B)]. The labeled cells were all relatively small.

Previous studies of the location, axonal conduction velocity, and soma size of relay cells in the cat LGN
suggest that the large-soma cells labeled from layer IVab are Y cells; that the smaller cells labeled from
layer IVc are X cells; and that the small cells labeled in the deep C laminae from superficial layers of the
cortex are W-class relay cells. These results therefore closely complement those of Ferster and LeVay
(1978), shown in Figs. 8.10 and 8.11. [Reproduced with kind permission of *Experimental Brain Research.*]

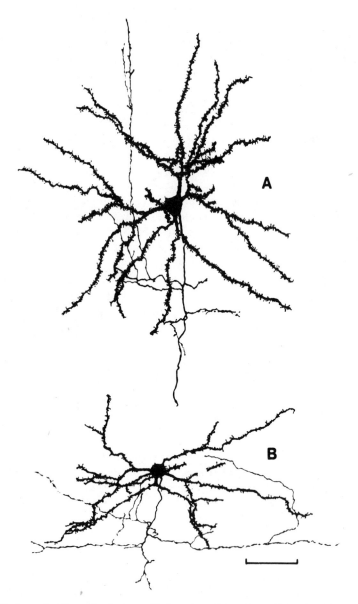

Figure 8.13. Layer 4 stellate cells of cat area 17. Drawings of the spiny stellate cells of layer 4 of the cat, made from Golgi-impregnated cortex of a 5-week-old animal (from Lund *et al.*, 1979). Cell A is a large, spiny stellate cell from layer 4A, the region of termination of Y-cell axons. Cell B is a spiny stellate cell of layer 4B, the region of termination of the axons of X cells. In both cases, the dendrites of the cells are confined to the same sublayer as the soma, implying that cell A receives Y-cell input and cell B X input, with little opportunity for either cell to receive both X and Y input. Scale bar represents 50 μm. [Reproduced with kind permission of the *Journal of Comparative Neurology*.]

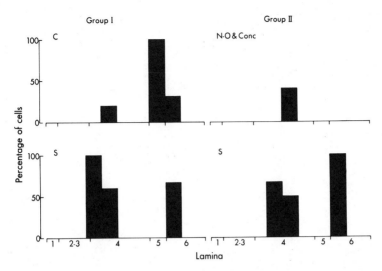

Figure 8.14. Laminar distribution of cell "types" in cat area 17. These histograms, from Bullier and Henry (1979c), show the laminar distribution of cells in area 17 of the cat, classified according to the scheme of Henry (1977), and also according to their afferent input. Group I cells are considered to receive Y input and group II cells X input. S-Type group I cells are found in the upper part of layer 4, where Y axons terminate, and S-type group II cells in the lower part of layer 4, where X axons terminate. S-Type cells of both groups are also found in layer 6 where collaterals of both X and Y axons terminate. Bullier and Henry also found evidence that C-type cells and nonoriented (N-O) or concentric cells may also receive X or Y input. The C cells are found in layers 4, 5, and 6 and the nonoriented cells in layer 4. [Reproduced from the *Journal of Neurophysiology* with kind permission of the American Physiological Society.]

layers in being less responsive to high velocities of stimulus movement, and having smaller receptive fields, as well as showing end-inhibition (a property which would have led Camarda and Rizzolatti to term them *hypercomplex*). Leventhal and Hirsch (1980) suggested that W-type geniculate relay cells may provide substantial input to these cells and, consistent with this, the Golgi analysis of Lund *et al.* (1979) shows that the apical dendrites of many pyramidal cells whose somas are in layers 2 and 3 extend into layer 1. Dreher *et al.* (1980) have provided latency-analysis evidence that W-cell axons provide the major input to a proportion of area 17 cells (Fig. 8.4), and noted further that such cells were commonly recorded in the first 500–700 μm of an electrode penetration. Clearly, W-input cells in area 17 are likely to be superficially located, and these workers may all have contributed to their detection.

Analyses of connections between cells in different laminae of cat visual cortex (Lund *et al.*, 1979; Gilbert and Wiesel, 1979) provide little clear evidence as to whether W-, X-, and Y-cell activities converge as cortical cells synapse with one another. or remain more or less separate. The circuits outlined in Figs. 8.15 and 8.16 are very partial; they omit, for example. the nonspiny stellate cells, which are believed to be inhibitory in function. Nevertheless, it seems clear from these figures that some pyramidal cells in layers 2, 3, 5, and 6 could receive the convergent activities of W, X, and Y cells. Further work, both physiological and anatomical, is needed to test for specificity

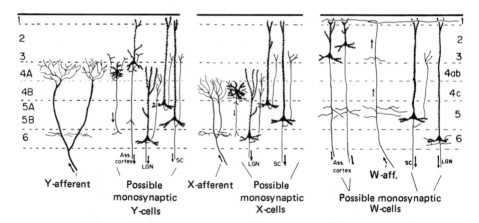

Figure 8.15. Schematic diagram of possible W-, X-, and Y-cell pathways in cat striate cortex (area 17). These diagrams rely on many studies, particularly Ferster and LeVay (1978), Leventhal (1979), and Lund *et al.* (1979). They show schematic cross-sections of cat area 17, with the layers numbered at the left following Lund *et al.* (1979) and at the right following Ferster and LeVay (1978). Only pyramidal and spiny stellate cells are represented.

Left: A Y-cell afferent to area 17 is shown at the left. It is of large caliber and terminates principally in layer 4A (4ab), and also in layer 6. Several neurons are drawn that are in a position to be directly postsynaptic to the afferent. They include stellate cells in layer 4A and pyramidal cells whose somas are located both more superficially and more deeply, in layers 5 and 6. Middle: An X-cell afferent is shown at the left. It is of lesser caliber and terminates principally in layer 4B (4c) and also in layer 6. Several neurons are drawn that might be directly postsynaptic to the afferent. They include stellate cells in layer 4B, and several cells with somas in layers 5 and 6. Note that the pyramidal cells are also shown in the diagram at the left. They are in a position to receive input from either Y or X afferents, or from both. Right: A W-cell axon is represented at the center. It is of fine caliber and terminates principally in layer 1, but also in layers 3 and 5. Several neurons are represented that might be directly postsynaptic to this axon. They include pyramidal cells with somas both above and deep to layer 4.

or convergence of afferents to these cells. So far, physiological studies (discussed above) suggest considerable specificity of input to individual cortical cells, but all techniques have their limitations and much work remains to be done. It is relevant to note that the comparable interlaminar connections within monkey area 17 seem more segregated into X- and Y-cell pathways than those described so far in the cat (Section 8.2.1).

Laminar distribution of cortical output cells: Cells in cat area 17 send axons to many sites, including the LGN, the pulvinar, pontine nuclei, and SC (subcortical projections), surrounding areas of the cortex in the same hemisphere (associational projections), and the cortex of the opposite hemisphere (commissural projections). There is now substantial evidence that the cells giving rise to these different projections are located in distinct layers of areas 17; thus, the efferent as well as the afferent mechanisms of the visual cortex vary as a function of cortical layering.

Early evidence of laminar segregation of cortical cells related to their axonal projections came from Toyama and co-workers' (1969) report (Fig. 7.8) that, in area 18 of the cat, cells that project to subcortical sites are found in lamina 5 and the upper part of lamina 6, while associational cells (projecting to surrounding cortical areas) are restricted to layers 2 and 3 and commissural cells to layer 2. Palmer and Rosenquist (1974) provided physiological evidence that cells projecting to the SC are restricted to layer 5 in all of areas 17, 18, and 19; this confirms and expands one element of Toyama and co-workers' scheme. Correspondingly, Holländer (1974) showed by the retrograde transport of HRP from an injection into the SC, that the SC receives input from pyramidal cells in layer 5 of all of areas 17, 18, 19, the Clare–Bishop area, and the lateral bank of the suprasylvian sulcus. Glickstein and Whitteridge (1976) and Shoumura (1973) reported that commissural cells in area 18 are located in layer 3 (rather than layer 2). Magalhaes-Castro *et al.* (1975*b*) confirmed the restriction of collicular projecting cells to layer 5 in areas 17, 18, and 19. Gilbert and Kelly (1975) confirmed several of these findings and added the important observation that pyramidal cells in layer 6 project back to the LGN and that, as with the cortical projection to the SC, this is true for areas 18 and 19, as well as 17. Lund *et al.* (1979) have confirmed these several

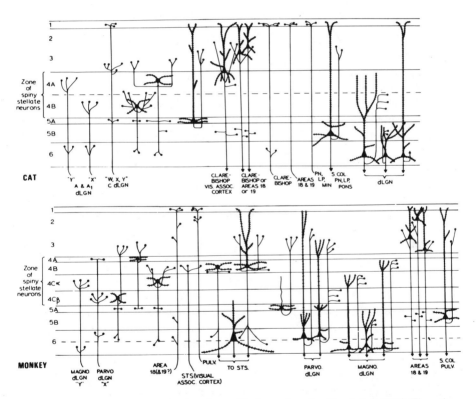

Figure 8.16. Diagrams comparing circuitry of areas 17 of cat and monkey (Lund *et al.*, 1979). The authors comment that the drawings are incomplete (they show, for example, only pyramidal and spiny stellate cells), inaccurate, and subject to modification. For example, the "W, X, Y" input to layer 1 of cat area 17 is now known to comprise only W-cell input (see Section 8.1.3). Nevertheless, they provide a valuable reference diagram for what is known of the interconnections within area 17 in these two species. Abbreviations: PN, posterior nucleus of Rioch (1929); LP, lateral posterior nucleus; STS, cortex around the superior temporal sulcus. [Reproduced with kind permission of the *Journal of Comparative Neurology*.]

findings and incorporated them into a valuable comparison of the organization of area 17 in the cat and monkey (Fig. 8.16).

This general pattern, in which layers 2 and 3 contain commissural and associational cells and layers 5 and 6 subcortically projecting cells, is found also in monkey visual cortex (Section 8.2.1) and may hold also for projections of cat visual cortex not yet extensively tested. Albus and Donate-Oliver (1977) and Gibson *et al.* (1978), for example, report that cells in area 18 that project to the pons are located in layer 5, together with other subcortically projecting cells. The general pattern is relevant to the parallel processing analysis of the visual cortex because there is evidence from several studies (e.g., Hoffmann, 1973; Leventhal and Hirsch, 1978) that Y-cell activity is prominent in cells in the deeper layers of area 17 (which project subcortically), while X- and W-cell activity is more dominant in layers 2 and 3 (which project to other areas of the cerebral cortex).

8.1.4. Corticofugal Projections of Areas 17, 18, and 19

In several ways just discussed, the laminar organization of corticofugal projections is very similar between areas 17, 18, and 19. In all three areas, for example, corticotectal cells are located in lamina 5 and corticogeniculate cells in layer 6. As reviewed in Chapter 6, Section 6.1.8, however, these three areas project to different target zones within the LGN, forming a strikingly reciprocal connection with the regions of the LGN from which their afferents arise. There is thus a considerable degree of parallel wiring in the corticogeniculate, as well as geniculocortical pathway.

As noted in Chapter 7, Section 7.2.3, a similar specificity may not exist in the projections of the visual cortex to the SC.

8.2. PRIMATE VISUAL CORTEX: PROCESSING OF GENICULATE INPUT

Whereas, in the cat, the relay cells of the LGN project in substantial numbers to several cortical areas (17, 18, 19, and the LSA), they project in primates principally to area 17. Indeed, until recently it was believed that in primates area 17 was the only cortical area to which the dLGN projects. With increasing confidence, however, recent studies have reported evidence that small numbers of neurons in the dLGN send axons to the prestriate cortex (Wong-Riley, 1976; Hendrickson *et al.*, 1978; Yukie *et al.*, 1979; Yoshida and Benevento, 1981). Yoshida and Benevento, for example, found that injections of HRP into the crown of the precruciate gyrus, which lies anterior to the striate cortex in the macaque monkey, label a small number of cells within the dLGN, in an interlaminar region between layers 3 and 4. In Yukie and co-workers' data too, the dLGN cells that appeared to project to the prestriate cortex were relatively few in number; they were located both in interlaminar zones, and within the laminae of the nucleus. These observations establish an interesting exception to previous understanding, but little is known of the functional significance of the geniculate projection to the prestriate cortex. The cells involved are so few in number that area 17 can still be considered to be, by a long margin, the principal cortical area processing the visual information relayed by the dLGN of the primate. The following paragraphs trace evidence that within primate area 17, different populations of cortical cells are organized to process the activities of X- and Y-like relay and ganglion cells, in parallel.

8.2.1. Parallel Organization of Area 17

As discussed in Chapter 6, Section 6.2, the LGN of primates shows a characteristic division into magno- and parvocellular laminae that, respectively, contain Y- and X-like relay cells. Hubel and Wiesel (1972) showed that relay cells in the magno- and parvocellular laminae terminate in different layers of the visual cortex, the cells in the parvocellular laminae (i.e., the X-like cells) terminating in layers 4A and 4Cβ, and the magnocellular (Y-like) cells in layer 4Cα (i.e., between the two layers of X-cell terminals). Tigges *et al.* (1977) and Hendrickson *et al.* (1978) presented autoradiographic evidence that in both New- and Old-World monkeys, geniculate relay cells terminate in layer 6 as well as layer 4. Further, Hendrickson and co-workers found evidence that the X-like relay cells in the Old-World rhesus monkey terminate in the upper part of layer 6 and that (less certainly) Y-like cells terminate in the deeper part. Moreover, the Golgi analyses of Lund and Boothe (1975) and Lund *et al.* (1979) provided evidence of a striking separation of the connections of X- and Y-like cells within area 17. For example, they showed (Figs. 8.16–8.19) that different populations of spiny stellate cells are confined, both soma and dendrites, to each of layers 4A, 4Cα, and 4Cβ, making it extremely likely that X- and Y-like afferents reach different groups of stellate cells. Further, the parallel wiring may persist both to the next cortical synapse and into the output of area 17. The stellate cells in layers 4A and 4Cβ (X-input) send axons to layer 3, where they are in a position to synapse on the apical dendrites of pyramidal cells whose somas are in the upper part of layer 6 and which project back to the parvocellular (X-like) part of the LGN. The same layer 6 cells presumably receive direct input from the X-like terminals demonstrated in the upper part of layer 6 by Hendrickson *et al.* (1978). Conversely, the axons of stellate cells in layer 4Cα (Y-input) spread widely within that layer and in layer 4B, where they are in a position to synapse on the apical dendrites of pyramidal cells whose somas are in the lower part of layer 6 and whose axons project back to the magnocellular (Y-like) part of the LGN. The same layer 6 cells may receive input from the Y-like afferents that Hendrickson *et al.* (1978) suggested may terminate there. In short, it may be possible to trace pathways of X- and Y-like cells through monkey area 17 into the corticogeniculate projection. Taking the analysis one step further, Lund *et al.* (1979) observed that pyramidal cells in upper layer 6 (X-system cells projecting to the X-like component of the LGN) send axon collaterals to layer 4Cβ, the major site of X-afferent termination. Conversely, the Y-system pyramidal cells in lower layer 6, which project to the Y-like component of the LGN, send axon collaterals to layer 4Cα, the principal site of Y-afferent termination. Thus, there may be a considerable separation of the pathways followed by X- and Y-like cells within area 17.

The above analyses do not establish a rigid, two-stream wiring of primate visual cortex, for several reasons. First, they do not demonstrate that the stellate cell axons actually synapse onto the apical dendrites of layer 6 pyramidal cells in the manner pos-

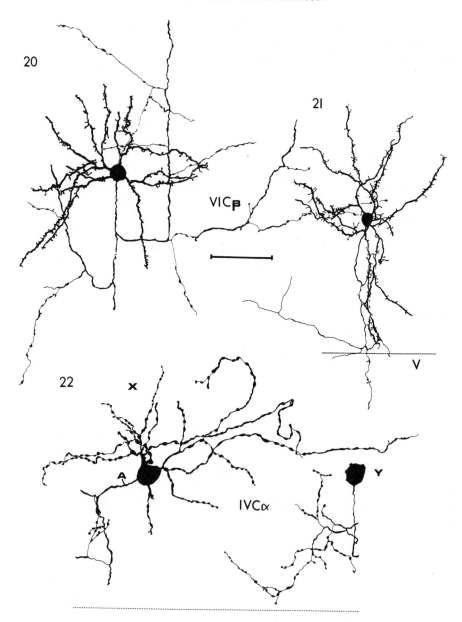

Figure 8.17. Stellate cells from layer 4 of monkey area 17, as seen in Golgi-impregnated material (from Lund, 1973). Cells 20 and 21 are stellate neurons from layer 4Cβ (X-like input). The cells shown at the bottom (labeled 22) are from layer 4Cα (Y-like input). The drawing at the left shows the dendritic formation of such a cell and part of its axon (labeled A). The drawing at the right shows the full axon morphology of a similar cell.

As in the cat, the dendrites of these cells seem confined to the layer in which their somas are found, providing little opportunity of the convergence of X- and Y-cell activities onto the same stellate cells. [Reproduced with kind permission of the *Journal of Comparative Neurology.*]

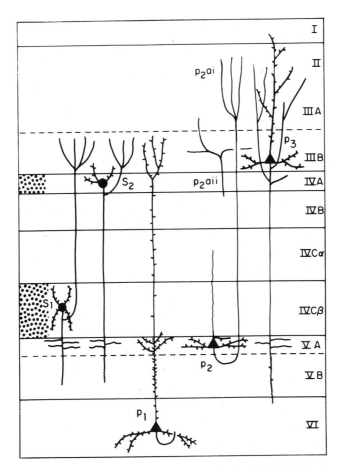

Figure 8.18. X pathways in monkey area 17 (from Lund and Boothe, 1975). The X-cell axons terminate in layers IVA and IVCβ (shown by dots), and influence separate groups of stellate cells. Axons of both stellate groups (S1 and S2) pass to layers IIIB and VA where they activate pyramidal cells whose dendrites spread in those regions, for example p1, p2, and p3. [Reproduced with kind permission of the *Journal of Comparative Neurology*.]

tulated above, or that if they do they provide more than a minor excitatory drive to those cells. Second, the same analyses show that both X- and Y-input stellate cells send axon collaterals to layer 5A, where convergence of the two systems may (or may not) occur. Third, both X- and Y-input stellate cells send axon collaterals to layer 3, where they may (or may not) converge on pyramidal cells of these layers. Nevertheless, the analyses do seem to provide evidence that, at least in certain parts of cortical circuitry, the activities of different groups of ganglion cells are likely to be processed by different groups of cortical cells.

No evidence is yet available of a W-like input to the visual cortex, relayed through the LGN, as has been described in the cat. One component of monkey LGN has yet to

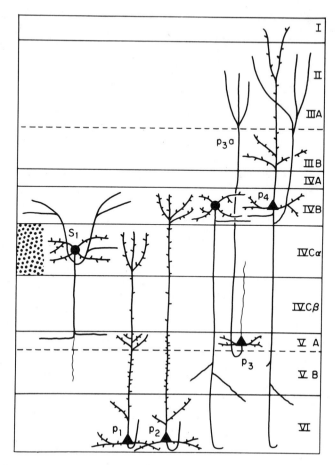

Figure 8.19. Y pathways in monkey area 17 (from Lund and Boothe, 1975). Y-like afferents terminate in layer IVCα (as indicated by dots) in relation to a group of stellate cells found in that layer (S1). The axons of these stellate cells pass to layers IVB and VA, where they are in a position to activate certain types of pyramidal cells (p1, p2, p4), whose dendrites spread in these layers. The dendrites of some pyramidal cells, such as p1, spread in layer IVCα, and so may be directly activated by Y-like afferents. The cells that seem likely to be activated by X- and Y-like afferents seem largely separate (compare this with the previous figure). However, some cells, such as p4 in this figure, may well receive the converging activities of both X- and Y-like cells. [Reproduced with kind permission of the *Journal of Comparative Neurology.*]

be characterized in terms of its ganglion cell input, viz., the S laminae. As noted in Chapter 6, Section 6.2, cells of these laminae have been shown to receive input from both eyes and to project to area 17, but their afferent ganglion cells, and their termination in area 17 have yet to be described.

In a physiological study, Dow (1974) described five classes of cells in area 17 of the monkey, distinguishing them by their stimulus-selectivity. Class I, for instance, included cells that lacked orientation- or direction-selectivity, having approximately circular receptive fields; some showed color-selectivity. Class III cells were strongly selective for the orientation of a line stimulus, and also for stimuli moving very slowly. Class V cells were markedly less selective for orientation or velocity, and had ON-OFF receptive fields similar to those of Y-input "complex" cells in cat visual cortex. Dow noted that different groups of cells seemed particularly responsive to different components of a visual stimulus, suggesting that "striate cortex performs several functions in parallel."

Moreover, Dow noted that the cell types he distinguished had different laminar distributions (Fig. 8.20), class I cells, for instance, concentrating in layer 4 and above, class V cells in layer 4 and below. To some extent, therefore, he observed some receptive field correlates of the laminar position of a cell, and some evidence that Y-cell properties (large receptive fields, phasic responses to stationary contrast stimuli, and responsiveness to fast stimulus movements) are common among cells in the deeper layers (5 and 6), while X-cell properties (small receptive fields, selectivity for slow stimulus movement) seem to dominate the upper layers (2 and 3). Bullier and Henry (1980) confirmed and extended several of these findings. For example, they also observed that cells lacking orientation- and direction-selectivity tend to be common in layer 4. Of particular interest in the present context, Bullier and Henry used electrical stimulation techniques to identify the afferents reaching individual cortical cells as slow-conducting (X-like) or fast-conducting (Y-like). In three ways, their results support the view that different cortical cells process the activities of X- and Y-like ganglion and relay cells separately, and in

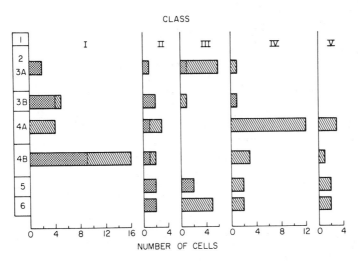

Figure 8.20. Physiological analysis of monkey area 17. Laminar distribution of different types of cells in monkey area 17, distinguished by receptive field properties (from Dow, 1974). Cells with few stimulus specificities (class I) are found principally in layer 4, but also in layers 3 and 2. Orientation-selective cells, resembling simple cells of other classifications (class II), are widely distributed across the thickness of the cortex. Other cells showing marked orientation specificities (class III) are found principally above or below layer 4, but not in layer 4. Cells showing direction-selectivity (class IV) are found principally in layer 4A, while ON-OFF-cells with some properties reminiscent of Y cells (class V) are found in layers 4, 5, and 6. [Reproduced from the *Journal of Neurophysiology* with kind permission of the American Physiological Society.]

parallel. First, Bullier and Henry found that only a proportion of cells in monkey area 17 could be shown (by the electrical stimulation technique) to be monosynaptically excited by geniculate relay cells; but that among those cells, some received X-like input and others Y-like input. They did not report evidence of cells receiving both X and Y input. Second, X- and Y-input cells tended to be found in the laminae in which X and Y afferents terminate. For example, X-input cells were found in layer 4Cβ and Y-input cells in layer 4Cα. Third, some of the receptive field properties of cortical cells matched those of the geniculate cells that appeared to provide their input. For example, cortical cells with color-coding properties received X input, and cortical cells with large receptive fields, transient response to stationary contrast stimuli, and relatively strong responses to fast-moving visual stimuli received their input from Y-like relay cells.

Malpeli *et al.* (1981) have recently described an analysis of X- and Y-cell input to monkey area 17 that took advantage of the segregation of X- and Y-class relay cells to different parts of the LGN. They injected transmission-blocking drugs into the parvocellular (X) or magnocellular (Y) parts of the LGN, and studied the responses of individual neurons in area 17. Having classified the cortical cells into "simple" and "complex" types according to whether the regions of the receptive field responsive to the two edges of a light slit were separate or overlapping, they tested whether the responsiveness of the cell was affected by the drug injections into the LGN. Many cells, both "simple" and "complex," lost responsiveness when either part of the LGN was injected, suggesting that they receive input from either X- or Y-class relay cells, and laying a basis for the parallel processing of X and Y activity in area 17. On the other hand, many cells, both "simple" and "complex," were affected by injection of either part of the LGN, suggesting that they receive convergent X- and Y-cell input.

Overall, these findings seem to me to provide strong evidence that different groups of cells in monkey visual cortex are involved in the processing of X- and Y-cell activity. Conversely, Bullier and Henry (1980) and Malpeli *et al.* (1981) note, in common with previous investigators of this problem, that X and Y "channels" of activity may well converge on individual cells within area 17. The elucidation of that convergence is of major importance to future analysis of cortical processing of visual information.

As in the cat, the axonal destinations of cortical cells vary with their laminar position, cells in layers 2 and 3 projecting to other cortical areas (areas 18 and 19, the superior temporal gyrus, and the visual cortex of the other hemisphere) and cells in layers 5 and 6 projecting to subcortical sites (LGN, SC, inferior pulvinar) (Wong-Riley, 1974; Lund and Boothe, 1975; Lund *et al.*, 1975, 1979). However, two exceptions to this general pattern should be noted. First, Lund *et al.* (1975; Fig. 8.21) found evidence that some cells in layer 4B (the part of layer 4 that does not receive geniculate terminals) project to the superior temporal sulcus. Second, Lund *et al.* (1979) suggest (Fig. 8.16) that some large pyramidal cells found in layer 5 of area 17 project to the superior temporal gyrus of the same hemisphere. [Rockland and Pandya (1979) have provided evidence that, as a general rule, some corticocortical association fibers do arise from cells of layers 5 and 6, particularly in caudally directed projections. Because area 17 is at the caudal pole of the hemisphere, however, this trend would not generate exceptions to the general pattern just discussed for area 17.]

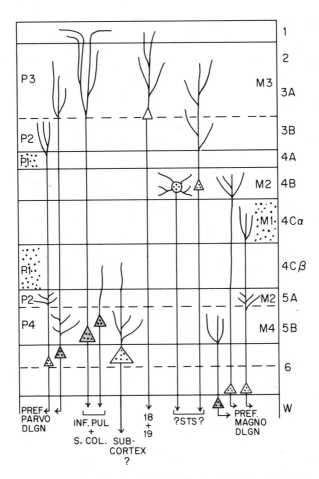

Figure 8.21. Summary diagram from Lund *et al.* (1975) showing the laminar distribution of cortical efferent cells in monkey area 17. Cells projecting to other cortical areas [18, 19, and the superior temporal gyrus (STS)] are found in layers 3 and 4B. Cells projecting subcortically are found in layers 5 and 6. In particular, cells projecting to the SC and pulvinar are found in layer 5B, while cells projecting to the LGN are found in layer 6. Note that cells in the upper part of layer 6 project to the parvocellular (X) part of the LGN, while cells in the deeper part of layer 6 project to the magnocellular (Y) part of the LGN. [Reproduced with kind permission of the *Journal of Comparative Neurology.*]

8.2.2. Organization of the Prestriate Cortex

Studies of the prestriate cortex in the monkey (Baizer *et al.*, 1977; Zeki, 1978*a,b;* van Essen and Zeki, 1978) provide considerable support for what van Essen and Zeki (1978) term "the notion of a functional division of labour within the prestriate cortex," i.e., for the idea that different groups of neurons in the prestriate cortex process different aspects of the visual image simultaneously, in parallel.

Baizer *et al.* (1977) studied area 18 neurons in the awake, behaving monkey and noted that different groups of cells were sensitive to the color, size, and direction of movement of stimuli; they suggested that different groups of cells might be concerned with each parameter of the stimulus, and be operating in parallel. van Essen and Zeki (1978) provided anatomical and physiological evidence that area 18 of Brodmann (which together with area 19 comprises the prestriate cortex) can be divided into several topographically distinct areas (which they termed *V2, V3, V3A,* and *V4*), each containing a separate, but not necessarily complete representation of the visual field, and each containing neurons concerned with a different parameter of the retinal image (color, movement, orientation, etc.), indicating a very striking separation of neurons with different stimulus specificities.

The above discussion of the organization of the striate and prestriate cortex has largely assumed that these areas should be understood as processing information reaching them from the LGN. There is some support for this assumption. Schiller and Malpeli (1977*b*), for example, reported that the visual responsiveness of area 18 neurons is totally, yet reversibly abolished by cooling area 17. Nevertheless, the prestriate cortex has recently been shown to receive some input directly from the LGN (see Chapter 6, Section 6.2.2), and there is also considerable evidence that both areas 17 and 18 receive substantial visual input from thalamic areas other than the LGN, and that this input is important for the visual behavior of the animal. Some aspects of this extrageniculate input to the cerebral cortex are considered, for both monkey and cat, in the following section. That evidence suggests that the geniculocortical pathway, itself comprising two or more parallel "channels" of neurons, operates in parallel with another, major system of neurons involving the midbrain, the posterior complex of thalamic nuclei, and the visual cortex.

8.3. CORTICAL AFFERENTS FROM EXTRAGENICULATE SOURCES

The earliest technique used to trace thalamic inputs to the visual cortex was the mapping of retrograde degeneration of thalamic neurons following destruction of the visual cortex. The most dramatic degeneration is seen in the LGN, but degeneration was also detected in the pulvinar (Le Gros Clark and Northfield, 1937; Chow, 1950), indicating that this nucleus also sends axons to the visual cortex. In 1961, Altman and Carpenter observed a pathway from the SC of the cat to the lateral posterior nucleus (LP), which is closely adjacent to the pulvinar, and raised the idea of a "second" visual pathway from the retina to the visual cortex, via the SC, LP, and pulvinar.

One intriguing but perhaps unresolvable question raised by this description is that formulated by Diamond and Hall (1969), which might be paraphrased: "Are the extrageniculate afferents to the striate cortex a late development in phylogeny, connecting the visual cortex to newly developed parts of the thalamus, Or conversely, is the geniculocortical pathway the newer development, superimposed on a phylogenetically older, less direct, and less specifically organized pathway?" Diamond and Hall argue strongly for the latter view, on the basis of the organization of the visual pathways in "prototype" mammals. Their position is not invulnerable to criticism; it is not clear, for example, that the brain of the hedgehog (their mammalian prototype) is in fact prototypical, rather than the product of an independent line of evolution. Nevertheless, their work drew renewed attention to the existence of extrageniculate inputs to the visual cortex, via the midbrain and pulvinar, in parallel with the retinogeniculocortical pathway so prominent in many species.

8.3.1. Sources of Extrageniculate Afferents in the Cat

Since Altman and Carpenter's (1961) report, many studies have contributed to our knowledge of the connections (especially with the cerebral cortex) of the posterior thalamic complex in the cat and monkey. In the cat, my attempt to piece together the findings of all these studies was not successful. This was, I think, partly because different workers divided up that complex in different ways (into, for example, pulvinar, posterior, and suprageniculate nuclei, some with medial, lateral, and inferior parts). The divisions were largely based on the cytoarchitectural appearance of the complex in Nissl-stained sections of the thalamus and, because the region generally lacks a sharp differentiation into groups of neurons, the divisions suggested by different authors often seemed (to me) incongruent. The following brief description relies on Berson and Graybiel's (1978) division of the posterior thalamic complex, which was based on the afferent inputs to the complex and the pattern of its cortical projections. The scheme matches closely that suggested by Updyke (1977) on the basis of the cortical afferents to the complex.

Berson and Graybiel suggest at least three "pathways," shown very schematically in Fig. 8.22, by which retinal activity can reach the visual areas of the cerebral cortex, via components of the posterior thalamic complex. They suggest dividing the complex into four adjacent regions, with the pulvinar nucleus (Pul) most laterally and dorsally, and the lateral (LPl) and medial (LPm) parts of the lateral posterior nucleus and the suprageniculate zone (Sg-L) located successively more ventrally and medially. One of the three pathways involves the more lateral part of the pulvinar nucleus, which receives

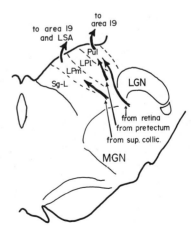

Figure 8.22. Extrageniculate pathways to the visual cortex in the cat. This diagram of a coronal section of cat thalamus shows the lateral (LGN) and medial (MGN) geniculate nuclei and the four zones into which Berson and Graybiel (1978) subdivided the posterior thalamic complex of nuclei that lie above and medial to the LGN. These divisions are the pulvinar nucleus (Pul), lateral (LP1) and medial (LPm) parts of the lateral posterior nucleus, and a region (Sg-L) at the confluence of the suprageniculate and lateral posterior nuclei.

Three pathways are shown by which retinal activity can reach the visual areas of cortex without traversing the LGN.

direct retinal input. This is the "retinal recipient zone" (RRZ) of the pulvinar discussed in Chapter 6, Section 6.6. Cells in that region project to area 19, area 21a and to the LSA. Although in current terminology the RRZ lies outside the LGN, and therefore deserves mention in the present discussion, it is relevant that the LGN was originally delimited on cytoarchitectural grounds, rather than on the basis of its connections. The RRZ lies immediately adjacent to what is currently recognized as the LGN and may, on the basis of its connections, come to be considered part of the LGN (Mason, 1979; Leventhal *et al.*, 1980; Guillery *et al.*, 1980).

The remaining, more medial part of the pulvinar nucleus receives a strong projection from the pretectal region of the midbrain (to which the retina projects directly), and its cells project to area 19, forming the second extrageniculate pathway to the visual cortex. Further medial, in LPm, is a region that receives a strong projection from the SC, and whose cells project to area 19 and the LSA, forming the third pathway. (The LPl zone, between Pul and LPm, is a corticorecipient zone, receiving a strong input from area 17.)

Many other connections of components of the posterior thalamic complex have been described, and important patterns have emerged, such as the reciprocal corticothalamic projections described by Updyke (1977). To trace them in any detail would, however, take us far from the main thrust of this section, which has been to summarize evidence, in the cat, of pathways lying outside the LGN by which retinal activity can reach the cerebral cortex, in parallel with geniculocortical pathways.

8.3.2. Sources of Extrageniculate Afferents in the Monkey

Extrageniculate pathways by which retinal activity can reach the visual cortex have been described in both New- and Old-World monkeys. One noticeable difference between the monkey and the cat is that, in the monkey, some of the cortical projections of this pathway reach area 17, though terminating in layers distinct from the geniculate afferents.

In both the squirrel monkey (Mathers, 1971) and the rhesus monkey (Benevento and Fallon, 1975; Partlow *et al.*, 1977), the SC has been shown to project to the inferior pulvinar nucleus. Winfield *et al.* (1975) and Benevento and Rezak (1975) demonstrated that this region of the thalamus projects to area 17, and noted that the projection appears to reach different layers of areas 17 and 18; specifically, pulvinar afferents appeared to terminate in layers 1 and 6 of area 17, and layer 4 of area 18. Rezak and Benevento (1979) repeated their earlier study of the rhesus macaque using autoradiography, rather than degeneration techniques. They confirmed their earlier conclusion that the inferior pulvinar nucleus projects to layers 1 and 6 of area 17, and found evidence of projections to layers 1, 3, and 4 of area 18. Very similar observations were reported for the pig-tailed macaque by Ogren (1977). Subsequently, Ogren and Hendrickson (1977) presented evidence that in the crab-eating macaque, pulvinar afferents terminate in layers 1 and 2 of area 17 and also reach layers 1, 3, and 4 of area 18. Ogren and Hendrickson (1976) and Ogren (1977) showed evidence of reciprocal connections between both lateral and inferior pulvinar nuclei and area 17 in both rhesus and squirrel monkeys, and Wong-Riley (1977) showed similar reciprocal connections between both lateral and inferior pulvinar and the prestriate cortex. Curcio and Harting (1978) reported that the layers of termination of pulvinar afferents in area 18 of the squirrel monkey are very similar to those described for the rhesus (i.e., layers 1, 3, and 4). Wong-Riley suggested that each of those two parts of the pulvinar (lateral and inferior) may act as a "vital subcortical visual centre mediating a thalamo-prestriate pathway, which runs parallel to the primary geniculo-cortical pathway."

The designation of the geniculocortical pathway as "primary" is, as discussed above, a moot point, and the studies just canvassed do not add to the resolution of the question raised at the beginning of the section, of the phylogenetic "primacy" of the geniculate or extrageniculate pathways to the visual cortex. They have added much, however, to our knowledge of extrageniculate inputs to the visual cortex. Since the pathway is topographically organized in the monkey (Partlow *et al.,* 1977; Benevento and Davis, 1977), as well as the cat, it may well provide a substrate for spatially organized visual behavior. Further, as Benevento and Rezak (1976) commented, the pathway appears to be present in the opossum as well as cats and primates, and segments of the pathway have been traced in hamsters (Schneider, 1969), the bush baby (Raczkowski and Diamond, 1978), the lemur *Microcebus* (Cooper *et al.,* 1979), and the rat (Olavarria, 1979). The pathway appears therefore to be a common feature of the mammalian brain.

8.3.3. Functional Significance of the "Second" Visual Pathway

The studies and writings of Klüver (1942), Trevarthen (1968), Schneider (1969), Diamond and Hall (1969), Humphrey (1974), Weiskrantz (1978), and many others on the behavioral capabilities of extrageniculate visual pathways have all involved the assumption that these pathways function in parallel with the retinogeniculocortical pathway. As Webster (1973) has commented, an assumption that different components of the brain can function in parallel is necessary (or else under test) in most assessments of the effects of brain lesions on behavior. In Chapter 11, evidence is discussed, much of it from behavioral studies of the effects of lesions to parts of the visual pathways, concerning the role that extrageniculate visual pathways may play in the visual behavior of cats and primates.

8.4. MODELS OF NEURONAL PROCESSING WITHIN THE STRIATE CORTEX: AN ARGUMENT AGAINST SERIAL PROCESSING

Despite its considerable detail, the discussion of the circuitry of the visual cortex in Sections 8.1.2, 8.1.3, and 8.2.1 above neglects many aspects of current knowledge about neural processing within the striate cortex. This neglect stems from the fact that a good deal of the extensive literature on this problem has not considered the different (Y/X/W) components of the input to the visual cortex, and therefore is not closely relevant to my thesis. Two series of studies can be traced, however, which seek to combine ideas of intracortical processing with understanding of the functional components of the input to the striate cortex. In both series, it has been assumed that intracortical processing is serial in mechanism; different groups of neurons are envisaged as processing visual information in turn, like the major components of a hi-fi system. The flow of information is in one direction.

The following discussion concerns those two series of studies and (though it was not my intention when I embarked on this section) ends by questioning the usefulness of serial processing as a model for intracortical processing, both in the visual cortex and in the neocortex in general. This is a more radical stand than I have taken previously, for instance in Stone *et al.* (1979). There we argued that "of course" some form of serial processing must be going on in the visual cortex; the problem was to figure out a viable hypothesis of its mechanism. Here I argue that some other mechanism of intracortical processing must be considered, one that takes into account, as a serial model it seems to me cannot, the extensive interconnections between the different layers of the visual cortex.

Before entering on that discussion, I would like to stress [as previously (Stone, 1972; Stone *et al.*, 1979)] that parallel and serial mechanisms of neuronal processing are not in principle mutually exclusive. Different groups of cortical cells could be devoted to processing in parallel the activities of different sorts of ganglion cells; yet within those groups, neurons could be connected in series with each other. The additional argument developed here is that neither parallel nor serial mechanisms can be assumed without evidence; that when the available evidence is considered, the idea of serial processing within the visual cortex can be seen to have been proposed, tested, and found wanting.

8.4.1. The Simple/Complex Model of Serial Intracortical Processing

The best-known approach to the analysis of intracortical processing was developed, much earlier than the parallel processing analysis, by Hubel and Wiesel (1962, 1965, 1968). They envisaged three classes of cortical neurons; cells of one class ("simple" cells) receive input from the LGN and pass it to cells of a second class ("complex" cells), which process it further and send it on to cells of the third class ("hypercomplex" cells). Further, areas 17, 18, and 19 were envisaged as in a hierarchical relationship, information being processed first by area 17, then by area 18, and then area 19.

Several criticisms of Hubel and Wiesel's model are set out in Stone *et al.* (1979). That review was in press before Gilbert and Wiesel's (1979) restatement of the simple/complex/hypercomplex model became available; the following comments concern that formulation.

Gilbert and Wiesel's model of intracortical processing is shown schematically in Fig. 8.23. They argue that the first stage of cortical organization is effected by the stellate cells of lamina 4. Their dendrites are confined to layer 4, where many geniculate afferents terminate. Gilbert and Wiesel presented evidence, based on a spectacular demonstration of cell structure using HRP, and in good agreement with the Golgi analysis of Lund *et al.* (1979), that many stellate cells in layer 4ab, and perhaps in layer 4c also, have substantial axonal terminations in layer 2 + 3. They argue that since virtually all layer 4 cells have "simple" receptive fields, while layer 2 + 3 cells have "complex" fields, the "complex" receptive field properties of layer 2 + 3 cells result from a convergence on them of several "simple" cells, as Hubel and Wiesel originally suggested.

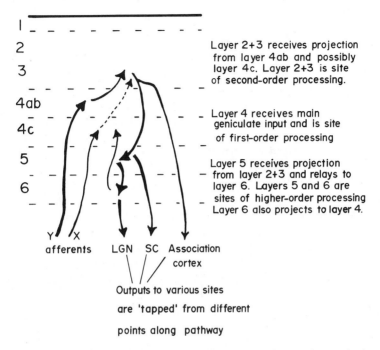

Figure 8.23. Schematic diagram of Gilbert and Wiesel's (1979) model of neuronal processing in area 17 of the cat (my drawing). Further discussion in text.

Thus, layer 2 + 3 cells perform a "second-order" processing of visual information passed on from first-order "simple" cells in layer 4.

Gilbert and Wiesel also confirmed earlier evidence that layer 2 + 3 cells send axons to layer 5, and presented new evidence that layer 5 cells project substantially to layer 6. This serial relay, they suggest, accounts for the "complex" receptive fields of cells in layers 5 and 6. Gilbert and Wiesel confirmed Ferster and LeVay's (1978) report that X- and Y-cell afferents terminate in separate strata of layer 4, as well as the evidence of Hoffmann and Stone (1971), Stone and Dreher (1973), and Leventhal and Hirsch (1977) that the receptive field properties of many cortical cells reflect whether they receive X- or Y-cell input. They do not discuss whether X- and Y-cell activities remain separate beyond layer 4, and do not consider the W-cell afferents that reach area 17 (Ferster and LeVay, 1978; Leventhal, 1979).

Gilbert and Wiesel's model leaves out the hypercomplex stage of Hubel and Wiesel's original proposal, as well as the idea of a hierarchical relationship between areas 17, 18, and 19. It is consequently a less ambitious model, and Gilbert and Wiesel were able to refer to, or themselves produced, evidence of the reality of every step of their circuit. Even so, it seems vulnerable to three criticisms:

1. Consideration of the full range of geniculocortical afferents: In addition to the X- and Y-cell input to area 17, the axons of W-class relay cells also terminate there (Chapter 6, Section 6.1.7, and Section 8.1.3, Figs. 8.10–8.12), particularly in layer 1 (where they presumably contact apical dendrites of pyramidal cells, many with somas in layers 2 and 3, and others with somas in layers 5 and 6), but also in layers 3 and 5. Moreover, the terminals of Y-cell axons spread into lamina 3, and collaterals of X- and Y-cell axons terminate in layer 6. As argued elsewhere (Stone *et al.,* 1979), this distribution of terminals means that some cells in all cortical layers are in a position to receive direct geniculate input. Gilbert and Wiesel's model thus seems to oversimplify the situation by equating first-order cells with layer 4 cells. Recently, for example, Peters *et al.* (1979) were able to demonstrate in their Golgi–EM studies in the rat that geniculate axons terminate on the dendrites of pyramidal cells in layers 3 and 5, as well as on layer 4 stellate cells. In a more limited study, Hornung and Garey (1980) found evidence of direct geniculate afferents reaching pyramidal cells in layer 3 as well as stellate cells in layer 4; and several authors, most recently Bullier and Henry (1979*c*), have provided physiological evidence for the presence of monosynaptically activated cells in layers 5 and 6. By all criteria so far applied, it seems likely that some cells in most or all of the layers of the visual cortex receive direct geniculate input.

This last conclusion does not imply that geniculate afferents terminate nonspecifically throughout the thickness of area 17; considerable specificity is apparent, particularly when the W, X, and Y components of the geniculate projection are considered separately. The conclusion does imply that the pattern of geniculate termination in the visual cortex is better understood when (1) the functional groupings among geniculocortical afferents are brought into the analysis and (2) it is recognized that cells with somas in one layer of the cortex may receive substantial synaptic contact on parts of their dendrites that extend into other layers. This view is close to that developed by White (1978, 1979) in his analysis of thalamic input to somatosensory cortex. White (1978) showed that afferents from the somatosensory nucleus of the thalamus (the ventrobasal nucleus) terminate on a number of cell types in layers 3, 4, and 5 of the somatosensory cortex and suggested that the different cell types are likely to be processing thalamic input in parallel. In considering the developmental mechanisms that might produce this pattern of thalamocortical connections, Peters and White (discussed in White, 1979) suggested that afferents are guided to the developing cortex by "mechanical" factors such as a glial scaffolding. Then, "having reached their sites of termination thalamocortical axons synapse on every available neuronal element capable of forming [the postsynaptic element in] asymmetrical [excitatory] synapses. . . ." White suggests that the distribution of thalamic afferents to a variety of cell types in different layers of the cerebral cortex, and the developmental mechanisms involved, may be common to different cortical areas, and to a range of species. The visual cortex would, in this respect, closely resemble other cortical areas.

2. Consideration of the full range of laminar connections within area 17: The circuit to which Gilbert and Wiesel drew attention (Fig. 8.23) does not include several components of cortical circuitry with quite different implications for cortical organization. For example, central to Gilbert and Wiesel's model is the proposal that layer 4 cells are first-order cells, and layer 2 + 3 cells are second-order. Yet consideration of the full known range of cortical connections does not support this sharp distinction.

Consider layer 4 cells. Most are stellate in form but some are pyramidal (Lund *et al.,* 1979). There is wide agreement that these cells receive excitatory synapses from the axons of geniculate relay cells. However, the same cells also receive substantial input from nearby cells. For example, the axons of both major varieties of layer 4 stellate cells (spiny and nonspiny) send branches that terminate in layer 4, presumably on other stellate cells. In addition, many pyramidal cells, particularly those in lamina 6, send axons to layer 4, which presumably terminate on the stellate cells there. If, as seems likely, the synapses formed by pyramidal and spiny stellate cells are excitatory, while those formed by nonspiny stellate cells are inhibitory, then layer 4 cells receive substantial excitatory and inhibitory input from other cortical cells. Layer 4 cells are thus both first-order and second-order cells; both the geniculate and the intracortical inputs reaching them appear substantial.

Consider then the cells in layer 2 + 3. These are principally pyramidal in form, although nonspiny stellate cells are also present (Lund *et al.*, 1979). The apical dendrites of many pyramidal cells extend to layer 1, where some presumably receive synapses from the W-class axons that Ferster and LeVay (1978) and Leventhal (1979) demonstrated terminate there. Further, the Y-cell axons terminating in layer 4ab extend into lamina 3, and Peters *et al.* (1979) showed that, in the rat, the basal and even apical dendrites of some layer 3 pyramids receive direct synapses from geniculocortical afferents (Fig. 8.24). LeVay and Ferster (1978) and Leventhal (1979) also found evidence of W-cell terminations in layer 3. Further, Dreher *et al.* (1980) have reported physiological evidence of direct W-cell input to many cells in these laminae. It appears likely

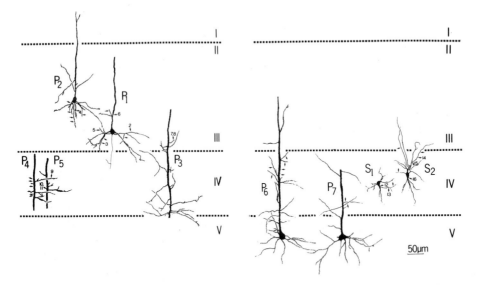

Figure 8.24. Target cells of geniculate afferents in the rat. Diagram from Peters *et al.* (1979) showing results from their analysis of geniculate synapses on cells identified in Golgi-impregnated material. After making lesions in the LGN of the rat, these workers used the electron microscope to identify terminals originating from LGN relay cells (by their degeneration), and to locate the synapses they formed on the dendrites of cells in the visual cortex. The morphology of the cells studied was identified by Golgi-impregnation. This study provides very direct evidence that LGN axons terminate on basal and apical dendrites of pyramidal cells in layer III (P1 and P2), on the dendrites of stellate cells in layer IV (S1, S2), and on the apical dendrites of pyramidal cells in layer V (P3–P7). The arrows show the locations of identified degenerating axon terminals. [Reproduced with kind permission of Chapman & Hall.]

therefore that many cells in layer 2 + 3 are first-order in the sense that they receive substantial, direct geniculate input. On the other hand, as Lund *et al.* (1979) and Gilbert and Wiesel have shown, spiny stellate cells of layer 4ab (which receive Y-cell input) send substantial axon collaterals to layer 2 +3, as do pyramidal cells of layer 5A and B (Lund *et al.,* 1979). Hence, layer 2 + 3 cells are also postsynaptic to other cortical cells (this is the feature of their circuitry that Gilbert and Wiesel emphasized) and many are likely to prove both first- and second-order.

A similar argument holds for cells of layers 5 and 6. These are principally pyramidal cells, and (as already noted) may receive direct geniculate input at a number of sites. On the other hand, the same cells are in a position to receive substantial input from cortical cells in several other layers. Gilbert and Wiesel noted, for example, the substantial input that passes from layer 2 + 3 to layer 5 and from layer 5 to layer 6; but especially considering the long apical dendrites, the cells may receive several other inputs from cortical cells.

The report of Meyer and Albus (1981) suggests a further component of cortical circuitry that seems to go against the serial processing interpretation developed by Gilbert and Wiesel. Meyer and Albus reported evidence that many stellate cells in layer 4 of area 17 send axons out of this area, to area 18. Previously, it had been considered that cortical output derives from pyramidal cells; stellate cells were presumed to receive geniculate input and pass on to other "secondary" neurons in area 17. In short, it seems that when the full range of demonstrated cortical connections is considered, it is hard to draw a compelling distinction between first-order and second-order cortical neurons.

3. Physiological evidence of monosynaptic "complex" cells: Gilbert and Wiesel's model, like its predecessor, seems quite at odds with evidence that many cells that have receptive field properties that would put them in the "complex" category receive strong monosynaptic input. To date, this evidence has been reported in at least five studies (Hoffmann and Stone, 1971; Stone and Dreher, 1973; Toyama *et al.,* 1973; Singer *et al.,* 1975; Bullier and Henry, 1979*a,b,c*), and has not yet been challenged. Toyama and co-workers' evidence is shown in Fig. 8.8.

It is, of course, possible that a circuit like that emphasized by Gilbert and Wiesel in the cat is functional in monkey visual cortex. So far, early evidence of geniculate input to layer 1 of monkey area 17, from degeneration studies, has not been confirmed by autoradiography, so that many cells in the superficial laminae may not receive direct geniculate input. It seems to be established, however (see Section 8.2.1), that layer 6 receives direct input, as well as layer 4, and the Golgi analyses available suggest that the apical dendrites of pyramidal cells in layers 5 and 6 extend into layer 4 where they may receive input from geniculate axons. Moreover, the interconnections between cortical layers are as profuse as in the cat. Overall, a stage-by-stage serial processing of information within the visual cortex seems as unlikely in the monkey as in the cat.

8.4.2. Synaptic Latencies: A Second Line of Evidence for Serial Processing?

A number of studies have examined the responses of cortical cells to volleys of impulses elicited in the afferents to the cortex by brief electrical stimuli (see Section 8.1.2). All such studies have noted that latencies indicative of monosynaptic activation are found in only a proportion of cortical cells sampled. Some cells respond at latencies that suggest, particularly when the analysis includes two or more stimulating sites along the pathway, that the cells are activated via two or more cortical synapses. Clearly, such cells could be viewed as second or later stages in a serially organized set of neurons.

Two arguments limit the force of this suggestion. One is technical: Electrically evoked volleys are a useful but not foolproof means of distinguishing the thalamic afferents to cortical cells. For example, many X-input cells in both the LGN and the visual cortex are not responsive to electrically evoked afferent volleys, probably because of inhibition of those cells by the volley of activity elicited by the same stimulus in the faster-conducting Y-cell system. As a consequence, many X-input cells might respond only to polysynaptic influences reaching them after the Y-evoked inhibition has fallen away. Moreover, only one study (Dreher *et al.*, 1980) has detected the physiological impact of the W-cell afferents that anatomical studies have shown to reach layers 3, 5, and especially 1. These afferents must synapse on some cortical cells including, presumably, many pyramidal cells of layer 2 + 3 [the layer where Singer *et al.* (1975), Toyama *et al.* (1977*b*) and Bullier and Henry (1979*c*) have suggested that di- or polysynaptically activated cells are most frequent]. Until this problem is sorted out, we cannot estimate what proportion of cells in layer 2 + 3 is monosynaptically contacted by geniculate afferents. The proportion seems certain to be higher than physiological studies indicate.

The second argument stems from a consideration of what is meant by serial or higher-order processing. Even assuming (as several reports indicate) that many cortical cells receive their major excitatory input only disynaptically, does it follow that they can usefully be regarded as "higher-order" processors? As Bullier and Henry (1979*b*) stress, following earlier developments of the same point, the receptive fields of cells they considered higher-order on the basis of the analysis of synaptic latencies do not appear markedly more complex than, or even different from, those of putative first-order cells. So far there are only very limited receptive field correlates of the first-order/second-order grouping of neurons suggested by synaptic analysis. Moreover, the circuitry discussed in (3) above still makes the pattern of connection between cortical layers resemble a loop or "feedback" circuit (Fig. 8.25), rather than a serial, stage-by-stage layout, and the synaptic analysis just discussed does not help distinguish between these two possibilities.

In summary, positive findings reached by synaptic analysis, such as the identification of unambiguously monosynaptic latencies, seem reliable; negative or apparently exclusive findings (such as the lack of monosynaptic input above and below layer 4) require further testing, by other techniques. The evidence for monosynaptic input to cells above layer 4, to take a topical example, seems to be mounting.

8.4.3. Summary

To summarize Section 8.4, it seems to me that what we know of the organization of the striate cortex goes against the operation of serial processing mechanisms within in. The variety and laminar spread of the cell types contacted by geniculate afferents seem too great, and the interconnections between cortical layers seem too extensive, to be compatible with a serial or hierarchical ground-plan for intracortical circuitry. Conversely, the data do provide support for the existence of parallel-wired mechanisms of cortical processing.

Although more radical than a previous critique of serial processing (Stone *et al.*,

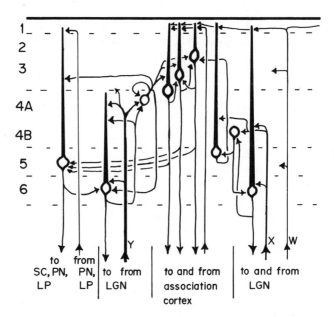

Figure 8.25. Loop circuits predominate in cat area 17. Schematic diagrams of the circuits followed by Y-, X-, and W-cell activity in cat visual cortex, as indicated by current knowledge of cortical circuitry. Note that although the three classes of axons seem to contact different groups of cortical cells initially, their activities may converge fairly quickly. The point of this diagram is that in no case do the circuits resemble a stage-by-stage serial processing mechanism. Rather, feedback or loop circuits seem to predominate.

The diagram seeks to emphasize two types of reciprocal connection. First, the different layers of the cortex are reciprocally connected, as for example layers 4 and 6, and layers 3 and 5. Layers 4 and 5 are interconnected both directly and via potential relays in layer 3. There are also instances where a layer seems to contain reciprocal connections within it; many stellate cells of layer 4, for example, send axons out of the layer, but the same axons send collaterals back to layer 4. Second, several layers are reciprocally connected with distant sites. For example, cells in layer 6 receive afferents from the LGN directly, as well as indirectly via layers 4, 3, and 5; they project back to the LGN. Cells in layer 5 project to the SC and to nongeniculate parts of the thalamus (PN, LP); those cells may receive direct input from afferents from some of these sites. Those afferents terminate in layer 1, to which the apical dendrites of many layer 5 pyramidal cells reach. Similarly, the cells that project to association areas of the cortex in the same hemisphere (principally pyramidal cells of layers 2 and 3) may receive direct input from cells in those areas, whose axons terminate in layer 1 of area 17.

1979), the present thesis is only a limited extension of the ideas of White (1978, 1979), which stemmed from his analysis of the somatosensory cortex. He used Golgi–EM techniques to demonstrate that ventrobasal thalamic afferents reach several types of neurons in layers 3, 4, and 5. He notes that "any ... hierarchical processing of thalamic input must occur in conjunction with the parallel ... processing of thalamic input ..." and that "neurons at several different hierarchical levels receive input directly from the thalamus." My own view goes beyond White's in arguing that the notion that neurons at different levels of a hierarchy receive primary input is at odds with the idea of a hierarchy; that the original evidence for a hierarchy (i.e., the simple/complex/hypercomplex classification of cortical cells, which was also the basis of White's adoption of the idea) has been substantially weakened by subsequent work; and that we need to develop quite new concepts to characterize intracortical circuitry, concepts that can accommodate recently established elements of the connections between cortical neurons.

8.5. FUTURE WORK: THE IMPORTANCE OF THE CLASSIFICATION AND TERMINOLOGY USED FOR CORTICAL CELLS

The issue raised in the previous section is not trivial: an established and widely accepted view of intracortical organization seems no longer valid. Conversely, the evidence supporting a parallel processing analysis of the visual cortex seems to me substantial. Even so, it would for several reasons be inappropriate now to urge the wholesale adoption of the latter view in place of the former. First, the parallel processing analysis is already being tested in many laboratories and several sensory systems; if it proves a powerful analysis (as I believe it will), that will be established by experimental work, not advocacy. Second, the evidence against the usefulness of the serial processing model is limited and the history of science is replete with ideas that have reasserted their usefulness; serial and parallel mechanisms of cortical circuitry are not in principle incompatible. Third, and conversely, the parallel processing model of the visual cortex may be the most powerful analysis available, but it too will no doubt prove limited and inadequate. We need to treat all models of cortical organization with both skepticism and respect: sufficient skepticism that we are always actively testing our current ideas and always willing to change them in the face of new evidence; and sufficient respect that we remain willing to reconsider any body of ideas, should the evidence demand it.

I do not therefore urge a single, dramatic solution to the question of the organization of intracortical circuitry; that solution will likely emerge from continued experimental testing of current ideas. What seems to me important is to maintain a conceptual framework that encourages that steady development. It is that imperative that leads me, again, to a consideration of the classification and terminology that neurobiologists apply in their study of cortical cells.

Issues of classification and terminology have always been central to the study of cortical organization, and Mann's (1979) review of the classification and naming of neurons in the somatosensory cortex shows that the problems that have arisen are in no way unique to the visual system. The discussion below does not go beyond analyses of the general problem of classification and terminology of nerve cells published elsewhere (Tyner, 1975; Rowe and Stone, 1977, 1979, 1980a,b; Mann, 1979) or the specific discussion of this problem for the visual cortex presented in Stone et al. (1979). Those analyses are critical of the simple/complex(/hypercomplex) terminology for the reason argued by Henry (1977): that terminology assumes a model of cortical organization that should be under test. The continuing survival of the simple/complex model in the face of overwhelming evidence against it has required the immunity to test that the terminology provides.

The same analyses lead, however, to a corresponding criticism of Henry's (1977) proposal of five groups of cortical cells, each defined by certain "key" properties of their receptive field. As Henry notes, the use of alphabetical terminology avoids any presuppositions about cortical organization, so that models of that organization can be tested. On the other hand, the defining of cell groups in terms of key physical features involves the presupposition that those features of the receptive fields are of particular importance, and other properties of the cell (such as its input and onward projection) are of little or no importance, in determining cell groupings. Moreover, these assumptions are made immune to test by the use of definitions, with the result that the adequacy of the cell groupings cannot be tested by the subsequent accumulation of more knowledge of their properties. The groupings are rigidly fixed.

We (Stone *et al.*, 1979) suggested that the W/X/Y terminology developed for retinal ganglion cells and geniculate relay cells can be extended successfully to the visual cortex, provided that the groups of cortical cells distinguished be established by description on the basis of many of their properties, rather than by definition in terms of any one or two. This would allow continued testing of both the model of cortical organization being considered, and the groupings of cells employed. Correspondingly, it would allow flexibility in the use and development of terminology. The approach allows identification, for example, of distinct populations of cortical cells processing the activities of different groups of retinal ganglion cells, and their onward projections. Such populations of cortical cells can then be termed W, X, and Y-class cortical cells. On the other hand, for analysis of a stage of cortical organization at which the activities of W, X, and/or Y cells had substantially converged on the same cells [perhaps in the inferotemporal cortex; see, for example, the study of Fuster and Jervey (1981)], this analysis may not be useful, and alternative ideas and terms would presumably have to be developed. Again, if it became clear that one group of cortical cells (say the Y cells) was serving a variety of functions that could not usefully be considered as part of an overall function, terminology would have to be developed to encompass that diversity. If, to take another possibility, one of the W, X, or Y components of the input to the visual cortex was shown to provide only a modulatory influence on cortical cells, that too would have to be reflected in the classification and terminology. But to the extent that different groups of cortical cells are indeed processing separately the activities of W, X, and Y cells, it may prove useful to regard such groups as the cortical components of systems of W, X, and Y cells, each system encompassing groups of cells at the various visual centers.

There are many inadequacies in such a scheme, but I believe they stem from the incompleteness of our understanding of the visual cortex. For example, the scheme offers no ready model of the cortical basis of perception; but that reflects only the real limitations of our understanding. It deliberately offers no fixed scheme for classifying cortical cells, both because our knowledge of these cells is incomplete, and because of the resulting need to retain flexibility in the assimilation of new evidence. What the scheme does offer is a classification and terminology that take into account, as no previous scheme has done, important aspects of the organization of the retinocortical pathways, and yet allow the constant testing and development of concepts of (1) models of cortical organization and (2) the functional groupings of cortical cells.

The testing of models of cortical organization has already led, for example, to fresh ideas of parallel processing mechanisms within the psychophysics of vision (Chapter 11). The potential value of interaction between visual psychophysics and neurobiology is well recognized; the realization of that potential requires the use of classifications and terminologies for visual neurons that are not only empirically based, but also empirically testable. To the extent that we lose sight of the need for testability, our classifications will become no more than ways of explaining old ideas. Made testable, a classification of cortical neurons can be a continuing source of new and better ideas of the neuronal basis of vision.

On the Understanding of Retinal Topography

A "Two-Axis" Model of Mammalian Retina

9

9.1. THE PROBLEM: THE VARIABILITY OF RETINAL TOPOGRAPHY

A remarkable feature of mammalian retina is the constancy of its organization across its thickness. In the retinas of all mammals, three layers of nerve cells are recognized, separated by plexiform (synaptic) layers. The receptor cells lie on the outer surface of the retina, with the tips of their outer segments enclosed by cells of the pigment epithelium; bipolar, amacrine, interplexiform, and horizontal cells form the middle layer, and ganglion cells and their axons form the innermost layers of the retina.

By contrast, there is considerable diversity among mammalian species in the structure of the retina along its other two axes, viz., its width and height. The variation is particularly well documented in the ganglion cell layer, in which a fovea may be present or absent, a visual streak may be a dominant feature or barely detectable, and the populations of ganglion cells that project to different sides of the brain may vary widely in their relative numbers, and in the degree to which they are spatially segregated. Moreover, recent studies have provided evidence of long-unsuspected differences in the properties of ganglion cells between areas of the retina nasal and temporal to the area centralis or fovea, and between the visual streak and other regions of peripheral retina. It has been suggested that these latter differences are related to the different functional roles and phylogenetic histories of these different areas of the retina. Regional variations have also been described in other layers of the retina, adding to the range of variation requiring explanation.

In this chapter, I attempt to provide a conceptual framework or "model" within which this variation in retinal topography can better be understood. The model relies on a variety of observations, but particularly on patterns of retinal distribution and naso-temporal division of the different classes of ganglion cells that have been described in various mammals. In brief summary, it is argued (1) that a basic pattern of retinal topography can be discerned that is common to a wide range of mammals and (2) that the considerable variation in topography between mammalian species can be fruitfully regarded as adaptations of that basic pattern. The model is, however, restricted to mammalian retina. In many fish, amphibians, and reptiles, retinal cells proliferate constantly throughout life in a zone at the edge of the retina (whereas in mammals the proliferation ceases in fetal life or infancy), and the effect of this steady proliferation on retinal topography has still to be described. In birds, the development in many species of two separate foveas in each retina, and the absence of any retinal projection to the ipsilateral side of the brain, suggest a quite separate line of development of retinal topography. Much needs to be learnt of retinal topography and ganglion cell groupings in nonmammalian vertebrates (and no doubt in mammals as well) before a model of retinal topography applicable to all vertebrates can be attempted. Nevertheless, the model suggested here has (I would argue) two principal useful features:

1. It provides a single framework for understanding the variety of retinal topography found among mammals, and relating it to their phylogenetic history and visual behavior.
2. It makes substantive generalizations and predictions about major features of retinal topography, whose testing should both assess the model and advance understanding of retinal organization.

9.2. A TWO-AXIS MODEL OF THE TOPOGRAPHY OF MAMMALIAN RETINA

The model is represented diagrammatically in Fig. 9.1, and the legend to that figure describes some of its detail. Its major premises and postulates are:

1. The topography of mammalian retina can be usefully regarded as organized around two axes, the approximately horizontal axis of the visual streak, and the approximately vertical axis of nasotemporal division. The former axis follows the length of the visual streak; this is an elongated region of specialized retinal structure that extends across much of the width of the retina. The latter axis runs approximately vertically across the zone of transition between the nasal region of the retina, from which ganglion cells project to the contralateral hemisphere of the brain, and the temporal region of the retina, to which ipsilaterally projecting ganglion cells are restricted.
2. The visual streak and the nasotemporal division of the retina represent different aspects of its function and phylogenetic development. The streak specialization seems appropriate to allow the animal to scan large parts of its visual field without eye movements; while the different laterality of projection of nasal and temporal areas of the retina seems related to the function of frontalized, binocular vision.

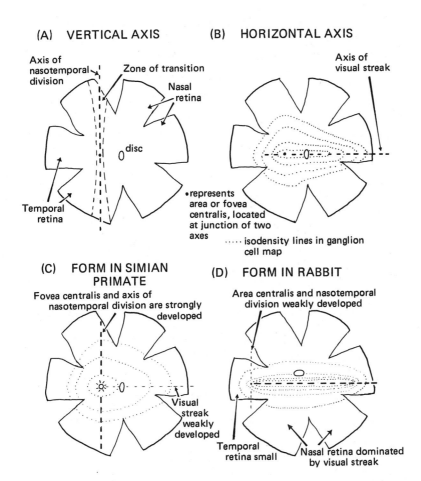

(A) VERTICAL AXIS

Axis of nasotemporal division

Zone of transition

Nasal retina

disc

Temporal retina

• represents area or fovea centralis, located at junction of two axes

(B) HORIZONTAL AXIS

Axis of visual streak

···· isodensity lines in ganglion cell map

(C) FORM IN SIMIAN PRIMATE

Fovea centralis and axis of nasotemporal division are strongly developed

Visual streak weakly developed

(D) FORM IN RABBIT

Area centralis and nasotemporal division weakly developed

Temporal retina small

Nasal retina dominated by visual streak

Figure 9.1. Schematic diagram of two-axis model of retinal topography in mammals. The retina is shown as it appears in a flattened whole-mount preparation; short radial cuts are made at intervals around the perimeter of the retina, so that it can be flattened. (A) The axis of nasotemporal division is approximately vertical. It runs down the middle of a zone of transition between nasal and temporal regions of the retina, delineated by the laterality of projection of their ganglion cells (Section 9.3). The axis crosses the area or fovea centralis, and the zone of transition is narrowest near the area or fovea. (B) The axis of the visual streak is approximately horizontal, but in some species it may not be precisely either horizontal or straight (Section 9.4). It is best developed in the retina nasal to the axis of nasotemporal division; it can be discerned in the patterns of distribution of ganglion cells (as here), of cones, and of horizontal cells. (C) In simian primates, the vertical axis is strongly developed and the fovea highly specialized. The streak is weakly developed and can be detected only by measurement and experiment. (D) In the rabbit, the streak is massively developed. The area centralis is only weakly developed and is found close to the temporal margin of the retina, i.e., temporal retina is very small. An axis of nasotemporal division can be demonstrated experimentally.

3. The area centralis is a region of specialized structure located at the intersection of the above two axes. Although commonly identified as a localized region of high ganglion cell density and considered to subserve high-resolution vision, the area centralis is only weakly developed in some species (which presumably rely on other retinal specializations for high-resolution vision). Because the area centralis is consistently found astride the axis of nasotemporal division, its common function among mammals may be as the retinal fixation point in binocular vision.

4. Because the area centralis and visual streak subserve distinct visual functions, their development may vary widely and independently between species with distinct evolutionary histories.

5. Conversely, the ubiquity among mammals of a number of retinal specializations suggests that a single pattern of retinal topography developed early in the phylogenetic history of mammals. The model proposed is a best guess at that basic pattern of retinal topography.

9.3. THE VERTICAL AXIS: THE NASOTEMPORAL DIVISION OF THE RETINA

9.3.1. Historical Note: The Nasotemporal Division of Human Retina

Query 15: Are not the Species of Objects seen with both Eyes united where the optick Nerves meet before they come into the Brain, the Fibres on the right side of both nerves uniting there, and after union going thence into the Brain in the Nerve which is on the right side of the Head, and the Fibres on the left side of both Nerves uniting in the same place, and after union going into the Brain in the Nerve which is on the left side of the Head, and these two Nerves meeting in the Brain in such a manner that their Fibres make but one entire Species or Picture, half of which on the right side of the Sensorium comes from the right side of both Eyes through the right side of both optick Nerves to the place where both Nerves meet, and from thence on the right side of the Head into the Brain, and the other half on the left side of the Sensorium comes in like manner from the left side of both Eyes. For the optick Nerves of such Animals as look the same way with both Eyes (as of Men, Dogs, Sheep, Oxen, c.) meet before they come into the Brain, but the optick Nerves of such Animals as do not look the same way with both Eyes (as of Fishes, and of the Chameleon,) do not meet, if I am rightly inform'd. [Isaac Newton, *Opticks* (4th ed., 1730), Book Three, Part I]

In probably all mammals, a minority of the fibers of the optic nerve do not decussate (cross the midline) as they run to the brain, but stay uncrossed and terminate in the ipsilateral side of the brain. This partial decussation of fibers of the optic nerve, and the related nasotemporal division of the retina, were first recognized in the human visual system. Polyak (1957) has traced the discovery of this pattern, noting that it was not known in the 17th century, for example when Descartes wrote his *Traité de l'homme* (Fig. 9.2A), but was understood in basic outline by the middle of the 18th century (Fig. 9.2B). Polyak attributes to Newton the earliest speculation (quoted above) that the chiasm represents a region of partial decussation. An empirical basis for the notion came, early in the 19th century, from anatomical studies and cases of homonymous hemianopias (matching half-blindness of each eye) caused by destruction of the occipital lobe of

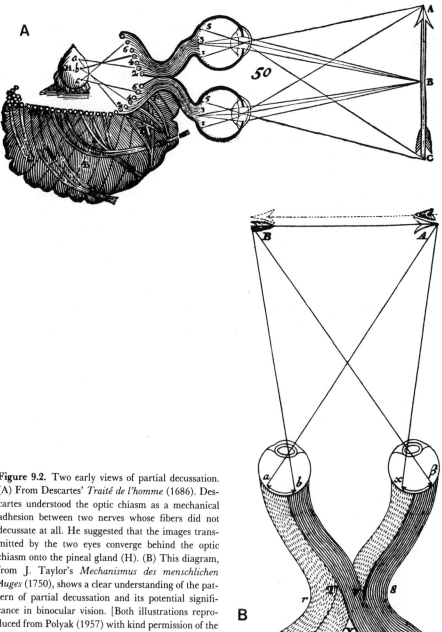

Figure 9.2. Two early views of partial decussation. (A) From Descartes' *Traité de l'homme* (1686). Descartes understood the optic chiasm as a mechanical adhesion between two nerves whose fibers did not decussate at all. He suggested that the images transmitted by the two eyes converge behind the optic chiasm onto the pineal gland (H). (B) This diagram, from J. Taylor's *Mechanismus des menschlichen Auges* (1750), shows a clear understanding of the pattern of partial decussation and its potential significance in binocular vision. [Both illustrations reproduced from Polyak (1957) with kind permission of the University of Chicago Press; copyright 1957 by the University of Chicago.]

one hemisphere. This early understanding was reinforced by extensive clinical experience of the effects of penetrating wounds of the occipital lobe on the visual fields, as assessed by perimetry, particularly in the aftermath of war (for a recent treatise, see Teuber *et al.*, 1960). Postmortem anatomical study of the retina following disease has added to the understanding gained from perimetry; the relatively recent studies of Gartner (1951), Kupfer (1953), and Hoyt *et al.* (1972) can be cited here, showing continuing modern interest in this basic puzzle of the human visual pathway. Hoyt and co-workers' study of the fiber bundles surviving in cases of cerebral hemiatrophy (Fig. 9.3) provides a particularly valuable view of the separation of human retina into nasal (contralaterally projecting) and temporal (ipsilaterally projecting) regions. In particular, these two areas are seen to join along a vertical axis that crosses the fovea.

Despite this work, however, several problems remain unresolved concerning the nasotemporal division of human retina:

1. How sharp is the separation of nasal and temporal regions of the retina? Linskz (1952) argued that the separation cannot be extremely sharp, because of the intrinsic scatter in neural wiring. Ogle (1962) argued that the persistence of Panum's area across the vertical meridian of the visual field means that there must be some overlap of ipsi- and contralaterally projecting areas of the retina along their border. More recently, students of the neural mechanisms of binocular depth discrimination (stereopsis) have argued (Fig. 9.4) that the high stereoacuity of humans in the region immediately in front of and behind the fixation point requires an overlap of the areas of the retina projecting to the two hemispheres, at least in the region of the fovea. Such an overlap could be produced by commissural connections between the visual cortices of the two hemispheres, or by a bilateral projection from the foveal region of the retina. Stone *et al.* (1973) and Bunt *et al.* (1977) have provided evidence of such a bilateral projection in the monkey.

2. Is the foveal region distinct from other regions of the retina in that all parts of it are represented in both hemispheres? The clinical observation of "macular sparing" (i.e., of vision spared at the fovea in an otherwise-blind hemifield), which is particularly common following cortical lesions, led to this suggestion. It is argued in Section 9.3.2 that, at least in the monkey and probably in the human and other primates as well, the fovea is indeed bilaterally represented, but that the region of bilateral representation extends all along the border between nasal and temporal retina. The bilateral representation of the fovea does not then provide a basis for macular sparing.

Considerable attention has been given in recent years to understanding the pattern of partial decussation in other mammals, in which an experimental approach is possible. The studies that are relevant to this section are those in which an attempt was made to plot the distribution over the retina of ganglion cells that project to the different hemispheres. When this is done, a division of the retina into nasal and temporal areas, according to the laterality of projection of the ganglion cells, can be usefully made. Overall, the nasotemporal division of the retina in nonprimates seems less sharp and precise than in primates; and this lack of precision has proved an intriguing problem to untangle.

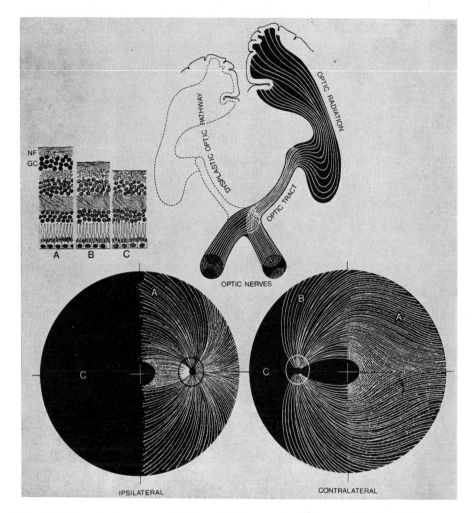

Figure 9.3. The axis of nasotemporal division of the human retina. Perimetry of the visual fields in cases of homonymous hemianopias had established that this axis is vertically oriented and crosses the fovea. These drawings of the pattern of axon bundles in the fiber layer of the retina in a case of dysplasia of one hemisphere provide an anatomical demonstration of the axis. The original legend reads:

"Pattern of distribution of nerve fibres (white lines) in the ipsilateral and contralateral fundi viewed from the front. The dysplastic side of the brain, the involved portion of the visual pathway, and the corresponding (blind) homonymous hemiretinae contain no visual fibres. Contrast the pattern of nerve fibres entering the ipsilateral disc (located in the 'seeing' hemiretina) with the pattern of the nerve fibres entering the contralateral disc (located in the 'blind' hemiretina).

The normal (A) and hypoplastic zones (B and C) of the retinae are indicated in each fundus (below) and in cross-sectional diagrams of the retinal layers (above, left):

(A) Normal hemiretina;

(B) Hypoplastic retina in contralateral fundus that contains no ganglion cells but is traversed by nerve fibres from the seeing hemiretina;

(C) Hypoplastic retina in both fundi that contains neither ganglion cells nor nerve fibres."

[From Hoyt *et al.* (1972). Reproduced with kind permission of the Editor of the *British Journal of Ophthalmology*.]

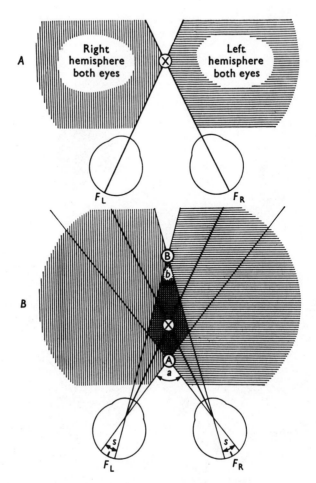

Figure 9.4. Blakemore's (1969) suggestion of the possible role of a strip of bilateral representation in mechanisms of binocular vision. Blakemore explained the diagrams as follows:

"A. Assuming that partial decussation exactly divides the retinae through the foveae, F_L and F_R, the region of space shaded with vertical lines is entirely represented in the right hemisphere, since it projects upon nasal retina in the left eye and temporal in the right. The horizontally striped area is likewise represented in the left hemisphere. The areas of space nearer than the fixation point, X, and beyond it should project to separate hemispheres through the two eyes.

B. Now if a central vertical strip of retina of angular width s is represented in both hemispheres, the striped regions of . . . A are expanded to include the area round the fixation point which projects to both hemispheres through both eyes."

In the "clean split" situation diagrammed in (A), the regions of the visual field in which stereoacuity is greatest (the regions just in front and beyond the fixation point) would not be represented binocularly in either hemisphere; yet binocular representation of an image in at least one hemisphere is believed necessary for stereopsis. The strip of bilateral representation solves this problem, by providing for binocular representation of these regions in both hemispheres. [Reproduced with kind permission of the *Journal of Physiology (London)*.]

9.3.2. The Nasotemporal Division of the Retina in the Monkey

The pattern of nasotemporal division of the retina in the monkey has been described by Stone *et al.* (1973), Bunt and Minckler (1977), and Bunt *et al.* (1977). These studies showed that, as expected from earlier observations (e.g., Gartner, 1951; Kupfer, 1953), the ganglion cells in the retina temporal to the fovea project to the ipsilateral side of the brain, while those nasal to the fovea project to the centralateral side. Both groups of workers went on to show that, when the areas of the retina containing ipsi- and contralaterally projecting ganglion cells were delineated in whole-mount retinas, each area is bounded by a vertical line that extends across the fovea (Fig. 9.5). In

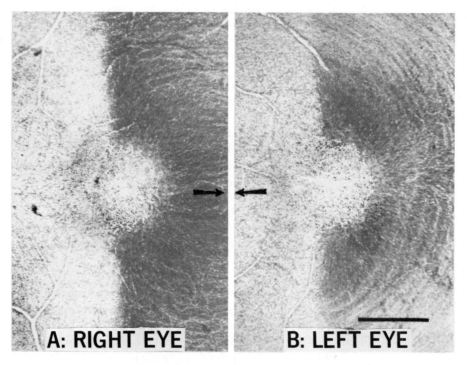

Figure 9.5. Relation of the axis of nasotemporal division to the fovea. The right optic tract of this monkey was sectioned and several months later whole-mount preparations of the retinas were made (from Stone *et al.*, 1973). (A) In the right retina, ganglion cells temporal to the fovea have degenerated and disappeared; the affected region of the retina (on the left of the photomicrograph) appears light, with scattered cells. Cells are present in large numbers above, below, and nasal to the fovea; these regions appear dark and crowded with cells. The arrow points nasally, toward the optic disc. The line separating light and dark areas seems sharp and vertical, and crosses the fovea (the light, circular formation at center) but slightly to the temporal side of its center. (B) The situation in the left eye of the same animal. Here, cells have disappeared from the retina nasal to the fovea (the arrow again points nasally, toward the optic disc), and the line between light and dark areas of the retina passes slightly nasal to the center of the fovea.

The following Fig. 9.6 shows our interpretation of these findings. [Reproduced with kind permission of the *Journal of Comparative Neurology*.]

each case, the line did not bisect the fovea but passed slightly (50–100 μm) to one side of the center. For example, the line that forms the edge of the nasal region of the retina containing contralaterally projecting cells passes slightly temporal to the center of the fovea (Fig. 9.5A), so that large numbers of contralaterally projecting cells are found around most of the foveal margin. Conversely, the line limiting the temporal region containing ipsilaterally projecting cells passes slightly nasal to the center of the fovea (Fig. 9.5B), so that, again, most of the foveal margin appears to contain large numbers of ipsilaterally projecting cells. Both groups of workers interpreted this observation as indicating that, within a vertically oriented strip of the retina about 100–200 μm wide (equivalent to about 1 deg of visual angle) and centered on the fovea, ipsi- and contralaterally projecting ganglion cells intermingle. Stone and co-workers' demonstration of this "median strip of overlap" of nasal and temporal areas of the retina is illustrated in Fig. 9.6. This strip of overlap would seem to match well the expectations of Linskz (1952), Ogle (1962), and other investigators of binocular vision discussed in the previous section.

The existence and dimensions of this strip of overlap have been confirmed in subsequent studies (e.g., DeMonasterio, 1978a). However, four qualifications to the above conclusions must also be noted. First, ganglion cells subserving the center of the fovea are displaced radially from the receptors that connect to them; it is this displacement that forms the foveal pit. As a consequence of it, the above conclusions may not hold for the fovea. The present results show considerable specificity in the movement of ganglion cells by which the fovea is formed; with few exceptions (demonstrated by Bunt and co-workers and discussed further below), contralaterally projecting cells move to the nasal margin of the fovea and ipsilaterally projecting cells to the temporal margin. Nevertheless, the presence of a 1-deg-wide "median strip of overlap" suggested in the above studies requires the assumption that each ganglion cell lies directly internal to its input receptors, and that assumption cannot be made at the fovea. It seems natural to assume that functionally the strip of overlap extends across the fovea without discontinuity, as it does across the area centralis of the cat retina (Section 9.3.3), but this remains an assumption.

Second, Bunt and Co-workers used the sensitive technique of histochemical localization of retrogradely transported HRP and found evidence that a small minority of cells located on the temporal edge of the fovea project contralaterally (i.e., the "wrong" way), while a similarly small minority of cells on the nasal margin of the fovea project ipsilaterally (also the "wrong" way). This pattern is illustrated in Fig. 9.7; these cells were not detected by Stone et al. (1973), who used a technique involving retrograde degeneration. Bunt and co-workers suggest that these cells create a "widening of the 1 deg of vertical overlap to a total of 3 deg at the fovea," hence providing a structural basis for macular sparing. However, two considerations seem to make this suggestion unlikely. On the one hand, as Bunt and co-workers note, most clinical reports of macular sparing refer to a region of spared vision much wider than the 1.5 deg of sparing that might be explained by Bunt and co-workers' observation. On the other hand, these cells can explain macular sparing only if they are located directly internal to their input receptors. As already noted, however, ganglion cells in the foveal region seem to be displaced radially from the receptors that connect to them, along the radius of the foveal pit. The cells at issue might be an exception to this pattern, but if they are not, then

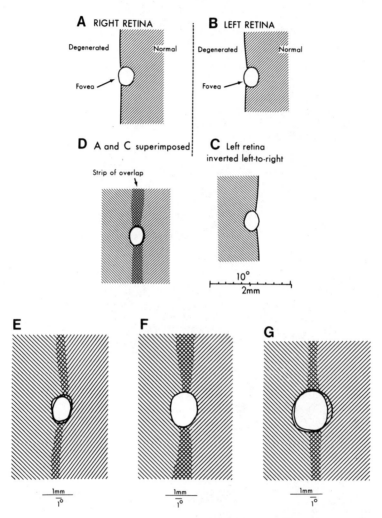

Figure 9.6. Argument for a strip of overlap in primate retina. The upper figure shows (A and B) schematic diagrams of the foveal regions of the retinas in Fig. 9.5. The fovea is outlined and the region of the retina containing ganglion cells is shaded. If (B) is reversed left-to-right, the diagram in (C) is obtained. Now (A) shows the distribution of contralaterally projecting cells around the fovea of a right retina and (C) shows the distribution of ipsilaterally projecting cells, also in a right retina. When (A) and (C) are superimposed, we obtain the diagram in (D). A narrow strip of retina is delineated that is vertically oriented and centered on the fovea. In that strip, ipsi- and contralaterally projecting ganglion cells intermingle. The axis of naso-temporal division can be envisaged as running down the center of this strip, crossing the center of the fovea.

The diagrams at the bottom (E, F, G) show similar strips delineated in three other monkeys. [All diagrams are from Stone *et al.* (1973). Reproduced with kind permission of the *Journal of Comparative Neurology*.]

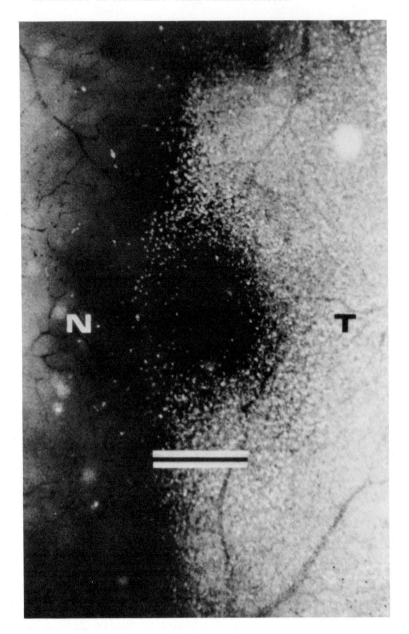

Figure 9.7. Demonstration of the nasotemporal division of the monkey retina by retrograde axonal transport. Bunt and Minckler (1977) injected HRP into one LGN of a monkey. The enzyme was transported retrogradely back along the axons of cells projecting to that LGN, to their somas. Those cells can be identified histochemically, by reacting the enzyme with a benzidine compound. Brown granules are deposited in cells that sent axons to the injected LGN. Under dark-field illumination, the cells containing HRP granules glow brightly, generating this beautiful picture of the foveal region of the left eye. The left (ipsilateral) LGN was injected and labeled cells are found temporal to (T), above, and below the fovea. This histochemical technique is very sensitive and detected a feature that was presumably present in Stone and co-workers'

they are likely to be connected to receptors closer to the center of the fovea than their somas. It is entirely feasible, and I would argue likely, that these cells are connected to receptors within 100 μm of the center of the fovea, i.e., within the median strip of overlap. If so, the strip of overlap is functionally no wider at the fovea than immediately above or below it, and Bunt and co-workers' observation does not provide a basis for macular sparing.

Third, Bunt and co-workers noted a very sparse scattering of ipsilaterally projecting ganglion cells up to 2 deg nasal to the strip of overlap and, symmetrically, an equally sparse scattering of contralaterally projecting cells up to 2 deg temporal to the strip. Further away, all cells project either ipsi- or contralaterally.

Fourth, the observations so far discussed concern the retina several (3 to 4) millimeters above or below the fovea. The retina extends at least 12 mm above and below the fovea, however, and the increasing scatter of ganglion cell somas further away from the fovea, and the distortion of the retina inevitable when it is flattened to make a whole mount, make it difficult to assess the width of the strip more than a few millimeters above or below the fovea. Some evidence of the width of the strip of overlap in the more peripheral regions comes from the study of Malpeli and Baker (1975). They used physiological techniques to study the retinotopic organization of the LGN of the monkey and (surprisingly in view of the anatomical work just discussed) found "no evidence of significant nasotemporal overlap in the central visual field." However, they did observe "negative azimuths" (which would indicate overlap) of up to 0.4 deg; the maximum predicted from the 1-deg-wide strip of overlap seen anatomically is 0.5 deg. Malpeli and Baker noted, however, that their maximum error of plotting might be as much as 0.3 deg, which may explain their conclusion that they had not observed significant overlap near the fovea. On the other hand, these workers noted negative azimuths of up to 6 deg in the far upper and lower regions of the visual field, and suggested that at corresponding distances from the fovea the strip of overlap becomes much wider. This widening could result from a decrease in the precision of the neural circuits in more peripheral parts of the retina, or from more positive considerations yet to be understood.

In general, however, the impression is gained that the nasotemporal division of monkey retina is very sharp (at least near the fovea), vertically oriented and centered on the fovea. A comparison of Figs. 9.3 and 9.5 suggests that the pattern in the human will prove to resemble closely that seen in the monkey. A comparison of Fig. 9.5 with Fig. 9.9, which is a summary diagram of the nasotemporal division in cat retina, brings out an important feature of the pattern in primates: a single pattern appears to be applicable to all groups of ganglion cells.

(1973) preparations, but which we missed. A small number of cells on the nasal side of the fovea are filled and presumably projected ipsilaterally; a similar number of cells on the nasal side of the fovea, detected at higher-power inspection, were not labeled, presumably because they projected contralaterally. A similar pattern was present in the right eye of the animal, with the labeling concentrating on the nasal side of the fovea.

Differing interpretations of the significance of this result are discussed in the text. [From Bunt and Minckler (1977), *Archives of Ophthalmology*, Vol. 95; copyright 1977 by the American Medical Association.]

9.3.3. The Nasotemporal Division of the Retina in the Common Cat

Ganser (1882) was the first to investigate the distribution in cat retina of cells projecting to different sides of the brain; he reported that ipsilaterally projecting ganglion cells are fewer in number than contralaterally projecting cells and are confined to temporal retina. In 1966, I reported an extension of Ganser's finding, made possible by the use of whole mounts rather than sections of the retina. One optic tract was sectioned in young (11-to-16-day-old) kittens causing, within several weeks, the apparently complete retrograde degeneration of the ganglion cells whose axons had been cut. The distribution of surviving ganglion cells was then studied in both retinas, and the following conclusions were suggested.

First, ipsilaterally projecting ganglion cells are almost all confined to a temporal region of the retina whose nasal boundary is an approximately vertical line that crosses the area centralis (Fig. 9.8A). Second, all ganglion cells in the region of the retina nasal to the area centralis project contralaterally, and an edge to this nasal area could also be seen (Fig. 9.8B), although it was less clear-cut than the edge in Fig. 9.8A. [In fact, an extremely small proportion ($< 0.1\%$) of ganglion cells in nasal retina apparently project ipsilaterally (Murakami *et al.*, 1982); their functional significance is not clear.] After mapping both retinas, I argued that a "median strip of overlap" could be delineated along the junction of nasal and temporal regions of the retina. The edges of the strip are the edges shown in Fig. 9.8A and B; the strip is about 200 μm wide (equivalent to about $0.9°$ visual angle), vertically oriented and centered on the area centralis. Within it, ipsi- and contralaterally projecting ganglion cells intermingle in about equal numbers. This strip seems to correspond closely to the $1°$-wide strip of overlap described in monkey retina (see preceding Section 9.3.2, Fig. 9.6). Third, a minority (about 25%) of ganglion cells in temporal retina project contralaterally. This is why the edge of the nasal region of the retina (Fig. 9.8B) does not appear sharp, and is a major point of difference between the visual pathways of cat and monkey.

Five substantial developments in the understanding of this pattern have emerged from subsequent work. The first followed the classification of ganglion cells. Stone and Fukuda (1974*b*) and Kirk *et al.* (1976*a,b*) showed that the pattern of nasotemporal division just described, and summarized in Fig. 9.9A, is a composite of distinct patterns found among W-, X-, and Y-class ganglion cells. Stone and Fukuda's analysis is shown in Fig. 9.9B; Kirk and co-workers' analysis is in close agreement and also provides evidence of distinct patterns among subgroups of W cells. Among X cells there is a quite precise pattern of nasotemporal division, very similar to that described for the whole ganglion cell population of the monkey. Ganglion cells in temporal retina project ipsilaterally, those in nasal retina project contralaterally; the ipsi- and contralaterally projecting cells intermingle within the $0.9°$-wide median strip of overlap. Among Y and W cells, all those found nasal to the area centralis project contralaterally, as do X cells. However, a proportion of both cell classes located temporal to the area centralis project contralaterally; specifically, a small minority (about 5%) of the Y cells and a distinct majority (about 60%) of the W cells project to the "wrong" (contralateral) side of the brain. Moreover, for Y and W cells, the region of transition from the projection pattern typical of nasal retina to that typical of temporal retina is centered, not on the area

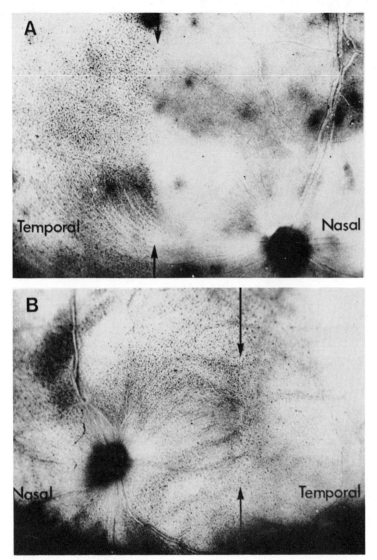

Figure 9.8. The nasotemporal division of the cat retina. These photographs (from Stone, 1966) show the appearance of the ganglion cell layer of the retina several weeks after section of the left optic tract in a young kitten. The retinas are still *in situ* in the posterior half of the eyeball; a Nissl stain (methylene blue) was applied after formalin fixation. In the right retina (shown at top) surviving ganglion cells are found temporal to an approximately vertical line (arrowed), that traverses the area centralis (the concentration of ganglion cells at lower center). The optic disc is in the lower right of the field.

In the left eye (B), ganglion cells survive in normal numbers nasal to the area centralis. Some surviving cells are found in temporal retina, but they comprise only a minority of the normal population. I suggested that a line of transition (arrowed) between normal and reduced numbers of cells could be delineated. This line is also approximately vertical and traverses the area centralis. [Reproduced with kind permission of the *Journal of Comparative Neurology.*]

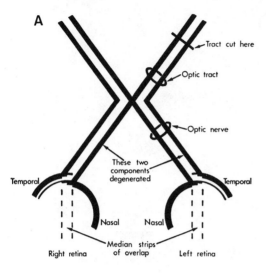

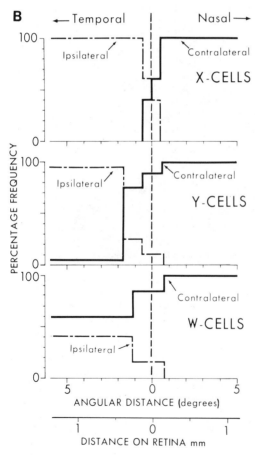

Figure 9.9. W-, X-, and Y-cell contributions to the pattern of nasotemporal division of the cat retina. (A) Schematic diagram of the nasotemporal division of the cat retina (from Stone, 1966; the diagram also shows some of the experimental design). The portions of each retina that project ipsi- and contralaterally are drawn separately. The variations in the thickness of the retinas indicate variations in the percentage of ganglion cells that belong to ipsi- and contralaterally projecting components. Note that the proportion of contralaterally projecting cells is 100% nasal to the area centralis. Across the narrow median strip of overlap, which is centered on the area centralis, that proportion drops sharply to 25%, and throughout temporal retina approximately 25% of cells project centralaterally. The remaining cells project ipsilaterally. Dimensions are not to scale. (B) These three graphs (from Stone and Fukuda, 1974*b*) show estimates of the contributions of X, Y, and W cells to the above pattern. For X cells there is a sharp transition from nasal retina (from which all project contralaterally) to temporal retina (from which all project ipsilaterally) and the transition is centered on the fovea.

For both Y and W cells, the zone of sharpest change in the laterality of their projection is centered not at the area centralis, but 200–300 μm temporal to the area centralis. About 5% of Y cells and 60% of W cells throughout temporal retina project contralaterally.

As discussed in the text, the median strip of overlap may be considerably wider above and below the area centralis. Also, the pattern shown for W cells has been shown to comprise distinct patterns found in the subgroups of W cells. [Reproduced with kind permission of the *Journal of Comparative Neurology*.]

centralis (as it is in X cells), but 0.2–0.3 mm more temporal. Stone and Fukuda (1974*b*) noted that among W cells in temporal retina, those termed *tonic* or (by Rowe and Stone, 1977) *W1 cells* mostly projected ipsilaterally, while those termed *phasic* or *W2 cells* mostly projected contralaterally. Kirk and co-workers provided a more detailed description of the laterality of projection of W-cell subclasses. These observations have been corroborated by the morphological study of Rowe and Dreher (1979, 1982*b*), who showed that small- and medium-soma γ cells (which may correspond to major subclasses of W cells) have distinct patterns of nasotemporal division, that of medium-soma cells resembling W1 cells, that of small-soma cells resembling W2 cells. For all subclasses, however, as for Y cells, the transition from nasal to temporal patterns of projection appears displaced temporally from the median strip of overlap.

Second, Sanderson and Sherman (1971), Cooper and Pettigrew (1979*b*), and Rowe and Dreher (1979, 1982*b*) have provided evidence of the nasotemporal division of the retina, considering just those cells projecting to the thalamus. In brief, they provided evidence that much of the pattern shown in Figs. 9.8 and 9.9 is formed by cells that project to the thalamus (see Fig. 9.10 and its legend for detail of Cooper and Pettigrew's experiment). The small-soma γ (W) cells throughout temporal retina that project contralaterally (the "wrong" way) apparently project to the midbrain; several studies (e.g., Harting and Guillery, 1976) have provided evidence that these cells project to the anterior pole of the contralateral SC. However, many of the medium-soma γ (also W) cells that project contralaterally reach the medial interlaminar component of the LGN, in the forebrain (Rowe and Dreher, 1979, 1982*b*), and presumably contribute to the overall pattern observed by Cooper and Pettigrew.

Third, Cooper and Pettigrew (1979*b*) have provided evidence that, as in the monkey (Section 9.3.2), the width of the median strip of overlap may increase considerably with distance above (and presumably below) the area centralis. They confirmed that its width is about 200 μm at the area centralis, but suggested that 3 mm above the area centralis it is fully 1.3 mm wide.

Fourth, the occurrence of contralaterally projecting cells in temporal retina is not unique to the cat. It has been observed in the rabbit (Provis, 1979), possum (Tancred and Rowe, 1979), fox (Rapaport *et al.,* 1979), rat (Cowey and Franzini, 1979), and opossum (Rowe *et al.,* 1981) and may be present in all nonprimate mammals (Lane *et al.,* 1971; Stone *et al.,* 1973).

Fifth, a quite different, apparently abnormal pattern of nasotemporal division has been reported in the Siamese cat; this is discussed in the following Section 9.3.4.

Some functional implications: The patterns illustrated in Fig. 9.9 have led to a number of interesting suggestions about the functions of different ganglion cell classes. For example, the nasotemporal division is sharpest among X cells, and only among X cells is the division centered on the area centralis and, presumably therefore, on the fixation point. Following the argument in Fig. 9.4, it seems reasonable to suggest that in the cat, X cells (rather than Y or W cells) subserve binocular depth discriminations close to the fixation point. Since the small receptive fields of X cells and their concentration at the area centralis make it likely that X cells are also important in high spatial resolution, it seems reasonable to suggest that the mechanisms of binocular depth discrimination and high-acuity vision both involve X cells and may be closely interrelated.

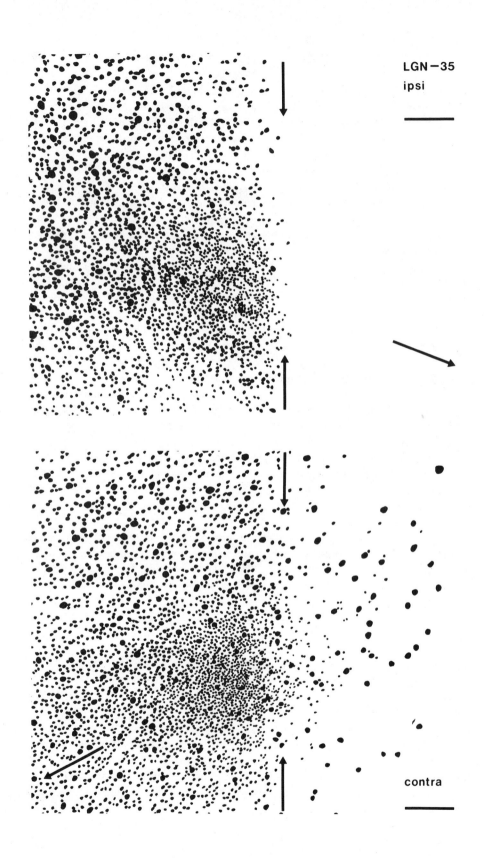

LGN−35
ipsi

contra

Interpretations have also been proposed of the patterns seen among the other ganglion cell classes. For example, the distinct pattern of nasotemporal division of Y cells, and certain of their receptive field properties, was the basis of Levick's (1977) suggestion that they subserve a particular role in the detection of objects located closer to the eyes than the fixation point. Again, many of the contralaterally projecting W cells located in temporal retina have been shown to project to the midbrain (see above). Because such contralateral projections to the midbrain are less prominent in mammals than in other vertebrate groups, the suggestion has been made (Stone, 1966; Mitzdorf and Singer, 1977) that W cells may be phylogenetically "older" than other ganglion cell classes. This suggestion has received little support, however, and recent work suggests that retinal projections to forebrain sites are phylogenetically very old, although in many species less prominent than the projection of each eye to the contralateral optic tectum.

To these ideas I add here that the division of the retina into nasal and temporal regions seems important for binocular vision, and is particularly sharp in species (such as some primates and carnivores) in which binocular fixation is a well developed component of visual behavior.

9.3.4. The Nasotemporal Division of the Retina in the Siamese Cat

The work of Guillery (1969a), Guillery and Kaas (1971), and Hubel and Wiesel (1971) drew attention to the presence in the Siamese cat of a major, apparently hereditary abnormality in the proportion of fibers of each optic nerve that decussate and pass to the contralateral side of the brain. The proportion of contralaterally projecting fibers is unusually high in the Siamese cat, about 80–87% (Stone et al., 1978) as against about 72% in the common cat [my own estimate from the data for Fig. 1 in Rowe and Stone (1976a), assuming that 25% of the ganglion cells temporal to the area centralis project contralaterally]. This high proportion is associated with marked disturbances in the pattern of lamination of the dLGN and in the retinotopic organization of the geniculocortical pathway.

Guillery (1969a) and subsequently Guillery and Kaas (1971) and Lund (1975) suggested that there must be a major abnormality in the nasotemporal division of Siamese retina, and studies from several laboratories have recently provided descriptions of this abnormality. Stone et al. (1976, 1978) and Marzi (1978, 1980) reported studies

---

←

Figure 9.10. The nasotemporal division for ganglion cells projecting to the thalamus in the cat. Cooper and Pettigrew (1979b) provided these superb diagrams of the distribution in the region of the area centralis of ganglion cells that project to the thalamus of one hemisphere. The cells were labeled histochemically following an HRP injection into the right LGN that also spread into surrounding thalamic nuclei; these are camera lucida drawings made from retinal whole mounts.

The pattern resembles that seen for the ipsi-/contra-analysis of all ganglion cells (Figs. 9.8 and 9.9). Ipsilaterally projecting cells (top drawing) are confined to a region of temporal retina bounded nasally by a vertical line (arrowed) that traverses the area centralis. Contralaterally projecting cells (lower drawing) in temporal retina include both large somas (α/Y cells) and small somas (γ/W cells). The principal differences between this pattern and that observed for all cells (Fig. 9.8) is that the contralaterally projecting small (presumably W) cells in temporal retina are confined to within 1 mm of the area centralis. The scale represents 200 μm. [Reproduced with kind permission of the *Journal of Comparative Neurology*.]

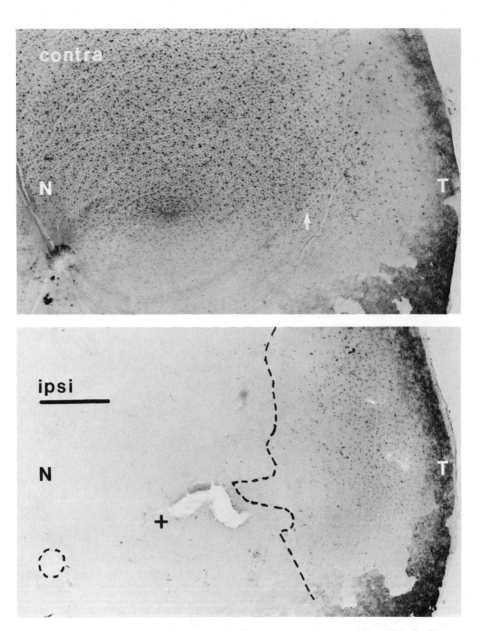

Figure 9.11. The nasotemporal division of the retina for thalamus-projecting ganglion cells in the Siamese cat. Regions of whole mounts of the retinas of a Siamese cat after injection of HRP into the right LGN (from Cooper and Pettigrew, 1979c). The upper photomicrograph shows that contralaterally projecting ganglion cells are found in massive numbers in temporal retina; the whole of the area centralis appears to project contralaterally. Conversely, the ipsilaterally projecting cells shown in the lower photomicrograph, taken from the other retina, are far fewer in number than in the common cat (Fig. 9.8) and, in this animal, are restricted to far temporal retina, temporal to the dashed line. Note that the lower photomicrograph is from the right retina of the animal; it is reversed left-to-right for easy comparison with the other retina.

Thus, the axis of nasotemporal division in this breed appears to be shifted well temporal to the area centralis. [Reproduced with kind permission of the *Journal of Comparative Neurology.*]

based on ganglion cell degeneration following optic tract section, Kirk *et al.* (1976*c*) reported a physiologically based study, Cooper and Pettigrew (1977, 1979*c*) reported a study of the nasotemporal division of the retinothalamic component of the Siamese visual pathway (Fig. 9.11), and Murakami *et al.* (1982) have provided both anatomical and physiological evidence of abnormalities in the topography of temporal retina in Siamese cats. These studies are in close agreement, and may be summarized as follows:

1. As previous work had predicted, many ganglion cells in the temporal retina project contralaterally, rather than ipsilaterally, as they would in the common cat. This is true for ganglion cells of all sizes and functional classes.

2. The zone of transition from the pattern of projection typical of nasal retina (all contralateral) to the pattern typical of temporal retina (most ipsilateral) is much wider in the Siamese cat. Stone and co-workers estimate the zone as 3 to 8 mm wide (varying between individuals) as against 0.5 mm or less in the common cat. Kirk and co-workers estimated that for two classes of ganglion cells (Y and X cells), the zone is 30 deg wide, or about 7 mm on the retina. Because of the width of the zone of transition, it is difficult to talk of a "median strip of overlap," in the sense used in the common cat and monkey. Nevertheless, the zone of transition does appear to be oriented approximately vertically.

3. The zone of transition is centered, not at the area centralis, as in the common cat, but rather a few millimeters temporal to the area centralis, which therefore projects almost entirely contralaterally. For example, Stone and co-workers estimate the temporal displacement of this zone of transition at 1.7 to 3.0 mm, variable between cats, and a similar range but somewhat larger displacement was observed by Cooper and Pettigrew (1979*c*). Kirk and co-workers estimated the displacement at 15 deg, about 3.5 mm on the retina.

4. The abnormality is more marked among Y cells than among the population of ganglion cells as a whole. For example, Stone and co-workers concluded that the center of the zone of transition for Y cells is centered still further temporal than for the population as a whole, about 3–5 mm as against 1.7–3 mm. Further, they estimated that the proportion of Y cells projecting ipsilaterally is lower than that of the population as a whole (4–6% as against 13–20%). Confirming this, Murakami *et al.* (1982) concluded from HRP and field potential studies that very few of the large-soma (Y) cells in temporal retina project to the ipsilateral LGN.

5. There is a gradual increase in the proportion of ipsilaterally projecting ganglion cells between the area centralis and the zone of transition; early studies had suggested some discontinuity in this pattern.

Clearly, the pattern of nasotemporal division of the retina in the Siamese cat is markedly different from and more variable than that of the common cat. In terms of the two-axis model being developed, the Siamese retina is abnormal in that the axis of nasotemporal division is severely blurred and is shifted away from the area centralis. It might be simplistic to attribute the squint, poor acuity (Blake and Antoinetti, 1976), and poor stereoacuity (Packwood and Gordon, 1975) of the Siamese cat directly to this blurring and shift, but it does seem apparent that normal binocularity and acuity of vision are incompatible with this arrangement.

9.3.5. The Nasotemporal Division of the Retina in Marsupials, the Rabbit, Rodents, and the Fox

Descriptions of the nasotemporal division of the retina are available for species drawn from several mammalian orders. The descriptions provide some indication of the features of the division that are general among mammals, and of the features found only in certain species or groups of species.

Possum and opossum: Modern marsupials, such as the Australian brush-tailed possum and the North American opossum, represent a line of phylogenetic development long separated from that of the placental mammals. Moreover, the possum is a member of one major subgroup of marsupials (the diprodonts) and the opossum is a member of the other subgroup (the polyprodonts) so that the histories of these two species have long been separate from each other. It is remarkable, therefore, to find close similarities between these species and the cat in the pattern of nasotemporal division of the retina. In the brush-tailed possum, Tancred and Rowe (1979) have shown that ipsilaterally projecting ganglion cells are confined to a region of the retina temporal to the area centralis; as in the cat, the nasal boundary of this region is a straight, approximately vertical line that passes across the area centralis (Fig. 9.12). All cells in nasal retina and a minority of cells in temporal retina project contralaterally. This minority is rather larger in the possum than in the cat (40% as against 25%) but, as in the cat, most of the contralaterally projecting ganglion cells in temporal retina are small-bodied and project to the midbrain. Thus, if the medium- and large-bodied ganglion cells were studied separately, it seems likely that for them a quite sharp pattern of nasotemporal division might be demonstrable, similar to that of X cells in the cat.

In the opossum, Rowe *et al.* (1981) have similarly shown that ipsilaterally projecting ganglion cells are largely confined to the retina temporal to the area centralis where they intermingle with many contralaterally projecting cells. They noted, moreover, that the ipsilaterally projecting ganglion cells in temporal retina are from particular size groups; as in the cat and possum, these cells are either large or medium-sized in soma diameter. Further, they have shown in subsequent work (personal communication) that (as in the monkey, cat, and possum) the nasal boundary of the region of the retina containing ipsilaterally projecting cells is approximately vertical and crosses the area centralis. It is intriguing that, despite the long phylogenetic separation of these species, these details of the pattern of nasotemporal division are common to them.

Rabbit: The rabbit has undergone a long period of phylogenetic development separate from other placental mammals. It has developed a highly differentiated and distinctive visual system, with laterally placed eyes whose retinal topography (Fig. 9.15) is dominated by a massive visual streak. It has only a small region of binocular visual field (Hughes, 1971) and (see Section 9.5.3 below) its area centralis is relatively poorly developed and inconspicuous. Moreover, only a small proportion of ganglion cells project ipsilaterally. Nevertheless, Provis (1979) described a specialization of the retina at the temporal end of the visual streak [the region that, as Hughes (1971) pointed out, the

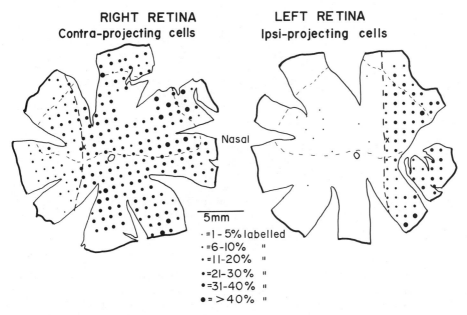

RIGHT RETINA
Contra-projecting cells

LEFT RETINA
Ipsi-projecting cells

Nasal

5mm

· =1 - 5% labelled
· =6-10%   "
· =11-20%   "
•=21-30%   "
●=31-40%   "
●= >40%   "

Figure 9.12. The nasotemporal division of the possum retina. The maps show the distribution of labeled ganglion cells in the right and left retinas of a possum, following injection of HRP into the left optic tract. The distribution is represented as the proportion of cells in local regions of the retina that show the HRP label. The lightly dashed line in each map represents the edge of the tapetum. As in the cat and monkey, the ipsilaterally projecting ganglion cells (those found labeled in the left retina) are, with very few exceptions, confined to a region of temporal retina bounded nasally by an approximately vertical line (heavily dashed) that traverses the area centralis. Virtually all cells nasal to the area centralis project contralaterally, and contralaterally projecting cells are also found throughout temporal retina (as in the cat, but not the monkey). As in the cat, such cells are small or large in soma size. The approximately vertical dashed line in the map at left represents the temporal limit of the region in which contralaterally projecting cells equaled or outnumbered ipsilaterally projecting cells. [Maps kindly provided by E. Tancred.]

rabbit uses for binocular vision], which might correspond to the area centralis of other mammals. The major feature of this specialization was a localized maximum in the density of large-bodied ganglion cells.

Provis and Watson (1981) have shown that the pattern of nasotemporal division in the rabbit is in many ways different from that seen in the cat. Only about 10% of all ganglion cells project ipsilaterally (Giolli and Guthrie, 1969), and there is little impression of even a blurred edge to the nasal region of the retina from which ganglion cells project contralaterally; such an edge is strikingly sharp in the primate (Figs. 9.3, 9.5A) and somewhat less clear in the cat (Fig. 9.8B). Moreover, the number of cells in nasal retina that project ipsilaterally seems considerably higher in the rabbit (1% of cells as against < 0. 1% in the cat). Further, Provis and Watson note that (by way of contrast with the cat and the possum) the cells in temporal retina that project contralaterally are mostly large or medium in soma size, rather than large or small. Despite these differences, one basic aspect at least of the nasotemporal division of the retina in the rabbit is

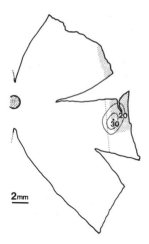

Figure 9.13. The nasotemporal division of the rabbit retina. This map shows the temporal half of the left retina of a pigmented rabbit. HRP had been injected into the left (ipsilateral) optic tract. Ipsilaterally projecting ganglion cells (i.e., those found labeled with HRP) were concentrated in the area shaded; note that this area is at the temporal margin of the retina and that its nasal border is an approximately vertical line. Some labeled cells were found nasal to this line; they were sparsely scattered and were confined to the retina temporal to the dotted line, which is also approximately vertical.

The isodensity lines labeled 20 and 30 show the distribution of large ganglion cells (labeled and unlabeled) at the temporal margin of the retina; this is the large-cell node that may (see text) correspond to the area or fovea centralis of other mammals. Note that the nasal border of the area containing ipsilaterally projecting ganglion cells crosses this node (the peak of the node is represented by the dot). [Reproduced with kind permission from Provis and Watson (1981).]

common to previously examined species. The ipsilaterally projecting ganglion cells are, with few exceptions, confined to an area of temporal retina whose nasal boundary is an approximately vertical line that crosses the presumed area centralis (Fig. 9.13). Thus, as in the monkey, cat, and possum, a vertical axis of nasotemporal division can be detected as the nasal boundary of the temporal region of the retina containing ipsilaterally projecting ganglion cells (Fig. 9.13).

Rat and mouse: Cowey and Franzini (1979) and Keens (1981) have provided a description of the location of ipsilaterally projecting ganglion cells in the rat. Such cells, which comprise approximately 20% of the overall population, are located (as in other mammals) in temporal retina. Moreover, Cowey and Perry's (1979) report indicates that, as in the cat, possum, and rabbit, some cells throughout temporal retina project contralaterally. Given these basic similarities, however, it is interesting to note three distinctive features. First, the nasal boundary to the region containing ipsilaterally projecting cells seems more poorly defined than in other species. Second, the boundary does not appear to be vertical; the region containing ipsilateral ganglion cells extends markedly into inferior retina, but not at all into superior retina, so that its nasal border, inasmuch as it can be delineated, tilts laterally at its top. Third (Keens, 1981; Fig. 9.14), the boundary does not cross the peak of ganglion cell density, which Fukuda (1977) identified as the area centralis, but runs 2–3 mm lateral to it.

Dräger and Olsen's (1980) report of the nasotemporal division of the mouse retina indicates a marked similarity with the rat. Again, features were noted that are present in many nonrodents. For example, ipsilaterally projecting cells were largely confined to a region at the temporal margin of the retina, with only a small minority of cells in nasal retina projecting ipsilaterally; a considerable proportion of cells throughout that temporal region project contralaterally, so that this temporal region projects bilaterally; and there is a size difference between cells projecting ipsi- and contralaterally from temporal retina (the latter are distinctly smaller), indicating that they may represent different functional classes of ganglion cells. In two important aspects, however, the pattern in the mouse, as in the rat, is distinct from that of other species so far described. First, the

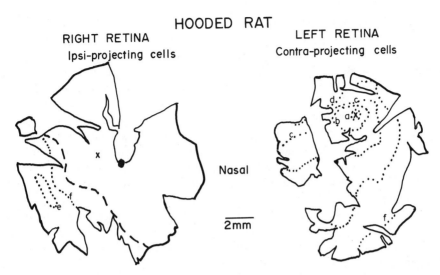

Figure 9.14. The nasotemporal division of the rat retina. The maps represent whole-mount preparations of the left and right retinas of a pigmented rat, following an injection of HRP into the right optic tract. The optic disc is shown as a filled circle (split in the left retina), and the region of peak ganglion cell density is indicated with a cross. The dotted lines represent isodensity lines for labeled ganglion cells, with values as follows: a, 5000/mm^2; b, 4000/mm^2; c, 3000/mm^2; d, 2000/mm^2; e, 1500/mm^2; f, 1000/mm^2.

Ganglion cells labeled with HRP were found throughout the retina contralateral to the injection (map on the right). In the retina ipsilateral to the injection, labeled ganglion cells were found concentrated in a crescent along the inferotemporal margin of the retina. Note that the inner boundary of this region (indicated by the dashed line) is not vertical, as it is in other mammals examined, and that it is located about 2 mm temporal and inferior to the region of peak cell density. In albino rats, ipsilaterally projecting ganglion cells were fewer in number and concentrated in an even narrower crescent, separated from the region of peak cell density by as much as 5 mm. [Maps kindly made available by J. S. Keens.]

region in which ipsilaterally projecting cells are found extends into inferior retina, so that the boundary to the region is not vertical, but strongly tilted laterally; indeed, the trend for ipsilaterally projecting cells to lie in inferior retina seems even more marked than in the rat. Second, Dräger and Olsen also note that the boundary does not cross the region of peak ganglion cell density; rather, as in the rat, the boundary runs temporal to and below the region of peak cell density.

These aspects of retinal topography may, however, prove quite consistent with the two-axis model of retinal topography proposed here; if so, the model is strengthened, and much learnt about the rodent retina, from the following considerations. First, the lack of sharpness to the pattern of nasotemporal division may reflect the poor development of binocular vision in this species (see Hale *et al.*, 1979, for discussion of this issue). Second, the inferior placement of the region containing ipsilaterally ganglion cells matches the above-the-horizon position of the binocular part of the rat's visual field (A. Hughes, 1977); this region of the retina seems, in the rodent as in other mammals, particularly related to binocular vision. Third, the separation of the line of nasotemporal division from the region of peak cell density may be evidence that this region is not the homolog of the area or fovea centralis of other species, but rather represents (as does the peak cell density of the rabbit) the crest of the visual streak. This issue is further discussed in Section 9.8.2.

Cowey and Perry (1979), Dräger and Olsen (1980), and Jeffery *et al.* (1981) comment on the possibility that, as suggested by Cunningham and Freeman (1977), a substantial proportion of ganglion cells in the rodent retina project to both hemispheres by means of a branching of their axons, one branch entering each optic tract. Jeffery and co-workers used a double-labeling technique to obtain some assessment of the proportion of double-projecting cells. They suggest that such cells are found only in the inferotemporal crescent of the retina in which ipsilaterally projecting cells concentrate; and that double-projecting cells comprise only 1% of ganglion cells in that region.

Fox: In the gray fox (Rapaport *et al.,* 1979), the pattern of nasotemporal division of the retina resembles fairly closely that seen in the cat. Ipsilaterally projecting cells are found only in temporal retina, and the nasal boundary of temporal retina is sharp, and runs vertically across the area centralis. Some cells in temporal retina project contralaterally; as in the cat, most of these cells belong either to small-soma or to large-soma ganglion cell groups. One distinct difference from the pattern found in the cat was noted. In the fox, virtually all large-soma ganglion cells throughout temporal retina project contralaterally, whereas in the cat only a small minority do so. The difference is quite striking, especially given the relatively close phylogenetic relationship of the two species.

9.3.6. Summary

In probably all mammals, a minority of ganglion cells project ipsilaterally, and such cells are almost entirely restricted to temporal retina. The completeness and sharpness of the division of the retina into nasal (contralaterally projecting) and temporal (all or partly ipsilaterally projecting) areas vary greatly between species, however, from the sharp, clear pattern seen in the monkey, to the pattern in, for example, the rabbit, in which the presence of a vertical axis of nasotemporal division can be detected only in the distribution of ipsilaterally projecting ganglion cells. Among the limited number of mammals studied, the sharpness and completeness of the division seem related to factors such as the degree of development of the area or fovea centralis, the degree of frontalization of the eyes, and the proportion of ganglion cells that project ipsilaterally.

In all mammals studied except the rodents (rat and mouse), the axis of nasotemporal division is vertically oriented and crosses the area centralis. In some mammals, possibly all nonprimates, the pattern of nasotemporal division differs among the different classes of ganglion cells. In the cat, for example, the pattern is sharpest and clearest among X cells (among which the pattern resembles that seen for all ganglion cells in the primate), and least developed among the W2 subgroup of W cells. In the possum, the nasotemporal division of the retina varies between ganglion cells of different sizes in a way remarkably similar to that of the cat; but in the rabbit and fox, a somewhat different correlation between cell size and the pattern of nasotemporal division has been reported.

Clearly, much remains to be learnt about the nasotemporal division of the retina, especially about its involvement of different classes of ganglion cells. That the pattern is fundamental to the organization of the visual pathway is indicated by the situation in the Siamese cat, in which a gross disturbance of the nasotemporal division is accompanied by marked abnormalities in the organization of the visual pathways, and deficiencies in visual performance.

9.4. THE HORIZONTAL AXIS: THE VISUAL STREAK

The term *visual streak* was used by Davis (1929) to refer to an elongated, horizontally oriented region of specialized structure in the rabbit retina. Confirming and expanding the observations of several earlier workers, Davis noted that ganglion cells are more numerous in the visual streak than above or below it, that the receptors appear longer, and that both inner and outer plexiform layers are relatively thick. The spatial layout of the streak is dramatically apparent in maps of the ganglion cell layer (Fig. 9.15), but comparable patterns have also been observed in the distribution of cones [e.g., by Hughes (1971) in the rabbit and Steinberg *et al.* (1973) in the cat, Fig. 9.20] and in the distribution of horizontal cells (Wässle *et al.,* 1978). The visual streak is thus not exclusively, or even predominantly, a specialization of the ganglion cell layer. It involves several of the neuronal classes of the retina.

Where the streak is strongly developed, it is often apparent in macroscopic features of the retina or fundus. In the rabbit, for example, a striking feature of the fundus of the eye is formed by the vascularized, horizontally spreading bundles of myelinated (and therefore opaque) ganglion cell axons that course along the upper edge of the concentration of ganglion cells. In species with a retinal circulation, the location of a strongly developed streak can be recognized by a watershed of blood vessels, centered on the ridge of the streak; for example, in the retina of the marsupial Tasmanian devil (Fig. 9.16). It is proposed in this section that:

1. A visual streak is present in most mammalian retinas examined, and may therefore have been developed at an early stage in the phylogenetic history of mammals.
2. The streak is mainly a feature of the region of the retina nasal to the axis of nasotemporal division. It is thus, at least principally, a feature of the crossed component of the visual pathway.
3. The streak may involve, principally or exclusively, certain functional groups of ganglion cells.
4. The streak is a specialization quite distinct from the area centralis and subserves a distinct function.

In short, the suggestion is made that the visual streak is fundamental to the organization of mammalian retina in two senses. First, it is an important parameter in the organization of retinal topography, and second, it may be part of the mammalian heritage. Expanding on this latter suggestion, it is possible (arguably likely) that the patterns of retinal topography found in extant mammals are derivatives of a pattern that was present in early mammals and whose history may extend into their reptilian ancestors. While the characteristics of that early mammalian pattern cannot in any practical way be determined, the widespread occurrence of a visual streak among extant mammals suggests that a streak was part of it. The strong development of the streak in, for exam-

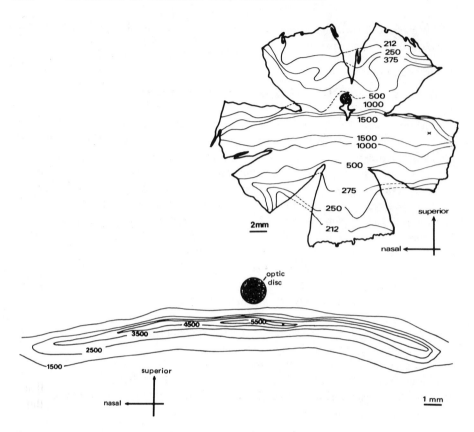

Figure 9.15. The distribution of ganglion cells in the rabbit retina. The upper map shows the outline of the left retina of a pigmented rabbit. The optic disc is shaded and the lines are isodensity lines, outlining the regions of the retina within which density was above a criterion level (expressed as cells/mm²). The cross in temporal retina marks the position of the maximum density of large ganglion cells in this retina. The isodensity lines are interrupted where estimates of ganglion cell density could not be made.

The upper diagram does not show the pattern of isodensity lines in the visual streak; this is shown separately in the lower diagram, for the same retina. The asterisk indicates the position of the maximum density of all ganglion cells. [Reproduced from Provis (1979) with kind permission of the *Journal of Comparative Neurology*.]

ple, the rabbit and its weak development in primates might then be regarded as evolutionary variants on a basic pattern.

An analogy might usefully be drawn with the five-digit extremity of the mammalian forelimb. That extremity is recognizable in all mammals, although its form and functional specialization differ extraordinarily between, for example, bats (in which the digits form part of the wing framework), ungulates (in which the digits form stubby toes), sea-dwelling mammals (in which the digits form the internal framework of a digit-

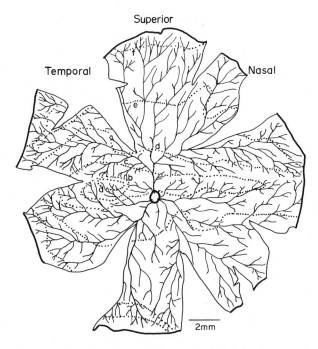

Superior

Temporal

Nasal

2mm

Figure 9.16. Distribution of ganglion cells in the Tasmanian devil retina. This carnivorous Australian marsupial has a strongly developed retinal circulation, similar to that found in many placental carnivores. The ganglion cell density map (shown superimposed on a representation of retinal blood vessels) shows a strongly developed visual streak and an identifiable area centralis. Note that the location of the streak matches closely a "watershed" of blood vessels in nasal retina. [Map kindly provided by Tancred.]

less flipper), and primates (in which the hand's manipulative ability may have been of great importance in the development of intelligence). I draw this analogy because the implications of point (1) above seem substantial. This view of the visual streak as a functional variant of a phylogenetically inherited character, rather than as a purely functional adaptation, has two important implications at least. First, it implies that at least some vestige of a visual streak is likely to be present in most mammals, even if in their particular ecological niches the specialization is of no use. Second, it implies that the function of the visual streak may vary between species; as with forelimb digits, the same structure might be adapted for quite distinct functions.

Descriptions of ganglion cell distribution are now available for many species of mammals; these are illustrated below or summarized in Section 9.7. The following three parts of the current Section 9.4 concern the visual streak of the rabbit, cat, monkey, and opossum. It is principally in the first three of these species that analysis of retinal topography has gone beyond the mapping of ganglion cell distribution to the analysis of the differential distribution of different classes of ganglion cells; and, fortunately for my argument, the degree of development of the visual streak varies widely among them, from its dominance in the rabbit retina to its relative insignificance in the monkey. Discussion of the opossum is included because it is the one species so far examined in which there seems to be no evidence of a visual streak specialization.

The following discussion is in good agreement with previous discussions and interpretations of the visual streak (e.g., Slonaker, 1897; Davis, 1929; Munk, 1970; Rodieck, 1973; A. Hughes, 1977). The interpretation of retinal topography suggested is close to that of A. Hughes (1977) in viewing variability in topography as the result of different evolutionary development; it adds the historical perspective that that variability may be imposed on a pattern of retinal topography common to most or all mammals.

9.4.1. The Visual Streak of the Rabbit Retina

Ganglion cell composition: Every published description of the ganglion cell layer of rabbit retina has made mention of the concentration of ganglion cells in the visual streak. Particularly when seen in Nissl-stained whole mounts, this is a prominent feature of the visual streak specialization. The analyses of both Hughes (1971) and Provis (1979) have shown that ganglion cell density is higher along the ridge of the streak than elsewhere in the retina, reaching values of 5000–6000 cells/mm^2. Cell density falls off rapidly with distance above or below the streak, but the ridge of high density spans much of the width of the retina. In some retinas, the isodensity lines that outline the visual streak extend all the way to the temporal edge of the retina, and in others (compare Figs. 4 and 5 in Provis, 1979) the streak seems to stop distinctly short of the temporal edge. The point of maximum ganglion cell density is found toward the middle of the streak, and perhaps because of this several authors have viewed the streak as a modified area centralis. (In many species, it is at the area centralis that ganglion cell density reaches its maximum.) Provis (1979) noted, however, that the location of maximum cell density is difficult to specify, because the density gradient along the streak is quite shallow. Moreover, the location of the maximum is quite variable between individual rabbits, whereas in species with a clearly identifiable area centralis, the location of the density maximum is both sharp and constant. Hughes noted that, because of the lateral placement of the rabbits' eyes (the optic axes of the eyes are directed almost directly sideways from the head), the rabbit has a smaller binocular visual field than many other mammals. That is, the only part of its very wide visual field that is simultaneously seen by both retinas is a segment approximately 30 deg wide, directly in front of the head. This binocular part of the visual field is imaged on the retina at the temporal end of the visual streak; and both Hughes and Provis described a ganglion cell specialization at the temporal end of the streak that may correspond to the area centralis. In Provis' analysis, this region contains a localized peak in the density of a class of large-bodied ganglion cells, already discussed in Section 9.3.5. Hughes observed a "bulging" of the isodensity lines at the temporal end of the streak; Provis did not confirm this observation.

Both Hughes and Provis observed that the mean soma size of ganglion cells, and the range of their soma sizes, are smaller in the visual streak than in upper or lower retina. This could result from a "packing constraint," i.e., because of the need to crowd many ganglion cells in a small area. Alternatively, it could result from the concentration in the visual streak of intrinsically small-bodied ganglion cells. Evidence from other species suggests that at densities of 5000–6000/mm^2, a packing constraint probably is operating; on the other hand, evidence has been advanced (see Section 9.4.2) that in the cat, small-bodied ganglion cells do concentrate in the streak, independent of ganglion cell density.

Evidence that ganglion cells with particular properties are relatively frequent in the rabbit visual streak comes from the physiological study reported by Levick (1967), in which he compared the receptive field properties of ganglion cells in the streak with those of cells in upper and lower retina. Levick described several types of receptive fields not encountered elsewhere in the retina (uniformity detectors, local edge detectors, orientation-sensitive cells). Such cells may also be present above and below the streak, but apparently they are particularly numerous in the streak. Interestingly, their receptive fields resemble those of the W group of ganglion cells described in the cat retina that, as discussed in the next section, appear to concentrate in the visual streak. Correspondingly, Rapaport et al. (1981) reported evidence that the morphology of rabbit ganglion cells does not vary with their position with respect to the visual streak. Thus, the cells in the visual streak are small because they are members of a small-bodied class of cells, and not simply because of the density at which they are pooled together.

Other cell types: Davis (1929) noted that the receptors are longer in the visual streak region of the rabbit retina, and that the inner and outer nuclear layers are considerably thicker, implying that greater numbers of receptors, bipolars, and amacrine cells are present. Hughes (1971) noted a strong concentration of cone-type receptors in the visual streak.

Orientation of the streak: All observers (for example, Slonaker, 1897; Davis, 1929; Hughes, 1971; Provis, 1979) agree that when the eye and head are in a normal position of rest, the visual streak is oriented horizontally.

9.4.2. The Visual Streak of the Cat Retina

Early investigators of the cat retina differed as to whether its specialized area is "round" (Chievitz, 1889) or a horizontally elongated "bandlike" region (Slonaker, 1897). Present understanding is that both a localized "round" area centralis and a horizontally elongated visual streak are present. The streak is apparent in a map of the overall distribution of ganglion cells (Fig. 9.17) as a horizontal elongation of the isodensity lines [or "arm of high density" of Stone (1965)], particularly apparent in the retina nasal to the area centralis. Isodensity lines closer to the area centralis are more circular, and in an early description of this pattern (Stone, 1965), I commented that this change from relatively elongated to relatively circular lines "suggests a continuous quantitative change in organization between the area centralis and the periphery."

The major features of the map in Fig. 9.17 were confirmed by Hughes (1975), by Rowe and Stone (1976a) who analyzed the ganglion cell population of the visual streak, and by Stone (1978). Many analyses of the features of the ganglion cell map of cat retina have been reported; at this stage, I would argue that the overall pattern of ganglion cell distribution reflects the superimposition of (at least) two independent topographical specializations (the area centralis and visual streak). Further, the "change in organization" suggested in the above quotation can now be understood as a change with position in the retina in the relative prominence of the area centralis and the visual streak.

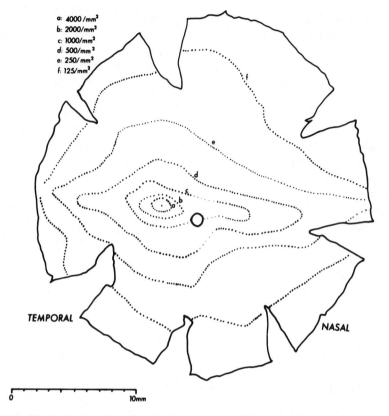

Figure 9.17. The distribution of ganglion cells in the cat retina. The original legend reads:
"Map of ganglion cell distribution made from a cresyl-violet stained whole mount. Counts were made under a microscope. No correction has been applied for shrinkage. The dotted lines are isodensity lines and mark out areas of retina within which ganglion cell density was higher than certain criterion levels, as indicated at top left. The fine dotted lengths in the 1,000/sq. mm line (c) near the optic disc were interpolated across regions where density could not be measured, because of the thickness of the axon layer. The region of peak density (i.e., the centre of the area centralis) is indicated by a dot . . . . Peak density was approximately 8,600/sq. mm." [From Rowe and Stone (1976a). Reproduced with kind permission of the *Journal of Comparative Neurology.*]

Ganglion cell composition: Evidence has been reported (Rowe and Stone, 1976a; Stone and Keens, 1978, 1980) that the ganglion cells found in the visual streak exhibit much or all of the range of ganglion cell size, morphology, receptive field properties, and central connections found in the cat retina. However, certain classes of ganglion cells seem to concentrate there, in particular small-bodied ganglion cells with receptive field properties typical of W cells (Rowe and Stone, 1976a). Rowe and Stone's two principal lines of evidence for this concentration may be summarized as follows:

1. Ganglion cells with small somas are particularly frequent in the visual streak; this effect is not due to the density of ganglion cells in the streak (Fig. 9.18).

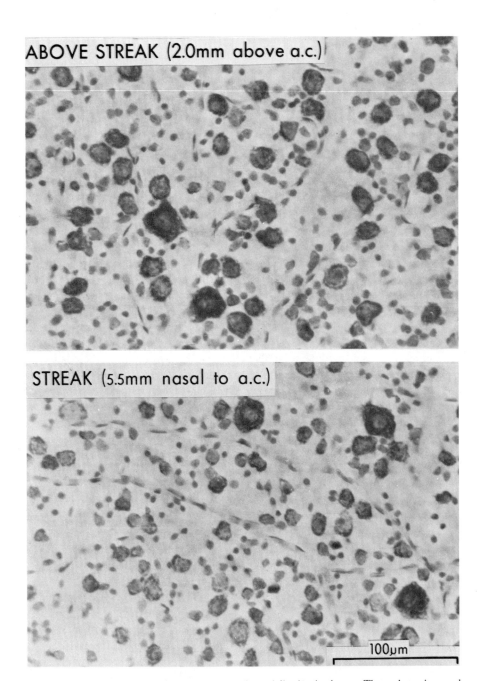

Figure 9.18. "Equidensity" analysis of visual streak specialization in the cat. These photomicrographs (from Rowe and Stone, 1976a) provide a comparison between areas of the cat retina in the visual streak (lower) and out of it (upper), with similar ganglion cell densities. The point of the comparison is that the prevalence of small-bodied ganglion cells in the visual streak cannot simply be attributed to the density of ganglion cells there. Rowe and Stone argue that small-bodied W cells are particularly numerous in the streak. [Reproduced with kind permission of the *Journal of Comparative Neurology.*]

2. W-Class ganglion cells were encountered more frequently in the streak than elsewhere, both relatively (i.e., as a proportion of the W, X, and Y cells encountered) and absolutely (i.e., as the number of cells encountered per set number of electrode penetrations of the retina).

Rowe and Stone concluded that both major subgroups of W cells that they distinguished [the "tonic" and "phasic" subgroups of Stone and Fukuda (1974a), later termed *W1* and *W2 cells* by Rowe and Stone (1977)] contribute to the concentration of ganglion cells in the visual streak, their relative frequency in the streak being as high or higher than in nonstreak peripheral retina. One subgroup of the W2 cells (cells with ON-OFF receptive fields) seemed particularly numerous in the streak. Y cells appear to make a small contribution to the streak. The analyses of Wässle *et al.* (1975) and Stone (1978) show that the absolute frequency of large-soma cells is higher in the streak than in peripheral nonstreak areas of the retina, so that a visual streak pattern is apparent in a map of the retinal distribution of these cells (Fig. 9.19). On the other hand, my own analysis (Stone, 1978) suggested that the relative frequency of Y cells is distinctly lower in the streak than above or below it (2.8–4.4% in the streak, variable between individuals, as against 4.2–6.9% above or below it).

Hughes (1981), on the basis of a "modal analysis" of the soma sizes of ganglion cells at different positions in the cat retina, has challenged the view that W cells concentrate in the visual streak, arguing that except at the area centralis, the relative proportions of the different ganglion cell classes do not change with retinal position. However, his analysis does not take into account the recently described medium-soma γ-cells, and assumes that all medium-sized somas are the somas of β-cells. Other recent evidence, moreover, supports the earlier view. In particular, Ballas *et al.* (1981) present evidence that the cells that project to the nucleus of the optic tract in the pretectum are predominantly small-soma γ cells and concentrate in the visual streak.

Current evidence neither suggests nor excludes the possibility that X cells contribute to the streak; in Rowe and Stone's (1976a) data, the relative frequency of X cells is much lower in the streak than elsewhere, and their absolute encounter rates do not show a significant increase in the visual streak. [Rodieck's (1979) conclusion that only ON-OFF W cells have been shown to concentrate in the visual streak was apparently based on a consideration of only the relative encounter rates we reported, rather than of the absolute encounter rates, or of the implications of the relative encounter rates, in the presence of substantial changes in ganglion cell density.]

In summary, all varieties of W cells appear to contribute to the concentration of ganglion cells in the visual streak (although not all to the same degree), and Y cells also make a small contribution. This evidence led Rowe and Stone (1976a; Fig. 9.21B) to suggest that the isodensity lines in Fig. 9.17 (which shows a map of the distribution of all ganglion cell classes in cat retina) might comprise two distinct but spatially superimposed components: a distribution of X cells described by circular isodensity lines and a distribution of W cells described by horizontally elongated lines. The same diagram also summarizes Rowe and Stone's view that the major contribution to the visual streak in the ganglion cell layer comes from small-bodied W cells.

Receptors and horizontal cells: As already noted above, maps of the distribution of cones (Steinberg *et al.,* 1973; Fig. 9.20) and of class A horizontal cells (Wässle *et al.,*

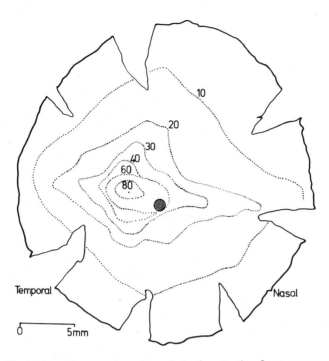

Figure 9.19. Distribution of large-soma ganglion cells in the cat retina. Large-soma ganglion cells are considered to be the α cells identified in Golgi studies and the Y cells identified physiologically. They comprise less than 10% of the total ganglion cell population. A map of their distribution over the retina, such as that shown here (from Stone, 1978), shows that both the area centralis and visual streak specializations can be identified. I reported evidence, however, that the visual streak specialization is less well developed in Y cells than in the population as a whole. At the area centralis, moreover, there is a very localized region of reduced density of large cells; the reduction is too localized to be represented on this map. The single dot here represents the peak density of all cells.

The maximum density of large cells found in this retina was 212/mm², in an annular region with outer diameter of 400 μm and inner diameter of 200 μm, centered on the peak density of all cells. In the central circle of diameter 200 μm, the density of large cells was 159/mm²; outside the annulus of maximum large-cell censity, density fell monotonically toward the periphery, as shown. The total number of large cells was 7100 (rounded to the nearest 100). [Reproduced with kind permission of the *Journal of Comparative Neurology*.]

1978) show that these two cell classes at least concentrate along the visual streak. Thus, the streak cannot be viewed as exclusively or even primarily a specialization of the ganglion cell layer.

Is there a streak in temporal retina?: Available evidence is limited but suggests that the isodensity line for ganglion cells (both for the total population and for small-soma cells considered separately), for cones, and for horizontal cells are all horizontally elongated in the retina temporal to the area centralis. The elongation is much weaker than in nasal retina, but seems consistent. Rowe and Stone (1976a) provided limited evidence that, as in the nasal part of the streak, small-soma cells are more frequent in the temporal part of the streak than elsewhere in peripheral retina. I would stress, how-

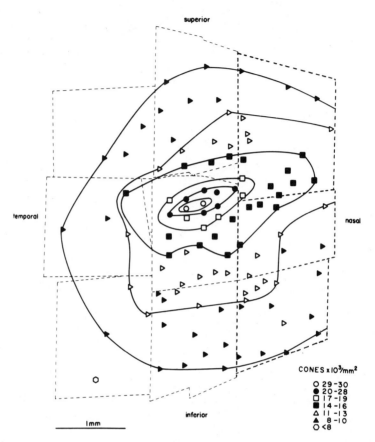

Figure 9.20. The distribution of cones in the cat retina. Although this map does not show the full extent of nasal retina, in which the visual streak is best developed, the isodensity lines obtained when cone densities are counted are horizontally elongated, in a manner similar to that seen for ganglion and horizontal cells. The density of cones reaches a maximum at the area centralis of about 27,000/mm², several times the maximum density of ganglion cells. The cone/ganglion cell ratio generally increases away from the area centralis. [From Steinberg *et al.* (1973). Reproduced with kind permission of the *Journal of Comparative Neurology.*]

ever, that the temporal end of the visual streak seems considerably less distinct than the nasal end. Stone and Keens' (1980) reassessment of the distribution of small- and medium-soma ganglion cells (Fig. 9.21C) emphasizes the point. They concluded that the elongated isodensity lines that describe the distribution of small-soma cells extend far less into temporal retina than Rowe and Stone (1976a) had envisaged.

    Orientation of the streak: There is some disagreement as to the orientation of the visual streak in the conscious cat. In my own maps, it was clear that the streak is approximately horizontal, but it was also apparent that without control for the possible distortion of the retina during flattening, and knowledge of the orientation of the optic disc and area centralis in the conscious, unparalyzed cat, its exact orientation could not be

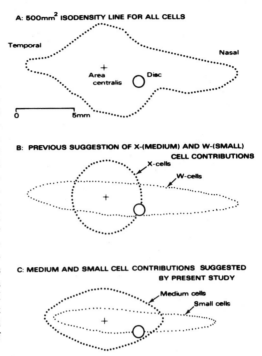

Figure 9.21. Contributions of small- and medium-soma ganglion cells to ganglion cell map. The diagram at the top reproduces the 500/mm² isodensity line in the map of the distribution of all ganglion cells in the cat retina shown in Fig. 9.17. Rowe and Stone (1976a) suggested that medium-soma X cells and small-soma W cells make different contributions to that line, as shown in the middle diagram; their particular point was the concentration of small ganglion cells in the visual streak. Stone and Keens (1980) subsequently suggested the pattern shown in the lower diagram. This differs from Rowe and Stone's suggestion in two ways. First, Stone and Keens noted recent evidence that many medium-soma ganglion cells are probably not the somas of X cells; second, they noted that the visual streak seems less marked in temporal than in nasal retina. [Reproduced with kind permission of the *Journal of Comparative Neurology*.]

specified. Hughes (1975) considered these and other factors and concluded that the visual streak of cat retina is straight and, in the conscious cat, horizontal in orientation. On the other hand, McIlwain (1977b) reported evidence from studies of the living eye that the visual streak, or at least the portion of it nasal to the area centralis, runs at right angles to the long axis of the slit pupil of the eye. He also confirmed previous reports that the pupils of the two eyes of normally active cats are not vertical and parallel to each other, but incline toward each other, being closer at the top. The mean angle between the long axes of the two pupils was estimated at 13.6° by McIlwain (1977b), at 14° by Olson and Freeman (1978), and at 8.7° by Cooper and Pettigrew (1979a). This result implies that the streak is not usually held horizontal, but with its nasal end tilted a few degrees downwards. Cooper and Pettigrew (1979b) added the further suggestion that the visual streak in the cat may not be straight. They suggested that it might be useful to regard the streak as comprising nasal and temporal arms, which diverge by a few degrees. Thus, these studies raise the possibility that the visual streak in the cat, and possibly many other species, may not be either horizontal or straight.

9.4.3. Evidence of a Visual Streak in the Monkey and Other Primates

The horizontal elongation of ganglion cell isodensity lines by which the visual streak is recognized in maps of the rabbit and cat retinas (Figs. 9.15 and 9.17) is less marked in the monkey or human (Fig. 9.22), but nevertheless seems clear and consistent.

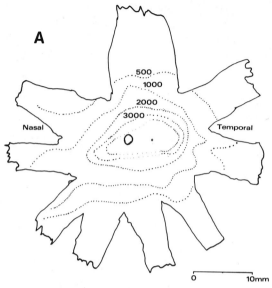

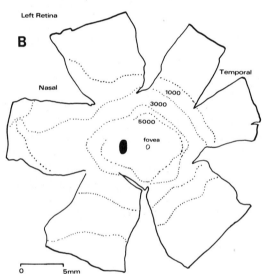

Figure 9.22. Maps of ganglion cell distribution in human and monkey. (A) Map of the distribution of ganglion cells in a human retina. Isodensity lines are not shown close to the fovea, because the multilayering of ganglion cells there prevented the counting of ganglion cells. In peripheral retina, the isodensity lines are generally elongated horizontally. (B) Map of the distribution of ganglion cells in the retina of a crab-eating macaque monkey *(M. irus)*. As in the map above, ganglion cell density could not be estimated close to the fovea. The isodensity lines are distinctly elongated horizontally, especially in nasal retina. As in the human, the elongation is less marked than in either the cat (Fig. 9.17) or the rabbit (Fig. 9.15). [From Stone and Johnston (1981). Reproduced with kind permission of the *Journal of Comparative Neurology.*]

Although A. Hughes (1977) concluded that ganglion cell distribution in the macaque monkey is "quite radially symmetrical," suggesting that no visual streak is present, the isodensity lines in maps of the macaque monkey (van Buren, 1963; Stone and Johnston, 1981), of the owl monkey (Webb and Kaas, 1976), of the human (Stone and Johnston, 1981) and of the prosimian bush baby (DeBruyn *et al.*, 1980; Stone and Johnston, 1981) are all elongated horizontally, at least away from the fovea. Two arguments support the suggestion that this elongation reflects the presence of a visual streak.

First, when soma sizes of ganglion cells were measured, in an analysis comparable to that reported by Rowe and Stone (1976a) for the cat, mean soma size of ganglion cells was found to be relatively small at the nasal end of the isodensity lines (Stone and Johnston, 1981), as it is in the cat. The analysis showed that the small size of ganglion cell somas is independent of cell density (Fig. 9.23), and is due to a relatively high proportion of small-soma ganglion cells in the population found at the nasal end of the isodensity lines.

Second, in the retinas of chronically decorticate monkeys, the massive loss of ganglion cells caused by retrograde transneuronal degeneration of ganglion cells projecting to the thalamus (van Buren, 1963; Cowey, 1974; Weller *et al.*, 1979) is most marked at the foveal region (confirming previous reports). The surviving ganglion cells are all small in soma size (15 μm or less in diameter), and form a more marked streak than found among the population of ganglion cells as a whole (Fig. 9.24). Note that the 500/mm^2 isodensity line in this map is very elongated in nasal retina, but not in temporal

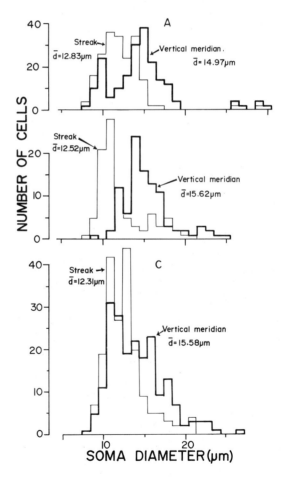

Figure 9.23. Evidence of the high proportion of small ganglion cells in the visual streak. In each of the cat (A), monkey (B), and opossum (C), frequency/soma size histograms were constructed for samples of ganglion cells in the nasal part of the visual streak, and in a region of similar ganglion cell density directly above the area or fovea centralis. Such areas of a cat retina are shown in Fig. 9.18. In all three species, mean soma size is less in the visual streak sample, and the histograms suggest that in the visual streak, fewer cells are found in the middle part of the soma range, and more at the small end. [Data from Rowe and Stone (1976a), Stone and Johnston (1981), and, with kind permission, from E. Tancred.]

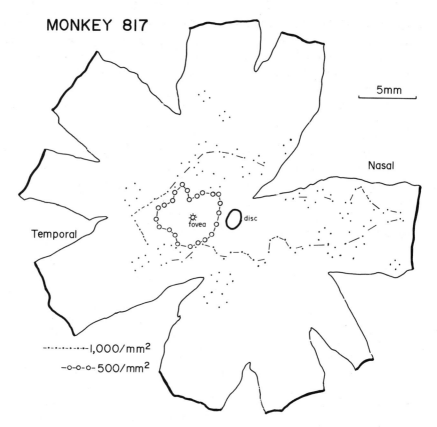

MONKEY 817

5mm

Nasal

fovea   disc

Temporal

----------1,000/mm²

-o-o-o- 500/mm²

Figure 9.24. Further evidence of a visual streak in primates. Map of the distribution of ganglion cells in the retina of a rhesus monkey, which had survived 5 years after a bilateral occipital lobectomy. Each closed dot represents the border between two sampling areas, in one of which the density of surviving ganglion cells was 500/mm² or greater, and in the other was less than 500/mm². The open circles represent similar borders for a density criterion of 1000/mm². There was a massive loss of ganglion cells (over 90% in the region of the fovea, up to 50% in peripheral retina), confirming earlier reports. The surviving cells were all small in soma size, and the 500/mm² line seems to form a distinct visual streak, particularly in nasal retina.

Presumably, these cells survived because they projected to sites other than the LGN, which degenerates following the lesion to the occipital cortex. The results suggest that in the monkey, as in the cat, there is a tendency for small-soma ganglion cells to concentrate in the visual streak and project to the SC. [The retina was kindly provided by Drs. T. and P. Pasik.]

retina. If it does represent the visual streak, then in the macaque monkey the streak does not extend into temporal retina. The cells that survived presumably did so because they projected to the midbrain, or perhaps because they projected to both mid- and forebrain and were sustained by an axon branch passing to the SC or other midbrain or brain-stem site. They may correspond to the small-soma W cells that appear to concentrate in the cat visual streak and are also known to project predominantly to the SC (Sections 2.3.3.7, 9.4.2).

9.4.4. The Opossum: An Exception?

The one mammalian species so far reported in which there appears to be no sign of a visual streak is the American opossum. In both South and North American species of opossum (Hokoc and Oswaldo-Cruz, 1979; Rapaport *et al.*, 1981*b*), the isodensity lines in a map of ganglion cell distribution are only slightly elongated horizontally (Fig. 9.25), and Rapaport and co-workers' (1981*b*) analysis showed no evidence of a concentration of a particular size group of ganglion cells in the region where a streak might be anticipated. This apparent lack of a streak is not a feature general to marsupials, many of which have a strongly developed streak. It is possible that a visual streak specialization may be present in other layers of the retina, or that a streak specialization appears transiently during development in this species but does not survive into adulthood. It is also possible that the opposum will prove to be exception to the general occurrence of a streak in mammalian retinas. It has been considered a rather primitive mammal, with a poorly differentiated visual system (Rapaport *et al.*, 1981*b*); the lack of a streak in this species raises the possibility that this feature was not present in some early mammals.

9.4.5. Summary

The visual streak specialization of the retina is dominant in the rabbit retina, clear in the cat, and demonstrable in the macaque. The species examined span a considerable part of the mammalian range, suggesting that the visual streak is a consistent feature of mammalian retina.

9.5. THE FOVEA OR AREA CENTRALIS: A SPECIALIZATION AT THE JUNCTION OF THE TWO AXES

The presence in many mammals of a very localized, strongly specialized region of the retina has been recognized for many decades. Chievitz (1889) traced reports of the human fovea or macula lutea as long as a century prior to his own work, and Slonaker (1897), Polyak (1957), A. Hughes (1977), and Provis (1979) have all traced and discussed the history of its discovery. Chievitz quotes H. Müller's (1861) observation that "in mammals at least an area centralis is present which resembles in structure the yellow spot and is recognizable by the similar course of the central bundles." This section describes the area or fovea centralis of the monkey, cat, and rabbit. Its purpose is not to review the literature on the morphology of this specialization, which has been intensively studied in the monkey and cat, but to set out a case for the presence of such an area in all mammalian retinas, as part of the overall scheme of retinal topography being proposed.

Before doing so, however, some comment seems necessary on a persistent problem of terminology raised by the term *area* or *fovea centralis* and, equally, by *visual streak*. Müller seemed to introduce the term *area centralis* rather than *fovea centralis* or *macula lutea* because many mammals have neither a foveal specialization nor the yellow pig-

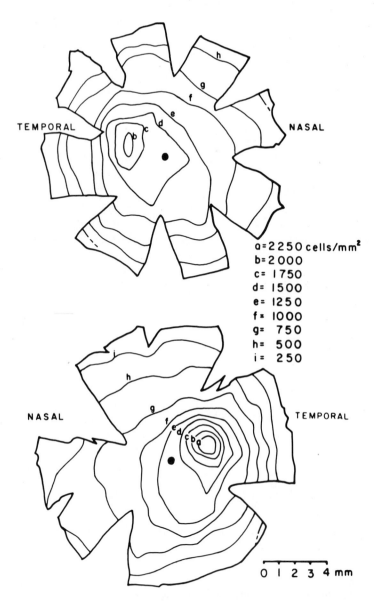

Figure 9.25. Maps of the distribution of ganglion cells in the right (upper) and left (lower) retinas of the North American opossum [*Didelphis virginiana* (from Rapaport *et al.,* 1981*b*)]. The outlines represent whole-mounted retinas, with the optic discs represented as solid spots. Isodensity lines are drawn to delineate areas within which ganglion cell density is higher than the value indicated at each line. Density reaches a maximum at the area centralis, located a few disc diameters temporal to the area centralis. The highest density of ganglion cells observed in either retina was 2750/mm^2, at the area centralis of the left retina.

Note that the isodensity lines are only slightly elongated horizontally. In this species, and in the South American opossum [*Didelphis marsupialis* (Hokoc and Oswaldo-Cruz, 1979)], there is little evidence of a visual streak. [Reproduced with kind permission of the *Journal of Comparative Neurology.*]

ment found at the fovea in some primates, but nevertheless have an apparently corresponding specialization of the retina. The term *area* is more vague than *fovea* (which means "pit") or *macula* ("spot"), and its use has avoided a number of problems. However, *centralis* does create a difficulty, for even in the human or macaque monkey, the fovea is not exactly central in the retina but, as in all mammals, is nearer the temporal margin of the retina than the nasal. In species with laterally placed eyes (such as the rabbit), this asymmetry is very marked, and the term *centralis* seems particularly inappropriate. A similar problem is raised by *streak*. It was named in the rabbit, where the specialization is so marked that the term makes sense. Perhaps, however, the very descriptiveness of the term has led investigators to conclude that primates have no streak, for the specialization is not obvious to macroscopic inspection, as it is in the rabbit, and if first studied in the monkey, would not have been termed a *streak*. It seems to me that these problems are of the sort that Tyner (1975) and Rowe and Stone (1977) have argued are commonly, perhaps inevitably, associated with the use of names for taxa (groups) in biological classifications that emphasize particular physical features of the objects being classified. The use of terms such as *area centralis* and *visual streak* is, in fact, part of an attempt to classify the sorts of regional specialization found in the retina. Ideally in such classifications, we need terms that are not only useful in summarizing our observations but also do not, by insisting on or emphasizing particular physical features, make it difficult to recognize the range of morphology found in such specializations among different species. Following Rowe and Stone's (1977) analysis, one should perhaps use nondescriptive terms (such as specialization *A* and specialization *B*, or perhaps some nondescriptive modification of present terms. A Hughes (1977), for example, uses the term *area retinae* instead of *area centralis*, and its vagueness does avoid much of the problem with *centralis*. For the time being, I retain the older terms (*centralis* and *streak*) while acknowledging the need for a reconsideration of terminology. One reason for this conservatism is that I suspect we will come to recognize retinal specializations other than the area centralis and visual streak (for example, the nasotemporal division of the retina discussed in Section 9.3 and the two specializations discussed in Section 9.6) that will also require classification and naming. A more radical revision of terminology might be more effective at that stage. In the meantime, the term *area retinae* has the disadvantage that, despite its vagueness, it implies that only one regional specialization is to be found in mammalian retina and thereby tends to perpetuate what seems to me to have been a confusion in earlier literature (though not in A. Hughes, 1977), namely, that the visual streak and area centralis are different modifications of the same specialization. The following discussions of monkey, cat, and rabbit retina confirm A. Hughes' (1977, p. 705) conclusion that all vertebrate retinas examined contain at least the rudiments of an area centralis.

9.5.1. The Fovea Centralis of the Monkey and Other Primates

Late in fetal life, the ganglion and bipolar layers at the foveal region of the monkey and human (and presumably most other primates as well) migrate a short distance radially from a point that becomes, in the adult, the center of the foveal specialization. The ganglion and bipolar cells pile up, surrounding a small, thin patch of the retina that is

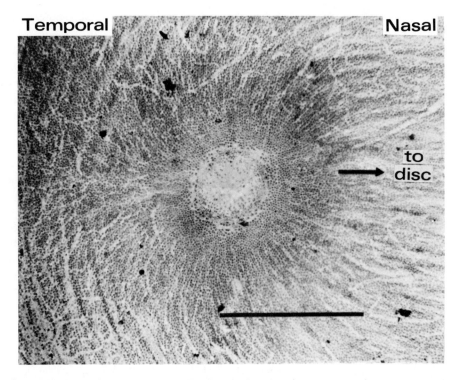

Figure 9.26. Fovea of the monkey retina. The fovea in the retina of a macaque monkey, seen in a whole mount. The floor of the foveal pit (the region cleared of ganglion and bipolar cells) is circular in outline. The scale represents 100 μm.

free of ganglion and bipolar cells and forms the floor of a pit, the fovea. The result is a spectacular, very localized specialization of retinal structure (Fig. 9.26). In addition, there are associated specializations in the morphology and connections of ganglion, bipolar, and receptor cells, which have been the subject of many studies (for example, Polyak, 1941, 1957; Boycott and Dowling, 1969; DeMonasterio and Gouras, 1975). In brief and incomplete summary, the ganglion cells at the fovea are extremely numerous and densely packed [80,000–100,000/mm^2 around the fovea (van Buren, 1963; Rolls and Cowey, 1970)], several hundredfold more densely than in far peripheral retina. Their soma sizes are sharply reduced and the dendritic fields of many [the "midget" cells of Polyak (1941), the B cells of Leventhal *et al.* (1981)] are extremely restricted in their spread and are contacted by very few bipolar cells, perhaps only one. The number of ganglion cells at the fovea approaches the density of receptors, and the receptors are specialized in two ways: only cones are present and the cones are slimmer than in peripheral retina, as slim as rods, allowing them to be densely packed. Moreover, the bipolar cells of the region have restricted axonal and dendritic trees, and appear in many cases to connect a single cone to a single granglion cell.

In terms of visual capacity, it is established that the fovea is the region of maximal visual acuity in the retina, and also maximal stereoacuity and chromatic acuity. Because its receptor population consists entirely of cones, the fovea is insensitive at scotopic (dark) levels of illumination. The small saccades of fixation and smooth pursuit eye movements are well developed in primates, and seem particularly related to control of the direction of view of the fovea; and the foveal region of the retina dominates the representation of the retina in the LGN and visual cortex. Among the primates, the fovea is always located a few disc diameters temporal to the optic disc, usually on the same horizontal level as the disc, and is always found astride the axis of nasotemporal division (Fig. 9.6), and along the axis of the visual streak (Fig. 9.24).

9.5.2. The Area Centralis of the Cat

In the cat, ganglion cell density reaches a maximum of 8000–9000 cells/mm^2, much lower than the peak value found in the monkey. However, the maximum is equally restricted in its location (Fig. 9.27) and very constant in position from individual

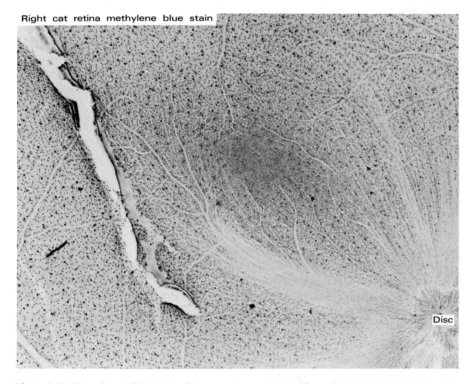

Right cat retina methylene blue stain

Disc

Figure 9.27. Central part of the retina of a cat, seen in a methylene blue-stained whole mount. It is a right retina, and the optic disc can be seen at the lower right, characterized by the convergence of fiber bundles onto it. The area centralis is apparent as a dense aggregation of cells above and temporal to the disc. In more peripheral retina, the ganglion cells are quite scattered and the large dark-staining somas of α (Y) cells can be distinguished. [From Stone (1965). Reproduced with kind permission of the *Journal of Comparative Neurology*.]

to individual. It is located approximately 3 mm temporal to and 1 mm above the optic disc, astride the axis of nasotemporal division (Fig. 9.9) and along the axis of the visual streak. At the area centralis, the soma sizes of at least two of the main groups of ganglion cells [the α and β cells of Boycott and Wässle (1974)] are reduced in size, and their dendritic fields spread less widely than in peripheral retina [Chapter 2, Sections 2.1.4 and 2.3.3; for a detailed summary of centroperipheral gradients, see Rowe and Stone (1980b)]. Rowe and Dreher (1982a) have recently described another specialization of β cells at the area centralis; viz., the somas of many β cells are displaced from their dendritic fields, the displacement being away from the center of the area centralis. The magnitude of the displacement is small compared to the size of the primate fovea, but it does serve to minimize the thickness of the ganglion cell layer and may represent a "primitive" form of fovea.

In physiological studies, X cells have been shown to be particularly numerous at the area centralis and to have smaller receptive fields than elsewhere in the retina. Y and W cells are also more numerous in absolute terms in the region of the area centralis (except that at the very center the density of Y cells shows a localized minimum), but their frequency relative to X cells is minimal at the area centralis. [For a review of these features of cat ganglion cells, see Rowe and Stone (1977, 1980b), Chapter 2, Section 2.3.3.5]. The density of cone-type receptors also reaches a maximum at the area centralis (Fig. 9.20), as does the density of type A horizontal cells (Boycott et al., 1978).

Behaviorally, the area centralis seems to be the region of the retina capable of maximal spatial acuity (Blake and Bellhorn, 1978), and its representation dominates the retinotopic organization of the LGN and area 17 and, to a lesser extent, of areas 18 and 19 as well. It appears to be the strongest specialization in the cat retina and seems, as earlier workers suggested, to correspond to the primate fovea.

9.5.3. Evidence of an Area Centralis in the Rabbit Retina

According to my present hypothesis, an area centralis should be present in the rabbit retina, and should be located in line with the visual streak, and on the axis of nasotemporal division. It should be located therefore (from Fig. 9.13) at the temporal end of the visual streak, about 3 mm from the temporal margin of the retina.

This is, as Hughes (1971) noted, the region that the rabbit uses for frontal, binocular vision, and it was here that Hughes noted an upward bulging in the isodensity lines of both ganglion cells and cones, which might represent an area centralis. This latter feature of the ganglion cell layer was not confirmed by Provis (1979), who nevertheless did observe a specialization in the ganglion cell layer at this position (Fig. 9.28). She reported that the density of ganglion cells with large (greater than 20-μm diameter) somas, which is less than 20 cells/mm^2 in most areas of the retina, reaches a maximum of 48–70/mm^2 at the temporal end of the streak. This maximum is quite separate from the maximum density of all ganglion cells, which is located along the ridge of the visual streak, several millimeters closer to the optic disc. Because, however, the large-cell maximum is localized and constant in position, and is located astride the axis of nasotemporal division, in the center of the part of the retina used for binocular vision, it seems to correspond to the area or fovea centralis of other mammals.

Figure 9.28. Evidence of an area centralis in the rabbit retina. This map (from Provis, 1979) shows that when the distribution of large-bodied ganglion cells is separately plotted, a distinct concentration of large cells is present at the temporal end of the visual streak. Its position is consistent between different individuals and it is located astride the axis of nasotemporal division.

The map is from the same retina as used for Fig. 9.15. The numbers represent the density of large ganglion cells per square millimeter. [Reproduced with kind permission of the *Journal of Comparative Neurology*.]

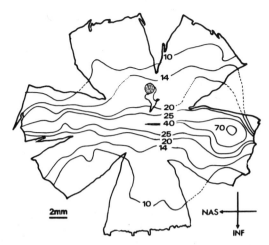

Choudhury (1981) has confirmed Provis' observation of a large-cell maximum at the temporal end of the visual streak. Oyster *et al.* (1981), on the other hand, were unable to detect it.

9.5.4. Summary

The fovea or area centralis specialization is dominant in the monkey, strongly developed in the cat, and demonstrable in the rabbit. I argue that when considering a range of species, this specialization cannot be reliably identified by the absolute value of ganglion cell density reached there, but is better identified in terms of the densities of ganglion cells of different classes, and by its position at the junction of the axes of naso-temporal division and of the visual streak. This argument follows the discussion of Provis (1979); it implies that without the sort of analysis she has provided for the rabbit retina, it would be premature to conclude that a retina lacks this specialization.

9.6. TWO REGIONAL SPECIALIZATIONS RELATED TO THE AXIS OF NASOTEMPORAL DIVISION

9.6.1. Nasal–Temporal Gradients in the Properties of Ganglion Cells

Regional variations in ganglion cell properties summarized: The properties of ganglion cells vary with position in the retina in at least three distinct ways. First, where the area centralis is well developed, there are strong gradients between central and peripheral retina in the soma size, dendritic field size, receptive field size, axonal conduction velocity, and other properties of many ganglion cells. In the cat, for example, these changes are very evident in the Y- and X-cell classes of ganglion cells and are reviewed in detail by Rowe and Stone (1980*b*) and in Chapter 2, Section 2.3.3. Descriptions of the morphology and physiology of monkey ganglion cells (Chapter 3, Section 3.1) make it clear that similar gradients are present among them.

Second, the properties of ganglion cells well away from the area centralis vary between the visual streak region and nonstreak peripheral retina. In the cat, as already noted, the major differences appear to be differences in the proportions of various classes of ganglion cells, W cells predominating in the visual streak. Rowe and Stone (1976a) did, however, find a small but distinct reduction in the receptive field sizes of ganglion cells between visual streak and nonstreak regions of peripheral retina, and Stone and Keens (1980) reported that in the visual streak the somas of presumed β- (X-) class ganglion cells are slightly smaller than in nonstreak areas. In the rabbit, where the streak is very strongly developed, the cell size gradients between streak and nonstreak periphery are much stronger (Section 9.4.1). Although a packing constraint seems likely to be operating in the rabbit streak because of the high density of cells there, evidence is also available (summarized in Section 9.4.1) that small-bodied ganglion cells in several ways comparable to cat γ cells concentrate in the visual streak of the rabbit.

A third gradient has been described between ganglion cells temporal to and nasal to the axis of nasotemporal division, as a difference in the soma size, axonal conduction velocities, and relative numbers of cells of particular classes. These nasal–temporal differences seem to be independent of ganglion cell density, and cells with intermediate properties have been described (in the cat) close to the axis of nasotemporal division. These differences may be just one aspect of the nasotemporal division of the retina; so far they have been described in three carnivores, three primates, four Australian marsupials, and the American marsupial opossum.

Specific observations: Following early brief comments by Stone et al. (1976, 1977), a more detailed set of observations was reported by Stone et al. (1980). They provided evidence that the X (β) and Y (α) cells found in the retina temporal to the axis of nasotemporal division have larger somas than those found in nasal retina (larger by about 10%). Further, they have slightly faster-conducting axons, a difference first noted by Bishop et al. (1953). Similar differences were not detected among W (γ) cells. Stone and co-workers noted further that within temporal or nasal retina, the properties they assessed were relatively constant; within 1 or 2 mm of the axis of nasotemporal division, however, cells with intermediate properties were found. Thus, there appeared to be a gradient in properties extending a few millimeters on either side of the axis of nasotemporal division. Stone and co-workers also noticed a consistent trend for cells with medium-sized somas (16- to 23-μm diameter) to be relatively frequent in temporal retina, and for cells with smaller somas to be most frequent in nasal retina. They were unable to establish, however, whether this reflected a high proportion in temporal retina (and a correspondingly low proportion in nasal retina) of medium-soma X (β) cells or of medium-soma W (γ) cells.

In Australian marsupials such as the possum, wallaby, Tasmanian devil, and (less certainly) the wombat (Tancred, 1981), in the fox (Rapaport et al., 1979), in the bush baby, monkey, and human [all primates (Stone and Johnston, 1981)], in the dog (Osmotherly, 1979), and in the rabbit (Provis, 1979), there is a consistent tendency for cells in the medium part of the soma size range for any particular animal to be relatively numerous in temporal retina, and for smaller ganglion cells to be less common. The trend is least marked in the rabbit and wombat, in which the area centralis is weakly developed and the area of temporal retina is small, and in the bush baby, possibly because of the small overall range of ganglion cell soma size found in this species. The trend is distinct in the carnivores (particularly the fox), in the other marsupials mentioned, and in the macaque monkey and human (Fig. 9.29). In the American opossum, Rowe *et al.* (1981) reported that the high proportion of medium-soma ganglion cells appears to be restricted to upper temporal retina, and they provide single-unit and field potential data corroborating their soma size analysis.

These results are both puzzling and intriguing. They are puzzling because they were unexpected, and because their functional significance seems obscure. They are intriguing in that they seem quite consistent across species with widely different evolutionary backgrounds, and may therefore reflect a feature in the visual system common to all mammals. That feature is presumably the partial decussation of the optic nerve, with the associated development of the nasotemporal division of the retina and of mechanisms of binocular depth discrimination. Several behavioral correlates of these trends have been described, but they seem more related to the central connections of these cells than to the slight variations observed in their properties. This behavioral evidence is considered in Section 9.8.3, and the suggestion is made that the variations observed in the physical properties of nerve cells may not all be specific adaptations to particular environmental pressures; some may be structural side effects of other major developments in brain structure, such as the development of partial decussation.

9.6.2. A Vertical Streak?

A. Hughes (1977) described in the ganglion cell density maps of several mammalian species an upward sweep of isodensity lines above the area centralis, noting that the feature is particularly prominent in herbivores, such as the goat, cow, deer, and rabbit. It seems unlikely, however, that the specialization is entirely restricted to this group of animals. DeBruyn *et al.* (1980) have suggested the presence of a similar specialization in the retina of a primate, the prosimian bush baby, and Fig. 9.30 (from Osmotherly, 1979) shows evidence of a similar feature in the dog retina. The ganglion cell density map in Fig. 9.30 was made from the retina of a blue cattle dog, and shows a clear area centralis and an approximately horizontal visual streak, and also an upward elongation of isodensity lines (arrowed), directly above the area centralis.

As to the function of this specialization, Hughes noted that, when an animal is fixating a region of ground several feet before it, this region of relatively high ganglion cell density would scan the area between the fixation point and the animal's feet. This region is arguably of special interest to grazing animals and to hunters (such as the dog) of small, scurrying prey.

Osmotherly reported evidence that this "vertical streak," like the horizontal streak, contains a high proportion of ganglion cells of a particular soma size. He suggested that, in peripheral regions of the dog retina, ganglion cells can be divided into four soma-size groups (8–15, 16–20, 21–25, > 25 μm), and found that the 16- to 20-μm-diameter group is commonest in both horizontal and vertical streaks.

Little else is known of this specialization. It differs from the other specializations considered here in that it seems to be detectable in only a proportion of the species examined. We need to know more about its occurrence and the classes of ganglion cells

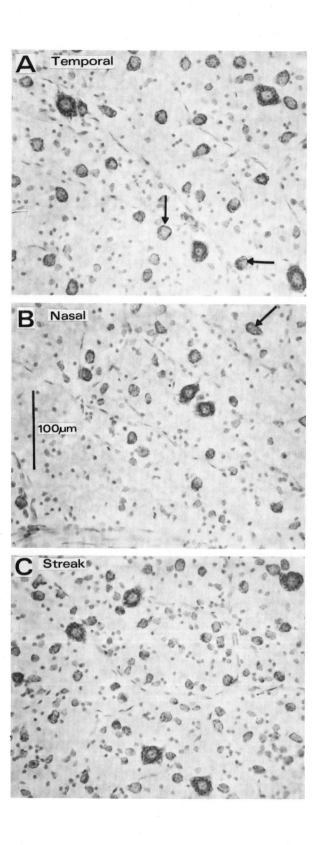

that concentrate there, to determine (for example) whether it is a specialization inde-pendent of the area centralis, or is better understood as an extension of the area centralis.

9.7. HOW GENERALLY CAN THE MODEL BE APPLIED?

Having proposed a model of the topography of mammalian retina, I must argue for its generality. In fact, arguments supporting its generality have already been advanced at several points in the chapter; they are summarized in the following paragraphs.

9.7.1. The Generality of the Area Centralis

In most mammalian retinas, an area centralis can be recognized as a localized region of high ganglion cell density located a few millimeters temporal to the optic disc. Although in some species difficult to detect by ophthalmoscopic observation of the fundus or even by inspection of serial sections of the retina, the area centralis can usually be identified by inspection or mapping of Nissl-stained retinal whole mounts. In a survey of nearly 100 mammalian species, A. Hughes (1977) recognized an area centralis in all but three (the rabbit, ground squirrel, and gerbil); I would extend Hughes' observations by arguing that an area centralis may be present in all mammals. Hughes commented, for example, that "the rabbit shows it [the absence of an area centralis] to best effect"; while in Section 9.5.3, it is argued from Provis' (1979) analysis that an area centralis can be recognized in that species. Even if Provis' analysis is accepted as compelling, it does not, of course, prove the general occurrence of an area centralis in all mammalian retinas, but it does suggest that, even in species where an area centralis is not strongly developed, a corresponding specialization may be present. The further general point can be made that the position of the area centralis within the retina is highly consistent: with few exceptions, the area centralis is located astride the axis of nasotemporal division. The exceptions are of interest; one is found in the Siamese cat (Section 9.7.4), the other (which may prove more apparent than real) is found in the rodent retina (Sections 9.3.5; 9.8.2).

9.7.2. The Generality of the Visual Streak

A. Hughes (1977) considered a visual streak absent from the retinas of many noc-turnal mammals (such as the hedgehog, guinea pig, mouse, weasel, and rat) and from

←──

Figure 9.29. Nasal–temporal differences in ganglion cell soma sizes in the cat retina. (A) and (B) provide a comparison of regions of the ganglion cell layer in peripheral cat retina temporal and nasal to the area centralis. The arrows point to medium-soma ganglion cells, which seem far more common in temporal retina, even though ganglion cell density is very similar in the two areas. Measurement and experimental techniques showed other differences discussed in the text.

(C) shows a region of the visual streak in nasal retina. Here, small-soma ganglion cells are even more frequent, and medium-soma ganglion cells less frequent, than in nasal retina (B). [From Stone *et al.* (1980). Reproduced with kind permission of the *Journal of Comparative Neurology*.]

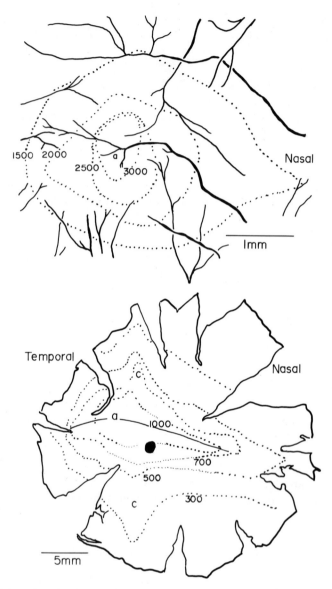

Figure 9.30. Maps of the distribution of ganglion cells in the dog retina. The lower map shows the distribution of ganglion cells in the peripheral regions of the right retina of a blue cattle dog. The isodensity lines close to the area centralis (a) are not shown. The arched line follows the ridge of the visual streak. Note the vertical elongation of isodensity lines above and below the area centralis. The upper map shows the distribution of ganglion cells in the region of the area centralis. The maximum density of 4545 cells/mm² was encountered at a. This maximum is distinctly lower than that found in the cat. The blood vessel that crosses the area centralis was a consistent feature of dog retinas; a similar vessel has been observed in cats only in the Siamese breed. Overall, the distribution of ganglion cells resembles that of the cat quite closely. Two distinct differences were noted by Osmotherly (1979). First, there was a tendency for the isodensity lines above and below the area centralis to be vertically elongated, suggesting a specialization of the retina in this region more marked than any yet described in the cat. Second, when the contribution of ganglion cells of

certain diurnal species (tree shrew, squirrel, squirrel glider, tree kangaroo, macaque monkey, apes and man). Section 9.4.3 presents evidence that in one of these species (the macaque monkey), the elongation of isodensity lines in peripheral retina does represent a visual streak. It seems fair to comment, moreover, that in the other of these species for which ganglion cell density maps have been published, with the exception only of the opossum (Section 9.4.4), the isodensity lines in peripheral retina are horizontally elongated; for example, in the rat (Fukuda, 1977; A. Hughes, 1977), in primates [including a prosimian, New- and Old-World monkeys, apes, and man (van Buren, 1963; Webb and Kaas, 1976; DeBruyn et al., 1980; Stone and Johnston, 1981)], in the guinea pig and gray squirrel (A. Hughes, 1977), and in the tree kangaroo (Hughes, 1974). It seems reasonable to hypothesize that analyses similar to that developed for the monkey would provide evidence of a visual streak specialization in the great majority of these species.

It should be added that the hypothesis that a visual streak is a general feature of the mammalian retina does not require or imply that the structural specializations that characterize the streak will be identical between species, varying only in degree of development. For example, in both the dog (Osmotherly, 1979) and the cow (Hebel and Holländer, 1979), the concentration of ganglion cells in the visual streak involves not the smallest-diameter class of ganglion cells (as it appears to do in the cat, rabbit, and monkey) but a class with medium-sized somas. Similarly, the hypothesis does not require that the function of the streak is identical between species.

9.7.3. The Generality of the Nasotemporal Division of the Retina

In all mammals so far investigated, the optic nerve decussates only partially and the ipsilaterally projecting ganglion cells are found concentrated in the region of the retina temporal to the area centralis. In most species studied (human, monkey, cat, possum, rabbit), the nasal boundary of this temporal region is vertical and crosses the area of fovea centralis. Particularly in primates, but also in the cat, a vertically oriented strip of the retina can be delineated, crossing the area or fovea centralis, across which the laterality of ganglion cell projection shifts. Nasal to the strip, virtually all ganglion cells project contralaterally; temporal to the strip, most or (in primates) all ganglion cells project ipsilaterally.

Important exceptions to this pattern are found in the Siamese cat, where the transition from contra- to ipsilateral projection patterns is more diffuse than in the common cat, and centered temporal to the area centralis; and in rodents in which the region containing ipsilaterally projecting ganglion cells lies in inferotemporal retina, with its nasal boundary well temporal to the presumed area centralis. The situation in the Siamese cat seems clearly to be a deleterious abnormality. The pattern seen in the mouse and rat may also represent an undeveloped, or degenerated form of the more common pattern, but this possibility requires further analysis.

---

different soma sizes to the visual streak was analyzed (four soma-size groups were distinguished), the group that seemed most common in the streak was that with small–medium somas, rather than the group with the smallest somas, as in the cat. [Map kindly supplied by S. Osmotherly.]

Where (as in all nonprimates) some cells in temporal retina project ipsilaterally and some contralaterally, it appears that different ganglion classes contribute differently to this pattern. In the cat for example, the β-class ganglion cells in temporal retina do not contribute to the contralateral projection, but some γ and α cells do contribute. Analogous patterns have been apparent in all species investigated, including the Siamese cat, rat, and mouse.

The differences between nasal and temporal areas of the retina in the properties of ganglion cells (Section 9.6.1) have also been detected in all species studied, and may be a general feature of mammalian retina.

9.7.4. The Siamese Cat: A Case to Prove Two Points

One concern I have had in developing the "two-axis" model of retinal topography set out in Section 9.2 is that its general applicability to mammals might result from its being so vague as to be untestable. The Siamese cat presents an exception among mammals, which seems to support the model in two ways: methodological and functional. The exception provides methodological support because the Siamese retina is a concrete example of how a retina might be constructed that did not fit the model. And it provides functional support for the model because the unusual pattern of nasotemporal division and the separation of the area centralis from the axis of nasotemporal division are accompanied (by comparison with the common cat) by structural abnormalities in the visual pathways and by functional deficiencies (loss of spatial acuity, squint; see Section 9.3.4). The general pattern of retinal topography summarized by the model thus seems important for the optimal visual performance of the animal.

9.8. FUNCTIONAL SIGNIFICANCE OF REGIONAL SPECIALIZATIONS

There is an extensive literature on the functional roles of the regional specializations of the retina, particularly the prominent features, such as the nasotemporal division, the visual streak, and the fovea or area centralis. I add here only a few comments relevant to the two-axis model I have tried to develop.

One comment is prefatory: there is no reason [following the arguments developed by Simpson (1961) and reviewed by Campbell and Hodos (1970)] to expect structures believed to be homologous between different species to be identical in function. For example, the visual streak of the possum may prove to be homologous to the streak of the rabbit, in the sense that both structures have developed from a structure found in their common ancestors. Their life styles are very different, however (one is an arboreal, nocturnal scavenger, the other a diurnal grazing species), and they may employ the streak specialization quite differently. Similarly, the fovea centralis of primates subserves high-resolution color vision; its presumed homolog in the cat, the area centralis, is capable of only about 1/10th of the spatial resolution of the human fovea and has only a very limited capacity to subserve color vision; while the presumed homolog of these structures in the rabbit may be capable of less spatial resolution than the visual streak in that species. The suggestion of homology is not weakened by this functional diversity, any more than (using my previous analogy) the homology between mammalian forelimbs requires that the dolphin's flipper be identical in function with the bat's wing. The basis for suggesting that the rabbit area centralis may be homologous to the primate fovea rests on properties such as its location relative to the horizontal and vertical axes of the retina, rather than on a commonality of function.

9.8.1. The Nasotemporal Division of the Retina

Even before the first known anatomical demonstrations of partial decussation, Newton (1730) had speculated that at the optic chiasm, some information from each eye passes to each hemisphere, so that each hemisphere can coordinate the information it receives from corresponding parts of the two retinas. This idea is now generally accepted and greatly developed empirically. The pattern of nasotemporal division described in Section 9.3, involving a narrow, vertically oriented "median strip of overlap" at the junction of nasal and temporal halves of the retina, is thus only a limited development of earlier ideas. Present understanding of the function of the strip in binocular depth discrimination (Fig. 9.4) can also be viewed as a development of the early idea that the different laterality of projection of nasal and temporal regions of the retina comprises a mechanism for coordination of the two eyes.

9.8.2. The Fovea or Area Centralis and the "Vertical Streak"

Where, as in primates and many nonprimates, the fovea or area centralis specialization is well developed, it seems clear that it comprises a mechanism for high-resolution, binocular vision. High levels of spatial resolution are attained by providing a large number of ganglion cells, each of which monitors only a small region of the retina (i.e., it has a small receptive field). Chromatic resolution (color vision) is added to that spatial acuity in primates by the development of specialized receptor pigments and retinal circuitry appropriate to provide color specificity in foveal ganglion cells, especially in the X-like ganglion cells, which tend to be most numerous there. Because this specialization is centered on the axis of nasotemporal division, the same cells are in a position to provide a mechanism for high levels of stereoacuity.

Where, by contrast, the area centralis is only weakly developed, it seems unlikely to subserve high levels of spatial resolution or stereoacuity. In the rabbit, for example, the concentration of ganglion cells at the putative area centralis is quite weak, and the animal often appears to inspect objects monocularly, using its strongly developed visual streak (van Hof and Lagers-van Haselen, 1973). Moreover, a strongly developed area centralis is associated with the development of eye movements (small saccades, smooth pursuit movements) needed for effective deployment of the area centralis; in the rabbit, small (< 1-deg amplitude) saccades and smooth pursuit eye movements appear to be absent (Collewijn, 1977). Nevertheless, in the rabbit, as in the cat and monkey, the area centralis is located in the middle of the binocular part of the retina and the rabbit does use its area centralis for fixation, perhaps preferentially (van Hof and Lagers-van Haselen, 1973). Collewijn (1977) suggested that the voluntary saccadic movements that he observed in rabbits provide "a rather global redirection of the visual field, possibly in particular of the binocular area" (which contains the area centralis).

Thus the area centralis is not in all species the retinal area capable of optimal spatial resolution, and is not the only part of the retina used for fixation. It does seem to be the one area of the retina used for binocular fixation. When a rabbit shifts from monocular fixation with its visual streak to binocular fixation with its area centralis, very different neuronal machinery is brought to bear on the image.

One possible exception to this generality must be noted, however; as with the exception to the rule that the vertical meridian crosses the area centralis (Section 9.3), it is found in a rodent. Tiao and Blakemore (1976) studied the topography of the retina in the golden hamster, and found a broad peak of maximum density located temporal to the optic disc, which they termed the *area centralis,* following common usage of the term in mammals, and in other rodents such as the rat (Fukuda, 1977). They noted that the *areae centrales* of the two eyes do not lie on corresponding points, and cannot therefore be used for binocular fixation. The same situation is found in the rat (Keens, 1981). This seems to me to reinforce the suggestion made above (Section 9.3.5) that the region of peak cell density in the rodent retina is not homologous with the area centralis of the cat or fovea centralis of the monkey, and may be better considered as the crest of the visual streak.

The "vertical streak" discussed in Section 9.6.2 would seem to provide, as A. Hughes (1977) has commented, some degree of retinal specialization for viewing the region between where an animal is fixating and its feet. Little is known of the types of ganglion cells present in this specialization, or of where they project in the brain (for example, is this region represented more strongly in the LGN of the thalamus or the SC of the midbrain?). It may prove to be an extension of the area centralis or a quite distinct specialization.

9.8.3. Nasal–Temporal Differences in Ganglion Cell Properties

In a discussion well ahead of its time, Walls (1953) noted histological differences between laminae of the cat LGN that subserve ipsi- and contralateral eyes and took the position ("until the results of experiments destroy it") that nasal and temporal areas of the retina make qualitatively different contributions to the "binocular image," at least in the cat. He reviewed many suggestions from clinical literature of differences in the functional capabilities of retinal regions nasal and temporal to the fovea, and hence to the axis of nasotemporal division, and predicted an accumulation of evidence of such differences. Further, he suggested that nasal–temporal differences in retinal function may reflect some basic process in the evolution of mammals, such as the development of substantial ipsilateral projections, and of stereoscopic vision. Nearly three decades later, Stone *et al.* (1980) summarized the still-limited evidence of nasal–temporal differences in the structure and function of the retina, and suggested that the differences might be common to all mammals. In addition to the specific differences discussed in Section 9.6.1, ipsilaterally projecting axons undergo orthograde degeneration more rapidly than those projecting contralaterally (Guillery, 1970). Moreover, the ipsilaterally innervated laminae of the LGN of the cat undergo transneuronal degeneration following enucleation of one eye more rapidly than the contralaterally innervated laminae (Guillery, 1973), and are more severely affected by binocular competition following monocular deprivation. This latter effect has been seen in both cat (Sireteanu and Hoffmann, 1979) and monkey (Headon and Powell, 1973; Hubel *et al.*, 1977).

Perhaps the most specific of the many lines of evidence of the functional significance discussed by Walls was Broendstrup's (1948) report of differences between nasal and temporal retina in their capacity for visual localization consequent to cataracts. His conclusions were recently confirmed by Moran and Gordon's (1982) report of a patient who had undergone surgery at age 19 for removal of a cataract present in one eye since birth. Assessment of the visual fields of the operated eye showed that the region of the visual field related to nasal retina was normal in extent, while there was a considerable loss of the region related to temporal retina. That is, the function of temporal retina was the more seriously affected by the cataract.

Particularly interesting evidence of nasal–temporal differences in visual function is provided by Sherman's (1977) report that, in bilaterally occipitally lobectomized cats, nasal retina shows clear residual function, while temporal retina does not. Although there is some disagreement concerning the magnitude of the ipsilateral projection to the midbrain in cats (discussed by Harting and Guillery, 1976), that projection appears to be much weaker than the contralateral projection. The ipsilateral projection to the SC in monkeys seems stronger than in the cat (Allman, 1977), but still less than the contralateral projection (Hubel *et al.*, 1975). Correspondingly, Sherman concluded from his behavioral studies that in the cat, temporal retina seems more dependent than does nasal retina on forebrain visual structures. It is tempting to relate this observation to the circumstance, that in both amphibians (e.g., Khalil and Szekely, 1976) and teleost fish (e.g., Springer and Landreth), ipsilateral projections from the retina pass to diencephalic centers and also to the pretectal area, but do not reach the optic tectum itself, whose massive retinal input is contralateral in origin. Apparently, wherever it has developed, the ipsilateral projection of the retina passes principally to the forebrain.

Marzi and di Stefano (1981) have recently looked again at human visual performance with these animal data in mind. They noted that, comparing visual performance using nasal and temporal regions of the retina, temporal retina provides slightly better acuity (assessed by letter recognition) for regions of the retina within 15 deg of the fovea, while nasal retina provides better acuity for targets beyond 15 deg. When the speed of the subjects' response (a key press) following presentation of a light stimulus was tested, nasal retina provided faster response times at all eccentricities. The functional implications of these differences would seem still to be rather unclear, however.

While these last considerations suggest a phylogenetic and behavioral framework for understanding nasal–temporal differences in visual function, they do not suggest why there should be subtle but consistent differences in ganglion cell properties between nasal and temporal areas of the retina. Perhaps, following Gould and Lewontin's (1979) article on the limits of adaptation as an explanation of the physical features of organisms, these subtle differences are not themselves functional adaptations, but minor structural consequences of major evolutionary changes, such as the development of an ipsilateral projection from the retina. Arguably, that explanation seems ad hoc, but it certainly cannot be excluded. There is no reason to assume that every bump on an animal or every variation within a population of neurons is an adaptation to a particular environmental pressure; and it is difficult (without being equally ad hoc) to suggest what else to make of (for example) the circumstance that the somas of cat α cells are, on the average, 10% larger in the temporal region of the cat retina than in the nasal region.

9.8.4. The Visual Streak

A number of discussions of the function of the visual streak are availble, for example in Slonaker (1897), Davis (1929), Munk (1970), Hughes (1971, 1977), Rodieck (1973), and Rowe and Stone (1976b). All these discussions, despite many differences of emphasis, seem to agree that the visual streak enables the animal to scan a wide segment of its visual field (usually considered to be the horizon) with a specialized retinal structure, without the specialized eye movements needed for scanning with a fovea or area centralis. The structural specializations at the streak might provide greater sensitivity to small, dim, slowly moving objects.

The only important new element that I would like to add to previous discussion is the hypothesis (argued in Section 9.7.2) that the visual streak is a general feature of mammalian retina. The point is not trivial, for the occurrence of a visual streak is often assumed to be a specific adaptation to a specific environment. But it is entirely possible that many mammals have a streak specialization in their retinas, not because they developed it to suit their current visual needs, but because they inherited it from a phylogenetic ancestor in whom it developed. The implications of this idea for the understanding of retinal topography are pursued in the following Section.

9.8.5. Retinal Topography: The Influence of Visual Environment and Phylogenetic Heritage

It is natural to interpret regional specializations in retinal structure as adaptations that are of survival value to the animal; and it is easy to construct intriguing parallels between retinal specialization and visual environment. The following are familiar examples: the need of that classic predator's victim, the rabbit, to be able to detect possible predators is met by its panoramic field of vision and highly developed visual streak; the need of a predator such as the cat for acute vision with good depth perception is met by its frontalized eyes and well-developed area centralis and nasotemporal division. Plains-dwelling kangaroos have a strongly developed streak, presumably to scan the ever-present horizon for natural enemies, while arboreal relatives, such as the tree kangaroo, have at best a weakly developed streak (Hughes, 1974).

It is as easy, however, to find counterexamples: The possum spends most of its waking hours climbing trees in the Australian bush at night, when there really is no horizon to be seen (though many strong verticals of tree trunks), yet has a much better developed streak than the tree kangaroo or, indeed, the cat. The wombat is remarkable for the enormous burrows it digs and lives in but has an exquisitely sharp visual streak; yet most rodents, many of which also live in holes, have weakly developed streaks. These counterexamples do not indicate that functional advantage has not been important in the evolutionary development of regional specializations of the retina, but they do seem to provide evidence that they are not a sufficient explanation of observed variations in topography.

What other factors might influence the topography of mammalian retina? I suggest two. One was foreshadowed at the end of the last Section 9.8.4, where I suggested that the regional specializations found in the retina of a particular species may not each be an adaptation to its visual needs; many may be part of the animal's inherited architecture. When a hand, a paw, a bat's wing, a hoof, and a dolphin's flipper are understood as adaptations of a common five-digit forelimb extremity, their common and unique features make a comprehensible pattern; but if one seeks to understand each bone as a specific adaptation, the mysteries multiply. In the same way, the regional specializations found in mammalian retinas seem better understood when viewed as adaptations of a common plan of the retina than when every feature of the retina is interpreted as a specific adaptation. It has been the purpose of this chapter to formulate a testable guess at that common plan.

The second factor that seems to me relevant to understanding retinal topography and its variations is the ontogeny of the retina. Little is known of the ontogeny of retinal topography in mammals, and my own guess is that analysis of, for instance, the mechanisms of development of the visual streak or area centralis will prove deeply relevant in the assessment of retinal topography, and of models such as that proposed here. The regional specializations of the retina are best understood, I am suggesting, not as a set of specific adaptations to particular environments, but as adaptations imposed on a basic ground-plan of the topography of the retina determined by its phylogenetic and developmental history.

9.8.6. Significance of the Two-Axis Model: A Ground Plan for the Topography of Mammalian Retina

Finally, I will state formally the suggestion made in the last paragraph of the preceding Section. It is widely accepted that the range of bodily structure found among mammals can be understood as variations on a basic pattern of morphology common to all mammals. Features of that plan include general features of vertebrates (such as the vertebral column, the development of two limb girdles and five-digit extremities and bilateral symmetry) and specifically mammalian features (such as the mode of gestation and the suckling of infants, and particular patterns of specialization of the cerebral cortex). The two-axis model set out in Section 9.2 may be a useful approximation to the basic pattern of retinal topography in mammals.

On the Understanding of the   10
Visual Pathways'
Dependence on the Visual
Environment

10.1. INTRODUCTION: THREE STARTING POINTS

A great deal has been learnt in recent years concerning the postnatal development of the visual pathways, and the dependence of that development on visual experience. Important concepts have emerged, such as the critical period of maturation, the plasticity of stimulus specificities of individual cortical neurons, and the competition of the two eyes for synaptic space on cortical cells. The purpose of this chapter is not, however, to review this extensive literature, but to argue that, particularly in the cat, our understanding of the impact of visual experience on the development of the visual pathways has been both deepened and simplified by analysis of that impact in terms of the functional groupings of ganglion cells.

10.1.1. Amblyopia: The Clinical Starting Point

The clinical condition of amblyopia, i.e., of functional blindness of an eye with no apparent abnormality of the eye or the visual pathways, has been recognized for many decades. Duke-Elder and Wybar (1973) trace understanding of amblyopia to the work of LeClerc in the late 18th century; Deller (1979) notes that the understanding of amblyopia as the result of the disuse of the affected eye stems from the work of Javal in the mid-19th century. Amblyopia is caused by disuse of the affected eye during the first few years of life, most commonly as a result of strabismus (incorrect alignment of the two eyes) or anisometropia (unequal refractive power of the two eyes), but also as the result of ptosis (drooping of the eyelid) or severe optic defects in the eye, such as a cataract. If the disuse is continued for the first several years, the loss of visual function is irreversible. The severity of the loss depends on the severity and duration of the cause, and the time of its occurrence. The earlier a problem occurs, the more sensitive the visual system is to it: a cataract present at birth is much more potentially damaging than an equivalent optical defect beginning in the third or fourth year of life. Similar disuse beginning after the first 5 years of life does not cause amblyopia at all and if the causative condition itself can be corrected, then visual function mediated by the affected eye is found to be unimpaired. Recently, attempts have been made (Banks *et al.,* 1975; Hohmann and Creutzfeldt, 1975) to determine the sensitive period of the human visual system for the development of binocular vision: both reports estimate that the sensitivity is greatest in early postnatal life and lasts to the third year of life, up to about 2½ years of age.

The condition of amblyopia indicates clearly that the normal development of part of the brain (the visual system) is dependent on specific environmental factors. People who are unable to view the world binocularly during infancy will never be able to experience stereopsis; without the experience of sharply focused retinal images during development, our ability to resolve or, in perceptual terms, to experience sharp images never develops. Moreover, the ophthalmologist's observation that amblyopia can be mitigated by patching or optically penalizing the child's functional eye, thus forcing the child to use his potentially amblyopic eye, demonstrated that part of the cause of amblyopia lies in the interaction of the two eyes during maturation. The neurobiological studies of the influence of visual experience on the visual pathways reviewed here have confirmed these insights and considerably expanded them. Their starting point was the clinical analysis of amblyopia.

10.1.2. The Philosophical Starting Point: Rationalism and Empiricism

There has been an enduring tension within Western philosophy since the 18th century between rationalism (an attempt to understand the world by the exercise of innate powers of the mind, particularly reason) and empiricism (which argues that understanding and knowledge are gained when our ideas, indeed our minds, are shaped by experience, and hence by the senses). Some idea of why I am discussing this tension as a starting point for the analysis of environmental effects on the maturation of the visual pathways is perhaps provided by the following two quotations, one each from a dominant figure of empiricism (the Englishman David Hume) and of rationalism (the Frenchman René Descartes).

Hume regarded sensation as the source of ideas and concepts.

> 'Tis certain, that the mind, in its perceptions, must begin somewhere: and that since the impressions precede their correspondent ideas, there must be some impression, which without any introduction make their appearance in the soul. As they depend upon natural physical causes, the examination of them would lead me too far from my present subject, into the sciences of anatomy and natural philosophy. [From *A Treatise of Human Nature*, 1897, p. 275]

Contrast this with Descartes' view that the intellect is the source of knowledge, the senses serving only as a distraction. The reality of the world, he argued, is "properly perceived" by the intellect, not the senses. Descartes believed that one can unambiguously understand one's own mind, especially if it is freed from the distractions of the senses.

> [Since] . . . there is nothing more easily or clearly apprehended than my own mind . . . I will now close my eyes, I will stop my ears. I will turn away my senses from their objects, I will even efface from my consciousness all the images of corporeal things . . . and thus holding converse only with myself . . . I will endeavour to obtain . . . a more intimate and familiar knowledge of myself. [From "Meditations on the First Philosophy" in Cantor's *Seventeenth Century Rationalism,* 1969]

This "turning away from the senses" was Descartes' route to his famous "cogito," the starting point of his attempt, after he had skeptically set aside all his previous experience, to understand the world by a process of deduction from indubitable truths such as "I think, therefore I must exist." Clearly, for Descartes the senses were a noisy distraction from his search for the source of certain knowledge; whereas for Hume the senses were that very source.

Both approaches have been tremendously influential on later thinkers, yet either pursued alone leads to absurdity. The empiricist philosophers laid the intellectual foundations of empirical science, whose success in the expansion of knowledge has far outstripped that of any other approach, and has been central to recent human history. Yet, as Hume expounded, pure reliance on the senses leads to an extraordinary, some would say disabling, degree of skepticism. There are many familiar examples of this skepticism, some stemming from Hume's analysis of causation, others from the analysis of whether any degree of reality can be established by the senses. On the other hand, Descartes' physics of the universe (his system of whorls and vortices) and Leibniz's "monadology," which are both cosmologies formulated by the exercise of pure reason by extraordinarily able intellects, seem absurd in the light of modern physics. Knowledge of the natural world cannot be acquired by the exercise of pure reason; gaining of that knowledge seems to require constant recourse to empirical experience and, therefore, to the senses. Kant's *Critique of Pure Reason* can be viewed as an attempt to understand the interaction of innate properties of the mind (Kant's "categories") with the information provided by the senses; and hence to reconcile the extraordinary success of empiricism with the absurdities of the skepticism to which, pursued alone, it leads.

No modern philosopher of science (I am less sure about all philosophers) seems currently to argue that observation alone is sufficient for the gaining of knowledge. Even the "naive" falsificationist [Lakatos' (1970) term] sees the rigorous application of reason (an innate property of the mind) as necessary in the empirical testing of ideas, which are themselves reached by intuition (another innate property). All philosophers of science seem to agree that the intellect interacts with experience: each is vulnerable to the other. Yet the rationalist–empiricist tension persists. Much of the influence exerted, and the criticism attracted, by Kuhn's "The Structure of Scientific Revolutions," for example, stemmed from his emphasis on the influence of psychological factors on the progress of scientific knowledge. Other philosophers, notably Polanyi *(Personal Knowledge),* emphasize the personal, "internal" nature of human knowledge.

What has the neurobiologists' work on the visual pathways contributed to this discussion? I think something unique. In neurobiology, as in biology as a whole, the empiricist–rationalist tension recurs in the debate over the determinants of the properties of behavior of animals and man: are those determinants hereditary or environmental? Hume (quoted above) foresaw the relevance of "anatomy and natural philosophy" to the problem of the source of knowledge; visual neurobiologists have begun to provide an account, based on intensive empirical analysis at the level of individual neurons, of the interaction between part of the brain (the visual pathways) and the external world. Their work is providing an empirical basis for assessing the hereditary and environmental determinants of the properties of the visual pathway. The account they are providing tells us not just that both heredity and environment contribute to the normal development of vision; most debates on this point end with that conclusion. More impressively, neurobiologists are beginning to delineate what components of vision are particularly determined by these two basic factors, and precisely where in the brain, and at what stage of its development, environment has its impact.

10.1.3. Animal Models of Amblyopia: The Neurobiologists' Starting Point

Modern neurobiology has contributed to the understanding of amblyopia and of the interdependence of the brain and environment, by developing animal models of amblyopia. Many current models can be traced to the early and continuing work of Wiesel and Hubel (1963a,b, 1965, 1974) in the cat and monkey. Their work did not encompass the functional groupings of ganglion cells (my principal concern in this chapter), but recent work has lain the basis for the analysis of the effects of visual experience and deprivation in terms of those groupings, and is therefore summarized briefly here.

Amblyopia in animals: Animal models of human amblyopia have been devloped in monkeys and cats. In monkeys, for example, von Noorden *et al.* (1970) showed that experimentally induced lid closure (which mimics the congenital cataracts or ptosis of the eyelid that cause amblyopia in humans) produces a severe form of amblyopia, shown in their tests as a loss of visual acuity, and in more general behavior as apparent blindness. The loss was irreversible if the lid closure was effected during the first 3 months of life, but appeared reversible if performed at an older age. von Noorden and Dowling (1970) further showed that surgically induced convergent strabismus (which mimics the convergent strabismus quite common in human childhood) can also produce a loss of acuity if it is induced early in postnatal life. Divergent strabismus did not produce an acuity loss, apparently because the animals learned to fixate with either eye, again mimicking the human experience.

In the cat, Dews and Wiesel (1970) showed that monocular lid suture causes a loss of acuity, of visual placing reactions, and of pattern discriminative ability in the deprived eye. These losses could be reversed by closing the experienced eye, and forcing the animal to use the "bad" eye, thus mimicking the patching or penalization procedures used clinically. The reversal was partial unless the experienced eye was closed within the first few months of life, but some improvement was obtained when the closure was done as late as 1 year of age. Sherman (1973), van Hof-van Duin (1977), and Heitlander and Hoffmann (1978) demonstrated a restriction of visual fields of monocularly and binocularly deprived cats. Hirsch (1972) and Muir and Mitchell (1973) found evidence that the loss of acuity was specific to the deprivation of experience; in kittens exposed only to vertical stripes, there was a loss of acuity for gratings of orthogonal orientation. Ikeda and Jacobson (1977) extended these observations to cats with a surgically induced squint, again showing a loss of acuity and a restriction of the visual field. Blake and Hirsch (1975) provided evidence of a sharp loss of binocular depth discrimination in cats reared with alternating monocular occlusion. Without the opportunity to use both eyes simultaneously, these cats do not develop the ability to use binocular clues to judge the relative depth of visual stimuli.

Given these animal conditions of amblyopia, neurophysiologists undertook to find the basis for the various components of amblyopia in the connections and properties of individual nerve cells, to determine the time course of the abnormal developments caused by visual deprivation and, thereby, to determine more closely the interdependence between visual experience and the maturation of the visual pathways.

Neurophysiological correlates of amblyopia: Neurophysiological correlates have been described for at least six components of amblyopia:

1. **Loss of acuity:** Evidence has been presented that the resolving power of individual X-class ganglion cells (the class considered likely to subserve high-resolution vision; Chapter 2, Section 2.3.3.1) is reduced by convergent strabismus (Ikeda and Wright, 1976), by optical penalization [chronic defocusing of the retinal image (Ikeda and Tremain, 1978)], or by lid suture (Lehmkuhle *et al.*, 1978, 1980*b*). Several studies have denied the occurrence of these effects; however, they are discussed in Section 10.3. Eggers and Blakemore (1978) have provided evidence that cells in the visual cortex of kittens reared with an artificial anisometropia (i.e., with one eye defocused) could nevertheless develop a receptive field in each eye. However, the contrast sensitivity of the cells responsive through the previously defocused eye was much poorer than of those driven through the normal eye. In monocularly deprived cats, there is a massive drop in the number of cortical cells that receive strong excitatory input from the deprived eye (Wiesel and Hubel, 1965), and this presumably also contributes to the acuity loss caused by monocular deprivation.

2. **Loss of pattern discrimination:** In binocularly deprived cats, many cortical cells remain visually responsive and, even in monocularly deprived animals, a proportion of cells remains responsive to the deprived eye. However, considerable abnormalities have been described in the stimulus specificities of these cells. It appears, although there is considerable disagreement in the literature on this point [see Pettigrew (1978), Blakemore (1978), and Movshon and van Sluyters (1981) for reviews], that some degree of orientation-selectivity and binocularity can be found in some cortical cells of the very young, visually inexperienced kitten. As might be expected, these selectivities are less well developed than in the adult. If the cat is raised without visual experience, moreover, a corresponding deficit in orientation-selectivity persists (Wiesel and Hubel, 1965; Pettigrew, 1974; Blakemore and van Sluyters, 1975; Buisseret and Imbert, 1976; Leventhal and Hirsch, 1977; Fregnac and Imbert, 1978). The development or survival of the property of direction-selectivity also appears dependent on normal visual experience (Daw and Wyatt, 1976; Singer, 1976). Similarly, as discussed below, the binocularity of cor-

tical cells is readily disturbed by visual deprivation, or any manipulation that prevents the animal from experiencing normal binocular vision. The stimulus specificities of neurons in monkey visual cortex also require normal experience if they are to persist in the adult (Wiesel and Hubel, 1974; Baker et al., 1974). If these stimulus specificities are important for the processing of visual information by the visual cortex, their absence or reduction in visually deprived animals may underlie the inability of amblyopic animals to perform pattern discrimination.

3. **Loss of binocular vision (without loss of acuity):** Perhaps the most common component of amblyopia is the loss of binocular depth discrimination and of its perceptual counterpart, stereopsis. The work of Hubel and Wiesel (1965) and Blake and Hirsch (1975) in the cat and Wiesel and Hubel (1974) and Baker et al. (1974) in the monkey, showed that the conditions that cause this loss in humans (squint, monocular or binocular occlusion of vision) sharply reduce the proportion of cortical neurons that receive strong excitatory input from both eyes (i.e., the proportion that "are binocular"). Indeed, Hubel and Wiesel (1965) used a technique of alternately occluding one eye of a kitten to show that where a kitten is deprived of binocular visual experience, but not of monocular experience by either eye, cortical neurons develop their normal range of specificities, except for binocularity. The conclusion seems compelling that binocular depth discrimination and stereopsis require a normal complement of binocularly wired cells.

4. **Loss of visual field:** The massive loss of visual field apparent in the monocularly deprived eye (Sherman, 1973; Heitlander and Hoffmann, 1978; but see van Hof-van Duin, 1977) seems explicable in terms of the failure of many geniculate cells to form effective synapses onto cortical cells. However, the loss of nasal parts of the visual fields in binocularly deprived cats does not have a ready correlate in available physiological results. The loss of nasal field in cats with convergent squint (Ikeda and Jacobson, 1977) has been related to a loss of functioning geniculate relay cells (Ikeda et al., 1977). In such animals, apparently, the ganglion cells in temporal retina fail to form effective synapses on their target cells in the LGN.

5. **Meridional amblyopia:** Mitchell et al. (1973) found evidence that humans with congenital astigmatism grow up with an uncorrectable loss of acuity for line or grating stimuli of the orientation for which the optics of their eye were defective. Muir and Mitchell (1973) demonstrated a similar condition in artificially reared kittens. A number of workers, particularly Blakemore and Cooper (1970), Hirsch and Spinelli (1970), Freeman and Pettigrew (1973), and Stryker et al. (1978), have shown that kittens that never experience lines of a particular orientation grow up lacking the normal complement of cortical cells tuned to respond to those lines. Similarly, monkeys reared with astigmatic refractive errors develop a meridional amblyopia (Harwerth et al., 1980).

6. **Recovery by reverse suturing:** The clinicians' experience that the development of amblyopia in a squinting eye can be prevented by patching the preferred eye, suggested that the cortical neurons affected by deprivation might recover some of their normal physiological properties under the same conditions. Chow and Stewart (1972), Blakemore and van Sluyters (1974), and Smith et al. (1978) have provided evidence that opening the eyelids of the deprived eye of a monocularly deprived cat, and suturing the experienced eye ("reverse suturing"), does facilitate recovery of the connections between

the initially deprived eye and the visual cortex. They further concluded that, as with human amblyopia, the extent of the recovery is greater the earlier the reverse suturing is done. In the monkey, Blakemore *et al.* (1978), Vital-Durand *et al.* (1978), and LeVay *et al.* (1979) have shown that reverse suturing can counteract the effects of monocular deprivation on the morphology of LGN cells and on the ocular dominance "stripes" of the visual cortex. Again, the earlier the reverse suturing is done, the more complete the reversal of effects.

The above is a very bare summary of the neurophysiological correlates of amblyopia. It fails to convey either the richness of the experimental work that has been directed to the understanding of the impact of visual experience on the development of the visual pathways, or the considerable controversies that enliven the recent literature. Pettigrew's (1978) review chapter seeks to resolve one of those controversies, and the articles by Blakemore (1978) and Movshon and van Sluyters (1981) also provide critical reviews. In addition, the 1979 volume edited by Freeman and the July 1979 issue of *Trends in Neurosciences* contain a wide range of articles summarizing many aspects of current neurobiological work in this area. An incisive review of many issues considered in this chapter was provided by Hirsch and Leventhal (1979); and Hickey's recent (1981) review provides a valuable discussion of the critical period of vulnerability of the visual system to the environment. To all these the reader is recommended.

10.1.4. Summary

This introduction has attempted to summarize the clinical, philosophical, and experimental framework of the work described below: the investigation of the effects of deprivation on the different functional groupings of ganglion cells, and of the brain centers to which they project. It is apparent that the problem of the dependence of the visual pathways on visual experience is older in clinical work than in neurobiology and older again in philosophy.

The clinical and experimental work, taken together, does allow some response to the question raised by the philosophers: does our understanding of the world depend on the innate properties of the brain, or on our sense impression? It seems clear that the visual pathways are genetically programmed at least to the extent that, at birth, the axons of retinal ganglion cells have found their target nuclei, including the LGN; that the axons of geniculate relay cells have reached the visual cortex; and that some cortical cells have started to develop stimulus specificities. Moreover, the functional organization of retinal ganglion cells is largely, though perhaps not completely, determined genetically. It seems, therefore, that sense impressions cannot, as Hume argued some must, "without any introduction make their appearance in the soul." Rather, even in newborn or older but visually inexperienced animals, there is an "introduction," provided by a highly, but still incompletely organized visual pathway. Moreover, the limits of visual capabilities that an animal develops postnatally still seem genetically determined: those limitations vary considerably between species, regardless of visual experience.

Yet the dependence of the visual pathways on early postnatal visual experience is very great: the development of the normal properties of neurons in the pathway, especially in the LGN and visual cortex, requires normal visual experience. Moreover, the loss of visual function and of the stimulus specificities of cortical cells matches the type of visual deprivation experienced. In short, the normal development of this part of the brain, and of the perceptual and higher mental processes for which it is necessary, require that the individual enjoy a normal visual environment during early postnatal life: the first stages and the limits of that development seem to be genetically determined.

10.2. EFFECTS OF VISUAL DEPRIVATION ON THE LGN OF THE CAT

It was in the LGN of the cat that the earliest observations were made that suggested that the X- and Y-cell components of the cat's visual pathway are differently affected by deprivation of visual experience. Sherman *et al.* (1972) observed a "loss" of Y-class relay cells from the population of cells recorded in the LGN of monocularly (MD) and binocularly deprived (BD) cats by microelectrode techniques. The "loss" (actually a decrease in the frequency of encounter of Y cells relative to X cells) was most severe in the laminae of the LGN of MD cats that receive input from the deprived eye; a less severe effect was noted throughout the LGN of BD animals. A further, apparently distinct abnormality was reported by Maffei and Fiorentini (1976), Ikeda and Wright (1976), Sireteanu and Hoffmann (1977), Hoffmann *et al.* (1978), and Lehmkuhle *et al.* (1978, 1980*a,b*); these reports all describe a loss in the spatial resolving power of individual relay cells of the LGN caused by visual deprivation of their input eye.

10.2.1. The "Loss" of Y Cells from the LGN of Visually Deprived Cats: Its Nature, Morphological Correlates, and Cause

The "loss" of Y cells: Sherman *et al.* (1972) studied the physiological properties (receptive field properties, afferent and axonal conduction velocities) of relay cells in cats that had been monocularly or binocularly deprived of pattern vision since early postnatal life by means of eyelid suture. They were able to identify the X and Y classes of relay cells described in other studies (Chapter 6, Section 6.1.1) and, as far as their tests indicated, the relay cells present were normal in properties: this confirmed Wiesel and Hubel's (1965) observation that visual deprivation does not affect the physiological properties of relay cells.

However, Sherman and co-workers did note a major abnormality in the relative numbers of X- and Y-class relay cells encountered. In normal cats, Y cells formed up to 50% of relay cells encountered in regions of the LGN subserving peripheral regions of the visual field: in MD cats, that proportion was normal in the laminae of the LGN subserving the nondeprived eye, but was reduced to 10% in laminae subserving the deprived eye. A less severe loss (to 20–30%) was observed in BD cats. [Our analysis did not include the W-class relay cells found in the C laminae of the LGN (Chapter 6, Section 6.1.2), at that stage still unrecognized.] The more severe "loss" observed in MD cats was apparently restricted to the binocular part of the affected laminae (i.e., to parts subserving regions of the visual field normally viewed by both eyes), while the less severe "loss" in BD cats seemed equally prominent in monocular and binocular parts of the nucleus.

This result (a "loss" of Y cells assessed by microelectrode techniques) has been confirmed in several subsequent studies, including Sherman and Stone (1973), Hoffmann and Cynader (1977), Sireteanu and Hoffmann (1977), Hoffmann and Holländer (1978), Kratz *et al.* (1979a), Eysel *et al.* (1979), and Geisert *et al.* (1980). One study (Shapley and So, 1980) has not confirmed it; these workers suggest that the numbers of Y and X cells are not affected by deprivation, but only their morphological size. As Shapley and So note, however, the sample of Y and X cells encountered in physiological experiments may depend critically on the tip size of the microelectrodes used. These workers used micropipette electrodes with resistances of 5–15 megaohms, as did Hoffmann *et al.* (1972). However, Shapley and So used a much more dilute salt solution within the electrode (approximately 0.15 M rather than 4 M), so that their micropipet electrodes must have had considerably larger tips, and presumably quite different sampling biases. Shapley and So report the same result with tungsten-in-glass electrodes, to which the above criticism does not apply. Further experiment, hopefully with other techniques, seems necessary to resolve this question; at the moment, the weight of evidence seems to me to support the view that visual deprivation causes a change in the properties of Y cells that decreases the frequency with which they are encountered in microelectrode sampling experiments.

Morphological correlates of the Y-cell "loss": Wiesel and Hubel (1965) had noted the greater severity of monocular (as against binocular) deprivation in its effects on the properties of cortical cells and had noted also that, in the LGN of MD cats, cells in the deprived laminae appeared pale and abnormally small; a similar effect was not detectable in the LGN of BD cats. They commented, confirming the clinical experience, that the damage to the visual system caused by closure of one eye "may not be caused by disuse, but may instead depend to a large extent on interaction between the two pathways."

That interaction has come to be called *binocular competition,* the idea being that LGN cells subserving the two eyes may be in competition for synaptic space on cortical cells. Guillery and Stelzner (1970) showed that the morphological deterioration (or failure to devleop) caused in LGN cells by closure of their input eye was confined to the binocular parts of the affected laminae (Fig. 10.1). Because cells in the monocular part of the LGN do not have to compete with relay cells innervated by the other eye for synaptic space in the cortex, they cannot be put at a competitive disadvantage by closure of the eye innervating them; correspondingly, these cells are unaffected morphologically. Guillery (1972) and Sherman *et al.* (1975b,c) went on to show that, even in the binocular region of a deprived lamina, the morphological effects of deprivation can be prevented if the cells with which they are competing are put at a similar or greater disadvantage. Specifically, they showed that cells in certain segments (which they termed *critical segments*) of lamina A contralateral to the sutured eye of an MD cat grow to apparently normal size, and include an apparently normal complement of Y cells, when cells in the corresponding part of lamina A1 are deafferented by a lesion to a localized area of the temporal retina of the nondeprived eye. Moreover, the animal is responsive, immediately after the deprived eye is opened, to visual stimuli presented in parts of the visual field of the deprived eye subserved by the "critical segments."

Following out this idea, three groups of workers tested another corollary of the binocular competition hypothesis: since (according to this hypothesis) relay cells in the

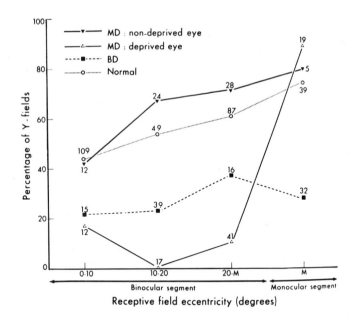

Figure 10.1. Early evidence that Y- and X-cell components of the visual pathway are differentially affected by deprivation of visual experience (from Sherman *et al.,* 1972). Graph shows the frequency of encounter of Y-class relay cells in the cat LGN, relative to X-class cells. In normal animals, the proportion increases with eccentricity (solid triangles). In MD cats (open triangles), the proportion of Y cells encountered is sharply reduced at eccentricities from the area centralis (0 on abscissa) to the monocular region of the LGN (M). In the monocular region, the proportion of Y cells was normal, so that the loss of Y cells appeared confined to the binocular segment of the LGN. In BD cats squares), the proportion of Y cells encountered was depressed in both segments of the LGN, but was never as severe as in the binocular segment of the MD LGN.

Thus, in both MD and BD cats, Y cells seem particularly severely affected by deprivation of visual experience. [Reproduced from the *Journal of Neurophysiology* with kind permission of the American Physiological Society.]

nondeprived laminae gain more than their normal share of cortical synapses, those cells should be somewhat larger than they would have been had the other eye not been deprived. This result was demonstrated by Wan and Cragg (1976), Hickey *et al.* (1977), and Hoffmann and Holländer (1978).

In BD cats, in which the loss of Y cells is less severe than in MD animals but is found in all segments of the LGN, a decrease in cell size has been reported (Hickey *et al.,* 1977) that is less severe than in monocular deprivation but is found throughout the LGN. However, Kratz *et al.* (1979*a*) found that in cats subjected to dark-rearing (a form of binocular deprivation), a loss of Y cells occurred that was not accompanied by a decrease in soma size.

The cause of the Y-cell "loss": There seems general agreement that the hypothesis of binocular competition can account for the abnormality in the number of Y cells encountered in deprived laminae of the LGN of MD cats. Implicit in this argument are the postulates that geniculate Y cells are in fact in competition with each other during early postnatal life, and that X cells are not similarly in competition during this period.

Two observations support these postulates. First, Leventhal and Hirsch (1977, 1978) found evidence that cortical cells receiving input from geniculate Y cells are binocular in animals reared without any visual experience, while cells receiving X input are almost entirely monocular. They suggest that among Y-input cells binocularity may be present at birth, making geniculate Y cells vulnerable to binocular competition in early postnatal life; while X-input cells may be monocular at birth, making geniculate X cells immune to such competition. Second, Sherman (1979a) has suggested that the relatively wide spread of the terminal arbors of the axons of geniculate Y cells in layer 4ab of area 17 (Section 8.1.3 and Fig. 8.10) may underlie a relatively high degree of interaction between Y-cell afferents from the two eyes. A further relevant observation was reported by Daniels et al. (1978), who presented evidence that the receptive field properties of geniculate Y cells are immature during much of the period of sensitivity of the cat's visual system to visual deprivation, while X cells mature early in the sensitive period: this prolonged immaturity may increase the relative vulnerability of Y cells to binocular competition.

Analysis of the differential sensitivity of Y and X cells to visual deprivation has thus sharpened our understanding of the dynamics of formation of geniculocortical synapses. Binocular competition and the different rates of connection and maturation of X and Y cells can account for several effects of monocular deprivation, but not for the less severe "loss" of Y cells from the LGN of BD cats. This implies a "direct" effect, still not understood, of deprivation on the development of geniculate Y cells.

The nature of the Y-cell "loss"—Shrinkage or true loss?: Sherman et al. (1972) noted that the "loss" of Y cells they reported on the basis of microelectrode recordings in the LGN of visually deprived cats could result either from a reduced number of functional Y cells ("true loss") or from a selective reduction in the size of Y cells relative to X cells ("shrinkage"), which would reduce the probability of Y cells being encountered by an electrode tip. They preferred the idea of true loss, but did not exclude shrinkage as a contributing or even principal factor. In the course of considerable subsequent work on the problem, evidence has emerged to support the prime importance of both mechanisms.

In brief summary, Hoffmann and Cynader (1977) noted, supporting the importance of shrinkage as the mechanism of Y-cell loss, that the sampling properties of microelectrodes are strong determinants of the proportions of Y cells encountered. Eysel et al. (1979) found that, although the proportion of Y cells is reduced by monocular deprivation among a sample of relay cells recorded in the LGN itself (Sherman and co-workers' original observation), their proportion is not reduced among recordings from their axons in the optic radiations. This implies that the number of functional Y cells is not reduced by monocular deprivation but rather that their somas have shrunken, while their axons have not. Further, Garey and Blakemore (1977) noted that the number of cells found labeled in the A laminae of the LGN following an injection of HRP into area 18 (thought to receive input only from Y-class relay cells) is not reduced by monocular deprivation (suggesting no true loss), but that the soma size of labeled cells is reduced (supporting shrinkage). Further, Hoffmann and Holländer (1978) also found evidence that the large (presumably Y) cells of deprived laminae of the LGN shrink more than smaller (presumably X) cells, explaining why there is a relative "loss" of Y cells when assessed by microelectrode techniques; and Eysel *et al.* (1979) noted that when the activities of X and Y classes of relay cells were assessed by the postsynaptic components of the field potentials generated in the LGN by an afferent volley, there was no evidence of relative loss of Y-cell activity.

On the other hand (against shrinkage and in favor of true loss), Hoffmann and Holländer (1978) noted that reverse suturing in MD cats (opening the initially closed eye and suturing the eyelids of the other, thus forcing the animal to rely on the initially deprived eye) can restore the number of Y cells identified physiologically to near-normal numbers, but does not cause any growth of the initially deprived cells. These cells remain at their abnormally small size, implying that the changes in the proportions of Y cells encountered are determined not by their soma size but by the number of Y cells that are functionally connected. The implication is weakened, however, by Dürsteler and co-workers' (1976) report that after reverse suturing, the initially deprived LGN cells grow substantially toward their normal size. LeVay and Ferster (1977) developed a hypothesis that LGN Y cells correspond to the class 1 relay cells described by Guillery (1966) in his Golgi study (Chapter 6, Section 6.1.6) and noted that monocular deprivation causes a reduction in the number of class 1 cells present. Lin and Sherman (1978) have reported, in contrast with Garey and Blakemore's (1977) finding mentioned above, that there is in MD cats a substantial reduction (compared to normal) in the numbers of relay cells labeled after injections of HRP into area 18 (Fig. 10.2). And subsequently, Kratz *et al.* (1979a) have noted that the proportion of Y cells is reduced in the LGN of dark-reared cats without any measurable reduction in the soma size of relay cells; in this situation, the "loss" of Y cells cannot be accounted for in terms of cell shrinkage.

Quite direct morphological support for a line loss of Y cells from the LGN of MD cats has emerged from the study of Sur *et al.* (1982) if the morphology of the axons of retinal Y cells. These axons normally branch to terminate in the A and C laminae (Bowling and Michael, 1980; Sur and Sherman, 1982). In MD cats many axons from Y cells in the deprived eye have very shrunken terminations, some not reaching the A laminae at all; while the terminals of X-cell axons were abnormally broad, perhaps reaching LGN cells which normally would have received Y-cell axons.

My own judgment, is that it is very likely that true loss of Y cells does occur: fewer relay cells seem to accept functional connections from retinal Y cells. Clearly, however, there is good evidence that Y cells may also remain functional and shrink. Presumably, both mechanisms operate; their relative prominence is still at issue.

10.2.2. An Abnormality of Geniculate X Cells in Visually Deprived Cats: A "Direct" Effect of Deprivation

Although Sherman *et al.* (1972) had confirmed Wiesel and Hubel's (1963a) conclusion that relay cells present in the LGN of visually deprived animals are in many ways normal in their physiological properties, several reports noted an overall loss in the spatial resolving power of relay cells consequent to visual deprivation. Ikeda and Wright (1976), Ikeda and Tremain (1978, 1979), Lehmkuhle *et al.* (1978, 1980a,b), and Chino *et al.* (1980) noted that the loss of resolving power seems most marked among X cells. In detail, the results of these groups of workers differ considerably. In particular,

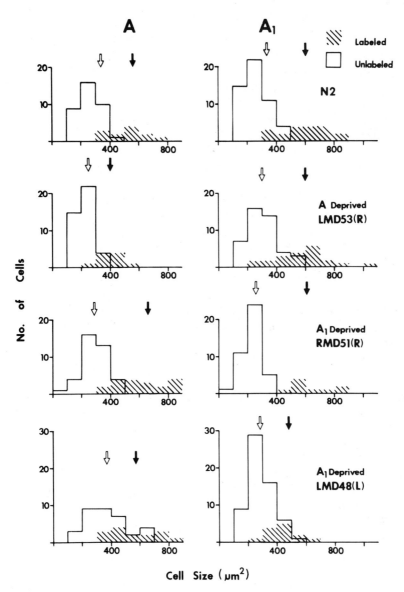

Figure 10.2. Evidence on the question whether Y cells in the cat LGN are reduced in size or number by visual deprivation. Lin and Sherman (1978) injected HRP into area 18 of MD cats, to identify the relay cells projecting there from the LGN. Confirming previous workers, they showed (top two histograms) that only a minority of LGN cells project to area 18 from laminae A and A1 of the LGN; those are large cells, confirming physiological evidence that only Y-class relay cells project to area 18 (Chapter 6, Section 6.1.7). The two histograms in the second row [LMD53(R)] show that when the eye providing input to lamina A is closed during early postnatal life, the number of cells projecting to area 18 is reduced (note smaller number of labeled cells in the left-hand histogram). The lower two rows of histograms show a symmetrical effect for layer A1.

The reduction in cell numbers labeled in layers driven by the deprived eye appeared significant statistically; also significant was the smaller size of labeled cells in deprived laminae. These data thus support the view that the effect of monocular deprivation on Y-class relay cells in the deprived laminae is caused by a reduction in both their number and their size. [Reproduced with kind permission of the *Journal of Comparative Neurology*.]

Ikeda and her co-workers found that among X cells, the loss of resolving power was confined to cells subserving the central 5° of the visual field (Fig. 10.3); they noted a similar but weaker effect among central Y cells. Chino *et al.* (1980) confirmed the loss of resolution among central X cells but reported no loss among Y cells. By contrast, Lehmkuhle and co-workers found the effect in X cells subserving all parts of the visual field, including the monocular segment. Further (see Section 10.3), Ikeda and co-workers concluded that the effect is determined by abnormalities present among X-class retinal ganglion cells, while the latter group of workers found the retina quite normal. In addition, Hoffmann *et al.* (1978) also noted a loss of resolving power among area centralis X cells, but in their results the effect was confined to layer A1 and was detectable, though weaker, among the remaining Y cells. To add to the variety of conclusions, Cleland *et al.* (1980) reported that in their experiments, lid suture had no effect on the resolving power of either X- or Y-class ganglion cells at any eccentricity.

As further illustration of the variety of effects found, Ikeda *et al.* (1977) reported that in the LGN of animals in which one eye was disadvantaged by surgically causing it to squint inwards, many cells in the LGN normally driven by temporal retina of the squinting eye fail to receive any functional retinal input at all. Yet again, Shapley and So (1980) and Derrington and Hawken (1981) reported that they could find no evidence

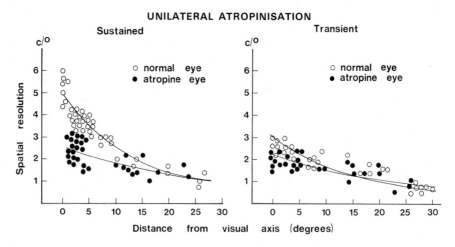

Figure 10.3. Evidence of an effect of visual deprivation on the spatial resolving power of geniculate relay cells (Ikeda and Tremain, 1978). These two graphs show the abilities of individual relay cells of the cat LGN to resolve grating stimuli. The graph at the left shows that among "sustained" cells (i.e., X cells), the resolving power of individual cells is greatest near the area centralis (or visual axis), and decreases steadily with distance from the visual axis. The open circles show data for normal cats, the closed circles for an eye that was defocused by atropinization during neonatal life. Note that the spatial resolving power of cells between 0 and 10 deg from the visual axis is sharply reduced by chronic defocusing. The graph at the right shows the equivalent data for transient- (i.e., Y-) class relay cells. Their resolving power is normally lower than that of sustained cells; it is only slightly reduced by chronic defocusing. [Reproduced with kind permission of the Editor of the *British Journal of Ophthalmology*.]

of an influence of deprivation by lid suture on the spatial resolution of either X- or Y-class relay cells in the LGN.

Presumably, some of these differences stem from the different techniques used to deprive the animals of visual experience. Ikeda and Wright (1976) and Ikeda and Tremain (1979) used surgically induced convergent squint of one eye, while Ikeda and Tremain (1978) used chronic defocusing of one eye. Lehmkuhle *et al.* (1978), on the other hand, used lid suture of one eye. One feature common to these different experiments and results is that the effects do not seem accountable in terms of binocular competition: a "direct" effect on the maturation of either ganglion cells or geniculate relay cells seems to be involved.

10.2.3. Other Evidence of a "Direct" Mechanism in Visual Deprivation

The effects of binocular lid suture and of dark-rearing cannot be explained in terms of a competitive interaction of the two eyes, since in these animals the two eyes were equally disadvantaged. Thus, the loss of Y cells reported by Sherman *et al.* (1972) and the reduction in cell size reported by Hickey *et al.* (1977) imply a non competitive or "direct" effect that is exerted relatively strongly on Y cells. In addition, Guillery's (1972) observation on "critical segments" was pursued by raising cats deprived of form vision by suture of one eye and enucleation of the other. If binocular competition were the sole determinant of visual development in MD cats, then (it was argued) the visual pathway and behavior mediated by the sutured eye should in such animals be normal, since the sutured eye would seem to have a competitive advantage over the other, enucleated eye. In fact, the visual fields of such an eye are abnormal (Sherman and Guillery, 1976), the response properties of cortical cells driven by that eye are abnormal (Kratz and Spear, 1976), and the LGN relay cells driven by that eye are slightly shrunken (Hickey *et al.,* 1977). These abnormalities are all less marked than seen for a sutured eye when the other eye was normal, and the cortical effects were less marked than seen in binocular deprivation. Clearly, however, there seems to be some effect of visual deprivation on all three parameters of visual system development, not explicable by binocular competition.

More recently, Anker and Cragg (1979) provided evidence that the "direct" and competitive mechanisms operating in monocular deprivation have different time courses. They studied kittens that were binocularly deprived by lid suture until they were 23 days of age, when one eye was opened. Over the next 26 days, the growth of cells in the still-deprived laminae of the LGN lagged behind that of cells in the laminae connected to the opened eye. The effect extended throughout the monocular as well as the binocular parts of the LGN, suggesting that the growth difference was not caused by binocular competition. Over the period 30–60 days after the one eye was opened, however, cells in the binocular segment of the LGN contralateral to the still-deprived eye became markedly smaller than those in the monocular segment, and the cell size difference between the two monocular segments disappeared. During this period, therefore, binocular competition seems to come into operation and to obscure the effects of the "direct" mechanism.

In summary then, there is substantial evidence of noncompetitive or "direct" effects of visual deprivation on the developing visual pathways in both BD and MD cats. The strong competitive mechanism that operates when one eye is more deprived than the other comes into play later than the "direct" mechanism and appears to "write over" the effects of the "direct" mechanism. In MD cats studied in adulthood, therefore, evidence is found principally of the competitive mechanism.

10.2.4. The Effect of Visual Deprivation on W-Class Relay Cells

Perhaps because W-class relay cells are encountered so infrequently in single-unit physiological experiments, the effects of visual deprivation on this class of cells have been reported only recently, and then only in terms of their soma morphology. Hickey (1980) measured the soma size of relay cells in the ventrally located laminae C1 and C2, which contain W-class relay cells, and reported a relatively weak influence of monocular deprivation. Specifically, Hickey noted that cells in layers A, A1, and C receiving input from the deprived eye were significantly (23–30%) smaller than in their nondeprived counterparts; for laminae C1 and C2, however, cells in deprived laminae were only slightly smaller (8 and 3%, respectively) than cells in nondeprived laminae and the differences were not statistically significant. Murakami and Wilson (1981), however, have subsequently reported a statistically significant result in the same experiment. Like Hickey, they studied soma sizes in the different laminae of the dLGN of MD cats. They confirm earlier observations (including Hickey's) of a significant reduction (25–35%) in the soma size of neurons in laminae A, A1, and C, but also observed a significant (28%) reduction in soma size in lamina C1. Murakami and Wilson confirm Hickey's observation that cells in lamina C2 undergo only a small (3%) and statistically insignificant reduction in soma size.

Overall therefore, the two studies seem in disagreement concerning the influence of deprivation on cells in layer C1. One possible explanation of the discrepancy is the longer survival of Murakami and Wilson's cats (140–256 days, as against the maximum survival time of 112 days used by Hickey); perhaps the neurons in lamina C1 are affected only slowly by visual deprivation. Hickey's conclusion that the W-cell system may be immune to visual deprivation may therefore need qualification.

The reason for the greater sensitivity of relay cells in lamina C1 (as against lamina C2) is not known. It seems an exaggerated form of the differences between laminae A1 and A in their reaction to visual deprivation (see Chapter 9, Section 9.8.3). Lamina A1 is affected more strongly than A and, also like C1, receives input from the ipsilateral eye. Even accepting this analogy, no clear reason has been established for the greater sensitivity to visual deprivation of ipsilaterally innervated laminae of the dLGN.

10.2.5. The Effect of Immobilizing the Eye

Brown and Salinger (1975) and Salinger *et al.* (1980) have tested the effect of depriving an eye of visual experience by immobilizing it. Movements of the eye and the consequent movements of the retinal image are a normal part of visual experience. Working in the cat, Salinger and co-workers found an unexpected effect of that deprivation on properties of dLGN neurons assessed physiologically. Whether the eye was immobilized by section of the nerves supplying the extraocular muscles (Brown and Salinger, 1975) or by detachment of those muscles from their insertion into the eyeball (Salinger *et al.*, 1980), these workers observed a reduction in the relative frequency with which X-class relay cells were encountered in microelectrode studies of the LGN.

Five aspects of the result are striking. First, it is the opposite of the effect caused by monocular deprivation effected by lid suture, suggesting that a quite different mechanism is operating. Second, it was found in adults, whose visual pathways seem unaffected by monocular deprivation by lid suture. Third, Salinger and co-workers suggest that the effect may be caused, not by a loss of sensory input, but by the loss of the normal kinesthetic input from the extraocular muscles. Fourth, the changes are found throughout the nucleus, even though only certain layers are deprived, supporting the idea that the mechanism involved is distinct from that operative in lid-suture deprivation. Fifth, the mechanism affects X- and Y-cell pathways differentially.

Berman *et al.* (1979) could find no comparable effects in the visual cortex, again suggesting that binocular competition, which is so important in monocular deprivation by lid suture, is not a factor. However, many exploratory and confirmatory experiments remain to be done on this effect, particularly to establish some morphological correlate of it. If the sensory input from the extraocular muscles can be confirmed as a determining factor, these experiments raise the possibility of major, unsuspected mechanisms operating in the control of geniculate function.

10.3. EFFECTS OF VISUAL DEPRIVATION ON CAT RETINAL GANGLION CELLS

Evidence presently available indicates that development of the functional groupings of retinal ganglion cells is not affected by visual deprivation. Although Ganz *et al.* (1968) reported a 40% drop in the amplitude of the b-wave of the ERG of visually deprived eyes, Sherman and Stone (1973) reported that the distinctive receptive field properties and axonal conduction velocities of W, X, and Y cells are as recognizable in the retinas of lid-sutured eyes as in retinas that have had normal visual experience; and that as seen in Nissl-stained retinal whole mounts, no obvious morphological abnormalities are produced in the ganglion cell layer by lid suture. Further, they noted no evidence, at least as between X and Y cells, of abnormalities in the proportions of the different cell classes present, such as appears in the LGN. Kratz *et al.* (1979b) confirmed and expanded these findings, and Thibos and Levick (1982) reported that when cats are raised with a severe artificial astigmatism, the grating resolution and orientation bias of retinal ganglion cells are unaffected. In another species, Daw and Wyatt (1974) reported that they could not modify the direction selectivities of rabbit retinal ganglion cells by selective exposure of young animals to moving patterns. The effects of deprivation on ganglion cell properties have not been examined in other species: as far as we know, therefore, the principal functional specificities of retinal ganglion cells are genetically determined and are not dependent on visual experience for their development.

The work of Ikeda and Wright (1976), Ikeda and Tremain (1978, 1979), and Ikeda *et al.* (1978) has, however, provided evidence that the spatial resolving power of area centralis X cells seems to depend for its full development on the retina being exposed to well-focused images during early postnatal life. Specifically, their data indicate that the spatial resolving power of X cells at the area centralis is reduced by visual deprivation produced by a chronic defocusing of the retinal image caused either by atropinization of the eye Ikeda and Tremain (1978) or by surgically induced strabismus (Ikeda and Tremain 1979; Ikeda *et al.*, 1978). Ikeda and Tremain suggest that the cause of this effect and of its limitation to the area centralis is as follows. The animal is deprived only of higher spatial frequencies; coarse patterns are still experienced. Only the X-class ganglion cells at the area centralis normally develop sensitivity to the higher spatial frequencies, and hence an abnormality is found only among this subgroup of X cells. Ikeda and Tremain's finding on this point has been corroborated by Chino *et al.* (1980). It should be noted, however, that Kratz *et al.* (1979*b*) could not demonstrate a similar abnormality among the retinal ganglion cells of cats monocularly deprived by lid suture (rather than by strabismus or atropinization) and concluded that the loss of resolving power that lid suture produces among LGN relay cells must be generated more centrally than the retina. Their conclusion was recently confirmed by Cleland *et al.* (1980). The discrepancy between these two results may somehow stem from the different techniques of deprivation used (atropinization/strabismus as against lid suture); otherwise, there is a clear disagreement in the literature on this point.

Despite this disagreement, however, there seems to be general agreement that the X cells at the area centralis of deprived retinas remain identifiable as X cells. The principal functional groupings of retinal ganglion cells, and many of their more specific properties as well, appear to be independent of visual experience for their development.

10.4. EFFECTS OF VISUAL DEPRIVATION ON THE SC OF THE CAT

The effects of visual deprivation on the properties of cells in the superficial gray stratum of the cat were investigated by Sterling and Wickelgren (1970) and were analyzed in terms of Y, X, and W cells by Hoffmann and Sherman (1974, 1975). Wickelgren and Sterling (1969*a,b*) had earlier shown that the normal receptive field properties of cells in the superficial gray stratum depend strongly on the input the colliculus receives from the visual cortex of the same hemisphere. Destruction of the visual cortex causes a loss of the cells' selectivity for moving stimuli and for the direction of their movement, and of their binocularity. This was confirmed by Rosenquist and Palmer (1971), who concluded that these effects are caused by destruction limited to area 17. Subsequently, Palmer and Rosenquist (1974) showed that the corticocollicular cells in area 17 are a fairly homogeneous group of layer 5 "complex" cells. They all respond well to fast-moving stimuli, and have large receptive fields, properties consistent with their receiving Y-cell input. Most are binocular and direction-selective, properties that fit well the idea that they provide the strong cortical influence on the colliculus described by Wickelgren and Sterling (1969*a,b*).

As discussed in Chapter 7, Section 7.2.2, Hoffmann (1972, 1973), working in the context of the Y/X/W classification of ganglion cells, described three principal input pathways from the retina to the colliculus: two direct pathways, comprising Y- and W-class ganglion cells projecting directly to the colliculus and an "indirect" Y pathway. This last pathway appeared to involved a Y-cell projection from the retina to the cortex, presumably via the LGN, and from the cortex to the colliculus. The corticocollicular cells described by Palmer and Rosenquist (1974) probably form the last link in this indirect path.

Hoffmann and Sherman (1974) reported that MD cats, the Y-indirect pathway to the colliculus from the deprived eye was almost undetectable, while the two direct pathways remained clearly demonstrable. Further, the effects were seen only in the binocular segment of the colliculus, just as the effect of monocular deprivation seemed largely confined to the binocular segment of the dLGN and of the visual cortex (Section 10.5). Hoffmann and Sherman concluded that the effects of monocular deprivation on the SC can be largely or fully explained in terms of the known breakdown of Y-cell transmission through deprived laminae of the dLGN.

Hoffmann and Sherman's model of the effects of monocular deprivation on the SC is shown in Figs. 10.4A–C. In addition, they noted some reduction in the overall number of cells driven by the deprived eye, though a lesser reduction than had been reported by Wickelgren and Sterling (1969*b*).

Hoffman and Sherman (1975) undertook a similar analysis of the effects of binocular deprivation (effected by lid suture) on the cat SC. They concluded (Figs. 10.4D and E) that the W-direct pathway is largely unaffected by this form of deprivation, that the Y-direct pathway seems somewhat diminished, and that the Y-indirect pathway is effectively absent. As a result, the SC of BD cats seems to be dominated by its retinal input, as Sterling and Wickelgren (1970) had earlier reported. Hoffmann and Sherman suggest that the effect of binocular as well as monocular deprivation on the colliculus may be secondary to their effects on the geniculostriate pathway.

10.5. EFFECTS OF VISUAL DEPRIVATION ON THE VISUAL CORTEX OF THE CAT

Many aspects of the influence of visual deprivation on the visual cortex have been investigaged without reference to the classification of retinal ganglion cells or geniculate relay cells. For three aspects of the cortical effects of visual deprivation, understanding was advanced when the analysis took into account the parallel organization of the retinogeniculocortical pathway.

10.5.1. Which Stimulus Selectivities Can Develop without Visual Experience?

We have seen that many properties of retinal ganglion cells and of geniculate relay cells can develop without visual experience, including their receptive field organization and at least some of the connections they form from the retina to the LGN and thence to the visual cortex. A substantial debate exists in the literature as to the experience-dependence of the stimulus specificities of cortical cells (viz., orientation- and direction-selectivity and binocularity), specificities that are, with few exceptions, not found in the retinal ganglion cells or geniculate relay cells. Briefly, several groups of workers have

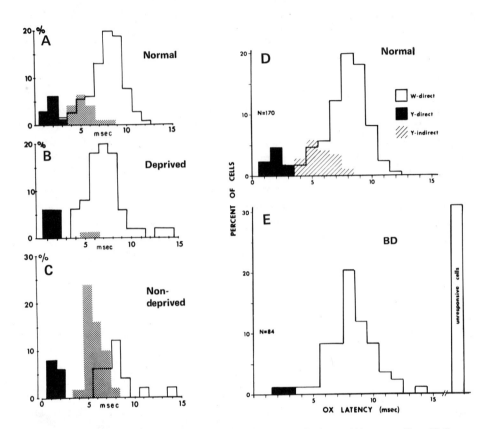

Figure 10.4. Effects of visual deprivation on the SC. The left-hand column of histograms (from Hoffmann and Sherman, 1974) shows frequency/latency histograms for cells in the SC. The latency value is the latency of the spike response of individual SC cells to a brief electrical stimulus delivered to the optic chiasm. Following Hoffmann (1973) (see Chapter 7, Section 7.2.2), these workers distinguished cells driven directly by retinal Y cells (black bars), cells driven directly by W cells (white bars), and cells driven indirectly (probably via the visual cortex) by Y cells (hatched bars). Comparison of data from a normal animal (A) with those from the SC contralateral to the sutured eye (B) shows that monocular deprivation caused reduction in the indirect Y pathway, i.e., in the pathway that involves the LGN and visual cortex. When data were taken from the SC receiving input from the nondeprived eye (C), no deficit in the Y-indirect pathway was apparent.

Similar data are shown in the right-hand column for the SC of normal (D) and BD (E) cats (from Hoffmann and Sherman, 1975). The effect of binocular deprivation was to eliminate the Y-indirect pathway, and to reduce the Y-direct pathway.

These workers suggested that the principal effect of monocular and binocular deprivation on the SC was secondary to the loss of Y-class relay cells from the LGN. [Reproduced from the *Journal of Neurophysiology* with kind permission of the American Physiological Society.]

stressed the ways in which the cells in the cortex of visually inexperienced (either binoc-ularly lid-sutured soon after birth or dark-reared) animals resemble those in the normal adult. Hubel and Wiesel (1963), for example, noted that the responses of cortical cells in inexperienced kittens were "strikingly similar" to those seen in adult cats. Their receptive fields could be classified as "simple" or "complex," and "receptive field ori-entation" was clearly apparent. Pettigrew (1974), on the other hand, developed distinc-tions (which seem very pertinent) between orientation- and direction-selectivity and between the responsiveness of a cell to stimulation of either eye ("binocularity") and the specificity found in many cortical cells for the relative positions of the two retinal images of a binocularly viewed target ("disparity-selectivity"). He concluded that, in kittens up to 4 weeks old, cortical cells lack both orientation- and disparity-specificity, although some are direction-selective. Furthermore, the lack of specificity persisted if the animals were kept inexperienced by binocular deprivation; by 5 weeks of age, fewer than 1% of cells showed orientation-selectivity, and none was disparity-selective. Again, Wiesel and Hubel (1965) reported that over half the neurons in BD cats are quite normal in prop-erties, while Fregnac and Imbert (1978) reported that the proportion of orientation-specific cells in dark-reared animals was only 4%, as against 69% in normally reared cats. These contrasting views are partly a matter of emphasis and partly a matter of experimental technique [as Pettigrew (1978) and Blakemore (1978) have suggested in reviewing the matter]. Studies by Blakemore and van Sluyters (1975), Buisseret and Imbert (1976), and Fregnac and Imbert (1978) on very young, visually inexperienced kittens seem agreed that a minority of cortical cells do show orientation-selectivity, but the proportion is much lower and the specificity often less developed than in the normal adult. By contrast, Sherk and Stryker's (1977) study supported Wiesel and Hubel's orig-inal finding that, in the BD kitten, a majority of nerve cells remain highly orientation-selective.

Some understanding of the genetically determined orientation-selectivity apparent in very young, visually inexperienced kittens has come from Blakemore and van Sluy-ters' (1975) suggestion that the orientation-selective cells they observed in such animals (which form a fairly homogeneous population of strongly monocular "simple" cells, located principally in layer 4) may receive input from geniculate X cells. In such cells, orientation-selectivity is innate; by implication, the development of orientation-selectivity in cells receiving Y input depends on visual experience. Leventhal and Hirsch (1977) have provided substantial evidence in support of this view (Fig. 10.5). In the dark-reared adult cats that these workers studied, nearly all orientation-selective cells had properties (small receptive fields, selectivity for slow-moving stimuli) that suggested that they receive X-cell input (see Chapter 8, Section 8.1.2). Among these cells, there was a strong tendency for their preferred orientations to group around the horizontal or vertical. Con-versely, most of the cells in the dark-reared cortex that lacked orientation-selectivity had relatively large receptive fields and responded well to fast-moving stimuli, suggesting that they receive input from geniculate Y cells. Fregnac and Imbert (1978) studied the cortex of dark-reared cats at a much younger stage than did Leventhal and Hirsch; nevertheless, their analysis is in close agreement. They propose a hypothesis of "differ-ential plasticity" of cortical neurons, in which, like Leventhal and Hirsch, they distin-guish two populations. Cells of one group are monocular at birth, but orientation-selec-

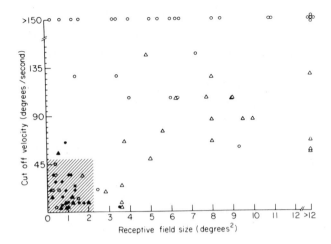

Figure 10.5. Evidence of a differential effect of visual deprivation on X-input and Y-input cortical cells. Leventhal and Hirsch (1977) measured many parameters of the visual reponsiveness of cells in area 17 of cat visual cortex, including two (receptive field size and "cutoff velocity") that they considered to reflect whether the cell received input from X-class or Y-class relay cells of the LGN. Cells receiving X input, they suggested, could be identified as those with small receptive fields and low cutoff velocities (i.e., they were unresponsive to fast-moving stimuli); in this graph, they are found in the lower left corner. Y-Input cells, by contrast, would appear toward the upper right corner.

This graph shows data for animals reared in the dark (data represented by circles) or with diffusing goggles (triangles). The closed symbols (both circles and triangles) represent cells that were orientation-selective; the open symbols represent cells insensitive to stimulus orientation. Note that solid symbols tend to congregate in the lower left-hand corner of the graph; this indicates that most orientation-selective cells have small receptive fields and low cutoff velocities (X-class properties). Thus, in animals with no pattern vision experience, it is only in X-class cortical cells that orientation-selectivity develops.

tive. That selectivity is not experience-dependent and the preferred orientations tend to be vertical or horizontal. These cells do require normal visual experience to develop normal levels of binocularity; they clearly correspond to Leventhal and Hirsch's X-input cells. Cells of the other group are binocular at birth, and require normal visual experience to develop or maintain orientation-selectivity; they clearly correspond to Leventhal and Hirsch's (1977) Y-input cells.

Leventhal and Hirsch (1980) have developed their earlier (1977) analysis by distinguishing a third group of cortical cells, viz., those with large receptive fields and poor responsiveness to fast-moving stimuli. Cells of this group, they suggest, are distinct from the small-field "slow" (presumably X-input) and large-field "fast" (presumably Y-input) cells distinguished in their earlier study. Leventhal and Hirsch suggest that this third group, which they found to concentrate in layers 2–4 of area 17, receives their main geniculate drive from W-class relay and ganglion cells (Chapter 8, Section 8.1.2). In dark-reared animals, these cells lack any specificity for stimulus orientation or direction, and their normal pattern of binocularity was to some degree disturbed, most cells tending to be dominated by one eye or the other.

In summary then, the question "which specificities require experience for their development?" seems best answered in terms of X, Y, and W cells. Cortical cells receiving X input are innately orientation-selective, but require experience to develop either binocularity or the disparity-selectivity to which Pettigrew (1974) drew attention. Cortical cells receiving Y input are binocular at birth (but not disparity-selective) and require experience to develop orientation-selectivity. Like Y-input cells, cortical neurons receiving W-cell input require experience to develop orientation-sensitivity; they also depend on experience for the development of normal levels of binocularity.

10.5.2. Determinants of Ocular Dominance

Leventhal and Hirsch's (1977, 1980) suggestion, following Blakemore and van Sluyters (1975), that X-input cortical cells are innately orientation-sensitive but require normal experience to develop binocularity, while Y-input cells are innately binocular but require normal experience to become orientation-sensitive, provides an explanation both for the distribution of ocular dominance in MD and BD cats and for the particular sensitivity of geniculate Y cells to monocular deprivation. First, binocular deprivation or dark-rearing reduces the proportion of binocular cells in area 17 (Kratz and Spear, 1976; Leventhal and Hirsch, 1977). The latter workers noted, moreover, that when X-input and Y-input cortical cells are considered separately, it is only among the X-input cells that binocularity is less commonly developed. Thus binocular deprivation does not "break up" the binocular connections of innately binocular cells; it acts only to prevent innately monocular cells from developing binocular inputs.

Second, as noted in the preceding section, the innate binocularity of Y-input cortical cells provides an explanation for the sensitivity of geniculate Y cells to monocular deprivation. Presumably, such cells are binocular because.they are contacted by Y-cell afferents that subserve both eyes and such afferents may be competing with each other for synaptic space. Because X-input cortical cells are innately monocular, binocular competition is not possible between X-class geniculate afferents subserving different eyes; and hence the "loss" of Y cells from the deprived laminae of the LGN of an MD cat (Section 10.2), and the morphological, physiological, and behavioral consequences of that "loss."

10.5.3. Eccentricity-Related Differences in the Modifiability of Orientation Stripes

Many workers, beginning with Hubel and Wiesel (1962), have shown that neighboring cells in cat and monkey visual cortex have similar orientation-specificities. A striking demonstration of a similar tendency in monkey visual cortex was provided by Hubel et al. (1978), who used the [^{14}C]deoxyglucose technique on animals exposed to high-contrast gratings of particular orientations. The technique shows that cells responsive to particular orientations are arranged in bands or stripes extending across the visual cortex, separated by less active cells. Stryker et al. (1977) have reported similar results in the cat.

Flood and Coleman (1979) demonstrated that in cats raised in striped cylinders, and thereby exposed visually only to horizontally or vertically oriented contours, the distribution of orientation-selective cells is abnormal. The "stripes" of cortical cells activated by contours of the orientation viewed during development are prominent, more so than in normal animals. The "stripes" of cells activated by contours orthogonal to that viewed during development are reduced below normal in prominence. Thus, the orientation "stripes" of the visual cortex seem modifiable by experience. Singer *et al.* (1981) also exposed kittens to only vertical or horizontal contours during rearing and showed a similar modification of ocular dominance stripes.

The point of Flood and Coleman's study that is of interest in the present context is that this modifiability is prominent only in regions of the cortex subserving peripheral parts of the visual field. The orientation stripes in regions of the cortex subserving central parts of the visual field were apparently undisturbed. Flood and Coleman related this eccentricity difference to Leventhal and Hirsch's (1977) finding that cortical cells that receive Y-cell input are (1) orientation-unselective in visually inexperienced cats and (2) more common in the cortex subserving peripheral retina. Perhaps orientation-selectivity can be modified by experience only among Y-input cortical cells; hence the greater modifiability of the peripheral cortex.

Recently, however, Singer *et al.* (1981) have reported a substantial study of the modifiability of orientation stripes in cat visual cortex, and their finding on this point was the converse of Flood and Coleman's: the disturbance of ocular dominance stripes was less rather than more severe in peripheral retina. This discrepancy, like several other issues in the study of environmental modification of the visual pathways, awaits clarification.

10.6. EFFECTS OF VISUAL DEPRIVATION IN OTHER SPECIES

In two other species, the monkey and the tree shrew, work has been done that indicates that the functional groupings of ganglion and geniculate relay cells may be important in understanding the environment-dependence of the visual pathways.

10.6.1. In the Monkey

LGN: There is limited evidence that X- and Y-like relay cells are differently affected by visual deprivation. The separation of X- and Y-class relay cells in the monkey LGN (the X cells forming the parvocellular layers of the nucleus and the Y cells the magnocellular layers; see Chapter 6, Section 6.2.1) makes it relatively easy to test the point. Several studies have reported that monocular deprivation in the monkey, effected either by lid suture or by surgically induced strabismus, causes the neurons of the affected laminae to fail to grow to their normal size (Headon and Powell, 1973; von Noorden, 1973; Baker *et al.*, 1974; Hubel *et al.*, 1977; von Noorden and Crawford, 1978; Vital-Durand *et al.*, 1978), and two studies (von Noorden and Crawford, 1978; Headon and Powell, 1978) noted that the magnocellular laminae are the more seriously affected. Headon and Powell (1978), for example, showed that in monkeys binocularly deprived of visual experience by lid suture at birth and examined at 65 days of age, the magnocellular laminae are distinctly more shrunken than parvocellular cells; however, by 376 days, the parvocellular laminae are equally affected. After deprivation of only one eye, however, a magno/parvo difference appeared that was more lasting; again, the magnocellular layers were the more seriously affected. At 376 days after monocular lid suture, cells in the deprived magnocellular laminae had shrunken 25–30%, while those in the parvocellular laminae had shrunken only 15–25%. No physiological correlates of these differences have been reported. Thus, the limited data available suggest that, in the monkey as in the cat, the Y-cell component of the dLGN is the more sensitive to visual deprivation.

Blakemore and Vital-Durand (1979) have described the properties of X-like relay cells of the monkey LGN, assessed physiologically. They were particularly concerned with the ability of individual X cells to resolve grating stimuli, describing a steady improvement over the first 200 days of postnatal life in the ability of cells to resolve fine gratings. The resolving power in individual X cells was only slightly reduced by suturing of the eyelids from birth. Blakemore and Vital-Durand concluded that the receptive field properties of X-like relay cells, or at least those important for spatial resolution, develop independently of patterned visual experience.

Visual cortex: One study (LeVay *et al.*, 1980) has reported evidence of a differential influence of deprivation on X- and Y-cell systems in the monkey. One portion of that study describes the effect of reverse suturing in MD monkeys on the pattern of termination of LGN relay cells. The terminals of X- and Y-class relay cells concentrate in layer 4 and are segregated in two ways. First, X- and Y-class cells terminate in different sublayers of layer 4 (Chapter 8, Section 8.1.3); this pattern is not disturbed by deprivation. Second, terminals of relay cells subserving different eyes terminate in alternating stripes of layer 4 (Hubel and Wiesel, 1972). This latter pattern is disturbed by deprivation (Hubel *et al.*, 1977), terminals of relay cells subserving the nondeprived eye forming abnormally wide stripes, those of relay cells subserving the deprived eye forming abnormally thin stripes. This disturbance appears to be the same for X- as for Y-class relay cells.

In animals in which the initially deprived eye opened, and the initially opened eye was closed ("reverse suturing"), an asymmetry between X-cell and Y-cell terminals was noted. When the reverse suturing was done at 3 weeks (initial suturing was done at 2–3 days), the terminals of the X-cell axons subserving the initially deprived eye responded dramatically, regrowing to fill more than their normal share of layer 4B, at the expense of the X-cell axons subserving the other eye. The Y-cell axons subserving the initially deprived eye did not respond, however, and remained restricted to abnormally narrow regions of layer 4Cα. When the reverse suturing was done at 6 weeks, the X-cell axons subserving the initially deprived eye still responded, though not as dramatically. They regrew to occupy an approximately normal share of layer 4B. As in the 3-week reverse suture, the Y-cell axon terminals did not respond. When the reverse suturing was delayed to 1 year of age, neither X- nor Y-cell axon terminals responded.

LeVay and co-workers noted partial correlates of these axon changes in the soma size of neurons in the parvo- and magnocellular layers of the LGN, and suggest that the termination of cortical plasticity occurs earlier for magno- (Y) than for parvocellular (X) axons.

10.6.2. In the Tree Shrew

The response of the X- and Y-cell components of the retinogeniculostriate pathway of the tree shrew seems very similar to that observed in the cat. Sherman *et al.* (1975*a*) reported that the relay cells of the LGN of the tree shrew can be classified into X- and

Y-cell classes, closely analogous in receptive field and conduction velocity differences to their namesakes in the cat. Further, they are intermingled within the nucleus (as in the cat), rather than segregated to different laminae (as in the monkey). Following monocular deprivation, Norton *et al.* (1977) found a selective loss of Y cells, restricted to the binocular parts of the nucleus. The X cells seemed unaffected by the tests applied, and the monocular segments of the deprived laminae had apparently normal complements of X and Y cells. Further, there was a corresponding loss of large-soma cells from the binocular segments of the LGN, and a loss of visual function in the deprived eye, apparently restricted to the binocular part of its visual field.

10.7. CONCLUSION: W, X, AND Y COMPONENTS OF THE NEURAL BASIS OF AMBLYOPIA

The recognition that the visual pathways comprise a number of parallel-wired "channels" or systems of neurons raised the possibility that the different systems might be differently sensitive to modification by experience. Such differences have now been reported in studies of the retina, LGN, SC, and visual cortex of the cat. The Y-cell system seems particularly sensitive, apparently because Y-class geniculate relay cells are wired binocularly to cortical cells at birth; they are therefore highly vulnerable to binocular competition during early life. Retinal effects of visual deprivation have been described only for X cells (with other studies reporting no effect even among X cells), and the effects of visual deprivation on the SC seem explicable in terms of the effects of deprivation on the geniculostriate pathway.

Lehmkuhle *et al.* (1980*b*) have recently proposed an interesting and potentially very valuable analysis of the contribution of X and Y cells to normal vision, and conversely to various classes of amblyopia. They note, in agreement with many workers, that X cells seem specialized to subserve high-resolution vision. Further, in amblyopia caused by defocusing of the retinal image, by astigmatism, or by anisometropia (which affect mainly the high-frequency components of the image), the loss of visual function is restricted to high spatial frequencies. This restriction occurs presumably because only X-cell development is affected. Conversely, other deprivation conditions, such as cataracts, eyelid suture (in cats), or, in humans, esotropic (convergent) strabismus, cause a "dense" amblyopia, which affects the perception of all spatial frequencies and at least in humans (Hess *et al.*, 1978) the spatial structure of the visual field. This, Lehmkuhle and co-workers suggest, occurs because such conditions affect Y-cell development as well, especially when the disuse is monocular. They suggest that Y cells are particularly important for the perception of low spatial frequencies; that (as argued in Section 10.2.1) Y cells are particularly sensitive to these forms of deprivation; and that the loss of low-spatial-frequency perception sharply reduces the visual system's ability to encode the spatial structure of a visual image.

One still unresolved question is whether the W-cell system is susceptible to modification by visual experience. Studies of the SC (Section 10.4) suggest that the W-cell projection to that center is unaffected by deprivation, and Leventhal and Hirsch (1980) have suggested that the W-cell system may prove to be relatively immune to environmental modification. If so, the W-cell system may then mediate much of the visual behavior of which visually deprived animals (and indeed humans) are capable. However, the question of which elements of the visual pathway are immune to deprivation of visual experience, and tend to make the structure of the brain purely innate, and which are experience-sensitive, making the structure and function of the brain a product of sensory input, is only partially settled.

It is already clear, however, that some properties of the visual pathways are determined primarily by genetic mechanisms, for example the functional groupings of retinal ganglion cells; and that other properties, such as the development of disparity-selectivity in cortical neurons, are critically dependent on early postnatal environment. Moreover, many remaining questions seem amenable to experiment; perhaps then the visual neurobiologist can provide an experimentally based answer to empiricist–rationalist debate discussed at the beginning of the chapter, at least as it applies to the visual system, showing which parts of that system are innate and which environment-dependent. The thrust of the present chapter is that a full understanding of the dependence of the visual pathways on visual experience requires an analysis in terms of the functional groupings of retinal ganglion cells, and of the subcomponents of the visual pathways to which they give rise.

On the Understanding of Visual Psychophysics and Behavior

11

11.1. INTRODUCTION

If, as I have argued at such length above, the arrangement of the mammalian visual pathways into parallel-wired systems of neurons is an important feature of their organization, then some evidence of that arrangement should be detectable in the visual performance of mammals; i.e., in their capacity, assessed by psychophysical or behavioral techniques, to respond to visual stimuli. That evidence is the subject of this chapter. The explanation of visual performance is perhaps the most important area of testing for any theory of the neuronal organization of the visual pathways; indeed, the explanation of visual performance in terms of neural mechanisms has long been a fundamental aim of the neurobiologist's study of vision. Earlier reviews of the correlation between visual performance and the parallel organization of the visual system have been provided by Breitmeyer and Ganz (1976), MacLeod (1978), Stone *et al.*(1979), and Lennie (1980a).

Psychophysical and behavioral correlates have been described for two sorts of parallel wiring in the mammalian visual system. First, two parallel neuronal pathways can be traced from the retina to the visual cortex (Chapter 8, Section 8.3). The more direct runs from the retina to the LGN to the striate cortex (the "RGC" path), the less direct runs from the retina to the SC to the posterolateral thalamus and thence to the visual cortex (the "RCTC" path). In Section 11.2, evidence is summarized that these two pathways form the basis of the focal/ambient division of vision functions. Second, there is evidence (discussed at length in Chapter 6, Sections 6.1—6.4, Chapter 8, Sections 8.1 and 8.2) that, within the RGC path different systems of neurons, such as X and Y cells, are connected in parallel between the retina and the cortex. In Section 11.3, evidence is presented that these systems of neurons within the RGC path form the neuronal basis of the dual-mechanism hypothesis of the perception of pattern and movement. A similar parallelism may exist in the RCTC path (Chapter 7, Section 7.2), but little is known of its psychophysical or behavioral correlates.

11.2. THE FOCAL/AMBIENT DIVISION OF VISUAL FUNCTION AND ITS NEURAL BASIS

11.2.1. The Focal/Ambient Division of Visual Function

A division of visual function into two principal mechanisms or "systems" has been suggested by several authors and seems of considerable value in understanding the contributions to visual processing of the different retinocortical pathways discussed above, and of the different (W/X/Y) systems of neurons present within those pathways. Trevarthen (1968) argued that visual function in primates can be usefully regarded as comprising two mechanisms, which he termed *focal* and *ambient* vision. Stone *et al.* (1979) adopted the same division, terming the categories of vision *foveal* and *ambient*. Trevarthen suggested that ambient vision is phylogenetically older than focal vision, being more prominent in prosimians, for example, than in monkeys, and includes all the functions that allow the prosimians to hunt successfully. This would seem to include the appreciation of visual space, and the orienting movements of the body, head, and eyes needed to move within it. Focal vision, on the other hand, includes the high-resolution vision mediated by the foveal specialization of the retina, and the various forms of manipulative, focal vision behavior dependent on the use of the fovea. Trevarthen suggests that focal vision is mediated by the visual centers of the forebrain such as the LGN and striate cortex while ambient vision is mediated by midbrain centers. Several later workers have proposed analogous divisions of visual function or have developed Trevarthen's concept. Diamond and Hall (1969), for instance, proposed that a pathway from the retina to the cerebral cortex via the midbrain and thalamus is present in all mammals. The presence of such a pathway, in addition to the RGC pathway, has been corroborated in a number of mammals (Chapter 8, Section 8.3); Diamond and Hall suggest that the RGC path develops relatively late in phylogeny, to serve specialized visual functions. Their idea envisages a cortical component to both pathways, in contrast to Trevarthen's suggestion that one system (ambient vision) might be entirely mediated by subcortical structures. Schneider (1969) proposed a "where is it"/"what is it" division of visual function in the hamster, the former mediated by the midbrain and the latter by forebrain (including cortical) visual centers; his hypothesis did not include a cortical component in the retinotectal path, but clearly envisaged two major components to visual function. Weiskrantz (1972, 1978) and Humphrey (1974) suggest that residual visual function in the destriate monkey, which includes considerable form and spatial vision (see following Section 11.2.3), corresponds to Trevarthen's "ambient vision". Humphrey, for example, commented that the destriate monkey appears capable of localizing objects in visual space, of differentiating figures from a background, and of "assessing how solid/empty space is momentarily structured around [its] body." Ambient vision in the monkey, it was suggested, is mediated by the RCTC pathway, while focal vision is mediated by the

RGC pathway. Weiskrantz argues, for example, that the visual association cortex normally integrates the activities of focal and ambient systems and comments that "perhaps this association cortex can build up a different, but nevertheless viable, world in the absence of the striate cortex with the information reaching it directly from the thalamus."

Ideas of the site and mode of interaction between ambient and focal vision have not been further developed. Two reports, however, provide evidence that in both primates (Bodis-Wollner *et al.*, 1977) and the cat (Sprague *et al.*, 1977), lesions that spare the striate cortex but involve the peristriate cortex have severe effects on visual performance, perhaps because they destroy the site of interaction of the two systems. I argue in the following paragraphs that ambient and focal systems of vision receive input from distinct ganglion cell populations: specifically, that X and Y cells (with X cells predominating) provide the input to the focal system, while W and Y cells (with W cells predominating) provide the input to the ambient system.

11.2.2. Evidence in Humans

In humans, the work that seems most relevant to a focal/ambient division of visual function has concerned the residual capabilities of individuals with lesions to the striate cortex. In terms of their subjective experience, these subjects are blind in the affected parts of their visual fields. Several recent studies suggest, however, confirming impressions expressed in earlier reports, that some residual visual function survives lesions to the visual cortex, but does not reach consciousness.

Pöppel *et al.* (1973) described observations on patients with large parts of their visual field rendered apparently blind by a cortical lesion. The patients were requested to direct their eyes toward a target presented in the blinded regions of their visual fields. The direction and amplitude of the eye movements were not accurate, but showed a significant, positive correlation with the parameters of movement needed accurately to locate the stimulus. Sanders *et al.* (1974) and Weiskrantz *et al.* (1974) examined a patient who had undergone surgery to one occipital lobe, resulting in a macular-splitting hemianopia. Although their subject could not perceive objects presented in the blind regions of his visual field, he showed considerable ability to reach out and touch a light source present in his blind hemifield, to distinguish a cross from a circle (provided they were large), to distinguish a horizontally from a vertically oriented bar (also provided they were large), and to detect the presence of a grating within a circular patch. Sanders and co-workers argue that considerable visual function remains in regions of the visual field blinded by a lesion to the visual cortex: they termed that residual function *blindsight*. Singer *et al.* (1977) reported evidence that local stimulation in peripheral regions of the visual field produces a localized modulation of visual threshold; and further that the modulation could be demonstrated in the "blind" regions of the visual fields of cortically blind subjects, even though they were unaware of the adapting stimulus. Singer and co-workers suggested that these modulations are a function of the retinotectal pathway related to the direction of visual attention. Barbur *et al.* (1980) have continued the analysis of residual function in humans. Their subject, with the visual cortex of one hemisphere damaged, showed a considerable ability to detect targets presented in the blind half of his visual fields. Systematic tests showed that his velocity sensitivity was little affected by the lesion, and that he could locate stimuli within the blind hemifield. However, he showed little ability to discriminate the shape or size of the stimuli used.

There are two striking features to these descriptions: First, after the loss of the striate cortex [which renders the major part of the retinogeniculocortical (RGC) pathway nonfunctional], the subject retains some capacity to direct his eyes and hands in visual space and to make limited discriminations of forms and contours, even though high-resolution vision is severely and permanently impaired. Second, the conscious appreciation of visual stimuli presented in the affected parts of the visual field was completely or largely lost; the RGC pathway seems of prime importance for conscious visual perception.

The idea seems compelling therefore that, in humans, the RGC pathway is necessary for at least conscious visual perception, and for high resolution of spatial patterns and thus for "focal" vision. The remaining components of the visual pathway seem able to serve functions which are part of "ambient" vision, such as the visual guidance of gaze and of limb movements, and limited pattern discrimination, as well as functions, such as the pupillary response to light, that are well established as subconscious reflexes.

What is known of the pathways which mediate ambient vision in humans? There is limited evidence that, in humans, the projection from the posterolateral thalamus to the prestriate cortex is of considerable importance. Bodis-Wollner *et al.* (1977), for example, describe a patient in which area 17 appeared, by the presence of cortically evoked potentials, to have survived a lesion to the prestriate cortex; the patient appeared more profoundly blind than other patients with lesions to the striate cortex that spared the prestriate cortex. Perenin and Jeannerod (1979) compared the visual performance of patients who had lost all the cerebral cortex of one hemisphere with that of patients (studied by others) whose lesions were more or less restricted to the striate cortex. They argue that the former patients have poorer form and spatial vision. The very limited functions that persist after large cortical lesions presumably depend on remaining structures such as the SC, or the accessory optic tract and its terminal nuclei.

11.2.3. Evidence in Monkeys

In monkeys, as in humans, lesions to the striate cortex cause a profound loss of visual function, but it has become clear that a good deal of visual ability remains, or can be recovered or learned. Following a number of early studies of the effects of cortical lesions in monkeys [for a brief but valuable summary, see Pasik and Pasik (1971)], Klüver (1942) provided a detailed account of this residual ability; his work indicated that destriate monkeys can localize objects in space, at least in the familiar environs of their cage, can detect very dim objects, and can respond to differences in luminous flux of stimulus objects, but not separately to their brightness, distance, or areas. Klüver found little evidence of form or color vision and concluded that loss of the striate cortex causes "a reduction in the number of properties which are effective in determining the responses of the normal monkey." On the important questions of form discrimination and the appreciation of spatial relationships within the visual field, Kluver concluded that "it appears questionable whether the monkey can achieve even primitive spatial organization after removal of the visual cortex."

Even so Klüver, in his 1942 paper, anticipated later work in noticing that destriate monkeys were able to respond to differences in the position of stimuli, and also in the following comments:

> For many decades it was thought that the bilateral occipital monkey is totally and permanently blind. . . . Closer observation, however, indicates that the animal only infrequently bumps into objects in its path if it . . . is not excited. . . . The role that visual cues may play. . . . in unfamiliar surroundings cannot be clearly determined from such [casual] observations. It is regrettable that the behavioral reactions in open space have not been studied more systematically.

Subsequent studies have borne out there observations. In particular, the Pasiks and co-workers [see Pasik and Pasik (1971) for a review; also Schilder *et al.* (1971, 1972)] demonstrated a residual ability in destriate monkeys to discriminate objects on the basis of brightness and shape. Subsequently, extensive reports of the spatial vision of the destriate monkey were provided by Weiskrantz (1972, 1978) and Humphrey (1974). Their reports describe the visual abilities of a rhesus monkey named Helen, who survived near-total, bilateral destruction of area 17. As with Klüver's animals, Helen's lesion included some prestriate cortex as well as striate, but her behavior was monitored over a longer period (8 years) and in a less restricted environment; Helen was commonly taken for walks, for example. Humphrey's description gives an impression of Helen's visual capacity.

> For 19 months following the operation nothing was seen in Helen's behavior to suggest any capacity for spatial vision.

Later, however, she began to turn her eyes to moving objects and reach out for them.

> At first she would look out and reach for objects only if they moved . . ., within a short time she became able to detect a flashing light source. . . . and finally a stationary black object against a light ground. . . . Her 'acuity' for detection progressively improved and. . . . she eventually had no trouble locating a black dot no more than 2 mm wide.

Helen learned to skirt obstacles relying apparently on vision, to pick up small tidbits from the floor (Fig. 11.1), to catch scurrying cockroaches, and even to judge distance when reaching for objects.

> It was clear that, given at least the experience of three-dimensional space, she was quickly developing a kind of three-dimensional spatial vision.

Weiskrantz and Humphrey both concluded that, after years of postoperative experience, Helen had developed something very like normal peripheral vision, perhaps akin to "ambient vision." Both conclude also that her ability to discriminate the shapes of small objects was severely and apparently permanently impaired (Fig. 11.2). To characterize the visual capabilities retained and lost by destriate monkeys, Weiskrantz (1972, 1978) developed a distinction between "noticing" and "examining" objects, one process involving the detection of objects, the other their identification. Weiskrantz noted earlier formulations of the same distinction, such as Trevarthen's (1968) "ambient" and "focal" components of vision, Schneider's (1969) suggestion that different parts of the visual pathway are used to determine "where" and "what" an object is, and Ingle's (1967) distinction between "orienting" and "evaluation." Weiskrantz suggests that destriate

Figure 11.1. Visual capacity of a destriate Monkey (from Humphrey, 1974). The drawings at the top, made from a cine film, show the monkey reaching directly and accurately for a small piece of chocolate placed on a table. The photograph shows the monkey proceeding between obstacles to reach very small tidbits (currants). [Reproduced with kind permission of *Perception*.]

monkeys like Helen are capable of noticing and locating objects, but not of examining them. We cannot know, of course, whether "focal vision" in the monkey includes conscious perception, but in other respects it seems comparable to the visual function lost in humans suffering lesions to the striate cortex.

The degree of visual recovery indicated by Humphrey's and Weiskrantz's reports has been confirmed in subsequent studies. For example, Feinberg *et al.* (1978) confirmed the ability of the destriate monkey to make accurate, visually guided reaching movements; Keating (1980) demonstrated the ability of destriate monkeys to detect and reach toward moving stimuli and to discriminate between stimuli moving at different speeds; and Dineen and Keating (personal communication) have demonstrated the ability of such animals to distinguish quite complex spatial patterns.

The degree of visual recovery now considered possible in destriate monkeys is greater than has ever been described in a human with a lesion to the striate cortex; for this, three reasons can be advanced. First, the monkey may be less dependent on the

Figure 11.2. Limitations on shape discrimination in a destriate monkey (from Humphrey, 1974). The monkey was able to distinguish the top pair of stimuli, a checkerboard pattern and a gray disc of identical mean luminance, when the checkerboard squares were 4 mm or more in side length. When that length was reduced to 2 mm, discrimination fell to 75%, and to chance at smaller dimensions. On the other hand, the animal failed in a series of 1000 trials to distinguish the second pair of stimuli, a triangle and a circle. She learned quickly to distinguish the two dots shown in the third row, one of which is surrounded by a circle. On the other hand, she failed to distinguish the stimuli shown at bottom, in which the circles are distinguished by the presence of a dot at the center of one. The monkey thus seemed capable of distinguishing large from small stimuli, but not of distinguishing patterns of similar size but different detail. [Reproduced with kind permission of *Perception*.]

striate cortex for visual function, although Helen's apparent lack of any spatial vision for many months after the lesion suggests considerable dependence, nevertheless. Second, the long period of Helen's survival, and the persistent and varied testing to which she was subject, may have been more effective in demonstrating residual function than the (necessarily) more limited testing applied to humans. Third, all the humans examined had suffered unilateral loss of the striate cortex, and therefore had half or more of their visual fields intact. They were not forced, as was the monkey, to rely entirely and for long periods on the residual visual capacity of the affected parts of their visual fields; presumably, the humans had less need to develop their residual visual capacities.

Considerable evidence is available concerning the visual pathways that might subserve the sort of residual vision seen in Helen. Destruction of the striate cortex causes the complete degeneration of the LGN and the loss, by transneuronal retrograde degeneration, of many ganglion cells as well (van Buren, 1963); the RGC path is totally destroyed. Cowey (1974) examined Helen's retinas, and confirmed van Buren's report of a massive loss of ganglion cells; he noted that the loss was particularly severe around the fovea, and that the remaining ganglion cells did not, as a consequence, form a prominent concentration around the fovea. My own observations on the retina of a destriate monkey add some further detail (see Chapter 9, Section 9.4.3 and Fig. 9.24). The remaining ganglion cells totaled about 140,000, perhaps 10% of the normal ganglion cell population. The cells concentrated weakly around the fovea and in peripheral retina appeared to concentrate in a visual streak. Cell loss was greatest around the fovea (over 95%), and considerably less (up to 60%) in peripheral retina. In peripheral retina, the loss was most marked in temporal retina and least marked in the visual streak region of nasal retina. Moreover, the ganglion cells remaining were predominantly small in soma size.

Thus, the ganglion cells that survive destruction of the striate cortex, and presumably are the cells that project to the midbrain and subserve all the vision apparent in destriate animals, resemble the W class of cat retinal ganglion cells in several ways (small soma size, concentration in the visual streak, projection to the midbrain).

Evidence on the importance for residual vision in destriate monkeys of the projection of the posterolateral thalamus to the prestriate cortex is problematical. Weiskrantz (1972) concluded that

> The information being exploited in the totally destriated brain is almost certainly received and processed by the posterior association cortex, because if this is also removed all but very crude discriminations seem to be impossible.

Poggio (1974), in his review, also concluded that the residual vision in destriate monkeys is abolished if the lesion is extended to the prestriate cortex. Further, Chalupa *et al.* (1976) have provided evidence that lesions to the posterolateral thalamus in monkeys produce substantial deficits in pattern discrimination. In their experiments, the effective lesions left the RGC pathway intact, but damaged the inferior component of the pulvinar, considered (Chapter 6, Section 6.6, and Chapter 8, Section 8.3.2) to be an important component of the RCTC pathway. And Solomon *et al.* (1981) have recently reported that the ability of destriate monkeys to visually locate and reach for objects is abolished by destruction of the SC. On the other hand, several reports (discussed in Zeki, 1969) suggested little or no deficit in visual performance following lesions restricted to the prestriate cortex; Zeki, reviewing this literature, argues that nevertheless some deficit is always caused by such lesions. Pasik and Pasik (1971) have provided evidence that some ability for form discrimination persists in monkeys as long as the accessory optic tract remains intact.

At present, therefore, it seems reasonable to suggest that the visual capabilities of the destriate monkey are mediated by a minority of ganglion cells that survives destruction of the striate cortex; and that these cells, which are relatively more numerous in the visual streak than elsewhere in the retina, are drawn principally from a particular class of ganglion cells, comparable to the W cells of the cat. It seems likely that a considerable portion (though not all) of the residual capacity for spatial and form vision involves the projection of the posterolateral thalamus to the prestriate cortex. It can thus be argued that ambient vision in monkeys is mediated, at least in part, by the RCTC pathway, beginning with a W-like class of retinal ganglion cells.

11.2.4. Evidence in Cats

In the cat, few workers have argued for a focal/ambient division of visual function. I think the reason for this is that the loss of visual function caused by lesions to the striate cortex is much less dramatic than in primates. Indeed, for many years the problem in the cat was to identify the visual functions performed by the striate cortex, since visual performance following damage to area 17 seemed remarkably good. In one respect, however, the cat is a valuable model for assessing the contributions of different classes of ganglion cells to visual function, because of the distinct projections of W, X and Y cells to the different areas of the visual cortex (Fig. 8.2). Recent studies of the effects on visual performance of ablations of these areas provide an interesting analogy between cat and primate.

Several workers have reported that destruction of area 17 in the cat has little effect on pattern discrimination (e.g., Doty, 1958, 1971). Pattern discrimination is much more severely affected, however, if the lesion includes areas 18 and 19, as well as 17 (Spear and Braun, 1969; Doty, 1971; Loop and Sherman, 1977; but see Winans, 1967). Sprague *et al.* (1977) reported a study that suggests a division of function between areas 17, 18, and 19 that was initially surprising (at least to me). After destruction of areas 17 and 18 in adult cats, the ability of the cats to relearn a range of visual discriminations (including horizontal versus vertical gratings, shape discrimination such as a circle versus a cross) remained at a high level, although in some individuals learning appeared slowed. Light/dark discriminations were quickly learned, and the visual fields of the animal appeared normal. This considerable level of visual performance was possible with the X-cell input and most of the Y-cell input to the cerebral cortex destroyed. Sprague and co-workers comment that

> The neuronal systems for perceiving and discriminating large planimetric patterns and forms, and for mediating visually guided behavior characteristic of this species lie outside of areas 17–18.

These "neuronal systems" may involve principally the W-cell system. The work of Kimura *et al.* (1980) and Dreher *et al.* (1980) has shown (Chapter 8, Section 8.1.1) that area 19 receives a substantial input from W-class relay cells in the LGN and, further (Kimura *et al.*, 1980), that the visual responsiveness of cells in area 19 is not disturbed, at least grossly, by cooling of area 17. Area 19 also receives input both from the pulvinar, which may relay W-cell activity directly from the retina (Itoh *et al.*, 1979; Leventhal *et al.*, 1980), and also from more medial regions of the posterolateral thalamus (See Fig. 8.22), which may be relaying activity from the SC and pretectum. Most of the retinal input to those latter structures also appears to be derived from the W class of retinal ganglion cells. Area 19 may, however, receive some Y-cell input via the medial interlaminar component of the LGN and via the SC and posterolateral thalamus.

It is of great interest, therefore, that in animals with areas 17 and 18 intact, but areas 19, 20, 21 and the LSA destroyed, Sprague *et al.* (1977) found a severe impairment of a form discrimination based on orientation or shape. Visual tracking and depth discrimination were also poor in such animals, but the extent of the visual fields appeared normal, and grating discriminations were relearned. By the tests these workers applied, therefore, visual performance in cats was more profoundly disturbed by lesions to areas 19, 20, 21 and the LSA than by lesions to areas 17 and 18. Sprague *et al.* (1981) developed this analysis further, arguing for a "double dissociation" of function between areas 17 and 18 on the one hand and areas 7, 19, 21 and the LSA on the other . Confirming Berkeley and Sprague (1979; Fig. 11.3), they noted that a lesion restricted to

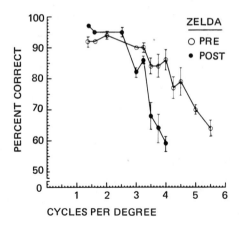

Figure 11.3. Loss of acuity in a destriate cat (from Berkeley and Sprague, 1979). The graph shows the performance of a cat in detecting the presence of a grating stimulus, as a function of the spatial frequency of the grating. Preoperatively, the animal could perform at above chance level for gratings as fine as 5–6 cycles/deg. After a lesion to areas 17 and 18, gratings could be detected only as fine as 4 cycles/deg. [Reproduced with kind permission of the *Journal of Comparative Neurology*.]

area 17 caused a "modest" (25—30%) elevation of threshold in grating acuity tasks, and sharper losses of orientation and vernier acuity, but did not affect learning of pattern discrimination when the discriminative cue was greater than threshold. This was also true of lesions that included area 18, as well as 17. Conversely, lesions that spared areas 17 and 18, but included areas 7, 19, 21 and the LSA, caused a sharp deterioration in the animals' postoperative performance on pattern discrimination tasks, but did not affect their performance on tests of spatial, angular, or vernier acuity. Lesions that included all these areas permanently eliminated all capacity for pattern discrimination. Sprague and co-workers conclude that the mechanisms for pattern recognition lie outside areas 17 and 18, and involve areas 7, 19, 21 and the LSA; while areas 17 and 18, and particularly area 17, determine the limits of acuity of the animal, both in acuity tasks and in pattern discrimination.

It is of interest in this context that in cats with retinal lesions restricted to the area centralis, behaviorally determined contrast-sensitivity functions shown a loss of resolution to high spatial frequencies (> 1 cycle/deg) but not to low frequencies (Blake and DiGianfilippo, 1980), with very little effect on general visuomotor behavior. To some extent, the visual function served by area 17 in the cat resembles the "focal vision" function suggested for area 17 in primates. Conversely, the considerable capacity for visual performance that, in the cat, survives the destruction of areas 17 and 18 may be analogous to the ambient vision function postulated in primates. The relatively mild impact of a lesion to the striate cortex indicates that the cat is less dependent than the monkey on the striate cortex (area 17) for visual performance, presumably because of the substantial projection of Y and W classes of cells in the LGN to area 19 and the LSA, and the RCTC pathways discussed in Chapter 8, Section 8.3. Perhaps, as Weiskrantz (1972) argued for the prestriate cortex in the monkey, these areas of cat visual cortex outside areas 17 and 18 are important for the integration of visual information reaching the cortex by different pathways.

11.2.5. Summary

In primates, a useful distinction can be drawn between the visual function (focal vision) that depends on the RGC pathway and is therefore lost following destruction of area 17, and the residual visual function (ambient vision) observable in the absence of the striate cortex, and apparently mediated by the RCTC pathway. The effects of lesions to the visual cortex in the cat provide a contrast with primates. In primates the visual loss suffered following a lesion to area 17 is very dramatic, whereas it required persistent testing to demonstrate the residual "ambient" visual function in destriate individuals. By contrast, residual vision following area 17 lesions in the cat is very obvious and substantial; it was the deficit caused by destruction of area 17 that was hard to demonstrate. Nevertheless, two analogies can be drawn between cat and primates that support the suggestion that focal and ambient vision can be distinguished in cats as well. First, in both cats and primates, the visual function subserved by area 17 involves high-resolution vision and, presumably, binocular depth discrimination. Second, area 17 receives input from similar classes of ganglion cells, the X cells of the cat and the X like cells of the monkey; while the retinal input to other areas of the visual cortex, which presumably mediate ambient vision, includes a large component of W or W-like cells. At least as a working hypothesis, it seems useful to suggest that the focal/ambient division of visual function can be related to the functional groupings of ganglion cells, and can be applied to nonprimate mammals, as well as to primates.

11.3. EVIDENCE OF DISTINCT X AND Y CONTRIBUTIONS TO FOCAL VISION

11.3.1. The Dual-Mechanism Hypothesis of the Perception of Form and Motion

Psychophysical studies of human visual performance have provided interesting evidence of separate neural mechanisms operating within focal vision. The functions subserved by these two mechanisms have been characterized in various ways: for example, as "spatial" as against "temporal" analyzers, or as "pattern detectors" as against "flicker" or "movement detectors". And the case has been argued that the spatial analyzers or pattern detectors correspond to the X-cell system of nerve cells, while the temporal analyzers or movement or flicker detectors correspond to the Y-cell system. The evidence (which seems to me considerable) in favor of this hypothesis is set out in this section. Some qualifications to and criticisms of the hypothesis are set out in the following Section 11.3.2.

In 1969, Keesey reported an experiment with stabilized retinal images. Her interest was to determine the degree to which the perceptual fading caused by image stabilization could be prevented by temporal modulation, i.e., by flashing the image on and off. She observed that flashing or flickering the stimulus (a bar-shaped patch of light) did indeed prolong its visibility; and she also noted the unexpected result that at flicker frequencies of 5–30 Hz, the perception of the flicker of the stimulus persisted after appreciation of its shape was lost. The importance of this finding for the present chapter is that it suggests that the perceptions of the temporal and spatial aspects of the stimulus (i.e., of its flicker and its shape) can be dissociated. This dissociation implies that different neural mechanisms may be involved in the perception of these two parameters.

Pursuant to this idea, Keesey (1972) reported experiments in which human subjects viewed a flickering bar pattern against a uniform field, and the brightness of the pattern was modulated sinusoidally with time. The subject was asked to set the amplitude of the brightness modulation so that he could just see the stimulus, and this amplitude setting was sought over a range of frequencies of brightness modulation. Keesey commented that

> Two categories of judgment emerged, 'flicker or no-flicker' and 'flickering bar or no flickering bar'.

At one setting of the amplitude of brightness modulation, the subject could just see the flicker of the stimulus; at another setting (usually higher), he could just see its shape. At flicker frequencies below about 30 Hz, the flicker of the bar stimulus was always detectable at a lower brightness setting than that required to see its shape. Further, Keesey showed that the relation between the threshold brightness of the stimulus and the flicker frequency differed for the two judgments. Keesey suggested that two different "mechanisms" may be involved, perhaps related to the distinct bipolar cell and amacrine cell responses that had recently been described in nonmammalian retinas. Finally, Keesey noted several prior descriptions of this phenomenon (e.g., by van Nes *et al.*, 1967); her 1972 report is one of the first to argue for the involvement of two separate neural mechanisms.

Tolhurst (1973) also argued for the presence of two mechanisms in human perception, one "movement-analyzing" and the other "pattern-analyzing," which seem closely analogous to the flicker- and shape-detecting mechanisms suggested by Keesey. Tolhurst confirmed Robson's (1966) observation that temporal modulation (flicker) increases the visibility of coarse but not fine grating stimuli, and also Blakemore and Campbell's (1969) observation of a "lowest adaptable spatial frequency channel." Blakemore and Campbell had provided evidence that a subject can "adapt" certain components of his visual system by fixedly viewing a stationary grating of a particular spatial frequency (say n cycles/deg.). His sensitivity to subsequently presented (test) gratings is decreased for a time, and the decrease is maximal for test gratings of spatial frequency n. In Blakemore and Campbell's data, this finding held for values of n of 3 or greater; for coarser gratings, the loss of sensitivity was still greatest at 3 cycles/deg. They postulated that the human visual system comprises independent "spatial channels," each maximally sensitive to grating of a particular spatial frequency, the "coarsest" channel being maximally sensitive to 3 cycles/deg.

Tolhurst's central observation was that when the adapting grating was moving instead of stationary, evidence could be gained of channels maximally sensitive to spatial frequencies much less than 3 cycles/deg. That is, there appear to be "coarse" channels in human vision that respond only to temporally changing (moving) stimuli, and "fine" channels that respond to stationary as well as moving stimuli. Tolhurst further reported, confirming Kulikowski (1971*b*), that when the latter channels are activated alone, whether by moving, flickering, or stationary stimuli, the stimulus appears stationary.

Tolhurst argued that these two sorts of channels may subserve "very different roles in perception," one coding the movement of stimulus, the other its shape. He noted that movement channels seem more sensitive at spatial frequencies less than 3 cycles/deg., while pattern channels seem more sensitive at frequencies greater than 3 cycles/deg. Further, he drew an interesting analogy between the properties of the two sorts of channels and of (respectively) the Y and X classes of cat retinal ganglion cells.

Kulikowski and Tolhurst (1973) expanded this work by further investigating a phenomenon described by Kulikowski (1971 *a,b*). Kulikowski (1971 *a*) had reported that the psychophysical threshold for the perception of flickering gratings depends upon whether they are presented and withdrawn (i.e., alternated with a screen of uniform luminance equal to the mean luminance of the grating—an "on–off modulation") or are "counterphased" (i.e., changed in phase by 180 deg., the black and white bars appearing to change places). The counterphase modulation appeared likely to provide the stronger stimulus, and indeed, at low spatial frequencies, the minimum contrast at which the grating was visible with counterphase modulation was half that required with on–off modulation. For gratings with spatial frequencies of 6 cycles/deg. or higher, however, thresholds for counterphase and on–off modulation did not differ (Fig. 11.4). Further, Kulikowski (1971*b*) noted that, at threshold contrast, coarse counterphased gratings appeared to flicker or to move, but that fine gratings (for which counterphase threshold was not lower) appeared stationary even though they were, in fact, flickering. The fine gratings appeared to flicker only when the grating contrast was increased above that required for detection of their presence.

Kulikowski and Tolhurst suggest that one neural mechanism is coding the flicker of their stimulus, another its pattern; when stimulated separately, the former mechanism gives rise to a perception of flicker, free of pattern (Keesey's observation), while the latter gives rise to a perception of pattern without flicker [Kulikowski's (1971 *b*) observation]. Kulikowski and Tolhurst noted that these two mechanisms (flicker- and pattern-detecting) may be comparable to the X and Y cell classes described in the cat, although they stress that the responses of cat X cells seem too transient to provide the exclusive input to the human pattern-detector mechanism.

King-Smith and Kulikowski (1975) pursued this analysis still further. Instead of grating stimuli, they employed small bar-shaped stimuli and a technique of "subthreshold summation" to explore the similarity of the flicker- and pattern-detecting mechanisms they had postulated to the receptive field properties of X and Y classes of ganglion cells. For example, they showed that the threshold contrast for the detection of a flickering bar (test stimulus) is affected by the presence nearby of flickering bars of subthreshold contrasts, i.e., of bars too dim to be seen (adapting bars). When the subject was asked to set the brightness of the test bar at threshold for seeing its shape, the interaction between test and adapting bars appeared linear; when the subject was asked to set the brightness of the test bar at threshold for seeing its flicker, the interaction was nonlinear. This difference in interaction could reflect the linear and nonlinear receptive field interactions seen in X and Y cells, respectively (Chapter 2, Section 2.1). Second,

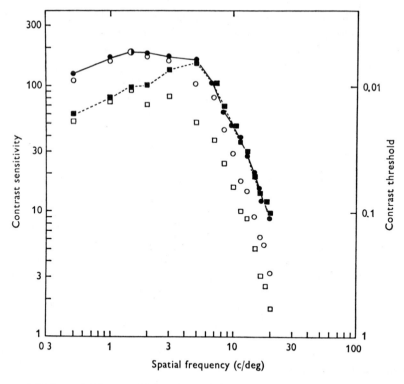

Figure 11.4 Evidence of different modulation thresholds for flicker and pattern detection (from Kulikowski and Tolhurst, 1973). The graph shows contrast-sensitivity functions for a human subject viewing a sinusoidal grating stimulus. The closed symbols show the functions obtained when the subject set the stimulus contrast so that the grating could just be detected, whether it appears first as a flicker or as a pattern. At spatial frequencies of the grating of the grating of 3 cycles/deg. or less, sensitivity was higher when the grating was modulated in a counterphase manner (black and white bars alternating positions, closed circles) than when alternated with a gray screen of equivalent luminance (closed squares). At spatial frequencies higher than 3 cycles/deg., this difference in sensitivity was not apparent.

The open symbols show the functions obtained when the subject set the threshold so that the 3.5Hz flicker of the stimulus was just visible, for counterphase modulation (open circles), and for on–off modulation (open squares). Now compare the functions plotted with open and closed squares (on/off presentation of the grating). For spatial frequencies below 3 cycles/deg., these functions do not differ, presumably because the subject was more sensitive to the flicker of the stimulus than to its pattern. At higher spatial frequencies, however, sensitivity to the flicker of the stimulus (open squares) was relatively low; i.e., the subject was more sensitive to the pattern of the stimulus than to its flicker. A converse trend was seen with the counterphased stimulus (compare open and closed circles); for low spatial frequencies, the subject was more sensitive to the flicker of the stimulus than to its pattern, and for high spatial frequencies was more sensitive to the pattern. Note moreover that where the subject was more sensitive to the pattern of the stimulus (at spatial frequencies greater than 3 cycles/deg., closed symbols), sensitivity was the same for the on/off and counterphased presentations of the grating (compare closed circles with closed squares).

These data suggest that two neural mechanisms are contributing to the obtained contrast-sensitivity functions. One is more sensitive at low spatial frequencies, is about twice as sensitive to counterphase as to on–off modulation, and, when stimulated alone, gives rise to a perception of flicker without pattern. The other is more sensitive to high spatial frequencies, is equally sensitive to counterphase and on–off modulation, and, when stimulated alone, gives rise to a perception of pattern without flicker. [Reproduced with kind permission of the *Journal of Physiology (London)*.]

the distance over which the interaction could be demonstrated differed between pattern detection and flicker detection, i.e., the adapting bars influenced pattern detection over relatively short distance (e.g., 10-min arc) but influenced flicker detection over distances twice as great (Fig. 11.5). This suggests that the cells subserving flicker detection have larger receptive fields; correspondingly, Y cells have larger receptive fields than X cells. Third, the "receptive fields" of both flicker and pattern detectors, as demonstrated by the same technique, showed that surround inhibition is stronger among the pattern detectors. This could reflect the relatively strong receptive field surrounds of X cells.

The hypothesis that separate mechanisms subserve the perception of the spatial and temporal aspects of a stimulus has been supported or invoked in many other studies. Nelson (1974), for example, argued that perceptual "switching" between the awareness of the movement and of the pattern of a stimulus may involve a switch in reliance between the Y and X cell components of retinal activity. Tolhurst (1975) used a two-stimulus paradigm to test the time course of adaption produced by a subthreshold grating stimulus; the time course was shorter (about 100 msec) when test and conditioning gratings were coarse, suggesting that the system sensitive to low spatial frequencies is "transient," and long (at least 100 msec) when the gratings were fine, suggesting that the system sensitive to high spatial frequencies is "sustained." Breitmeyer and Ganz (1976) and Legge (1978) discuss several other lines of psychophysical evidence that add to the concept of distinct mechanisms. The papers of Wilson (1978) and Wilson and Bergen (1979) also proposed the presence of "sustained" and "transient" mechanisms in human vision. Their work was distinct in using a spatially localized, aperiodic stimulus (a set of three spaced lines) rather than a grating stimulus, and they also employed two different time courses of stimulus presentation that they argue, preferentially acti-

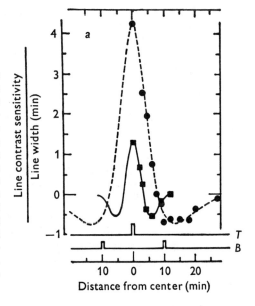

Figure 11.5. Spatial extent of functional units of flicker- and pattern-detecting mechanisms (from King-Smith and Kulikowski, 1975). The graph shows the effects of a subthreshold stimulus on the threshold for detection of a line stimulus. When the subject was concentrating on observing the flicker of the test bar, the function with closed circles and dashed lines was obtained. Interaction was facilitatory within a region about 10′ wide and was inhibitory when the separation of test and conditioning stimuli was wider. The function represented by closed squares and the continuous line was obtained when the subject set thresholds for pattern detection, i.e., for seeing the shape of the test bar. [Reproduced with kind permission of the *Journal of Physiology (London)*.]

vate one mechanism or the other. They suggest that the sustained and transient mechanisms may be related to the X and Y classes of ganglion cells described in the cat, and to corresponding groupings observed in the monkey. Subsequently, however, Wilson and Bergen (1979) argue for the operation of four mechanisms rather than two, and it should be noted that no ready physiological correlates seem available for the newly proposed mechanisms.

Thus, a number of workers have suggested the presence in the human visual system of distinct flicker- and pattern-detecting mechanisms that resemble properties of cat X and Y cells. However, the hypothesis of a correlation between mechanisms detected psychophysically in humans and ganglion cell classes described in the cat seeks to bridge two wide gaps, one between the perceptual and nonperceptual (i.e., between visual experience and the anatomy and physiology of single neurons), the other between species (human and cat). The hypothesis gains support, therefore, from three lines of evidence that help narrow those gaps.

First, Blake and Camisa (1977) used a behavioral technique to show strong parallels between cat and human perception of gratings. Thus, in the cat as in the human, the visibility of low-frequency gratings is enhanced if they are flickered rather than held stationary, while the visibility of fine gratings is reduced (Fig. 11.6). Further, the cat's threshold for perception of flickering gratings of low spatial frequency is halved (as it is in the human) when the flicker modulation is changed from on–off to counterphase. Blake and Camisa suggest that "the systematic shift in the cat's contrast sensitivity with temporal frequency reflects a shift in the relative involvement of X and Y cells in grating detection." Second, Kulikowski (1975) sought a correlate of pattern- and movement-detecting mechanisms in the mass potential evoked in human visual cortex by grating stimuli. He reported that the form of the response changed characteristically as the contrast of the grating was increased from the flicker threshold to the pattern threshold. Kulikowski argued that this difference is a pattern-evoked potential, and showed that it was selectively adapted by exposure to stationary grating stimuli. This observation thus provides evidence of the physiological activity of a pattern-detecting mechanism in human vision. Third, as argued in Chapter 3 (Section 3.1), there is now substantial evidence of classes of ganglion cells in the monkey retina, and of relay cells in the LGN, closely analogous to the X and Y cell groupings of the cat. Anatomically, the human visual system seems very similar to that of the monkey, making plausible the idea that the human system also comprises X-like and Y-like systems of cells. Recently, Harwerth *et al.* (1980) have provided evidence that the monkey's ability to discriminate grating stimuli of varying spatial frequency, duration, and contrast closely matches human performance. The tests they used with the monkeys had provided results in the human that suggested the operation of distinct "sustained" and "transient" mechanisms.

It seems clear, in summary, that the functional classification of ganglion cells has proved useful in understanding the components of focal vision. Considerable evidence is summarized here of psychophysical correlates of X and Y cell function, and additional evidence has been reviewed by Breitmeyer and Ganz (1976) and MacLeod (1978). It is also clear that these correlates are hypotheses that have yet to be extensively tested, and may need to be reinterpreted. Some of the evidence which might lead to reinterpretation is set out in the following section.

Figure 11.6. Contrast-sensitivity functions for grating detection in the cat (from Blake and Camisa, 1977). The graph shows the contrast of a sinusoidal grating stimulus presented to a cat, set at behaviorally determined thresholds. The test is equivalent to that shown in Fig. 11.4, for a human setting the threshold for detection of either flicker or pattern (closed symbols in Fig. 11.4). In the cat, as in the human, sensitivity is about twice as high for counterphase modulation (closed squares) as for on–off modulation (open squares), at the low end of the frequency range. At 4 cycles/deg., on the other hand, there is little detectable difference in the animals' threshold for counterphase and on–off presentation of the grating. Except for the relatively low abscissa values, the cat's detection of these stimuli follows the trends observed in the human. [Reproduced with kind permission of *Experimental Brain Research*.]

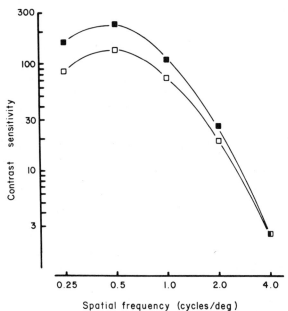

11.3.2. Qualifications

Several qualifications must be noted to the correlation discussed above between pattern- and movement-detecting mechanisms suggested from the psychophysical analysis of human vision and the functional groupings of ganglion cells.

First, Kulikowski (1971*b*, 1975) and Tolhurst (1975) have provided psychophysical evidence that, while pattern detection is mediated by an X-like system of cells at high spatial frequencies and by a Y-like system at low spatial frequencies, both systems are likely to contribute to the perception of patterns of intermediate spatial frequencies. Confirming this idea, Lehmkuhle *et al.* (1980*a,b*) have argued that the Y-cell system makes an important contribution to the perception of the low-spatial-frequency components of patterns. Second, there is physiological (Hoffman *et al.*, 1972; Singer and Bedworth, 1973) and psychophysical (Camisa *et al.*, 1977; Tolhurst and Barfield, 1978) evidence of an inhibitory interaction between X- and Y-cell systems, which may contribute to their function. This would qualify the notion that the cell systems function quite independently. Third, there is some disagreement as to which cells might subserve the two mechanisms discussed above. The disagreement seems principally to arise from studies of the visual cortex, in most of which the afferent input to the cortical neurons was not characterized. For example, Maffei and Fiorentini (1973) and Bisti *et al.* (1977) argue that the properties of "simple" and "complex" cells of cat visual cortex suggest that the

> X-chain ending up in the simple cells is devoted to the analyses of spatial information, while the Y-pathway (ending in complex cells) is involved in other visual functions, for example movement perception.

On the other hand, Ikeda and Wright (1975) saw no correlation between "simple" and "complex" categories of cells and the psychophysically distinguished mechanisms. They reported, however, that when cortical cells were classified as sustained or transient (each group including both "simple" and "complex" cells), the sustained (probably X-input) cells seemed primarily suited for the analysis of stimulus shape and transient (probably Y-input) cells for the analysis of movement or flicker. This difference in conclusion (concerning the involvement of different cortical cell "types") may spring from the different criteria these workers used to distinguish "simple" from "complex" cells. Both groups seem agreed that different cortical cell types may be processing different aspects of the visual stimulus, in parallel; i.e., that a basis can be found in the physiology of cortical cells for the movement- and shape-analyzing mechanisms described in human vision by psychophysicists, and that the basis is related to the properties of different classes of retinal ganglion cells.

Lennie (1980a) has advanced both psychophysical and physiological evidence against the view that independent mechanisms are operating in the human visual system for the detection of spatial and temporal aspects of a stimulus. He provides compelling physiological evidence that the differences in axonal conduction velocity between X and Y cells cannot account for latency differences that have been suggested between the responses of sustained and transient mechanisms (Breitmeyer and Ganz, 1976). Ikeda and Wright (1972a) had provided similar evidence some years before (although their interpreation was more optimistic for the presence of distinct sustained and transient mechanisms); as a consequence, the difference in axonal conduction velocity between X and Y cells has not been relied on for the correlations between parallel neuronal pathways and visual function argued above, or previously (Stone et al., 1979). In another physiological experiment, Lennie confirmed that X cells are more sensitive than Y cells to relatively high-spatial-frequency gratings (0.1–1 cycle/deg. in his data), but found that Y cells are only slightly more sensitive than X cells to stimuli of low spatial frequency, and only slightly more sensitive to stimuli of high temporal frequency. This latter point, he argues, goes against the view that Y cells subserve the threshold detection of coarse gratings, or of rapidly flickering stimuli. Further, in psychophysical experiments, Lennie demonstrated that, in a forced-choice experimental paradigm, a subject's ability to judge the spatial frequency of a grating of threshold contrast was equally good for gratings of low and high spatial frequency. The work of Kulikowski (1971a,b) and of Tolhurst (1973) had suggested (see above) that, at threshold, the spatial frequency of the coarser gratings would be difficult to detect.

Lennie's observations raise important questions, and may lead to a reevaluation of the dual-mechanism hypothesis, but they too are subject to other interpretations. Apparently, Lennie recorded only from ganglion cells in peripheral retina. It seems possible, from my own experience likely, that X cells at the area centralis will have significantly poorer sensitivity to coarse gratings and to fast-flickering stimuli than peripherally located cells, so that the threshold differences required for the dual-mechanism hypothesis may well exist; the point certainly seems worth testing. His psychophysical data do not directly address the results of Kulikowski and Tolhurst, or provide alternative explanations for them. Rather, they show that with a different experimental paradigm a different result is obtained. Kulikowski and Tolhurst relied on the subject's interpretation of his own conscious experience ("set the contrast so that you can just see the stimulus"); the forced-choice technique used by Lennie obliges the subject to choose whether or not a stimulus presented had a certain spatial frequency or other property, and then examines the proportion of correct choices as a function of stimulus conditions. The two techniques may not be equivalent tests of visual performance. For example, a cortically "blind" human cannot consciously see stimuli presented in affected regions of his visual field, and in tests that rely on conscious experience the individual seems quite blind. Put in a forced-choice situation, however, the same patient can make visually guided responses to stimuli presented in the affected visual field (Section 11.2.2); the blindness no longer appears total. The findings are not contradictory; each tells us something about aspects of cortical blindness. Perhaps in the cat as well, these two techniques are testing the operation of distinct components of the visual pathways.

11.3.3. The Oblique Effect: A Psychophysical Correlate of X-Cell Function

In several mammalian species, it has been demonstrated that the visual system is more sensitive to contours oriented vertically or horizontally than to those oriented obliquely. Several lines of evidence indicate (Boltz et al., 1979; Stone et al., 1979) that this dependence of sensitivity on orientation, which has been termed *orientational anisotropy,* or the *oblique effect* (Appelle, 1972), reflects the properties of the X-cell system.

For example, in humans the effect is maximal for a stimulus viewed by the fovea, and decreases progressively with retinal eccentricity (Mansfield, 1974; Berkeley et al., 1975). Thus, by analogy with the cat and monkey, in which X (or X-like) cells are particularly frequent at the area or fovea centralis, the magnitude of the oblique effect may be related to the distribution of X cells. Further, the oblique effect is demonstrable with fine, but not with coarse, gratings (Berkeley et al., 1975), suggesting that it is mediated by cells with small receptive fields (such as X cells), and it is not demonstrable with rapidly flickering stimuli (Camisa et al., 1977), which are effective stimuli for Y cells. Finally, Leventhal and Hirsch (1977) have described a group of cells in cat visual cortex that, judged by their receptive field properties, appear to receive input from retinal X cells. Among those cells, and not among other groups of cortical cells, orientation preferences are anisotropic; more cells are sensitive to vertical or horizontal contours than to oblique contours. The orientational anisotropy of these cells adds considerable weight to the idea that the oblique effect is mediated specifically by the X cell system. The oblique effect, then, is further evidence of the potential functional independence of the X cell system of neurons within the visual pathways.

11.4 THE NEURAL REPRESENTATION OF VISUAL PERCEPTION: A COMMENT

The work discussed in this and the other chapters of Part III bears on the neural representation of visual perception, a still-intractable problem of sensory biology. Barlow (1972) has set out perhaps the most articulate form of one view of neural representation, which takes as its starting point the idea that neurons in the visual pathway act as "feature detectors," each detecting the time and position of occurrence of a unique, invariant visual stimulus. Barlow argues that the "trigger features" that neurons detect in the visual environment are matched by evolutionary and developmental processes to the features of the animal's environment to which it needs to be responsive; that in the regions of the cerebral cortex processing the output of the visual cortex (which contains many millions of nerve cells), the number of neurons handling visual information reduces over successive stages to the order of 1000; that the range of firing rates found in individual nerve cells represents the degree of certainty of occurrence of the trigger features; and that activity in these high-level neurons is the neural representation of visual perception.

With others (Stone and Freeman, 1973; Rowe and Stone, 1977, 1979), I have argued a skeptical view of one of the fundamentals of Barlow's thesis, viz., the concept that visual neurons act as feature detectors. The stimulus selectivity apparent in many visual neurons does reduce the range of stimulus pattern to which they are responsive, but never reduces the range to a single unique pattern. A cortical neuron that is both orientation- and direction-selective will respond to line stimuli over a range of orientations (typically 30 deg. or more either side of the optimal); and such cells commonly respond well to spot stimuli either held stationary and flashing, or moving over a range of directions (commonly 90 deg. or more either side of the optimal). In the general case, a cell's response is less than optimal and the brain presumably must refer to other neurons with nearby receptive fields to determine which parameter(s) of the stimulus is (are) nonoptimal. And probably in all cases an array of neurons is needed to encode the size and shape of stimulus. On empirical grounds, then, visual neurons (at least as far into the pathway as area 17) do not seem to act as feature detectors. An epistemological criticism of the feature detector concept (as essentialist in methodology and untestable in formulation) has been set out elsewhere (Rowe and Stone, 1977, 1979) and still seems valid. Uttal (1971) adds the criticism (considered in more detail in the following chapter) that ideas of the neural representation of visual perception in terms of the activity of single cells are a by-product of the technology of single-cell neurophysiology. Barlow acknowledges that problem clearly, but does not change his view on the usefulness of feature detection as a characterization of neuronal coding. He argues rather that without some articulate hypothesis of brain function, we may win experimental battles, but lose the "war" on the brain's mysteries.

To extend Barlow's analogy, however, it seems to me unlikely that we can win this war simply by a determination to fight it; every neurobiologist for several generations has dreamed of solving the problem of neural representation of perception or of motor commands. Without a clear idea of the next conceptual breakthrough, it may be more fruitful to examine some of our present assumptions, in case their weakness is part of our problem. One alternative to Barlow's view is to regard visual perception as represented, not in the activity of a limited number of neurons turned to particular stimulus features, but rather in the activity of networks of neurons, whose coding capacities involve extensive interconnections with other neurons. Uttal (1977) argues persuasively the necessity of this sort of approach; the problem seems to be that the idea of neural networks is fairly undifferentiated, so that there is no clear way of testing or developing it. As Uttal (1977) puts it, we lack the methodology and technology to analyze neural networks; as a consequence, few compelling ideas are available of the coding properties of realistically neural networks. The suggestion that neurons function in networks therefore does little to alleviate the lack of a conceptual framework which was a major stimulus to Barlow's (1979) speculative analysis of cortical function.

Even so, it seems to me that we do not entirely lack a systematic, conceptual framework within which to understand brain function. The theory of evolution provided a systematics for understanding the morphology, ecology, development and behavior of animals; and, since nervous systems are parts of animals, the theory of evolution is of great value in understanding properties of the brain, as comparative neurobiologists have long been aware. It seems to me that a second type of systematics is emerging by reference to which we can relate properties of a visual neuron to the properties of the visual system. In this chapter, I have argued, following several previous authors, for the recognition of correlates between major components of visual perception (such as ambient and focal vision, movement and pattern vision), and systems of nerve cells, such as the Y-, X- and W-cell systems. Perhaps such correlations between systems of visual neurons and parts of the perceptual process can be viewed as a useful first step toward a "systematics" of the function of visual neurons. Put another way [and this argument is pursued one further step in the next chapter (Section 12.4)], knowledge of the major neuronal groupings within the visual pathway may be an important step toward an understanding of the coding properties of individual neurons.

Extensions and Limits of the   12
Parallel Processing Analysis

So far my thesis has concerned the visual pathways of mammals, summarizing the emergence and impact of a "parallel processing" view of their organization. In the first two sections of this final chapter, I will argue that the same idea (that a sensory system may comprise several parallel-wired subsystems) is equally valuable in another major sensory pathway, the somatosensory, and may prove of general value in sensory biology. There are of course problems as well as advantages in writing with such a specific aim, and to minimize one of those problems (that of exaggeration) I state in Section 12.3 the difficulties and limits of the parallel processing analysis, as I understand them. The analysis is not free of internal problems, nor is it a total explanation of the visual (or any other) pathway. In the final section, however, a more positive claim is argued: that, despite its weaknesses, the parallel processing analysis provides a starting point for a systematic analysis and understanding of sensory pathways that may help transcend some of the problems which limit the current ambit of sensory neurobiology.

12.1. PARALLEL PROCESSING IN THE SOMATOSENSORY PATHWAYS

The history of the parallel processing analysis of sensory pathways begins with studies on the somatosensory system, for it was in the analysis of peripheral nerve function that the idea arose that a single nerve might comprise subgroups of nerve fibers, each subgroup subserving a distinct component of the nerve's overall function. Brindley (1977), for example, traces to R. W. Volkmann and J. Müller the idea, formulated in the middle of the 19th century, that "for every kind of sensation there is a special kind of end organ and special kind of peripheral nerve fiber ... and that when peripheral nerve fibres of a given kind are stimulated, by whatever means, the sensation is always of the appropriate kind." This "doctrine of specific nerve energies" implies (though it was many decades before the implication was investigated) that the central somatosensory pathways are organized to keep these different neural pathways separate at least until their activities reach the region of the brain responsible for conscious sensation; i.e., that the pathways are organized to process different sensory streams separately, in parallel. Clinical and experimental support for the idea came from observations of the effects of trauma and of drugs on peripheral nerves. Gasser and Erlanger (1929) reviewed prior work on this problem, noting evidence that the sensation conveyed by a peripheral nerve could be fractionated. When pressure was applied to a nerve, for example, "sensations disappear in the order: contact, cold, warmth, pain." Conversely, when a drug such as cocaine was applied to the nerve, these different sensations were blocked in the opposite order.

These observations seemed to provide strong support for the idea that different groups of fibers, of different "chemical affinities" and sensitivity to trauma, mediate different aspects of sensation; and Gasser and Erlanger sought, using the newly developed cathode-ray oscillograph, to observe whether these hypothesized groups could be identified as conduction velocity groupings. Their results were dramatic. A peripheral nerve of a frog was stimulated electrically, and the compound action potential generated by the stimulus was recorded at some distance along the nerve. The potential had three or four components distinguished by their latency, each representing a different conduction velocity group of fibers within the nerve. The shortest-latency (fastest-velocity) group was labeled the α wave, and successively later waves were termed β and γ. When a pressure of 25 pounds per square inch was applied to a cuff around the nerve between the stimulating and recording sites,

> The α wave seemed to melt away from the fluorescent figure of the action potential on the screen of the tube.

When the pressure was applied more gradually, the components of the field potential could be seen to fail successively, beginning with the shortest-latency (fastest-conducting) component. Conversely, when cocaine was applied, the components fell away in the opposite order. The correlation with clinical and behavioral results was clear; conduction velocity groupings of the peripheral nerves seemed to match, at least approximately, the submodalities of somatic sensation. Touch fibers seemed to have the highest conduction velocity, pain the slowest, and temperature fibers to be intermediate in velocity. These correlations have been considerably developed and corroborated by subsequent work. The pathways followed in the brain by different components of somatic sensation have been traced, and shown to be organized, to a large extent (though certainly not completely) to process the activities of different somatosensory afferents separately, in parallel.

It was perhaps natural that the idea of parallel processing should have begun with somatosensory nerves: first, because some of the different components of somatic sensation (temperature, pain, vibration, touch) are recognizable in everyday experience, whereas quite sophisticated tests have proved necessary for a similar analysis of visual sensation; second, because somatosensory nerves are far more accessible than the optic nerve; and third, because in somatic sensation each subgroup of axons begins with a distinct class of receptor (with the apparent exception of thermal and nociceptive afferents, which both begin with naked endings), whereas in vision the two recognized receptor classes (rods and cones) provide input to all major groups of ganglion cells.

12.1.1. Classification of Somatosensory Afferents

By contrast with visual receptors, which in mammals form two classes (rods and cones), somatosensory receptors form numerous classes, with widely differing morphologies: In addition, many of these receptors are found in a variety of very different tissues [e.g., in hairy and smooth (glabrous) skin, in muscles, tendons, and ligaments, and in the serous membranes and mucosae of organs] and show different specializations in each. Somatosensory receptors also show much of the variation in properties seen in visual receptors. For example, they subserve a variety of conscious sensations and nonconscious processes (reflexes); and even within one tissue they are subject to considerable systematic variation, for example the different density of innervation of the skin of the palm and upper limb.

Perhaps because of this variety, an overall classification of somatosensory receptors, with higher- and lower-level categories and taxa (such as that discussed for visual system neurons in Chapter 2), has not been developed. Several classifications of somatosensory afferents have been proposed, however, but these overlap with, compete with, or complement each other. Thus, somatosensory receptors can be grouped by their location, into cutaneous and "deep" classes; or by the location of their stimuli, into extero-, intero-, and proprioceptors; or by the nature of the stimuli to which they are most sensitive, into mechano-, noci-, and thermoreceptors; by their morphology, into those with free endings and those with corpuscular endings, each comprising subtypes; or by the time course of their response to a steadily applied stimulus, into rapidly and slowly adapting types; or by the conduction velocity of their axons, into A and C fibers. All these groupings are in use, with well-developed subgroupings. What is not available is a single classification that attempts to encompass all this variety in a systematic way.

Perhaps, given the variety of somatosensory receptors, it would be inappropriate ever to seek such a classification. Yet the common embryological origin of the peripheral somatosensory nerves from the neural crest, and the close grouping of the thalamic nuclei and cortical areas involved in representing the different components of somatic sensation, suggest that, at least in a broad sense, the somatosensory pathways can be regarded as a single system. A single classification of receptors may not then be an inappropriate goal. I am not about to attempt it, however, and two schemes are discussed below. The first stems from conduction velocity groupings found among somatosensory afferents; the second stems from the idea of "arrays" of receptors. The mechanoreceptors of the skin, for instance, might be considered to subserve a particular part of somatosensory function, as might the receptors that signal muscle tension and joint position, or pain receptors; many authors have found it useful to consider such groupings of receptors as "functional arrays." Together, these descriptions give some impression of the variety of receptors present in the body tissues and of afferent fibers present in somatosensory nerves.

Grouping by conduction velocity: Erlanger and Gasser (1937) grouped the fibers (both afferent and efferent) of peripheral nerves into three groups (A, B, and C fibers), according to their conduction velocity (A fibers fastest-conducting, C fibers slowest). Only the B-group axons (myelinated fibers, conduction velocities 3–15 m/sec) proved to be functionally homogeneous; they are autonomic preganglionic efferent fibers, and are therefore not considered further here. The A fibers are also myelinated, with the largest diameters (up to 20 μ) and highest conduction velocities (up to 90 m/sec). The A group was further subdivided into α, β, γ and δ subgroups, again according to conduction velocity. The γ subgroup proved entirely efferent; some of the α and β fibers were also efferent. The afferent fibers in the A group were thus found in the A α, A β, and A γ subgroups; subsequently, Lloyd (1943) termed these groups I, II, and III, respectively. Group I fibers are the largest and fastest-conducting, and comprise two subgroups, both originating from "deep" structures: Group Ia fibers carry information from stretch receptors in muscle spindles, and group Ib fibers from tension receptors in muscle tendons. The Group II fibers are somewhat thinner and slower (5–15 μ, 20–90 m/sec), and carry information from most of the mechanoreceptors of the skin and deep tissues, including the "flower-spray" receptors in muscle spindles. The group III fibers are thinner and slower again (1–5 μ, 12–30 m/sec) and carry information from some hair follicles and pain receptors, and from receptors in the walls of blood vessels.

The class C fibers are relatively very thin (0.2–1.5 μ) and are also unmyelinated; both factors contribute to their slow conduction velocity (0.3–1.6 m/sec). The afferent fibers among them [termed group IV by Lloyd (1943)] carry much of the pain and temperature sensation from the body, and also some information from mechanoreceptors.

Functional arrays of receptors: There are thus, as early clinical observations had suggested, clear but inexact correlations between the sensations mediated by somatosensory afferents and their caliber and conduction velocity. Subsequent more detailed work on "functional arrays" of receptors has, however, shown considerable variety of receptor morphology and function, with little or no related variation in the conduction velocity of the afferent axons involved. For example, Merkel discs and Ruffini terminals are morphologically distinct endings formed by cutaneous afferents; yet physiologically they are similar in several ways, both responding in a maintained (slowly adapting) way to pressure on the skin, and having similar conduction velocities (in the group II range). Pacinian corpuscles are also cutaneous mechanoreceptors, but with a distinct morphology and particular physiological properties. They are extremely sensitive, responding to very slight depression of the overlying skin. They respond only very briefly to maintained pressure and are capable of following rapidly vibrating stimuli, yet the conduction velocity of their axons is also in the group II range. Indeed, most of the seven or eight types of cutaneous mechanoreceptors recognized have axonal velocities in the group II range; the exceptions are mechanoreceptors with group III and C velocities, which have been identified in the monkey.

Considering mechanosensory receptors found in deep tissues (muscles, tendons, ligaments, joint capsules) as a separate "array,' a further set of receptor specializations can be distinguished, such as the encapsulated endings of joint receptors, the Golgi tendon organs, and the annulospiral and flower-spray endings found in muscle spindles; but also including receptors, such as the Ruffini endings, found in the skin. Axons from Golgi tendon organs and annulospiral endings in muscle spindles endings form the group I conduction velocity group; the axons of other deeply located mechanoreceptors fall in group II and III.

Finally, several authors discuss "small-caliber" somatosensory afferent fibers as a group, including both the group III (thin, myelinated) and C (thin, unmyelinated) axon groups. Most of the latter carry information from nociceptors, nerve endings that appear to respond preferentially to tissue damage and whose activity presumably gives rise to pain; some, however, respond preferentially to warming of the skin ("warmth" fibers). Most of the group III fibers derive from receptors sensitive to cooling of the skin (cold fibers); others, however, appear to stem from nociceptors.

In summary, there exist a wide variety of receptors whose function is to transduce stimuli impinging on body tissues into electrical activity of sensory nerve fibers. Moreover, most of the various afferent types are particularly sensitive to one form of stimulation (their "adequate stimulus"), and activity in such fibers (however elicited) seems to give rise to perception of the adequate stimulus. Indeed, one of the compelling pieces of evidence that the activities of different classes of afferents are processed through the somatosensory pathways "in parallel" is the everyday experience that we can separate in our conscious experience the sorts of stimuli (touch, vibration, pain, temperature changes) to which the afferent fibers seem selectively sensitive. Evidence of that parallel processing is now available for all the major stages of the somatosensory pathways (spinal cord and brain stem, thalamus and necocortex).

12.1.2. Channeling of Submodalities through Spinal Cord and Brain Stem

Are submodalities channeled through different spinal cord fiber systems?: Somatosensory pathways ascending from the spinal cord and brain stem to the thalamus (and thence to the somatosensory cortex) were for many years described as comprising two major systems, distinct in their connections and in the submodalities they mediated. The dorsal-column/lemniscus system was thought to transmit "epicritic" sensation (Head's term to denote touch, vibration, proprioception) via the dorsal columns of white matter, the dorsal column nuclei (gracile and cuneate) of the medulla and their output pathway, the medial lemniscus, to the ventroposterior nucleus of the thalamus. The "anterolateral" or "spinothalamic" system was thought to transmit "protopathic" sensation (pain, temperature) via the anterior and lateral columns of spinal white matter, and their continuation through the brain stem (the spinal lemniscus), to the ventrobasal and the nearby posterior nucleus of the thalamus. To some extent at least, analogous components were recognized within the trigeminothalamic pathways transmitting somatic sensation from the head. The two spinal cord pathways are still recognized and considered functionally distinct by most authors, but increasingly sensitive anatomical and physiological analysis has led to a more sophisticated understanding of their distinct contribution to somatic perception; several simple views of the division of function between them have provided untenable.

Webster (1977) and Brown and Gordon (1977) reviewed the spinal cord circuitry of the somatosensory pathways; the studies on which their reviews were based concern principally the cat, monkey, and human. In his summary (Fig. 12.1), Webster still recognizes the dorsal column and anterolateral systems of fibers, but stresses several recent

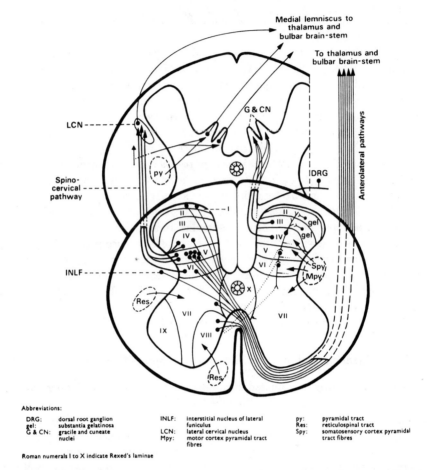

Figure 12.1. Webster's (1977) schematic diagram of the three major sensory pathways ascending to the thalamus from the spinal cord, and carrying somatosensory information. They are the anterolateral (spinothalamic) pathways, the medial lemniscal system, and the spinocervical pathway. [Reproduced from the *British Medical Bulletin* with kind permission of the Medical Department of the British Council.]

advances in understanding. First, much detail has been learnt of the circuitry of these two pathways; for example, the presence in the dorsal column system of axons originating from the dorsal horn, as well as directly from spinal nerves; and the location of cells contributing to the anterolateral (spinothalamic) pathways (principally in Rexed's lamina V, but also in laminae I, IV, VI, and even VII and VIII). Second, a third pathway has been recognized, the spinocervical pathway. Its axons arise from cells in the dorsal horn (laminae III, IV, and V) but, unlike the fibers of the anterolateral system, do not cross the midline but ascend to terminate in the lateral cervical nucleus, found at the upper end of the spinal cord. Axons of cells in this nucleus then cross the midline and join the medial lemniscus to reach the thalamus. Third, lesion experiments, and studies of the properties of single cells in each system, do not support a sharp "epicritic"/ "protopathic" division of function among these three systems of fibers, or between any two of them. While it still seems agreed that many fibers in the spinothalamic tracts are specifically transmitting nociceptive information, many other fibers of the tract respond to cutaneous stimuli, and those originating from laminae VII and VIII (which are numerous only in the lumbar part of the cord) are responsive to muscle and joint stimulation. Conversely, while most fibers in the spinocervical tract are responsive to light mechanical stimulation of the skin, many are also responsive to noxious (painful) stimuli. On the other hand, it has been recognized that, while proprioceptive information from the forelimb does traverse the dorsal columns and the cuneate nucleus, proprioceptive fibers reaching the thalmus from the hindlimb do not traverse the dorsal columns at all. Rather, proprioceptive information from the hindlimb reaches the thalamus via collaterals from the dorsal spinocerebeller tract that reach nucleus Z (located just rostral to the nucleus gracilis), which relays their activity to the thalamus.

Webster notes that these developments, in addition to introducing new and important findings such as the different paths taken by proprioception from hind- and forelimb, make largely untenable any simple division of function between the different somatosensory pathways ascending from the spinal cord. Some concentration of C-fiber function in the spinothalamic pathways is still apparent, and a similar concentration of C-fiber function is apparent in the trigeminothalamic pathways that transmit somatic sensation from the head (via the caudal part of the spinal nucleus of the trigeminal nerve); and similarly, the dorsal column fibers deal largely, as always thought, with cutaneous sensation and proprioception. Yet it is clear, Webster argues, that there is considerable overlap of function between the three major pathways he considers.

Brown and Gordon's analysis leads to the same view. Nevertheless, it seems reasonable to argue, the recognition that there is no simple division of function between the major tracts of the somatosensory pathways of the spinal cord does not rule out a parallel processing pattern of organization in the somatosensory pathways, for two reasons. First, as Brown and Gordon (1977) argue with insight, there are clear functional differences between the dorsal column, spinothalamic, and spinocervical pathways that are not expressed in terms of the modalities served by each. For example, many fibers in the dorsal column system can reasonably be considered "pure lines" in that they transmit information from just one peripheral receptor, or at most from a restricted group of cutaneous mechanoreceptors. Moreover, the receptive fields of individual cells in this system show considerable surround inhibition, which seems likely to increase the spatial resolution of each fiber. Conversely, convergence of sensory input is more common in fibers of the spinocervical tract, with single cells showing interaction of noxious and cutaneous sensation; while tissue damage and probably temperature variation are particularly common adequate stimuli of spinothalamic fibers. Brown and Gordon suggest that this difference in functional emphasis may be highly important:

> We see the question of parallel paths very much in this light. Each system does its particular part of the sensory task better than any other; but up to a point the nervous system can learn to use each system fairly well for a different purpose.

Second, even if there is overlap in the functions of major pathways within the somatosensory system, there may still be a strict, parallel connection of fibers subserving different submodalities within each system.

Some of the neuroanatomical basis of the separate processing of activity in different afferent classes has been summarized recently by Brown (1981). He reviews evidence of the morphology of single afferent fibers, traced with HRP techniques. The fibers traced were principally group A fibers, being cutaneous mechanoreceptors of various sorts. Pain, temperature, and smaller-caliber mechanoreceptor afferents were not studied; the technique involves intracellular recording from the axons, allowing both physiological identification of the fiber and injection of it with HRP, but all but eliminating the chance of studying thin oxons. Individual fibers commonly branch after entering the spinal cord. As their branches ascend or descend, they form a column of terminals in the dorsal horn (Fig. 12.2). The columns may be continuous or periodic along the length of the cord; Brown stresses, however, that different afferent types form distinct terminal patterns in the dorsal horn, and may therefore form quite distinct connections with the cells whose axons form the ascending tracts. The hair follicle afferents, for example, form a continuous column of terminals principally in lamina III; afferents from rapidly adapting Krause receptors form discontinous, slab like terminal patches in laminae III and IV; each Pacinian afferent forms a continuous column of terminals in laminae III and IV and a second, discontinuous column in laminae V and VI; the slowly adapting type I afferents form a discontinuous string of almost spheroidal terminal fields in laminae III and IV; and the rapidly adapting type II afferents form thin "slabs" of terminals at intervals along the dorsal horn, in layers III to VI.

These observations concern only terminal arborizations of certain primary afferents but, Brown argues, the terminal patterns formed by different afferents, although not spatially separate, seem specifically organized and distinct from each other. These patterns may form part of the basis of the distinct patterns of connections formed by different afferent types through the spinal cord to the thalamus and neocortex.

Parallel and convergent connections within brain stem centers: Studies of the stimulus specificities of single cells in brain stem centers concerned with somatic sensation, particularly the dorsal column (gracile and cuneate) nuclei, provide evidence of both parallel and convergent patterns of organization. Many studies (summarized in Brown and Gordon, 1977) have concluded that cells of these nuclei receive input from only one receptor class, or at least from a very limited range of receptors with closely similar properties, such as the cutaneous mechanoreceptors. Since these cells project to the thalamus, they would provide a basis for the "parallel processing" through these nuclei of particular submodalities. This parallelism is only slightly qualified by evidence of inhibitory interaction between different submodalities, such as the inhibitory influence of activity in large-caliber (cutaneous, proprioceptive) afferents on transmission in small-fiber (pain) pathways, postulated in the "gate-control" theory of pain transmission through the dorsal horn (Melzak and Wall, 1963). Such interaction seems to reinforce rather than dilute the significance of the separation between paths.

Other evidence, however, points to considerable convergence at the brain stem level. Brown and Gordon (1977), for instance, note evidence of excitation of relay cells in the cuneate and gracile nuclei by afferents from the dorsolateral fascicle. This extra input is cutaneous, however, and topographically in register with the input to the gracile and cuneate nuclei from the dorsal column, and could perhaps be considered to create only a small exception to a general pattern of parallel processing within the dorsal column nuclei. More substantial evidence of convergence of different sorts of somatosensory

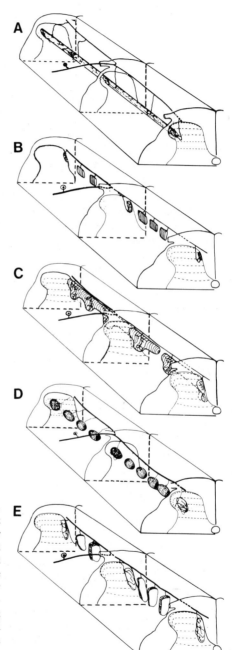

Figure 12.2. Parallel wiring in the dorsal horn. Brown's (1981) summary of the patterns of afferent termination of different classes of mechanoreceptor fibers in the dorsal horn of the spinal cord. Brown emphasizes the differences between different afferent classes in the patterns of terminals, as described in the text. (A) Hair follicle afferent; (B) Krause rapidly adapting afferent from smooth skin; (C) Pacinian corpuscle afferent; (D) slowly adapting type I afferent; (E) slowly adapting type II afferent. [Reproduced from *Trends in Neurosciences* with kind permission of Elsevier/North-Holland Biomedical Press.]

information was provided by Angaut-Petit (1975), who noted that, although a majority of cells in the gracile nucleus of the cat appear to be modality specific (responding only to gentle cutaneous stimuli such as hair displacement or light tapping), a substantial (31%) minority responded to both cutaneous and noxious stimuli. She suggested that this polymodal convergence may operate to allow some "gate-control" of pain transmission through the brain stem, as had previously been postulated for the dorsal horn. Dostrovsky et al., (1978) noted evidence that cells in the gracile nucleus of the cat may have two distinct components to their receptive field, one a very localized component indicative of very specific input to the cells, and the other a "widefield" input, making the cells responsive to receptors located over a wide region of skin. They suggested that the latter component may be attributable to "errors in connection" not normally effective in the adult. Rigamonti and Hancock (1978) reported evidence that visceral and cutaneous inputs converge on a substantial number of cells in the dorsal column nuclei of cats. The function of this convergence seemed unclear; Rigamonti and Hancock comment that the dorsal column nuclei should not be regarded as relaying only proprio- and exteroceptive activity. Further, Millar (1979) reported evidence of convergence of cutaneous and proprioceptive afferents (joint, cutaneous, and muscle) onto single neurons in the cuneate nucleus of the cat.

These instances of convergence of different submodalities are substantial; and since each demonstration required a specific experimental paradigm, it is possible that instances of convergence will multiply as experimental possibilities are explored. How significantly do they qualify the idea that different submodalities of somatic sensation are processed in parallel? This can be assessed by examining evidence of the organization of higher centers, in particular of the somatosensory areas of the thalamus and cerebral cortex.

12.1.3. The Parallel Organization of Somatosensory Thalamus

Somatosensory information is relayed to the cerebral cortex by the ventroposterior nucleus (VP) of the thalamus; because several subdivisions are recognized within the VP, it is also called the *ventrobasal* (VB) *complex*. From the considerable literature on the organization of this nucleus, I would draw three strands, which together indicate that the nucleus is very specifically organized to process different submodalities of somatic sensation in parallel.

First, many investigators have concluded that individual VP neurons are excited by just one receptor class, via specific spinal cord pathways. In 1952, Mountcastle and Henneman provided the first descriptions of the response properties of single cells in the VP, noting that each cell seemed stimulus-specific; i.e., it responded only to one sort of stimulus, such as touch/pressure, or vibration, or limb position. Moreover, their receptive fields were clearly localized to small areas of skin. Dykes (1981), reviewing recent evidence on the organization of VP, concludes that, subsequent debates notwithstanding, the view argued by Mountcastle and Henneman and by Mountcastle (1957) is fundamentally correct: different VP neurons relay different components (or submodalities) of somatic sensation.

Second, neurons relaying different submodalities tend to congregate in different components of the VP. Dykes *et al.* (1981) summarize previous evidence of submodality-specific areas within the VP in both cat and monkey and describe new observations in the squirrel monkey. They conclude that the VP, or more specifically its large lateral segment (VPL) that deals with sensory information from the body, receives input from cutaneous mechanoreceptors; specifically from both slowly and rapidly adapting touch and pressure receptors. Input from "deep" structures, for example from muscle spindles and Golgi tendon organs, is processed by a group of cells (for which no name is suggested) located just above the VPL. A nucleus located below the VPL, designated VPI (ventroposterior inferior), appears to process selectively the activities of Pacinian corpuscles (cutaneous mechanoreceptors sensitive to high-frequency vibration). Dykes and co-workers note further that, within the VPL, different subregions deal with the activities of rapidly and slowly adapting touch receptors; they were unable, however, to obtain a clear description of the overall pattern of separation of these subregions. Descriptions of a segregation of submodality-specific areas within the VP region of the cat thalamus are less complete (Dykes *et al.*, 1980a; Dykes, 1981) but suggest a segregation remarkably similar to that in the monkey.

Third, as discussed in Section 12.1.4 below, the different submodality-specific regions of somatosensory thalamus project to different regions of the cerebral cortex, so that the separate, parallel relay of submodalities continues within the cortex.

Understanding of the thalamic relay of pain and temperature information is less developed. The intralaminar group of thalamic nuclei has been considered important for pain sensation (Webster, 1977), and the posterior thalamic nuclei (designated PO; situated just behind the VPL) may also be involved (Webster, 1977). Tanji *et al.* (1978) found evidence that the PO project to the third somatosensory area of cortex, but the evidence for this pathway and for its involvement in pain perception seems relatively incomplete. From the experimentalist's point of view, there is another interesting analogy here with the visual system: in both systems, the fine-caliber pathways have been the most difficult to study and the last to be elucidated. Thin-fiber/small-cell systems are harder to record from and harder to stain and locate, and the tracing of nociceptive (pain) pathways is particularly difficult because their adequate stimulus (tissue destruction) cannot simply be applied repeatedly as the pathway is explored.

The evidence just discussed of modality-specific subregions within somatosensory thalamus has several important implications. It implies, for example, that there must be several different representations of the body within somatosensory thalamus, presumably one in each modality-specific subregion; and several authors comment that the representations seem discontinuous in ways hard to interpret. Most importantly for my present thesis, the evidence implies a degree of modality-specificity within the somatosensory thalamus that seems surprising in the light of the sensory convergence described in the dorsal column nuclei (previous section). Perhaps more systematic experimentation would reveal a similar degree of convergence in the thalamus. By present evidence, however, the tendency for different submodalities to engage different relay cells in distinct components of the VP region of the thalamus is striking. Parallel organization seems as prominent in the VP as in the LGN.

12.1.4. The Parallel Organization of Somatosensory Cortex

In discussing in Chapter 8 the organization of visual the cortex, I stressed two features as evidence that it is organized to process the activities of different ganglion cell classes separately, in parallel: (1) the presence in many species of several adjacent but separate areas of the visual cortex, each receiving a different component of the retinal output, and (2) the parallel circuitry of neurons within area 17. The same two features are as clearly apparent in the organization of the somatosensory cortex.

The areas of the somatosensory cortex: Historically, the delineation of the "somatosensory" cortical areas was part of attempts to show specificity of function within the neocortex. Since Marshall *et al.* (1941) located the somatosensory cortex in the cat and monkey, a rich literature of experimental work has grown concerning the delineation and organization of the somatosensory cortex, much as it has for the visual cortex. The spatial resolution of physiological mapping techniques has steadily improved, and modern anatomical techniques have proved sufficiently precise to complement them. In both sensory systems, it has emerged that the sensory field (the visual field, or the parts of the body) is represented several times over within the cerebral cortex; i.e., several separate areas of the neocortex have been delineated in each of which some aspect of body sensation is represented, for the whole body.

In both cat and monkey, the several areas of the somatosensory cortex are found in two or three separate groups known as S(omatosensory) I, II, and III. [As far as I can tell, the significance of the numerals is purely historical; SI was delineated first and SIII, where recognized, last. They do not (certainly should not) imply that the areas are related in any hierarchical sequence.] In the cat, SI is found in the posterior sigmoid gyrus; in the monkey, it is found in the postcentral gyrus. In both monkey and cat, SI comprises four cytoarchitectonic divisions (Dykes, 1978; Merzenich *et al.*, 1978; Jones and Porter, 1980). Most anteriorly, adjacent to the motor cortex, is area 3a; its cytoarchitecture shows the relatively strong development of the granular layers characteristic of the sensory cortex, but also some prominent layer V pyramidal cells typical of the motor cortex. Often considered traditional between motor and sensory functions, area 3a has been shown to receive input principally from group I afferents (from muscle stretch receptors and Golgi tendon organs), via a specific subregion of the somatosensory thalamus (see previous section). Just posterior to it (the areas are found in the same anteroposterior sequence in cat and monkey) is area 3b; this is unambiguously granular sensory cortex, with a prominent, much-studied representation of the body surface (i.e., of cutaneous sensation) relayed to it through the VPL (Dykes *et al.*, 1980*b*, 1981; Sur *et al.*, 1981. Moreover, within area 3b of both cat (Dykes *et al.*, 1980*b*) and monkey (Sur *et al.*, 1981), the activities of slowly and rapidly adapting cutaneous mechanoreceptors are processed by different groups of cells, which are arranged in strips or bands, and presumably receive input from the corresponding subregions of the VPL.

Posterior to area 3b, two further representations of the body are found in areas 1 and 2; the former receives cutaneous input and the latter input from deep tissues and muscles. The input to areas 1 and 2 may be formed by branches of axons reaching, respectively, areas 3b and 3a.

In the cat, the separate somatosensory area termed SII is located in the anterior ectosylvian gyrus. It receives its own input direct from the VP of the thalamus. It appears to receive input from cutaneous mechanoreceptors that also project to SI (Bennett et al., 1980). However, the study of Ferrington and Rowe (1980) indicates that the coding of information from a single receptor type (Pacinian corpuscles) may differ between SI and SII. In the monkey, SII is located at the foot of the postcentral gyrus, extending into the depths of the lateral fissure. In both species, it is possible that more than one representation of the body may be found within SII, but less is known of this area than of SI. Still less is known of SIII, located in the cat in area 5a of the parietal lobe; indeed, some authors seem skeptical and refrain from using the term. This area, and a comparable area 5 in parietal lobe of the monkey, receive input from the PO, some of which receive input from the spinothalamic tracts. This area too should perhaps be counted as part of the neocortex specifically related to somatosensory function.

The purpose of these multiple representations of bodily sensation is not known; but it is clear that a number of areas of the neocortex receive parallel input from the VP, or from other closely adjacent nuclei such as the ventrointermediate, ventroinferior, and posterior that, like VP, receive direct input from ascending somatosensory tracts. Further, at least in the squirrel monkey (Lin et al., 1980), each cortical area projects back, reciprocally, to the region of the VP that projects to it. Again, as in the visual system (Chapter 8, Section 8.1.4), the parallel processing of different components of the afferent stream seems to extend into corticothalamic pathways.

The cortical areas just discussed were differentiated principally on the basis of their cytoarchitecture and connections, assessed both anatomically and physiologically. A clear behavioral correlate of the difference between two cortical areas has been reported by Carlson (1981). She studied sensory deficits produced in macaque monkeys by lesions restricted to area 1 (which receives principally cutaneous input) and area 2 (input from deep structures, such as joints). The lesions were restricted to the region of representation of the hand and, correspondingly, the tasks involved sensory discrimination by the hand. Animals with lesions to area 1 showed lasting deficits in their ability to discriminate texture, but no loss in their discrimination of the shape or curvature of felt objects. Conversely, animals with lesions in area 2 showed little deficit on texture discrimination, but lasting deficits in the discrimination of shape and curvature. These results match well evidence of differential sensory input to areas 1 and 2; and they clearly support the notion that different regions of somatosensory cortex normally function in parallel, making different contributions to sensory function.

For more detailed discussion of the organization of the somatosensory cortex, the reader is referred to recent reviews such as those by Dykes (1978, 1981), White (1979), McCloskey (1979), Jones and Porter (1980), and Merzenich and Kaas (1980). For my present thesis, one more feature of the organization of the somatosensory cortex deserves discussion: the circuitry of the somatosensory cortex. And in this feature as well, somatosensory and visual areas of the neocortex show close similarity.

Parallel organization of the intrinsic circuitry of the somatosensory cortex: In Chapter 8 (Section 8.1.3), I argued that, in many respects, the intrinsic circuitry of the striate cortex seems organized for the parallel processing of different components of its afferent input. An important part of that argument stemmed from White's (1978, 1979) analysis of the termination of thalamic afferents in the somatosensory cortex of the mouse.

White's observation was this. He lesioned the ventrobasal complex, causing the axons coming from that nucleus to degenerate. A few days after the lesion, the degenerating terminals of those axons could be identified in the somatosensory cortex, under the electron microscope. Before being prepared for electron microscopy, however, the tissue was put through a Golgi impregnation procedure that allowed a proportion of the neurons in the somatosensory cortex to be visualized and characterized by their soma location and dendritic morphology. The identified neurons were then "de-impregnated" before being embedded, sectioned, and examined under the electron microscope. White could then assess the morphology of those cortical cells that did (and did not) receive synapses from afferents from the VP. For practical reasons, he examined only layers III, IV, and V; degenerating terminals (hence terminals of thalamic afferents) were found principally in layer IV, but also in the adjacent part of layer III. This pattern was in agreement with earlier reports, and the neuron types identified had also been described previously. The new observation was that thalamic terminals formed synapses on all dendritic processes present in the region where they terminated; specifically (Fig. 12.3) on the dendrites of spiny and nonspiny stellate cells, on the basal dendrites of pyramidal cells in layer III, on the apical dendrites of layer V pyramidal cells, and on the dendrites of a nonspiny bipolar-type cell. Thus, the cells receiving direct thalamic input are diverse in morphology and in the layer in which their somas are found; too diverse, White argues, to be compatible with any theory of cortical circuitry that holds that the thalamic input first reaches layer IV, and is then successively processed in deeper and more superficial layers. Moreover, as in the visual cortex, the onward projections of cells in the somatosensory cortex vary with the laminar position of their somas; layer III pyramidal cells, for example, project predominantly to other cortical areas in the same or opposite hemisphere, while layer V pyramidal cells project subcortically. It has not been established for the somatosensory cortex whether, as in area 17 of the visual cortex, different components of the afferent input reach different cortical layers. Even without a laminar organization of that sort, however, there is already substantial evidence that different groups of cells in the somatosensory cortex are organized to process sensory input in parallel, and relay their activity to different structures.

12.1.5. Summary

A comparison of the sort attempted here, between somatosensory and visual pathways, is perhaps best summarized in terms of similarities and differences. Principal among the similarities (relevant to a discussion of parallel processing) are:

1. The specificity with which individual thalamic relay cells receive input from one class of afferents.
2. The channeling of different components of somatic sensory input through largely distinct components of the thalamus to distinct cortical areas.

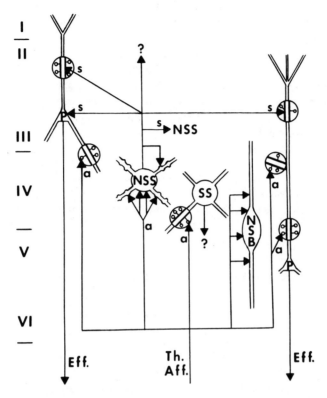

Figure 12.3. Summary diagram from White's (1979) review of the distribution of thalamic terminals in the somatosensory cortex. Note that thalamic afferents (Th. aff.) terminate on the dendrites of all cell types with dendrites in layer IV; specifically on the basal dendrites of pyramidal cells (P) with somas in layer III, on the apical dendrites of pyramidal cells with somas in layer V, on the somas and dendrites of nonspiny stellate (NSS) and bipolar (NSB) cells, and on the dendrites of spiny stellate cells (SS). [Reproduced with kind permission of Elsevier/North-Holland Biomedical Press.]

3. The similarities (discussed in the immediately preceding section) between somatosensory and visual areas of the neocortex in their intrinsic circuitry.

These three features seem fundamental evidence of parallel processing; both systems can be usefully analyzed in these terms. In the present context, four differences between visual and somatosensory systems seem to stand out:

1. There is a far greater variety of receptors in the somatosensory system, distributed in a variety of tissues and showing a range of specializations in each.
2. Each somatic receptor has its own axon, and at least major subgroups of afferents are transmitted separately to the somatosensory cortex. In the visual system by contrast, the two major classes of receptors (rods and cones) converge on all

ganglion cells; the sorting out of different pathways within the visual system begins at the ganglion cell level.

3. The somatosensory system forms at least three anatomically separate pathways extending from the dorsal roots to the thalamus (the spinocervical, spinothalamic, and dorsal column/lemniscal); whereas the visual pathway to the thalamus forms just a single anatomical entity (the optic tract).

4. Considerable excitatory convergence of different submodalities of somatic sensation has been described at the brain stem level; corresponding convergence has yet to be described in the visual pathways.

These differences (especially 1, 2, and 3) seem to emphasize the organization of the somatosensory pathways into separate, parallel channels processing different components of somatic sensation. In short, the somatosensory system appears as clearly organized for parallel processing as the visual system.

12.2. NOTES ON OTHER SENSORY PATHWAYS

12.2.1. The Auditory Pathways

Spoendlin's (1969, 1970, 1972) studies of the innervation of the cochlea have provided clear evidence of parallel innervation of the two classes of auditory receptor cells found in the organ of Corti, the inner and outer hair cells. His analysis has not, however, been pursued into the central auditory pathways. Conversely, several recent studies have suggested that subdivisions of the central auditory nuclei of the brain stem and forebrain are connected in parallel (Merzenich and Kaas, 1980; Andersen et al., 1980; Phillips and Irvine, 1981), but have tended to consider the output of the cochlea as homogeneous (Merzenich and Kaas, 1980). These two sets of findings have yet to be related to each other.

At the peripheral end of the auditory pathway, the hair cells of the cochlea (the sensory receptors of audition) seem divisible into two groups, inner and outer. The inner hair cells are fewer in number, and form a single row along the inner margin of the basilar membrane (which coils around the cochlear spiral). The outer hair cells are found on the outer part of the basilar membrane; they form four or five rows and hence are considerably more numerous (12,000 as against about 3500 inner hair cells in man). It is not known whether inner and outer hair cells make different contributions to auditory function (analogous perhaps to the different functional roles of the rod and cone receptors of the eye), but it is tempting to speculate that they do, for they are innervated by different fibers of the auditory nerve.

Spoendlin (1969, 1970) described two types (I and II) of sensory cells in the spiral ganglion (which contains the cell bodies of auditory nerve fibers). Most (approximately 95%) were of type I, with relatively large somas covered by a myelin sheath. A small minority of the cells lacked the myelin sheath, were relatively small in soma size, and contained fewer mitochondria and less rough endoplasmic reticulum. Spoendlin showed that these different cell types innervate separate groups of hair cells; specifically that type I cells innervate inner hair cells, while type II cells innervate outer hair cells. The result is remarkable in the present context because it indicates the presence of at least two major classes of auditory afferents, carrying information from different groups of hair cells. The result was also surprising because the outer hair cells outnumber the inner by about 3: 1, yet are innervated by a 5% minority of cochlear nerve cells. Spoendlin argued that each type I cell innervates just a single inner hair cell, while the axon of a single type II cell spreads along the basilar membrane to innervate many outer hair cells. These distinct innervation patterns would seem to emphasize the likelihood that inner and outer hair cells subserve different functional roles. Spoendlin (1972) noted further that a very small (0.5%) minority of auditory nerve fibers are very thick and innervate a row of approximately 10 inner hair cells.

Deol and Gluecksohn-Waelsch (1979) have described evidence of a similar pattern of innervation of hair cells in the mouse cochlea. They studied a mouse with an inherent and apparently specific degeneration of inner hair cells. The outer hair cells are apparently unaffected; and, as in the cat, they are considerably more numerous than inner hair cells. Nevertheless, most of the neurons in the spiral ganglion degenerated after birth. Presumably, the surviving minority innervate the outer hair cells. Perhaps reflecting the large proportion of afferent fibers devoted to the inner hair cells, the mutant mice have very poor hearing. The tests of hearing applied were not, however, sufficiently sensitive to indicate the auditory capacities or role of outer hair cells.

In short, there appear to be three classes of sensory afferents in the auditory nerve of the cat, and similar groupings may be present in other mammals. In the cat, each type I cell innervates an inner hair cell; each type II cell innervates a considerable number of outer hair cells; while each of the tiny proportion of very thick axons (whose somas are not yet identified) innervates a number of inner hair cells. Presumably, the different fiber types respond differently to auditory stimuli, though little evidence is available on this point. The analogy between these three afferent types in the auditory nerve and the functional groups of ganglion cell axons recognized in the optic nerve may be partly fortuitous, but it seems clear that auditory afferents may form distinct groups, and the recognition of such groups should provide, as G. H. Bishop commented in the context of the optic nerve, a valuable starting point in the further analysis of the pathway. However, this intriguing line of evidence of parallel organization stops at this point.

Centrally [see Merzenich and Kaas (1980) for a review], within the cochlear nuclei, the nuclei of the lateral lemniscus, the thalamic nucleus specialized for audition (the medial geniculate nucleus), and the auditory areas of the neocortex, marked subdivisions have been described. For example, the "primary" auditory area of cat cortex and the anterior auditory field receive direct, cochleotopically organized input from one particular division of the medial geniculate nucleus, in parallel with each other; and the "secondary" area of the auditory cortex receives a parallel, less clearly organized input from a different segment of the medial geniculate nucleus.

This parallel processing analysis of auditory pathways is continuing to attract attention, e.g., from Phillips and Irvine (1981), and there is thus much to suggest that the auditory pathways too comprise parallel-wired subdivisions. However, the extent of the analogy to be drawn with the parallel organization of the visual or somatosensory pathways remains to be assessed. That assessment may require a better understanding of the relation between Spoendlin's analysis of cochlear innervation, and more recent ideas of parallel organization of the central auditory pathways.

12.2.2. The Chemical Senses

Taste and olfaction provide an interesting contrast to the senses already discussed, for their stimuli are much harder to characterize. The stimulus "domain" has no spatial dimension and there is debate over whether different submodalities can be distinguished. The receptors have not yet been shown to form distinct morphological classes; and the physical basis of observed differences in sensitivity is not understood. For neither system is there evidence of conduction velocity groupings among the afferent fibers nor, a fortiori, of distinct central projections of different classes of afferents.

In the analysis of taste perception, several authors have proposed a typology of tastes, suggesting four "primary" tastes: sweet, sour, salty, and bitter. That division has been both assumed (e.g., Pfaffmann *et al.,* 1976) and challenged (Schiffman and Erickson, 1980; Erickson *et al.,* 1980). Without agreement on the submodalities of taste, or independent evidence based, for example, on neuronal connections within the pathways subserving taste, a viable hypothesis of parallel organization of taste pathways cannot be formulated.

In olfaction too, there is no agreed classification of receptors by either their physiology or their morphology, or of afferent fibers by their conduction velocity or response properties. In the context of central pathways, however, ideas of parallel processing have been put forward. Keverne (1978), for example, pointed out that in nonprimate mammals, a distinct patch of nasal mucosa can be identified, the vomeronasal organ. Two olfactory pathways can then be traced: one from the nasal mucosa proper, via the olfactory bulb to the prepyriform cortex, and then to the thalamus (medial dorsal nucleus) and neocortex (the orbitofrontal cortex); the second from the vomeronasal organ, via an accessory olfactory bulb to the amygdala and thence via the stria terminalis to the hypothalamus. These two pathways are analogous. Keverne argues, to the forebrain and midbrain projections of the retina; they establish potentially independent branches of the olfactory system. Takagi (1979) confirmed several aspects of Keverne's analysis but argued that, even in primates (which lack the accessory olfactory path) two distinct olfactory pathways can be distinguished within the forebrain. Olfactory information, he suggests, follows a common path through the olfactory bulb to the prepyriform cortex, the area traditionally recognized as the primary olfactory cortex. Thence two pathways arise; one passes through the hypothalamus to reach a lateroposterior part of the orbitofrontal cortex, while the other traverses the medial dorsal nucleus of thalamus to reach a distinct but adjacent region of the orbitofrontal cortex. Takagi describes evidence from ablation studies that indicates that the former of these pathways (the "transhypothalamic") is important for olfactory discrimination. He notes that the function of the other, "transthalamic" pathway remains to be defined; and comments that the evidence of "dual olfactory pathways" in the primate forebrain provides a challenge for future work.

Enough has been said, I think, to make clear that the idea that sensory systems comprise subsystems operating in parallel has proved widely useful in sensory neurobiology. Is there some significance to this broad applicability of the idea? I believe so; and I suggest in Sections 12.3 and 12.4 that the analysis of sensory systems in terms of the behavior of subsystems of neurons, whatever its limitations, is a conceptual advance on previous attempts to explain sensory processes in terms either of the properties of individual neurons or, at the other extreme, of schematic neuronal "networks."

12.3. IDEAS AND THEIR LIMITS

Ideas are fragile, easily broken by new evidence; yet without them the scientist is impotent. Perforce we must learn to deal with ideas, and to do so with both skepticism and respect: skepticism to enable scientific progress, and respect to ensure both the commitment essential for experimental work and the tolerance necessary for fruitful scientific debate. It is important to know the limits as well as the potential of our ideas.

Viewed then with skepticism, the parallel analysis of the visual (and other sensory) pathways has clear failings. Like most scientific ideas, it is imperfect and limited in its applicability (see, for example, Lennie, 1980a); it remains vulnerable to new evidence; and arguably it does little to bridge the gap between perceptual processes and their physical basis. Perhaps its most serious deficiency, however, is that while our perception of the visual world seems to involve the combination of many parameters (spatial, temporal, chromatic, volitional), the parallel processing analysis traces only the separateness of the neuronal pathways that seem to subserve different parameters. It tells us, as yet, nothing of the basis of their combination.

Viewing the parallel processing analysis, on the other hand, with respect, it seems a reasonable response to the criticism just formulated that no other available analysis tells us any more of the neural basis of the synthetic processes of visual perception. Conversely, the parallel processing analysis has proved more powerful than any other in understanding the range of cell morphology, connections, and physiology apparent among retinal ganglion cells and among neurons in the visual centers of the brain; it provides a more powerful framework than otherwise available for comparative studies that seek to assess the phylogenetic development of the visual pathways, and for understanding the neural basis both of certain perceptual phenomena, and of the environment-dependence of the ontogenetic development of vision; and the analysis has proved both powerful and relevant in the study of other sensory pathways. Even with its limitations, it is part of present understanding of mammalian sensory pathways.

Yet I would argue one further and substantial value for the parallel- processing analysis; that it offers a way of approaching the complexity of sensory pathways in terms of the properties of neuronal "systems." This possibility was broached at the end of the last chapter. It seems of such importance for progress in our field, yet so little recognized, that I would end this monograph with its advocacy.

12.4. "PARAMETRIC SYSTEMATICS": AN APPROACH TO THE UNDERSTANDING OF SENSORY PATHWAYS

12.4.1. Two Terms: Reductionist and "Holistic"

Historians of science tell us that the scientific revolution that has determined so much of the intellectual and material framework of modern Western society began with (among many other things) a willingness among natural philosophers to be satisfied with partial explanations. The early history of study of the lens of the eye, as discussed in particular by Polyak (1957) and Crombie (1964), will serve as an example.

Early philosophers concerned with anatomy of the human body, for example the renowned Alhazen, had considered at length how the eye functions. Despite considerable knowledge of optics (reading glasses, for example, had been in use since at least the 14th century), they had resisted the idea that the lens acts as a lens, preferring to interpret it as the receptive part of the eye, or as both an irradiator of energy and a receptor, in one. Part of their resistance seemed to stem from the quite correct understanding that if the lens were a lens, it would form an image of the world on the back of the eye that was physically inverted, both upside down and left to right. Yet this is not how the world appears. Accepting that the lens is an optical device seemed to create a problem of understanding; it made an overall understanding of vision more difficult. Beginning in the 16th century, however, empirically based work established the optical function of the lens. Felix Platerus (1536–1614, Professor of Anatomy at Basel) experimented on dogs to gain evidence that the lens was not receptive. He showed that destruction of the suspensory ligament of the lens (which was supposed to transmit activity to the retina and optic nerve) did not cause blindness. Platerus concluded that the role of the lens was purely physical, indicating that the retina must be the receptive part of the eye. Kepler found Platerus' conclusion convincing and (in his *Dioptrice*, 1611) relied on it in developing his pioneering account of the optics of the eye that assumed both (1) that an image of the visual world is formed on the back of the eye and (2) [the corollary of (1)] that that image is inverted. Subsequent experimental work on excised animal eyes, reported during the 17th century, demonstrated both these points directly. This work formed the beginnings of the experimental analysis of the visual pathways; but, as Crombie stresses, it required that scientists be satisfied (as they had not previously been) to carry on with major problems unsolved, and to analyze the eye as a mechanism reducible to its parts, even though the understanding of the whole visual system was apparently not advanced, or even seemed imperiled.

Much of the success of biological science has stemmed, Crombie argues, from the willingness of scientists to treat animals as mechanisms and to pursue the analysis of parts without demanding, in the short or medium term, an explanation of the whole. The success of "reductionist" science needs no exposition here.

Yet scientists often are not content with "reductionist" analyses of subareas of their field. They then seek to formulate an overall understanding of their field of work, a unifying cosmology in terms of which the interrelations of the many mechanisms they study can better be understood. In biology, the theory of the evolution of species is a spectacular example of an overall, unifying theory based, not on a mechanistic analysis of animals (the mechanism of evolution was in fact one of its unknowns), but on a broad survey of the physical variation present in animals. There seems no particular word for such ideas; borrowing from others, though not I fear from the Oxford English Dictionary, I will use the term *holistic* to refer to scientific ideas that seek to establish an overall framework for their field.

Now much of the experimental work I have tried to summarize in Chapters 1–3 and 6–11 was wholly or partly mechanistic in its thrust. Investigators traced connections, analyzed synaptic connectivity in terms of response latencies, explained a cell's receptive field size in terms of its dendritic field size, and so on. Yet the same sort of experimental work can be done with a different thrust. For example, one could undertake an experiment to answer the question "Is the size of the center region of a ganglion cell's receptive field determined by its dendritic field"? This would tell us more of the mechanism of receptive field organization; it would be a "reductionist" experiment. Conversely, one could, with very similar techniques, answer the question "What ranges of receptive and dendritic field sizes are found among different ganglion cell classes"? A series of experiments of the latter sort would provide the parametric data base for a classification of ganglion cells that might lead to the wide-ranging sort of analysis reviewed in this book. And since that analysis has led to a reconsideration of the organization of visual centers of the brain and their interrelations, has raised issues of phylogenetic development, and has provided new evidence on the environment-dependence of the visual system and on the neural basis of visual perception, its thrust seems fairly described as "holistic."

12.4.2. The Technical Limits of Ideas

It is axiomatic that our ideas are influenced by our experiences; that we attempt to explain mysteries in terms of what we know. The "physiologists" of Aristotle's time explained the procreativity of animals in terms of the actions of elements such as "earth" and "air"; Harvey's successful hydraulic analysis of the circulation led to hydraulic explanations of nerve transmission; computer scientists sometimes speculate about brain function in terms of the operations of a complex computer. Such arguments rest on analogies that, though useful and intriguing, can never be logically compelling and often lead quickly to absurdity. In sensory physiology, as others have argued, the spectacular success of the microelectrode analysis of single-cell function has led to theories of brain function in terms of the properties of single cells; examples are Hubel and Wiesel's simple/complex/hypercomplex analysis of the visual cortex, and Barlow's (1979) neuron doctrine. The history of these formulations, in particular their dependence on a feature extraction analysis of receptive fields, was traced in Rowe and Stone (1980*a,b*) and in Chapters 1 and 2. Critiques of them have been formulated by Uttal (1971), McIlwain (1976), and Rowe and Stone (1980*b*), and in Chapter 11 (Section 11.4).

As Uttal (1977) points out, however, great difficulties are involved in developing an alternative approach. How can we formulate viable ideas about complex neuronal systems, which take into account the large numbers of neurons likely to be involved in even simple functions of the brain? The problem is partly technological. As Uttal comments:

> An approach to the study of vertebrate neuronal networks, comparable in its universality to the microelectrode attack on single cell function, is still missing. Furthermore, we do not yet have the mathematical tools necessary to handle the complicated interactions of this . . . problem. This difficulty . . . is a substantial block to progress in the search for the functional interconnections of the brain's various regions.

Uttal was nevertheless optimistic that the technology to deal with complex arrays of neurons would be developed. I would like here to move one step beyond Uttal's analysis: to argue that a start toward the analysis of complex neuronal systems can and has been made without a technological breakthrough, but rather by means of a conceptual change, to thinking of neurons as forming subsystems within systems. Ulinski (1980) essayed such a conceptual change, proposing that neural systems such as the visual system can be fruitfully regarded as comprising "nested" levels of organization (cellular, modular, and network levels). The present approach differs from Ulinski's in stressing the presence of parallel-wired systems of neurons extending from the cellular to the network levels. The approach can be traced back (Rowe and Stone, 1980a; Chapters 1 and 2) to "parametric" studies of visual neurons, beginning in the late 1930s; and to the growing understanding among neurobiologists [apparent, for example, in the articles by Tyner (1975) and Ulinski (1980)] that the properties of the nervous system at all levels require understanding in terms of their evolutionary history. It offers an approach to the complexity of sensory pathways that requires no technological breakthrough.

12.4.3. A "Parametric Systematics" for Sensory Biology

Central to the approach I am advocating is the surveying of large numbers of nerve cells in the system under study, to observe and catalog the variation found among them in their many properties (structural, physiological, pharmacological). On the basis of those observations, hypotheses are formulated concerning the functional groups of neurons present and their interactions. If the observational base is sufficiently wide, these hypotheses may be extended to concern the psychophysics of the sensory system, and its phylogenetic history. Although its name is new, this "new" view has so strong a basis in the history of our field that, on reflection, it seems neither new nor (still less) revolutionary.

There seem to be three major antecedents to the idea of "parametric systematics." The first is the willingness of investigators in the somatosensory pathways, from their earliest work, to think in terms of neuronal subsystems operating in parallel within the somatosensory system as a whole. These "subsystems" are considered to employ different receptors, afferents, and central pathways and nuclei, and to subserve different components of somatic sensations; for example, "epicritic" and "protopathic" sensations. The function of a neuron is then partly understood from its own particular properties, and partly from knowledge of the properties of the system and subsystem of which it is part. This idea was brought into visual physiology by Bishop (1933; see Chapter 1, Section 1.1) and in many ways its use has continued; terms such as *geniculo-striate system* or *accessory optic system* attest to that. Yet in considerations of information processing in the visual pathways, this systematic approach was largely replaced in the 1950s and 1960s by the consideration of neurons as individuals, with the suggestion that successively higher levels of neural processing are performed by fewer and fewer neurons; or with attempts to view neurons as operating in schematized networks.

The second antecedent is the zoologists' experience that understanding of the enormous variety in the physical properties of animals could be gained by ceasing to think of them as either individuals or as rigid "types," but rather as populations within which more or less physical variation persists. From surveys of the properties of animals, groups, trends, and aggregations could be discerned that eventually formed the data base of the theory of evolution; and that broad "holistic" theory provided a framework for understanding the variety, phylogeny, and even ontogeny of all animals.

The third antecedent idea is the parametric approach to cell classification discussed (and advocated) in Rowe and Stone (1977, 1979, 1980*a,b*) and in Chapters 1 and 2. In zoology, that approach provided the observational basis of the theory of evolution; in sensory neurobiology, it has led to the parallel processing analysis of sensory pathways.

The parallel processing analysis may then provide the beginnings of an understanding of sensory pathways in terms of the systems of nerve cells identifiable within them, and the different functional contributions of those subsystems. The evidence presented in preceding chapters that this analysis provides new insights into the phylogeny, environment-dependence, and adult circuitry of the brain, emphasizes its potential. The parallel processing analysis enjoys, I would suggest, four advantages over alternative ways of analyzing sensory pathways:

1. It incorporates knowledge of the biology of animals; that is, by seeking understanding of the nervous system in terms of a wide range of neuron properties, it views the nervous system and its component systems as parts of the animal and subject, like the whole animal, to gradual change and considerable variation.
2. It deals better than any other theory with the physical variation found among neurons.
3. Because it is polythetically or parametrically based, the parallel processing analysis is to a large extent independent of particular experimental techniques.
4. Because of its broad observational base, it may provide a way of transcending the technical limits to our concepts of the organization of sensory pathways.

I would stress that my advocacy of a "holistic" approach to the study of sensory pathways is in no way meant to be exclusive; "reductionist" analyses of the visual pathways are of at least equal importance for progress in the field. Nevertheless, the "parametric systematics" I here advocate seem to offer a more powerful approach than any current alternatives for establishing a broad understanding of the visual and other sensory pathways. This optimism needs, of course, to be tested and retested in the laboratory.

References

Albus, K., and F. Donate-Oliver, 1977, Cells of origin of the occipito-pontine projection in the cat: Functional properties and intracortical location, *Exp. Brain Res.* **28:**167–174.

Allman, J., 1977, Evolution of the visual system of the early primates, *Prog. Psychobiol. Physiol. Psychol.* **7:**1–53.

Altman, J., and M. B. Carpenter, 1961, Fiber projections of the superior colliculus in cats, *J. Comp. Neurol.* **116:**157–178.

Altman, J., and L. I. Malis, 1962, An electrophysiological study of the superior colliculus and visual cortex, *Exp. Neurol.* **5:**233–249.

Anderson, R. A., P. L. Knight, and M. M. Merzenich, 1980, The thalamocortical connections of AI, AII and the anterior auditory field (AAF) in the cat: Evidence for two largely segregated systems of connections, *J. Comp. Neurol.* **194:**663–701.

Angaut-Petit, D., 1975, The dorsal column system. II. Functional properties and bulbar relay of the postsynaptic fibres of the cat's fasciculus gracilis, *Exp. Brain Res.* **22:**471–493.

Anker, R., and B. Cragg, 1979, Cell growth in the monocular segment of the lateral geniculate nucleus following the opening or closing of one eye, *J. Comp. Neurol.* **188:**107–112.

Appelle, S., 1972, Perception and discrimination as a function of stimulus orientation: The 'oblique effect' in man and animals, *Psychol. Bull.* **78:**266–278.

Ayala, F. J., 1978, The mechanisms of evolution, *Sci. Am.* **239:**56–59.

Backstrom, A.-C., S. Hemila, and T. Reuter, 1978, Directional selectivity and colour coding in the frog retina, *Med. Biol.* **56:**72–83.

Baizer, J. S., D. L. Robinson, and B. Dow, 1977, Visual responses of area 18 neurons in awake, behaving monkey, *J. Neurophysiol.* **40:**1024–1037.

Baker, F. H., P. Grigg, and G. K. von Noorden, 1974, Effects of visual deprivation and strabismus on the response of neurons in the visual cortex of the monkey, including studies on the striate and prestriate cortex in the normal animal, *Brain Res.* **66:**185–208.

Ballas, I., K.-P. Hoffmann, and H.-J. Wagner, 1981, Retinal projection to the nucleus of the optic tract in the cat as revealed by retrograde transport of horseradish peroxidase, *Neurosci. Lett.* **26:**197–202.

Balme, D. M., 1975, Aristotle's use of differentiae in zoology, in: *Articles on Aristotle,* Vol. 1, *Science* (J. Barnes, M. Schofield, and R. Sorabji, eds.), Duckworth, London.

Banks, M. S., R. N. Aslin, and R. D. Letson, 1975, Sensitive period for the development of human binocular vision, *Science* **190:**675–677.

Barbur, J. L., K. H. Ruddock, and V. A. Waterfield, 1980, Human visual responses in the absence of the geniculo-calcarine pathway, *Brain* **103:**905–928.

Barlow, G. W., 1977, Modal action patterns, in: *How Animals Communicate* (F. Sebeok, ed.), pp. 98–314, Indiana University Press, Bloomington.

Barlow, H. B., 1953, Summation and inhibition in the frog's retina, *J. Physiol. (London)* **119:**69–88.

Barlow, H. B., 1961, Possible principles underlying the transformation of sensory messages, in: *Sensory Communication* (W. Rosenblith, ed.), pp. 217–234, MIT Press, Cambridge, Mass.

Barlow, H. B., 1972, Single units and sensation: A neuron doctrine for perceptual psychology, *Perception* **1:**371–394.

Barlow, H. B., 1979, Three theories of cortical function, in: *Developmental Neurobiology of Vision* (R. D. Freeman, ed.), Plenum Press, New York.

Barlow, H. B., and W. R. Levick, 1969, Changes in the maintained discharge with adaptation level in the cat retina, *J. Physiol. (London)* **202:**699–778.

Barlow, H. B., R. Fitzhugh, and S. W. Kuffler, 1958, Change of organisation in the receptive fields of the cat's retina during dark adaptation, *J. Physiol. (London)* **137**:338–354.

Barlow, H. B., R. M. Hill, and W. R. Levick, 1964, Retinal ganglion cells responding selectively to direction and speed of image motion in the rabbit, *J. Physiol. (London)* **173**:377–407.

Barnes, J., M. Schofield, and R. Sorabji (eds.), 1975, *Articles on Aristotle,* Gerald Duckworth and Co., London.

Barris, R. W., W. R. Ingram, and S. W. Ranson, 1935, Optic connections of the diencephalon and midbrain of the cat, *J. Comp. Neurol.* **62**:117–153.

Benevento, L. A., and B. Davis, 1977, Topographical projections of the prestriate cortex to the pulvinar nuclei in the macaque monkey: An autoradiographic study, *Exp. Brain Res.* **30**:405–424.

Benevento, L. A., and J. H. Fallon, 1975, The ascending projections of the superior colliculus in the rhesus monkey (*Macaca mulatta*), *J. Comp. Neurol.* **160**:339–362.

Benevento, L. A., and M. Rezak, 1975, Extrageniculate projections to layers VI and I of striate cortex (area 17) in the rhesus monkey (*Macaca mulatta*), *Brain Res.* **96**:51–55.

Benevento, L. A., and M. Rezak, 1976, The cortical projections of the inferior pulvinar and adjacent lateral pulvinar in the rhesus monkey (*Macaca mulatta*): An autoradiographic study, *Brain Res.* **108**:1–24.

Benevento, L. A., and K. Yoshida, 1981, The afferent and efferent organization of the lateral geniculoprestriate pathways in the macaque monkey, *J. Comp. Neurol.* **203**:455–474.

Bennett, R. E., D. G. Ferrington, and M. Rowe, 1980, Tactile neuron classes within second somatosensory area (SII) of the cat, *J. Neurophysiol.* **43**:292–309.

Berkely, M. A., and J. M. Sprague, 1979, Striate cortex and visual acuity functions of the cat, *J. Comp. Neurol.* **187**:679–702.

Berkley, M. A., F. Kitterle, and D. W. Watkins, 1975, Grating visibility as a function of orientation and retinal eccentricity, *Vision Res.* **15**:239–244.

Berman, N., and E. G. Jones, 1977, A retino-pulvinar projection in the cat. *Brain Res.* **134**:237–248.

Berman, N., H. E. Murphy, and W. L. Salinger, 1979, Monocular paralysis in the adult cat does not change cortical ocular dominance, *Brain Res.* **164**:290–293.

Berson, D. M., and A. M. Graybiel, 1978, Parallel thalamic zones in the LP–pulvinar complex of the cat identified by their afferent and efferent connections, *Brain Res.* **147**:139–148.

Bishop, G. H., 1933, Fiber groups in the optic nerve, *Am. J. Physiol.* **106**:460–470.

Bishop, G. H., 1959, The relation between nerve fiber size and sensory modality: Phylogenetic implications of the afferent innervation of cortex, *J. Nerv. Ment. Dis.* **128**:89–114.

Bishop, G. H., and M. H. Clare, 1955, Organisation and distribution of fibers in the optic tract of the cat. *J. Comp. Neurol.* **103**:269–304.

Bishop, G. H., and P. Heinbecker, 1930, Differentiation of axon types in visceral nerves by means of the potential record, *Am. J. Physiol.* **94**:170–200.

Bishop, G. H., and P. Heinbecker, 1932, A functional analysis of the cervical sympathetic nerve supply to the eye, *Am. J. Physiol.* **100**:519–532.

Bishop, G. H., and J. S. O'Leary, 1938, Potential records from the optic cortex of the cat, *J. Neurophysiol.* **1**:391–404.

Bishop, G. H., and J. S. O'Leary, 1940, Electrical activity of the lateral geniculate of cats following optic nerve stimuli, *J. Neurophysiol.* **3**:308–322.

Bishop, G. H., and J. S. O'Leary, 1942, Factors determining the form of the potential record in the vicinity of the synapse of the dorsal nucleus of the lateral geniculate body, *J. Cell. Comp. Physiol.* **19**:315–331.

Bishop, G. H., M. H. Clare, and W. M. Landau, 1969, Further analysis of fiber groups in the optic tract of the cat, *Exp. Neurol.* **24**:386–399.

Bishinop, P. O., and G. H. Henry, 1972, Striate neurons: Receptive field concepts, *Invest. Ophthalmol.* **11**:346–354.

Bishop, P. O., and J. G. MacLeod, 1954, Nature of potentials associated with synaptic transmission in lateral geniculate of the cat, *J. Neurophysiol.* **17**:387–414.

Bishop, P. O., D. Jeremy, and J. W. Lance, 1953, The optic nerve: Properties of a central tract, *J. Physiol. (London)* **121**:415–432.

Bisti, S., R. Clement, L. Maffei, and L. Mecacci, 1977, Spatial frequency and orientation tuning curves of visual neurones in the cat: Effects of mean luminance, *Exp. Brain Res.* **27**:335–345.

Blake, R., and D. N. Antoinetti, 1976, Abnormal visual resolution in the Siamese cat, *Science* **194**:109–110.

Blake, R., and R. Bellhorn, 1978, Visual acuity in cats with central lesions, *Vision Res.* **18**:15–18.

Blake, R., and J. M. Camisa, 1977, Temporal aspects of spatial vision in the cat, *Exp. Brain Res.* **28**:325–333.

Blake, R., and A. DiGianfilippo, 1980, Spatial vision in cats with selective neural deficits, *J. Neurophysiol.* **43**:1197–1205.

Blake, R., and H. V. B. Hirsch, 1975, Deficits in binocular depth perception in cats after alternating monocular deprivation, *Science* **190**:1114–1116.

Blakemore, C., 1969, Binocular depth discrimination and the nasotemporal division, *J. Physiol. (London)* **205**:471–497.

Blakemore, C., 1978, Maturation and modification in the developing visual system in: *Handbook of Sensory Physiology,* Vol. XIII (R. Held, ed.), Springer-Verlag, Berlin.

Blakemore, C., and F. W. Cambell, 1969, On the existence of neurones in the human visual system selectively sensitive to the orientation and size of retinal images, *J. Physiol. (London)* **203**:237–260.

Blakemore, C., and G. F. Cooper, 1970, Development of the brain depends on the visual environment, *Nature (London)* **228**:477–478.

Blakemore, C., and R. C. van Sluyters, 1974, Reversal of the physiological effects of monocular deprivation in kittens: Further evidence for a sensitive period, *J. Physiol. (London)* **237**:195–216.

Blakemore, C., and R. C. van Sluyters, 1975, Innate and environmental factors in the development of the kitten's visual cortex, *J. Physiol. (London)* **248**:663–716.

Blakemore, C., and F. Vital-Durand, 1979, Development of the neural basis of visual acuity in monkeys, *Trans. Ophthalmol. Soc. U.K.* **99**:363–368.

Blakemore, C., and F. Vital-Durand, 1981, Distribution of X- and Y-cells in the monkey's lateral geniculate nucleus, *J. Physiol. (London)* **320**:17–18P.

Blakemore, C., L. J. Garey, and F. Vital-Durand, 1978, The physiological effects of monocular deprivation and their reversal in the monkey's visual cortex, *J. Physiol. (London)* **283**:223–262.

Bock, W. J., 1974, Philosophical foundations of classical evolutionary classification, *Syst. Zool.* **22**:375–392.

Bodis-Wollner, I. G., A. Atkin, E. Raab, and M. Wolkstein, 1977, Visual association cortex and vision in man: Pattern evoked occipital potentials in a blind boy, *Science* **198**:629–631.

Boltz, R. L., R. S. Harwerth, and E. L. Smith, 1979, Orientation anisotropy of visual stimuli in rhesus monkey: A behavioural study, *Science* **205**:511–513.

Bowling, D. B., and C. M. Michael, 1980, Projection patterns of single physiologically identified optic tract fibres in the cat, *Nature (London)* **286**:899–902.

Boycott, B. B., and J. E. Dowling, 1969, Organization of the primate retina: Light microscopy, *Philos. Trans. R. Soc. London Ser. B* **255**:109–184.

Boycott, B. B., and H. Wässle, 1974, The morphological types of ganglion cells of the domestic cat's retina, *J. Physiol. (London)* **240**:397–419.

Boycott, B. B., L. Peichl, and H. Wässle, 1978, Morphological types of horizontal cell in the retina of the domestic cat, *Proc. R. Soc. London Ser. B* **203**:229–245.

Brauer, K., W. Schober, and G. Winkelmann, 1979, Two morphologically different types of retinal axon terminals in the rat's dorsal lateral geniculate nucleus and their relationships to X- and Y-channels, *Exp. Brain Res.* **36**:523–532.

Breitmeyer, B., and L. Ganz, 1976, Sustained and transient channels for theories of visual pattern masking, saccadic suppression and information processing, *Psychol. Rev.* **83**:1–36.

Brindley, G. S., 1977, Somatic and visceral sensory mechanisms: Introduction, *Brit. Med. Bull.* **33**:89–90.

Broendstrup, P., 1948, The functional and anatomical differences between the nasal and temporal parts of the retina, *Acta Ophthalmol.* **26**:351–361.

Brown, A. G., 1981, The terminations of cutaneous nerve fibres in the spinal cord, *Trends Neurosci.* **4**:64–67.

Brown, A. G., and G. Gordon, 1977, Subcortical mechanisms concerned in somatic sensation, *Brit. Med. Bull.* **33**:121–128.

Brown, D. L., and W. L. Salinger, 1975, Loss of X-cells in lateral geniculate nucleus with monocular paralysis: Neural plasticity in the adult cat, *Science* **189**:1011–1012.

Brown, J. E., 1965, Dendritic fields of retinal ganglion cells of the rat, *J. Neurophysiol.* **28:**1091–1100.

Brown, J. E., and D. Major, 1966, Cat retinal ganglion cell dendritic fields, *Exp. Neurol.* **15:**70–78.

Brown, J. E., and J. A. Rojas, 1965, Rat retinal ganglion cells: Receptive field organisation and maintained activity, *J. Neurophysiol.* **28:**1073–1090.

Buisseret, P., and M. Imbert, 1976, Visual cortical cells: Their developmental properties in normal and dark reared kittens, *J. Physiol. (London)* **255:**511–525.

Bullier, J., and G. H. Henry, 1979a, Ordinal position of neurons in cat striate cortex, *J. Neurophysiol.* **42:**1251–1263.

Bullier, J., and G. H. Henry, 1979b, Neural path taken by afferent streams in striate cortex of the cat, *J. Neurophysiol.* **42:**1264–1270.

Bullier, J., and G. H. Henry, 1979c, Laminar distribution of first-order neurons and afferent terminals in cat striate cortex, *J. Neurophysiol.* **42:**1271–1281.

Bullier, J., and G. H. Henry, 1980, Ordinal position and afferent input of neurons in monkey striate cortex, *J. Comp. Neurol.* **193:**913–936.

Bullier, J. H., and T. T. Norton, 1977, Receptive-field properties of X-, Y- and intermediate cells in the cat lateral geniculate nucleus, *Brain Res.* **121:**151–156.

Bullier, J., and T. T. Norton, 1979a, Comparison of receptive field properties of X and Y ganglion cells in X and Y lateral geniculate cells in the cat, *J. Neurophysiol.* **42:**274–291.

Bullier, J., and T. T. Norton, 1979b, X- and Y-relay cells in cat lateral geniculate nucleus: Quantitative analysis of receptive field properties and classification, *J. Neurophysiol.* **42:**244–273.

Bunt, A. H., 1976, Ramification patterns of ganglion cells dendrites in the retina of the albino rat, *Brain Res.* **103:**1–8.

Bunt, A. H., and D. S. Minckler, 1977, Foveal sparing, *Arch. Ophthalmol.* **95:**1445–1447.

Bunt, A. H., R. D. Lund, and J. S. Lund, 1974, Retrograde axonal transport of horseradish peroxidase by ganglion cells of the albino rat retina, *Brain Res.* **73:**215–228.

Bunt, A. H., A. E. Hendrickson, J. S. Lund, R. D. Lund, and A. F. Fuchs, 1975, Monkey retinal ganglion cells: Morphometric analysis and tracing of axonal projections, with a consideration of the peroxidase technique, *J. Comp. Neurol.* **164:**265–286.

Bunt, A. H., D. S. Minckler, and G. W. Johanson, 1977, Demonstration of bilateral projection of central retina of the monkey with horseradish peroxidase neuronography, *J. Comp. Neurol.* **171:**619–630.

Burrows, G. R., and W. R. Hayhow, 1971, The organization of the thalamo-cortical visual pathways in the cat, *Brain Behav. Evol.* **4:**220–272.

Cajal, S. R., 1893, La retine des vertebres, *La Cellule* **9:**17–257.

Cajal, S. R. 1911, Le lobe optique des vertebres inferieurs, in: *Histologie de l'Homme et des Vertebres.* A. Maloine, Paris.

Caldwell, J. H., and N. Daw, 1978, New Properties of rabbit retinal ganglion cells, *J. Physiol. (London)* **276:**257–276.

Camarda, R., and G. Rizzolatti, 1976, Receptive fields of cells in the superficial layers of the cat's area 17, *Exp. Brain Res.* **24:**423–427.

Camisa, J. M., R. Blake, and S. Lema, 1977, The effect of temporal modulation on the oblique effect in humans, *Perception* **6:**165–171.

Campbell, C. B. G., and W. Hodos, 1970, The concept of homology and the evolution of the nervous system *Brain Behav. Evol.* **3:**353–367.

Campos-Ortega, J. A., and W. R. Hayhow, 1970, A new lamination pattern in the lateral geniculate nucleus of primates, *Brain Res.* **20:**335–339.

Campos-Ortega, J. A., and W. R. Hayhow, 1971, A note on the connections and possible significance of Minkowski's "intermediare Zellgruppe" in the lateral geniculate body of cercopithecoid primates, *Brain Res.* **26:**177–183.

Campos-Ortega, J. A., W. R. Hayhow, and P. F. de V. Cluver, 1970, A note on the problem of retinal projections to the inferior pulvinar nucleus of primates, *Brain Res.* **22:**126–130.

Carlson, M., 1981, Characteristics of sensory deficits following lesions of Brodmann's areas 1 and 2 in the front central gyrus of *Macaca mulatta, Brain Res.* **204:**424–930.

Chalmers, A. F., 1976, *What Is This Thing Called Science?* University of Queensland Press, Brisbane.

Chalupa, L. M., and I. Thompson, 1980, Retinal ganglion cell projections to the superior colliculus of the hamster demonstrated by the horseradish peroxidase technique, *Neurosci. Lett.* **19:**13–19.

Chalupa, L. M., R. S. Coyle, and D. B. Lindsley, 1976, Effect of pulvinar lesions on visual pattern discrimination in monkeys, *J. Neurophysiol.* **39**:354–370.

Chang, H.-T., 1951, Triple conducting pathway in the visual system and trichromatic vision, *Ann. Psychol.* **50**:135–144.

Chang, H.-T., 1956, Fiber groups in primary optic pathway of cat, *J. Neurophysiol.* **19**:224–231.

Chievitz, J. H., 1889, Untersuchungen über die area centralis retinae, *Arch. Anat. Physiol. Leipzig, Anat. Abtheil. Suppl.* 139–396.

Chino, Y. M., M. S. Shansky, and D. I. Hamasaki, 1980, Development of receptive field properties of retinal ganglion cells in kittens raised with a convergent squint, *Exp. Brain Res.* **39**:313–320.

Choudhury, B. P., 1981, Ganglion cell distribution in the albino rat's retina, *Exp. Neurol.* **72**:638–644.

Chow, K. L., 1950, A retrograde degeneration study of the cortical projection field of the pulvinar in the monkey, *J. Comp. Neurol.* **93**:313–340.

Chow, K. L., and D. L. Stewart, 1972, Reversal of structural and functional effects of long term visual deprivation in cats, *Exp. Neurol.* **34**:409–433.

Citron, M. C., R. C. Emerson, and L. S. Ide, 1981, Spatial and temporal receptive-field analysis of the cat's geniculo-cortical pathway, *Vision Res.* **21**:385–396.

Clark, W. E. LeGros, 1932, A morphological study of the lateral geniculate body, *Brit. J. Ophthalmol.* **16**:264–284.

Clark, W. E. LeGros, 1940a, Anatomical basis of color vision, *Nature (London)* **146**:558–559.

Clark, W. E. LeGros, 1940b, The laminar organisation and cell content of the lateral geniculate body in the monkey, *J. Anat.* **75**:419–433.

Clark, W. E. LeGros, and D. W. C. Northfield, 1937, The cortical projection of the pulvinar in the macaque monkey, *Brain* **60**:126–142.

Cleland, B. G., and W. R. Levick, 1974a, Brisk and sluggish concentrically organised ganglion cells in the cat's retina, *J. Physiol. (London)* **240**:421–456.

Cleland, B. G., and W. R. Levick, 1974b, Properties of rarely encountered types of ganglion cells in the cat's retina and an overall classification, *J. Physiol. (London)* **240**:457–492.

Cleland, B. G., M. W. Dubin, and W. R. Levick, 1971, Sustained and transient neurones in the cat's retina and lateral geniculate nucleus, *J. Physiol. (London)* **217**:473–496.

Cleland, B. G., W. R. Levick, and K. J. Sanderson, 1973, Properties of sustained and transient ganglion cells in the cat's retina, *J. Physiol. (London)* **228**:649–680.

Cleland, B. G., R. Morstyn, H. Wagner, and W. R. Levick, 1975, Long latency input to lateral geniculate neurones of the cat, *Brain Res.* **91**:306–310.

Cleland, B. G., W. R. Levick, R. Morstyn, and H. G. Wagner, 1976, Lateral geniculate relay of slowly conducting retinal afferents to cat visual cortex, *J. Physiol. (London)* **255**:299–320.

Cleland, B. G., T. H. Harding, and U. Tulunay-Keesey, 1979, Visual resolution and receptive field size: Examination of two kinds of cat retinal ganglion cell, *Science* **205**:1015–1017.

Cleland, B. G., D. E. Mitchell, S. Gillard-Crewther, and D. P. Crewther, 1980, Visual resolution of retinal ganglion cells in monocularly deprived cats, *Brain Res.* **192**:261–266.

Collewijn, H., 1977, Eye and head movements in freely moving rabbits, *J. Physiol. (London)* **266**:471–496.

Cooper, H. M., H. Kennedy, M. Magnin, and F. Vital-Durand, 1979, Thalamic projections to area 17 in a prosimian primate, *Microcebus murinus, J. Comp. Neurol.* **187**:145–168.

Cooper, M. L., and J. D. Pettigrew, 1977, Nasotemporal division of retinothalamic pathways in normal and Siamese cats, *Soc. Neurosci. Abstr.* **2**:556.

Cooper, M., and J. D. Pettigrew, 1979a, A neurophysiological determination of the vertical horopter in the cat and owl, *J. Comp. Neurol.* **184**:1–26.

Cooper, M., and J. D. Pettigrew, 1979b, The decussation of the retinothalamic pathway in the cat with a note on the major meridians of the cat's eye, *J. Comp. Neurol.* **187**:285–312.

Cooper, M., and J. D. Pettigrew, 1979c, The retinothalamic pathways in Siamese cats, *J. Comp. Neurol.* **187**:313–348.

Cowey, A., 1974, Atrophy of retinal ganglion cells after removal of striate cortex in a rhesus monkey, *Perception* **3**:257–260.

Cowey, A., and C. Franzini, 1979, The retinal origin of uncrossed optic nerve fibres in rats and their role in visual discrimination, *Exp. Brain Res.* **35**:443–445.

Cowey, A., and V. H. Perry, 1979, The projection of temporal retina in rats, studied by retrograde transport of horseradish peroxidase, *Exp. Brain Res.* **35:**457–464.

Crombie, A. C. 1964, Early concepts of the senses and the mind, in: *Perception: Mechanisms and Models,* Readings from *Scientific American.* W. H. Freeman, San Francisco.

Cunningham, T. J., and J. A. Freeman, 1977, Bilateral ganglion cell branches in the normal rat: A demonstration with electrophysiological collision and cobalt tracing techniques, *J. Comp. Neurol.* **172:**165–176.

Curcio, C. A., and J. K. Harting, 1978, Organization of pulvinar afferents to area 18 in the squirrel monkey: Evidence for stripes, *Brain Res.* **143:**155–161.

Daniels, J. D., J. D. Pettigrew, and J. L. Norman, 1978, Development of single-neuron responses in the kitten's lateral geniculate nucleus, *J. Neurophysiol.* **41:**1373–1393.

Darwin, C. 1872, *The Origin of Species,* 6th ed., Murray, London.

Davis, F. A., 1929, The anatomy and histology of the eye and orbit of the rabbit, *Trans. Am. Ophthalmol. Soc.* **27:**401–441.

Daw, N. W., and A. L. Pearlman, 1970, Cat colour vision evidence for more than one cone process, *J. Physiol. (London)* **211:**125–137.

Daw, N, and H. J. Wyatt, 1974, Raising rabbits in a moving visual environment: An attempt to modify directional sensitivity in the retina, *J. Physiol. (London)* **240:**309–330.

Daw, N., and H. J. Wyatt, 1976, Kittens reared in a unidirectional environment: Evidence for a critical period, *J. Physiol. (London)* **257:**155–170.

DeBruyn, E., V. L. Wise, and V. A. Casagrande, 1980, The size and topographic arrangement of retinal ganglion cells in the galago, *Vision Res.* **20:**315–327.

Deller, M., 1979, Are amblyopic man and ape related? *Trends Neurosci.* **3:**216–218.

DeMonasterio, F. M., 1978*a*, Properties of concentrically organized X and Y ganglion cells of macaque retina, *J. Neurophysiol.* **41:**1394–1417.

DeMonasterio, F. M., 1978*b*, Properties of ganglion cells with atypical receptive-field organization in retina of macaques, *J. Neurophysiol.* **41:**1435–1448.

DeMonasterio, F. M., and P. Gouras, 1975, Functional properties of ganglion cells of the rhesus monkey retina, *J. Physiol. (London)* **251:**167–195.

DeMonasterio, F. M., P. Gouras, and D. J. Tolhurst, 1976, Spatial summation, response pattern and conduction velocity of ganglion cells of the rhesus monkey retina, *Vision Res.* **16:**674–678.

Deol, M. S., and S. Gluecksohn-Waelsch, 1979, The role of the inner hair cells in hearing, *Nature (London)* **278:**250–252.

Derrington, A. M., and M. J. Hawken, 1981, Spatial and temporal properties of cat geniculate neurones after prolonged deprivation, *J. Physiol. (London)* **304:**107–120.

Derrington, A. M., P. Lennie, and M. J. Wright, 1979, The mechanism of peripherally evoked response in retinal ganglion cells, *J. Physiol. (London)* **289:**299–310.

Descartes, R., 1686, Meditations on the first philosophy, in: *Seventeenth Century Rationalism* (I. Cantor, ed.) Ginn (Blaisdell), Boston, 1969.

Dews, P. B., and T. N. Wiesel, 1970, Consequences of monocular deprivation on visual behaviour in kittens, *J. Physiol. (London)* **206:**437–455.

Diamond, I. T., and W. C. Hall, 1969, Evolution of neocortex, *Science* **164:**251–262.

Dodt, E., 1956, Geschwindigkeit der Nervenleitung innerhalb der Netzhaut, *Experientia* **12:**34.

Donaldson, I. M. L., and J. R. G. Nash, 1975, The effect of a chronic lesion in cortical area 17 on the visual responses of units in area 18 of the cat, *J. Physiol. (London)* **245:**325–332.

Dostrovsky, J. O., S. Jabbur, and J. Millar, 1978, Neurones in cat gracile nucleus with both local and widefield inputs, *J. Physiol. (London)* **278:**365–375.

Doty, R. W. 1958, Potentials evoked in cat cerebral cortex by diffuse and by punctiform photic stimuli, *J. Neurophysiol.* **21:**437–464.

Doty, R. W., 1971, Survival of pattern vision after removal of striate cortex in the adult cat, *J. Comp. Neurol.* **143:**341–369.

Dow, B. M., 1974, Functional classes of cells and their laminar distribution in monkey visual cortex, *J. Neurophysiol.* **37:**927–946.

Dräger, U., and C. Olsen, 1980, Origins of crossed and uncrossed retinal projection in pigmented and albino mice, *J. Comp. Neurol.* **191:**383–412.

Dreher, B., and L. Cottee, 1975, Visual receptive-field properties of cells in area 18 before and after acute lesions in area 17, *J. Neurophysiol.* **38**:735-750.

Dreher, B., and K. J. Sanderson, 1973, Receptive field analysis: Responses to moving visual contours by single lateral geniculate neurones in the cat, *J. Physiol. (London)* **234**:95-118.

Dreher, B., and A. J. Sefton, 1975, Receptive field properties of cells in the cat's medial interlaminar nucleus (MIN), *Proc. Aust. Physiol. Pharmarcol. Soc.* **6**:209.

Dreher, B., and A. J. Sefton, 1979, Properties of neurones in cat's dorsal lateral geniculate nucleus: A comparison between medial interlaminar and laminated parts of the nucleus, *J. Comp. Neurol.* **183**:47-63.

Dreher, B., and J. M. S. Winterkorn, 1974, Receptive field properties of neurons in cat's cortical area 17 before and after acute lesions to area 18, *Proc. Aust. Physiol. Pharmacol. Soc.* **5**:63P.

Dreher, B., and B. Zernicki, 1969, Visual fixation reflex: Behavioural properties and neural mechanism, *Acta Biol. Exp. (Warsaw)* **29**:359-383.

Dreher, B., Y. Fukada, and R. W. Rodieck, 1976, Identification, classification and anatomical segregation of cells with X-like and Y-like properties in the lateral geniculate nucleus of old-world primates, *J. Physiol. (London)* **258**:433-452.

Dreher, B., P. Hale, and A. G. Leventhal, 1978, Correlation between receptive field properties and afferent conduction velocity of cells in cat visual cortex: Evidence for different thalamic inputs to areas 17, 18 and 19, *Proc. Aust. Physiol. Pharmacol. Soc.* **9**:60P.

Dreher, B., A. G. Leventhal, and P. T. Hale, 1980, Geniculate input to cat visual cortex: A comparison of area 19 with areas 17 and 18, *J. Neurophysiol.* **44**:804-826.

Duke-Elder, S., and K. Wybar, 1973, *System of Ophthalmology,* Vol. VII, *Ocular Motility and Strabismus,* Mosby, St. Louis.

Dürsteler, M. R., L. J. Garey, and J. A. Movshon, 1976, Reversal of the physiological effects of monocular deprivation in the kitten's lateral geniculate nucleus, *J. Physiol. (London)* **261**:175-184.

Dykes, R. W., 1978, The anatomy and physiology of the somatic sensory cortical regions, *Prog. Neurobiol.* **10**:33-88.

Dykes, R. W., 1983, Parallel processing of cutaneous information in the somatosensory cortex, *Brain Res. Rev.* (in press).

Dykes, R. W., C.-S. Lin, and R. Rehman, 1980a, Thalamocortical projection nuclei for modality-specific regions of somatosensory cortex: Demonstration of different projections for areas 3a and 3b of the cat, *Physiologist* **23**:73.

Dykes, R. W., D. D. Rasmusson, and P. B. Hoeltzell, 1980b, Organisation of primary somatosensory cortex in the cat, *J. Neurophysiol.* **43**:1527-1546.

Dykes, R. W., M. Sur, M. M. Merzenich, J. H. Kaas, and R. J. Nelson, 1981, Regional segregation of neurons responding to quickly adapting, slowly adapting, deep and Pacinian receptors within thalamic ventroposterior lateral and ventroposterior inferior nuclei in the squirrel monkey (*Saimiri sciureus*), *Neuroscience* **6**:1687-1692.

Eccles, J. C., 1975, Under the spell of the synapse, in: *The Neurosciences: Paths of Discovery* (F. G. Worden, J. P. Swazey, and G. Adelman, eds.), MIT Press, Cambridge, Mass.

Edwards, S. B., A. C. Rosenquist, and L. A. Palmer, 1974, An autoradiographic study of the ventral lateral geniculate projections in the cat, *Brain Res.* **72**:282-287.

Eggers, H. M., and C. Blakemore, 1978, Physiological basis of an isometric amblyopia, *Science* **201**:264-267.

Enroth-Cugell, C., and J. G. Robson, 1966, The contrast sensitivity of retinal ganglion cells of the cat, *J. Physiol. (London)* **187**:517-552.

Enroth-Cugell, C., and R. M. Shapley, 1973, Adaptation and dynamics of cat retinal ganglion cells, *J. Physiol. (London)* **233**:271-309.

Erickson, R. P., E. Covey, and G. Doetsch, 1980, Neuron and stimulus typologies in the rat gustatory system, *Brain Res.* **196**:513-519.

Erlanger, J., 1927, The interpretation of the action potential in cutaneous and muscle nerves, *Am. J. Physiol.* **82**:644-655.

Erlanger, J., and H. S. Gasser, 1937, *Electrical Signs of Nervous Activity,* University of Pennsylvania Press, Philadelphia.

Erlanger, J., G. H. Bishop, and H. S. Gasser, 1926, Experimental analysis of the simple action potential wave in nerve by the cathode ray oscillograph, *Am. J. Physiol.* **81**:537-572.

Evinger, C., and A. F. Fuchs, 1978, Saccadic, smooth-pursuit and optokinetic eye movements of the trained cat, *J. Physiol. (London)* **285:**209–229.

Ewert, J. P., 1974, Neuronal basis of visually guided behaviour, *Sci. Am.* **230:**34–42.

Ewert, J. P., and F. Hock, 1972, Movement-sensitive neurones in the toad's retina, *Exp. Brain Res.* **16:**41–59.

Eysel, U. T., O.-J. Grüsser, and K.-P. Hoffmann, 1979, Monocular deprivation and the signal transmission by X- and Y-neurons of the cat lateral geniculate nucleus, *Exp. Brain Res.* **34:**521–539.

Famiglietti, E. V., 1970, Dendro-dendritic synapses in the lateral geniculate nucleus of the cat, *Brain Res.* **20:**181–191.

Famiglietti, E. V., and H. Kolb, 1976, Structural basis for on- and off-centre responses in retinal ganglion cells, *Science* **194:**193–195.

Famiglietti, E. V., and A. Peters, 1972, The synaptic glomerulus and the intrinsic neuron in the dorsal lateral geniculate nucleus of the cat, *J. Comp. Neurol.* **144:**285–334.

Farmer, S. G. and R. W. Rodieck, 1982, Ganglion cells of the cat accessory optic system: Morphology and retinal topography, *J. Comp. Neurol.* **205:**190–198.

Feinberg, T. E., T. Pasik, and P. Pasik, 1978, Extrageniculostriate vision in the monkey. VI. Visually guided accurate reaching behaviour, *Brain Res.* **152:**422–428.

Ferrington, D. G., and M. J. Rowe, 1980, Differential contributions to coding of cutaneous vibratory information by cortical somatosensory areas I and II, *J. Neurophysiol.* **43:**310–331.

Ferster, D., and S. LeVay, 1978, The axonal arborisations of lateral geniculate neurons in the striate cortex of the cat, *J. Comp. Neurol.* **182:**923–944.

Feyerabend, P., 1975, Imre Lakatos, *Brit. J. Philos Sci.* **26:**1–18.

Finlay, B. L., P. H. Schiller, and S. F. Volman, 1976, Quantitative studies of single-cell properties in monkey striate cortex. IV. Corticotectal cells, *J. Neurophysiol.* **39:**1352–1361.

Flood, D. G., and P. D. Coleman, 1979, Demonstration of orientation columns of 14 C 2-deoxyglucose in a cat reared in a striped environment, *Brain Res.* **173:**538–542.

Foote, W. E., E. Taber-Pierce, and L. Edwards, 1978, Evidence for a retinal projection to the midbrain raphe of the cat, *Brain Res.* **156:**135–140.

Foucault, M., 1970, *The Order of Things: An Archaeology of the Human Sciences,* Tavistock, London.

Freeman, R. E. (ed.), 1979, *Developmental Neurobiology of Vision,* Plenum Press, New York.

Freeman, R. D., and J. D. Pettigrew, 1973, Alteration of visual cortex from environmental asymmetries, *Nature (London)* **246:**359–360.

Fregnac, Y., and M. Imbert, 1978, Early development of visual cortical cells in normal and dark-reared kittens: Relationship between orientation selectivity and ocular dominance, *J. Physiol. (London)* **278:**27–44.

Friedlander, M. J., C.-S. Lin, and S. M. Sherman, 1979, Structure of physiologically identified X and Y cells in the cat's lateral geniculate nucleus, *Science* **204:**1114–1117.

Friedlander, M. J., C.-S. Lin, L. R. Stanford, and S. M. Sherman, 1981, Morphology of functionally identified neurons in the lateral geniculate nucleus of the cat, *J. Neurophysiol.* **46:**80–130.

Fries, W., 1981, The projection from the lateral geniculate nucleus to the prestriate cortex in the macaque monkey, *Proc. R. Soc. London Ser. B* **213:**73–80.

Fukada, Y., 1971, Receptive field properties of cat optic nerve fibres with special reference to conduction velocity, *Vision Res.* **11:**209–226.

Fukada, Y., and H. Saito, 1971, The relationship between response characteristics to flicker stimulation and receptive field organisation in the cat's optic nerve fibres, *Vision Res.* **11:**227–240.

Fukada, Y., and H. Saito, 1972, Phasic and tonic cells in the cat's lateral geniculate nucleus, *Tohoku J. Exp. Med.* **106:**209–210.

Fukada, Y., K. Motokawa, A. C. Norton, and K. Tasaki, 1966, Functional significance of conduction velocity in the transfer of information in the optic nerve of the cat, *J. Neurophysiol.* **29:**698–714.

Fukuda, Y., 1973, Differentiation of principal cells of the rat lateral geniculate body into two groups: Fast and slow cells, *Exp. Brain Res.* **17:**242–260.

Fukuda, Y., 1977, A three group classification of rat retinal ganglion cells: Histological and physiological studies, *Brain Res.* **119:**327–344.

Fukuda, Y., and K. Iwama, 1978, Visual receptive field properties of single cells in the rat superior colliculus, *Jpn. J. Physiol.* **28:**385–400.

Fukuda, Y., and J. Stone, 1974, Retinal distribution and central projections of Y-, X- and W-cells of the cat's retina, *J. Neurophysiol.* **37:**749–772.

Fukuda, Y., and J. Stone, 1976, Evidence of differential inhibitory influences on X- and Y-type relay cells in the cat's lateral geniculate nucleus, *Brain Res.* **113:**188–196.

Fukuda, Y., and M. Sugitani, 1974, Cortical projections of two types of principal cells of the rat lateral geniculate body, *Brain Res.* **67:**157–161.

Fukuda, Y., D. A. Suzuki, and K. Iwama, 1978, Four group classification of the rat superior collicular cells responding to optic nerve stimulation, *Jpn. J. Physiol.* **28:**367–384.

Fukuda, Y., I. Sumitomo, M. Sugitani, and K. Iwama, 1979, Receptive field properties of cells in the dorsal part of the albino rat's lateral geniculate nucleus, *Jpn. J. Physiol.* **29:**283–307.

Fuster, J. M., and J. P. Jervey, 1981, Inferotemporal neurons distinguish and retain behaviorally relevant features of visual stimuli, *Science* **212:**952–954.

Ganser, S., 1882, Ueber die periphere und centrale Anordnung der Sehnervenfasern and uber das Corpus bigeminum anterius, *Arch. Psychiatr. Nervenkr.* **13:**341–381.

Ganz, L., M. Fitch, and J. A. Satterberg, 1968, The selective effect of visual deprivation on receptive field shape determined electrophysiologically, *Exp. Neurol.* **22:**614–637.

Garey, L. J., and C. Blakemore, 1977, The effects of monocular deprivation on different neuronal classes in the lateral geniculate nucleus of the cat, *Exp. Brain Res.* **28:**259–278.

Garey, L. J., and T. P. S. Powell, 1967, The projection of the lateral geniculate nucleus upon the cortex in the cat, *Proc. R. Soc. London Ser. B* **169:**107–126.

Garey, L. J., and T. P. S. Powell, 1968, The projection of the retina in the cat, *J. Anat.* **102:**189–222.

Garey, L. J., and T. P. S. Powell, 1971, An experimental study of the termination of the lateral geniculo-cortical pathway in the cat and monkey, *Proc. R. Soc. London Ser. B* **179:**41–63.

Gartner, S., 1951, Ocular pathology in the chiasmal syndrome, *Am. J. Ophthalmol.* **34:**593–596.

Gasser, H. S., and J. Erlanger, 1929, The role of fiber size in the establishment of a nerve block by pressure or cocaine, *Am. J. Physiol.* **88:**581–591.

Gaze, R. M., and J. Jacobson, 1963, 'Convexity detectors' in the frog's visual system, *J. Physiol. (London)* **169:**1–3P.

Gaze, R. M., and E. G. Keating, 1969, The depth distribution of visual units in the tectum of the frog following regeneration of the optic nerve, *J. Physiol. (London)* **200:**128–129P.

Geisert, E. E., 1976, The use of tritiated horseradish peroxidase for defining neuronal pathways: A new approach, *Brain Res.* **117:**130–135.

Geisert, E. E., 1980, Cortical projections of the lateral geniculate nucleus in the cat, *J. Comp. Neurol.* **190:**793–812.

Geisert, E. E., P. D. Spear, and A. Langetsmo, 1980, Return of Y-cells in the lateral geniculate nucleus of monocularly deprived cats, *Soc. Neurosci. Abstr.* **6:**269.3.

Ghiselin, M. T., 1969, The triumph of the Darwinian method, *Syst. Zool.* **16:**289–295.

Gibson, A., J. Baker, G. Mower, and M. Glickstein, 1978, Corticopontine cells in area 18 of the cat, *J. Neurophysiol.* **41:**484–495.

Gilbert, C. D., 1977, Laminar differences in receptive field properties in cat primary visual cortex, *J. Physiol. (London)* **268:**81–106.

Gilbert, C. D., and J. P. Kelly, 1975, The projections of cells in different layers of the cat's visual cortex, *J. Comp. Neurol.* **163:**81–107.

Gilbert, C. D., and T. N. Wiesel, 1979, Morphology and intracortical projections of functionally characterised neurones in the cat visual cortex, *Nature (London)* **280:**120–125.

Giolli, R. A., and M. D. Guthrie, 1969, The primary optic projections in the rabbit: An experimental degeneration study, *J. Comp. Neurol.* **136:**99–126.

Giolli, R. A., and L. C. Towns, 1980, A review of axon collateralization in the mammalian visual system, *Brain Behav. Evol.* **17:**364–390.

Glickstein, M., and D. Whitteridge, 1976, Degeneration of layer III pyramidal cells in area 18 following destruction of callosal input, *Philos. Trans. R. Soc. London Ser. B* **272:**487–536.

Glickstein, M., R. A. King, J. Miller, and M. Berkley, 1967, Cortical projections from the dorsal lateral geniculate nucleus of cats, *J. Comp. Neurol.* **130:**55–76.

Gould, S. J., and R. C. Lewontin, 1979, The spandrels of San Marco and the Panglossian paradigm: A critique of the adaptationist programme, *Proc. R. Soc. London Ser. B* **205:**581–598.

Gouras, P., 1969, Antidromic responses of orthodromically identified ganglion cells in monkey retina, *J. Physiol. (London)* **204**:407–419.

Granit, R., 1955, Centrifugal and antidromic effects on ganglion cells of retina, *J. Neurophysiol.* **18**:388–411.

Granit, R., and E. Marg, 1958, Conduction velocities in rabbit's optic nerve, *Am. J. Ophthalmol.* **46**:223–231.

Griffin, A. E., and W. Burke, 1974, The distribution and nerve fibre groups in the optic tract and lateral geniculate nucleus of *Macaca irus*, *Proc. Aust. Physiol. Pharmacol. Soc.* **5**:74P.

Grossman, A., A. R. Lieberman, and K. E. Webster, 1973, A Golgi study of the rat dorsal lateral geniculate nucleus, *J. Comp. Neurol.* **150**:441–466.

Grüsser-Cornehls, U., O. J. Grüsser, and T. H. Bullock, 1963, Unit responses in the frog's tectum to moving and non-moving visual stimuli, *Science* **141**:820–821.

Guillery, R. W., 1966, A study of Golgi preparations from the dorsal lateral geniculate nucleus of the cat, *J. Comp. Neurol.* **128**:21–50.

Guillery, R. W., 1969a, An abnormal retino-geniculate projection in Siamese cats, *Brain Res.* **14**:739–741.

Guillery, R. W., 1969b, A quantitative study of synaptic interconnections in the dorsal lateral geniculate nucleus of the cat, *Z. Zellforsch. Mikrosk, Anat.* **96**:39–48.

Guillery, R. W., 1970, The laminar distribution of retinal fibres in the dorsal lateral geniculate nucleus of the cat, *J. Comp. Neurol.* **138**:339–368.

Guillery, R. W., 1972, Binocular competition in the control of geniculate cell growth, *J. Comp. Neurol.* **144**:117–130.

Guillery, R. W., 1973, Quantitative studies of transneuronal atrophy in the dorsal lateral geniculate nucleus of cats and kittens, *J. Comp. Neurol.* **149**:423–438.

Guillery, R. W., and M. Colonnier, 1970, Synaptic patterns in the dorsal lateral geniculate nucleus of the monkey, *Z. Zellforsch. Mikrosk. Anat.* **103**:90–108.

Guillery, R. W., and D. J. Stelzner, 1970, The differential effects of unilateral lid closure on the monocular and binocular segments of the dorsal lateral geniculate nucleus in the cat, *J. Comp. Neurol.* **139**:413–422.

Guillery, R. W., and J. H. Kaas, 1971, A study of normal and congenitally abnormal retinogeniculate projections in cats, *J. Comp. Neurol.* **143**:73–100.

Guillery, R. W., and M. D. Oberdorfer, 1977, A study of fine and coarse retinofugal axons terminating in the geniculate C-laminae and in the medial interlaminar nucleus of the mink, *J. Comp. Neurol.* **176**:515–526.

Guillery, R. W., and G. L. Scott, 1971, Observations on synaptic patterns in the dorsal lateral geniculate nucleus of the cat: The C laminae and the perikaryal synapses, *Exp. Brain Res.* **12**:184–203.

Guillery, R. W., E. E. Geisert, E. H. Polley, and C. A. Mason, 1980, An analysis of the retinal afferents to the cat's medial interlaminar nucleus and to its rostral thalamic extension, the "geniculate wing," *J. Comp. Neurol.* **194**:117–142.

Guiloff, G. D., 1980, Class I retinal ganglion cells in the toad, *Bufo spinulosus*, *Vision Res.* **20**:549–553.

Haight, J. R., and L. Neylon, 1978, Morphological variation in the brain of the marsupial brush-tailed possum, *Trichosurus vulpecula*, *Brain Behav. Evol.* **15**:415–445.

Hale, P. T., and A. J. Sefton, 1978, A comparison of the visual and electrical response properties of cells in the dorsal and ventral lateral geniculate nuclei, *Brain Res.* **153**:591–595.

Hale, P. T., A. J. Sefton, and B. Dreher, 1976, Functional properties of relay cells of the rat dorsal lateral geniculate nucleus, *Proc. Aust. Physiol. Pharmacol. Soc.* **7**:48P.

Hale, P., A. J. Sefton, and B. Dreher, 1979, A correlation of receptive field properties with conduction velocity of cells in the rat's retino-geniculo-striate pathway, *Exp. Brain Res.* **35**:425–442.

Harting, J. K., and R. W. Guillery, 1976, Organization of retinocollicular pathways in the cat, *J. Comp. Neurol.* **166**:133–144.

Hartline, H. K., 1938, The response of single optic nerve fibres of the vertebrate eye to illumination of the retina, *Am. J. Physiol.* **121**:400–415.

Hartline, H. K., 1940a, The receptive fields of optic nerve fibres, *Am. J. Physiol.* **130**:690–699.

Hartline, H. K., 1940b, The effects of spatial summation in the retina on the excitation of the fibers of the optic nerve, *Am. J. Physiol.* **130**:700–711.

Harvey, A., 1980, The afferent connexions and laminar distribution of cells in area 18 of the cat, *J. Physiol. (London)* **302:**483–505.

Harwerth, R. S., R. L. Boltz, and E. L. Smith, 1980, Psychophysical evidence for sustained and transient channels in the monkey visual system, *Vision Res.* **20:**15–22.

Hassler, R., 1966, Comparative anatomy of the central visual system in day- and night-active primates, in: *Evolution of the Forebrain* (R. Hassler and H. Stephan, eds.), pp. 419–434, Thieme, Stuttgart.

Hayashi, Y., 1969, Recurrent collateral inhibition of visual cortical cells projecting to superior colliculus in cats, *Vision Res.* **9:**1367–1380.

Hayashi, Y., I. Sumitomo, and K. Iwama, 1967, Activation of lateral geniculate neurons by electrical stimulation of superior colliculus, *Jpn. J. Physiol.* **17:**638–651.

Hayhow, W. R., 1958, The cytoarchitecture of the lateral geniculate body in the cat in relation to the distribution of crossed and uncrossed optic fibers, *J. Comp. Neurol.* **110:**1–48.

Hayhow, W. R., 1959, An experimental study of the accessory optic fiber system in the cat, *J. Comp. Neurol.* **113:**281–314.

Hayhow, W. R., 1966, The accessory optic system in the marsupial phalanger, *Trichosurus vulpecula, J. Comp. Neurol.* **126:**653–672.

Hayhow, W. R., C. Webb, and A. Jervie, 1960, The accessory optic fibre system in the rat, *J. Comp. Neurol.* **115:**186–216.

Headon, M. P. and T. P. S. Powell, 1973, Cellular changes in the lateral geniculate nucleus of infant monkeys after suture of the eyelids, *J. Anat.* **116:**135–145.

Headon, M. P., and T. P. S. Powell, 1978, The effect of bilateral eye closure upon the lateral geniculate nucleus in infant monkeys, *Brain Res.* **143:**147–154.

Hebel, R., and H. Holländer, 1979, Size and distribution of ganglion cells in the bovine retina, *Vision Res.* **19:**667–674.

Hedreen, J., 1969, Patterns of axon terminal degeneration seen after optic nerve section in cat, *Anat. Rec.* **163:**198.

Heinbecker, P., G. H. Bishop, and J. O'Leary, 1933, Pain and touch fibers in peripheral nerves, *Arch. Neurol. Psychiatry* **29:**771–789.

Heinbecker, P., G. H. Bishop, and J. O'Leary, 1934, Analysis of sensation in terms of the nerve impulse, *Arch. Neurol. Psychiatry* **31:**34–53.

Heitlander, H., and K.-P. Hoffmann, 1978, The visual field of monocularly deprived cats after later closure or enucleation of the non-deprived eye, *Brain Res.* **145:**153–160.

Hendrickson, A. E., N. Wagoner, and W. M. Cowan, 1972, An autoradiographic and electron microscopic study of retino-hypothalamic connections, *Z. Zellforsch. Mikrosk. Anat.* **135:**1–26.

Hendrickson, A. E., J. R. Wilson, and M. P. Ogren, 1978, The neuroanatomical organization of pathways between the dorsal lateral geniculate nucleus and visual cortex in Old-world and New-world primates, *J. Comp. Neurol.* **182:**123–136.

Henry, G. H., 1977, Receptive field classes of cells in the striate cortex of the cat, *Brain Res.* **133:**1–28.

Henry, G. H., P. O. Bishop, and B. Dreher, 1974, Orientation, axis and direction as stimulus parameters for striate cells, *Vision Res.* **14:**767–777.

Henry, G. H., A. R. Harvey, and J. S. Lund, 1979, The afferent connections and laminar distribution of cells in the cat striate cortex, *J. Comp. Neurol.* **187:**725–744.

Hess, R., F. W. Campbell, and T. Greenhalgh, 1978, On the nature of the neural abnormality in human amblyopia: Neural sensitivity loss, *Pflügers Arch.* **377:**201–207.

Hickey, T. L., 1980, Development of the dorsal lateral geniculate nucleus in normal and visually deprived cats, *J. Comp. Neurol.* **189:**467–481.

Hickey, T. L., 1981, The developing visual system, *Trends Neurosci.* **4:**41–44.

Hickey, T. L., and R. W. Guillery, 1974, An autoradiographic study of retinogeniculate pathway in the cat and the fox, *J. Comp. Neurol.* **156:**239–254.

Hickey, T. L., and R. W. Guilery, 1979, Variability in the laminar patterns in the human lateral geniculate nucleus, *J. Comp. Neurol.* **183:**221–246.

Hickey, T. L., P. D. Spear, and K. E. Kratz, 1977, Quantitative studies of cell size in the cat's dorsal lateral geniculate nucleus following visual deprivation, *J. Comp. Neurol.* **172:**265–282.

Hirsch, H. V. B., 1972, Visual perception in cats after environmental surgery, *Exp. Brain Res.* **15:**405–423.

Hirsch, H. V. B., and A. G. Leventhal, 1978, Functional modification of the developing visual system, Chapt. 7, *Handbook of Sensory Physiology*, Vol. IX, *Development of Sensory Systems* (M. Jacobson, ed.), Springer-Verlag, Berlin.

Hirsch, H. V. B., and D. N. Spinelli, 1970, Visual experience modifies distribution of horizontally and vertically oriented receptive fields in cats, *Science* **168**:869–871.

Hochstein, S., and R. M. Shapley, 1976, Quantitative analysis of retinal ganglion cell classification, *J. Physiol. (London)* **262**:237–264.

Hoffmann, K.-P., 1972, The retinal input to the superior colliculus in the cat, *Invest. Ophthalmol.* **11**:467–473.

Hoffmann, K.-P., 1973, Conduction velocity pathways from retina to superior colliculus in the cat: A correlation with receptive filed properties, *J. Neurophysiol.* **36**:409–424.

Hoffman, K.-P., and M. Cynader, 1977, Functional aspects of plasticity in the visual system of adult cats after early monocular deprivation, *Philos. Trans. R. Soc. London Ser. B* **278**:411–424.

Hoffman, K.-P., and H. Holländer, 1978, Physiological changes in cells of the lateral geniculate nucleus in monocularly-deprived and reverse-sutured cat, *J. Comp. Neurol.* **177**:145–158.

Hoffmann, K.-P., and A. Schoppmann, 1975, Retinal input to direction selective cells in the nucleus tractus opticus of the cat, *Brain Res.* **99**:359–366.

Hoffmann, K.-P., and S. M. Sherman, 1974, Effects of early monocular deprivation on visual input to cat superior colliculus, *J. Neurophysiol.* **37**:1276–1286.

Hoffmann, K.-P., and S. M. Sherman, 1975, Effects of early binocular deprivation on visual input to cat superior colliculus, *J. Neurophysiol.* **38**:1049–1059.

Hoffmann, K.-P., and J. Stone, 1971, Conduction velocity of afferents to cat visual cortex: A correlation with cortical receptive field properties, *Brain Res.* **32**:460–466.

Hoffmann, K.-P., J. Stone, and S. M. Sherman, 1972, Relay of receptive field properties in the dorsal lateral geniculate nucleus of the cat, *J. Neurophysiol.* **35**:518–531.

Hoffmann, K.-P., K. Behrend, and A. Schoppmann, 1976, A direct afferent visual pathway from the nucleus of the optic tract to the inferior olive in the cat, *Brain Res.* **115**:150–153.

Hoffmann, K.-P., H. Heitlander, P. Lippert, and R. Sireteanu, 1978, Psychophysical and neurophysiological investigations of the effects of early visual deprivation in the cat, *Arch. Ital. Biol.* **116**:452–462.

Hohmann, A., and O. D. Creutzfeldt, 1975, Squint and the development of binocularity in humans, *Nature (London)* **254**:613–615.

Hokoc, J. N., and E. Oswald-Cruz, 1979, A regional specialization in the opossum's retina: Quantitative analysis of the ganglion cell layer, *J. Comp. Neurol.* **183**:385–396.

Holden, A. L., 1969, Receptive field properties of retinal cells and tectal cells in the pigeon, *J. Physiol. (London)* **201**:56–57P.

Holden, A. L., 1977a, Responses of directional ganglion cells in the pigeon retina, *J. Physiol. (London)* **270**:253–269.

Holden, A. L., 1977b, Concentric receptive fields of pigeon ganglion cells, *Vision Res.* **17**:545–554.

Holländer, H., 1974, On the origin of the corticotectal projections in the cat, *Exp. Brain Res.* **21**:433–439.

Holländer, H., 1978, Identification of cytoplasmic laminar bodies in neurons of the cat lateral geniculate nucleus by phase contrast microscopy, *Microsc. Acta.* **81**:131–135.

Holländer, H., and D. Sanides, 1976, The retinal projection to the ventral part of the lateral geniculate nucleus: An experimental study with silver impregnation methods and axoplasmic protein tracing, *Exp. Brain Res.* **26**:329–342.

Holländer, H., and H. Vanegas, 1977, The projection from the lateral geniculate nucleus onto the visual cortex in the cat: A quantitative study with horseradish peroxidase, *J. Comp. Neurol.* **173**:519–536.

Hornung, J. P., and L. J. Garey, 1980, A direct pathway from thalamus to visual callosal neurons in cat, *Exp. Brain Res.* **38**:121–123.

Hoyt, W. F., E. N. Rios-Montenegro, M. M. Behrens, and R. J. Eckelhoff, 1972, Homonymous hemioptic hypoplasia, *Brit. J. Ophthalmol.* **56**:537–545.

Hubel, D. H., 1975, An autoradiographic study of the retino-cortical projections in the tree shrew (*Tupaia glis*), *Brain Res.* **96**:41–50.

Hubel, D. H., and T. N. Wiesel, 1960, Receptive fields of optic nerve fibres in the spider monkey, *J. Physiol. (London)* **154**:572–580.

Hubel, D. H., and T. N. Wiesel, 1961, Integrative action in the cat's lateral geniculate body, *J. Physiol. (London)* **155:**385–398.

Hubel, D. H., and T. N. Wiesel, 1962, Receptive fields, binocular interaction and functional architecture in the cat's striate cortex, *J. Physiol. (London)* **160:**106–154.

Hubel, D. H., and T. N. Wiesel, 1963, Receptive fields of cells in striate cortex of very young, visually inexperienced kittens, *J. Neurophysiol.* **26:**994–1002.

Hubel, D. H., and T. N. Wiesel, 1965, Receptive fields and functional architecture in two non-striate visual areas 18 and 19 of the cat, *J. Neurophysiol.* **28:**229–289.

Hubel, D. H., and T. N. Wiesel, 1968, Receptive fields and functional architecture of monkey striate cortex, *J. Physiol. (London)* **195:**215–243.

Hubel, D. H., and T. N. Wiesel, 1971, Aberrant visual projections in the Siamese cat, *J. Physiol. (London)* **218:**33–62.

Hubel, D. H., and T. N. Wiesel, 1972, Laminar and columnar distribution of geniculo-cortical fibers in macaque monkeys, *J. Comp. Neurol.* **146:**421–450.

Hubel, D. H., S. LeVay, and T. N. Wiesel, 1975, Mode of termination of retinotectal fibers in macaque monkey: An autoradiographic study, *Brain Res.* **96:**25–40.

Hubel, D. H., T. N. Wiesel, and S. LeVay, 1977, Plasticity of ocular dominance columns in macaque striate cortex, *Philos. Trans. R. Soc. London Ser. B* **278:**377–410.

Hubel, D. H., T. N. Wiesel, and M. P. Stryker, 1978, Anatomical demonstration of orientation columns in macaque monkey, *J. Comp. Neurol.* **177:**361–380.

Hughes, A., 1971, Topographical relationships between the anatomy and physiology of the rabbit visual system, *Doc. Ophthalmol.* **30:**33–159.

Hughes, A., 1974, A comparison of retinal ganglion cell topography in the plains and tree kangaroo, *J. Physiol. (London)* **244:**61–63P.

Hughes, A., 1975, A quantitative analysis of the cat retinal ganglion cell topography, *J. Comp. Neurol.* **163:**107–128.

Hughes, A., 1977, The topography of vision in mammals of contrasting lifestyles: comparative optics and retinal organisation, in: *Handbook of Sensory Physiology,* Vol. VII/5 (F. Crescitelli, ed.), Springer-Verlag, Berlin.

Hughes, A., 1979, A rose by any other name: On 'Naming of Neurones' by Rowe and Stone, *Brain Behav. Evol.* **16:**52–64.

Hughes, A., 1981, Population magnitudes and distribution of the major modal classes of cat retinal ganglion cells as estimated from HRP filling and a systematic survey of the soma diameter spectra for classical neurones, *J. Comp. Neurol.* **197:**303–340.

Hughes, A., and D. Whitteridge, 1973, The receptive fields and topographical organization of goat retinal ganglion cells, *Vision Res.* **13:**1101–1114.

Hughes, C. P., and S. B. Ater, 1977, Receptive field properties in the ventral lateral geniculate nucleus of the cat, *Brain Res.* **132:**163–166.

Hughes, H. C., 1977, Anatomical and neurobehavioral investigations concerning the thalamo-cortical organization of the rat's visual system, *J. Comp. Neurol.* **175:**311–336.

Hultborn, H., K. Mori, and N. Tsukahara, 1978, The neuronal pathway subserving the pupillary light reflex, *Brain Res.* **159:**255–267.

Hume, D., 1897, *A Treatise of Human Nature,* 2nd ed., p. 275, Oxford University Press (Clarendon), London.

Humphrey, N. K., 1974, Vision in a monkey without striate cortex: A case study, *Perception* 3:241–255.

Huxley, J. S., 1940, *The New Systematics,* Oxford University Press, London.

Ikeda, H., and S. G. Jacobson, 1977, Nasal field loss in cats reared with convergent squint: Behavioral studies, *J. Physiol. (London)* **270:**367–381.

Ikeda, H., and K. E. Tremain, 1978, Amblyopia resulting from penalisation: Neurophysiological studies of kittens reared with atropinisation of one or both eyes, *Brit. J. Ophthalmol.* **62:**21–28.

Ikeda, H., and K. E. Tremain, 1979, Amblyopia occurs in retinal ganglion cells in cats reared with convergent squint without alternating fixation, *Exp. Brain Res.* **35:**559–582.

Ikeda, H., and M. J. Wright, 1972a, Receptive field organisation of 'sustained' and 'transient' retinal ganglion cells which subserve different functional roles, *J. Physiol. (London)* **227:**769–800.

Ikeda, H., and M. M. Wright, 1972b, Differential effects of refractive errors and receptive field organisation of central and peripheral ganglion cells, *Vision Res.* **12**:1465–1476.

Ikeda, H., and M. J. Wright, 1975, Spatial and temporal properties of 'sustained' and 'transient' neurones in area 17 of the cat's visual cortex, *Exp. Brain Res.* **22**:363–383.

Ikeda, H., and M. J. Wright, 1976, Properties of LGN cells in kittens reared with convergent squint: A neurophysiological demonstration of amblyopia, *Brain Res.* **25**:64–77.

Ikeda, H., G. T. Plant, and K. E. Tremain, 1977, Nasal field loss in kittens reared with convergent squint: Neurophysiological and morphological studies of the lateral geniculate nucleus, *J. Physiol. (London)* **270**:345–366.

Ikeda, H., K. E. Tremain, and G. Einon, 1978, Loss of spatial resolution of lateral geniculate nucleus neurones raised with convergent squint produced at different stages in development, *Exp. Brain Res.* **31**:207–220.

Illing, R.-B., 1980, Axonal bifurcation of cat retinal ganglion cells as demonstrated by retrograde double labelling with fluorescent dyes, *Neurosci. Lett.* **19**:125–130.

Ingle, D., 1967, Two visual mechanisms underlying the behavior of fish, *Psychol. Forsch.* **31**:44–51.

Itoh, K., N. Mizuno, T. Sugimoto, S. Nomura, Y, Nakamura, and A. Konishi, 1979, A cerebello-pulvino-cortical and a retino-pulvino-cortical pathway in the cat as revealed by the use of anterograde and retrograde transport of horseradish peroxidase, *J. Comp. Neurol.* **187**:349–358.

Itoh, K., M. Conley, and I. T. Diamond, 1981, Different distributions of large and small ganglion cells in the cat after HRP injections of single layers of the lateral geniculate body and the superior colliculus, *Brain Res.* **207**:147–152.

Jakiela, H. G., C. Enroth-Cugell, and R. Shapley, 1976, Adaptation and dynamics in X- and Y-cells of the cat retina, *Exp. Brain Res.* **24**:335–342.

Jeffery, G., A. Cowey, and H. G. J. M. Kuypers, 1981, Bifurcating retinal ganglion cell axons in the rat, demonstrated by retrograde double labelling, *Exp. Brain Res.* **44**:34–40.

Jones, E. G., and R. Porter, 1980, What is area 3a? *Brain Res. Rev.* **2**:1-3.

Jordan, H., and H. Holländer, 1974, The structure of the ventral part of the lateral geniculate nucleus, *J. Comp. Neurol.* **145**:259–271.

Jung, R., 1973, Visual perception and neurophysiology, in: *Handbook of Sensory Physiology,* Vol. VII/3A (R. Jung, ed.), Chap. 1, Springer-Verlag, Berlin.

Kaas, J. N., M. F. Huerta, J. T. Weber, and J. K. Harting, 1978, Patterns of retinal termination and laminar organization of the lateral geniculate nucleus of primates, *J. Comp. Neurol.* **182**:517–554.

Kalil, R., and I. Worden, 1978, Cytoplasmic laminated bodies in the lateral geniculate nucleus of normal and dark reared cats, *J. Comp. Neurol.* **178**:469–486.

Kawamura, K., and T. Konno, 1979, Various types of corticotectal neurones as demonstrated by means of retrograde axonal transport of horseradish peroxidase, *Exp. Brain Res.,* **35**:161–175.

Kawamura, S., J. M., Sprague, and K. Niimi, 1974, Cortical projections from the visual cortices to the thalamus, pretectum and superior colliculus in the cat, *J. Comp. Neurol.* **158**:339–362.

Kawamura, S., N. Fukushima, and S. Hattori, 1979, Topographical origin and ganglion cell type of the retino-pulvinar projection in the cat, *Brain Res.* **173**:419–429.

Kawamura, S., N. Fukushima, S. Hattori, and M. Kudo, 1980, Laminar segregation of cells of origin of ascending projections from the superficial layers of the superior colliculus in the cat, *Brain Res.* **184**:486–490.

Keating, E. G., 1980, Residual spatial vision in the monkey after removal of striate and preoccipital cortex, *Brain Res.* **187**:271–290.

Keating, M. J., and R. M. Gaze, 1970, Observations on the "surround" properties of the receptive fields of frog retinal ganglion cells, *Q. J. Exp, Physiol.* **55**:129–142.

Keens, J. S., 1981, Aspects of the retinal projection in rats, *Proc. Aust, Physiol. Pharmacol. Soc.* **12**: 163P.

Keesey, U. T., 1969, Visibility of a stabilized target as a function of frequency and amplitude of luminance variation, *J. Opt. Soc. Am.* **59**:604–610.

Keesey, U. T., 1972, Flicker and pattern detection: A comparison of threshold, *J. Opt. Soc. Am.* **62**:446–448.

Kelly, J. P., and C. D. Gilbert, 1975, The projections of different morphological types of ganglion cells in the cat retina, *J. Comp. Neurol.* **163**:65–80.

Keverne, E. B., 1978, Olfaction and taste—Dual systems for sensory processing, *Trends Neurosci.* **1**:32–34.

Khalil, S. H., and G. Szekely, 1976, The development of the ipsilateral retinothalamic projections in the *Xenopus* toad, *Acta Biol. Acad. Sci. Hung.* **27**:253–260.

Kimura, M., T. Shiida, K. Tanaka, and K. Toyama, 1980, Three classes of area 19 cells of the cat classified by their neuronal connectivity and photic responsivness, *Vision Res.* **20**:69–77.

King-Smith, P. E., and J. J. Kulikowski, 1975, Pattern and flicker detection analysed by subthreshold summation, *J. Physiol. (London)* **249**:519–548.

Kinston, W. J., M. A. Vadas, and P. O. Bishop, 1969, Multiple projections of the visual field to the medial portion of the dorsal lateral geniculate nucleus and the adjacent nuclei of the thalamus of the cat, *J. Comp. Neurol.* **136**:295–316.

Kerby, A. W., 1979, The effect of strychnine, bicuculline, and picrotoxin on X and Y cells in the cat retina, *J. Gen. Physiol.* **74**:71–84.

Kirby, A. W., and C. Enroth-Cugell, 1976, The involvement of gamma-aminobutyric acid in the ganglion cell receptive fields, *J. Gen. Physiol.* **68**:465–484.

Kirk, D. L., B. G. Cleland, H. Wässle, and W. R. Levick, 1975, Axonal conduction latencies of cat retinal ganglion cells in central and peripheral retina, *Exp. Brain Res.* **23**:85–90.

Kirk, D. L., B. G. Cleland, H. Wässle, and W. R. Levick, 1976a, Crossed and uncrossed representation of the visual field by brisk-sustained and brisk-transient cat retinal ganglion cells, *Vision Res.* **16**:225–231.

Kirk, D. L., B. G. Cleland, H. Wässle, and W. R. Levick, 1976b, The crossed or uncrossed destination of sluggish-concentric and non-concentric cat retinal ganglion cells, with an overall synthesis of the visual field representation, *Vision Res.* **16**:233–236.

Kirk, D., W. R. Levick, and H. J. Wagner, 1976c, Decussation of optic axons in Siamese cat, *Proc. Aust. Physiol. Pharmacol. Soc.* **7**:153P.

Kluver, H., 1942, Functional significance of the geniculostriate system, *Biol. Symp.* **7**:253–299.

Kolb, H., 1979, The inner plexiform layer in the retina of the cat: Electron microscopic observations, *J. Neurocytol.* **8**:295–329.

Kolb, H., R. Nelson, and A. Mariani, 1981, Amacrine cells, bipolar cells and ganglion cells of the cat retina: A Golgi study, *Vision Res.* **21**:1081–1114.

Kratz, K. E., and P. D. Spear, 1976, Effects of visual deprivation and alterations in binocular competition on responses of striate cortex neurons in the cat, *J. Comp. Neurol.* **170**:141–152.

Kratz, K. E., S. V. Webb, and S. M. Sherman, 1978, Studies of the cat's medial interlaminar nucleus: A subdivision of the dorsal lateral geniculate nucleus, *J. Comp. Neurol.* **181**:601–614.

Kratz, K. E., S. M. Sherman, and R. Kalil, 1979a, Lateral geniculate nucleus in dark-reared cats: Loss of Y-cells without changes in cell size, *Science* **203**:1353–1355.

Kratz, K. E., S. C. Mangel, S. Lehmkuhle, and S. M. Sherman, 1979b, Retinal X- and Y-cells in monocularly lid-sutured cat: Normality of spatial and temporal properties, *Brain Res.* **172**:545–551.

Kruger, J., and B. Fischer, 1973, Strong periphery effect in cat retinal ganglion cells: Excitatory responses in ON- and OFF-centre neurones to single grid displacements, *Exp. Brain Res.* **18**:316–318.

Kuffler, S. W., 1953, Discharge patterns and functional organisation of mammalian retina, *J. Neurophysiol.* **16**:37–38.

Kuffler, S. W., R. Fitzhugh, and H. B. Barlow, 1957, Maintained activity in the cat's retina in light and darkness, *J. Gen. Physiol.* **40**:683–702.

Kuhn, T. S., 1962, The structure of scientific revolutions, in: *International Encyclopedia of Unified Science*, Vol. 2, No. 2, 2nd ed., University of Chicago Press, Chicago.

Kuhn, T. S., 1970, Reflections on my critics, in: *Criticism and the Growth of Knowledge* (I. Lakatos and A. Musgrave, eds.) Cambridge University Press, London.

Kuhn, T. S. 1974, Second thoughts on paradigms, in: *The Structure of Scientific Theories* (I. Suppe, ed.), University of Illinois Press, Urbana.

Kulikowski, J. J., 1971a, Some stimulus parameters affecting spatial and temporal resolution of human vision, *Vision Res.* **11**:83–93.

Kulikowski, J. J., 1971b, Effects of eye movements on the contrast-sensitivity of spatio-temporal patterns, *Vision Res.* **11**:261–273.

Kulikowski, J. J., 1975, Separation of occipital potentials related to the detection of pattern and movement, in: *New Developments of Visual Evoked Potentials of the Human Brain* (J. E. Desmedt, ed.), Oxford University Press, London.

Kulikowski, J. J., and D. J. Tolhurst, 1973, Psychophysical evidence for sustained and transient detectors in human vision, *J. Physiol. (London)* **232**:149–162.

Kulikowski, J. J., P. O. Bishop, and H. Kato, 1979, Sustained and transient responses by cat striate cells to stationary flashing light and dark bars, *Brain Res.* **170**:362–367.

Kupfer, C., 1953, Retinal ganglion cell degeneration following chiasmal lesions in man, *Arch. Ophthalmol.* **70**:256–260.

Lakatos, I., 1970, Falsification and the methodology of scientific research programs, in: *Criticism and the Growth of Knowledge* (I. Lakatos and A. Musgrave, eds.), Cambridge University Press, London.

Lakatos, I., and A. Musgrave (eds.), 1970, *Criticism and the Growth of Knowledge,* Cambridge University Press, London.

Lane, R. H., J. M. Allman, and J. H. Kaas, 1971, Representation of the visual field in the superior colliculus of the grey squirrel (*Sciurus carolinensis*) and the tree shrew (*Tupaia glis*), *Brain Res.* **26**:277–292.

Laties, A. M., and J. M. Sprague, 1966, The projection of optic fibers to the visual centers in the cat, *J. Comp. Neurol.* **127**:35–70.

Lederman, R. J., and W. K. Noell, 1968, Fast-fiber system of rabbit optic nerve, *Vision Res.* **8**:1385–1398.

Lee, B. B., B. G. Cleland, and O. D. Creutzfeldt, 1977, The retinal input to cells in area 17 of the cat's cortex, *Exp. Brain Res.* **30**:527–538.

Lee, C. W. F., H. T. H., and, B. Dreher, 1982, Area 21a and posteromedial lateral suprasylvian (PMLS) of the cat visual cortex: one or two areas? Horseradish peroxidase study, *Proc. Aust. Physiol. Pharmacal. Soc.* **13**:195*P*.

Legg, C. R., and A. Cowey, 1977, The role of the ventral lateral geniculate nucleus and posterior thalamus in intensity discrimination in rats, *Brain Res.* **123**:261–273.

Legge, G. E., 1978, Sustained and transient mechanisms in human vision: Temporal and spatial properties, *Vision Res.* **18**:69–81.

Lehmkuhle, S., K. E. Kratz, S. C. Mangel, and S. M. Sherman, 1978, An effect of early monocular lid suture upon development of X-cells in the cat's lateral geniculate nucleus, *Brain Res.* **157**:346–350.

Lehmkuhle, S., K. E. Kratz, S. C. Mangel, and S. M. Sherman, 1980*a*, Spatial and temporal sensitivity of X- and Y-cells in dorsal lateral geniculate nucleus of the cat, *J. Neurophysiol.* **43**:520–541.

Lehmkuhle, S., K. E. Kratz, S. C. Mangel, and S. M. Sherman, 1980*b*, Effects of early monocular lid suture on spatial and temporal sensitivity of neurons in dorsal lateral geniculate nucleus of the cat, *J. Neurophysiol.* **43**:542–556.

Leicester, J., and J. Stone, 1967, Ganglion, amacrine and horizontal cells of the cat's retina, *Vision Res.* **7**:695–705.

Lennie, P., 1980*a*, Perceptual signs of parallel pathways, *Philos. Trans. R. Soc. London Ser. B* **290**:23–37.

Lennie, P., 1980*b*, Parallel visual pathways, *Vision Res.* **20**:561–594.

Lennox, M. A., 1957, The ON-responses to colored flash in single optic tract fibers of cat: Correlation with conduction velocity, *J. Neurophysiol.* **20**:70–84.

Lettvin, J. Y., H. R. Maturana, W. H. Pitts, and W. S. McCulloch, 1961, Two remarks on the visual system of the frog, in: *Sensory Communication* (W. Rosenblith, ed.), pp. 757–776, MIT Press, Cambridge, Mass.

LeVay, S., and D. Ferster, 1977, Relay cell classes in the lateral geniculate nucleus of the cat and the effects of visual deprivation, *J. Comp. Neurol.* **172**:563–584.

LeVay, S., and C. D. Gilbert, 1976, Laminar patterns of geniculocortical projection in the cat, *Brain Res.* **113**:1–19.

LeVay, S., T. N. Wiesel, and D. H. Hubel, 1979, Effects of reverse suture on ocular dominance columns in rhesus monkeys, *Soc. Neurosci. Abstr.* **5**:2670.

LeVay, S., T. N. Wiesel, and D. H. Hubel, 1980, The development of ocular dominance columns in normal and visually deprived monkeys, *J. Comp. Neurol.* **191**:1–52.

Leventhal, A. G., 1979, Evidence that the different classes of relay cells of the cat's lateral geniculate nucleus terminate in different layers of the striate cortex, *Exp. Brain Res.* **37**:349–372.

Leventhal, A. G., and H. V. B. Hirsch, 1977, Effects of early experience upon orientation sensitivity and binocularity of neurons in visual cortex of cats, *Proc. Natl. Acad. Sci. USA* **74**:1272–1276.

Leventhal, A. G., and H. V. B. Hirsch, 1978, Receptive field properties of neurons in different laminae of visual cortex of the cat, *J. Neurophysiol.* **41**:948–962.

Leventhal, A. G., and H. V. B. Hirsch, 1980, Receptive field properties of different classes of neurons in visual cortex of normal and dark-reared cats, *J. Neurophysiol.* **43**:1111–1132.

Leventhal, A. G., and J. S. Keens, 1978, Aspects of the topographic organisation of thalamic afferent to cortical areas 17, 18 and 19 of the cat, *Proc. Aust. Physiol. Pharmacol. Soc.* **9**:192P.

Leventhal, A. G., J. S. Keens, and I. Törk, 1979, Evidence for an extrageniculate relay from the retina to cortical areas 19 and Clare–Bishop in the cat, *Soc. Neurosci. Abstr.* **5**:793.

Leventhal, A. G., J. S. Keens, and I. Törk, 1980, The afferent ganglion cells and cortical projections of the retinal recipient zone (RRZ) of the cat's 'pulvinar complex,' *J. Comp. Neurol.* **194**:535–554.

Leventhal, A. G., R. W. Rodieck, and B. Dreher, 1981, Retinal ganglion cell classes in cat and Old-world monkey: Morphology and central projections, *Science* **213**:1139–1142.

Levick, W. R., 1967, Receptive fields and trigger features of ganglion cells in the visual streak of the rabbit's retina, *J. Physiol. (London)* **188**:285–307.

Levick, W. R., 1975, Form and function of cat retinal ganglion cells, *Nature (London)* **254**:659–662.

Levick, W. R., 1977, Participation of brisk-transient cells in binocular vision—A hypothesis, *Proc. Aust. Physiol. Pharmacol. Soc.* **8**:9–16.

Levick, W. R., C. W. Oyster, and D. L. Davis, 1967, Evidence that McIlwain's periphery effect is not a stray light artefact, *J. Neurophysiol.* **28**:555–559.

Lewontin, R. C., 1974, *The Genetic Basis of Evolutionary Change,* Columbia University Press, New York.

Lewontin, R. C., 1978, Adaptation, *Sci. Am.* **239**:212–230.

Lin, C.-S., and S. M. Sherman, 1978, Effects of early monocular eyelid suture upon development of relay cell classes in the cat's lateral geniculate nucleus, *J. Comp. Neurol.* **181**:809–832.

Lin, C.-S., M. M. Merzenich, M. Sur, and J. H. Kaas, 1980, Connections of areas 3b and 1 of the parietal somatosensory strip with the ventroposterior nucleus in the owl monkey, *J. Comp. Neurol.* **185**:355–372.

Linskz, A., 1952, *Physiology of the Eye,* Vol. 2, *Vision,* Grune & Stratton, New York.

Lloyd, D. C. P., 1943, Neuron patterns controlling transmission of ipsilateral hind limb reflexes in cat, *J. Neurophysiol.* **6**:293–315.

Loop, M. S., and S. M. Sherman, 1977, Visual discriminations of cats with cortical and tectal lesions, *J. Comp. Neurol.* **174**:79–88.

Lund, J. S., 1973, Organization of neurons in the visual cortex, area 17, of the monkey (*Macaca mulatta*), *J. Comp. Neurol.* **147**:455–496.

Lund, J. S., and R. G. Boothe, 1975, Interlaminar connections and pyramidal neuron organisation in the visual cortex, area 17, of the macaque monkey, *J. Comp. Neurol.* **159**:305–334.

Lund, J. S. R. D. Lund, A. E. Hendrickson, A. H. Bunt, and A. F. Fuchs, 1975, The origin of efferent pathways from the primary visual cortex, area 17, of the Macaque monkey as shown by retrograde transport of horseradish peroxidase, *J. Comp. Neurol.* **164**:287–304.

Lund, J. S., G. H. Henry, C. L. MacQueen, and A. R. Harvey, 1979, Anatomical organization of the primary visual cortex (area 17) of the cat: A comparison with area 17 of the macaque monkey, *J. Comp. Neurol.* **184**:599–618.

Lund, R. D., 1975, Variations in the laterality of central projections of retinal ganglion cells, *Exp. Eye Res.* **21**:193–203.

Lund, R. D., and T. J. Cunningham, 1972, Aspects of synaptic organization of the mammalian lateral geniculate body, *Invest. Ophthalmol.* **11**:291–301.

Lund, R. D., P. W. Land, and J. Boles, 1980, Normal and abnormal uncrossed retinotectal pathways in rats: An HRP study in adults, *J. Comp. Neurol.* **189**:711–720.

McCloskey, D. I., 1979, Kinesthetic sensibility, *Physiol. Rev.* **58**:763–820.

Maciewicz, R. J., 1975, Thalamic afferents to areas 17, 18 and 19 of cat cortex traced with horseradish peroxidase, *Brain Res.* **78**:139–143.

McIlwain, J. T., 1964, Large receptive fields and spatial transformation in the visual system, *Int. Rev. Physiol., Neurophysiol. II* **10**:223–248.

McIlwain, J. T., 1966, Some evidence concerning the physiological basis of the peripheral effect in the cat's retina, *Exp. Brain Res.* **1**:265–271.

McIlwain, J. T., 1973, Topographical relationship in projection from striate cortex to superior colliculus of the cat, *J. Neurophysiol.* **36:**690–701.

McIlwain, J. T., 1976, Large receptive fields and spatial transformations in the visual system, *Int. Rev. Physiol.* **10:**223–248.

McIlwain, J. T., 1977*a*, Topographical organization and convergence in corticotectal projections from areas 17, 18 and 19 in the cat, *J. Neurophysiol.* **40:**189–198.

McIlwain, J. T., 1977*b*, Orientation of slit pupil and visual streak in the eye of the cat, *J. Comp. Neurol.* **175:**337–344.

McIlwain, J. T., 1978*a*, Cat superior colliculus: Extracellular potentials related to W-cell synaptic action, *J. Neurophysiol.* **41:**1343–1358.

McIlwain, J. T., 1978*b*, Properties of cells projecting rostrally from the superficial layers of the cat's superior colliculus, *Brain Res.* **143:**445–457.

McIlwain, J. T., and H. L. Fields, 1971, Superior colliculus: Single unit responses to stimulation of visual cortex in the cat, *Science* **170:**1426–1428.

McIlwain, J. T., and R. B. Lufkin, 1976, Distribution of direct Y-cell inputs to the cat's superior colliculus: Are there spatial gradients? *Brain Res.* **103:**133–138.

McKeon, R. (ed.), 1941, *The Basic Works of Aristotle,* Random House, New York.

MacLeod, D. I., 1978, Visual sensitivity, *Annu. Rev. Psychol.* **29:**613–645.

Maffei, L., and A. Fiorentini, 1973, The visual cortex as a spatial frequency analyser, *Vision Res.* **13:**1255–1267.

Maffei, L., and A. Fiorentini, 1976, Monocular deprivation in kittens impairs the spatial resolution of geniculate neurones, *Science* **264:**754–755.

Magalhaes-Castro, H. H., L. A. Murata, and D. Magalhaes-Castro, 1975*a*, Cat retinal ganglion cell projections to the superior colliculus as shown by the horseradish peroxidase method, *Exp. Brain Res.* **25:**541–549.

Magalhaes-Castro, H. H., P. E. S. Saraiva, and D. Magalhaes-Castro, 1975*b*, Identification of corticotectal cells of the visual cortex of cats by means of horseradish peroxidase, *Brain Res.* **83:**474–479.

Magoun, H. W., and S. W. Ranson, 1935, The afferent path of the pupillary light reflex: A review of the literature, *Arch. Ophthalmol.* **13:**862–874.

Malpeli, J. G., and F. H. Baker, 1975, The representation of the visual field in the lateral geniculate nucleus of *Macaca mulatta, J. Comp, Neurol.* **161:**569–594.

Malpeli, J. G., P. H. Schiller, and C. L. Colby, 1981, Response properties of single cells in monkey striate cortex during reversible inactivation of individual lateral geniculate laminae, *J. Neurophysiol.* **46:**1102–1119.

Mann, M. D., 1979, Sets of neurons in somatic cerebral cortex and their ontogeny, *Brain Res. Rev.* **1:**3–46.

Mansfield, R. J. W., 1974, Neural basis of orientation perception in primate vision, *Science* **186:**1133–1135.

Marchiafava, P. L., and G. C. Pepeu, 1966, Electrophysiological study of tectal responses to optic nerve volley, *Arch. Ital. Biol.* **104:**406–420.

Marrocco, R. T., 1978, Conduction velocities of afferent input to superior colliculus in normal and decorticate monkeys, *Brain Res.* **140:**155–158.

Marrocco, R. T., and J. B. Brown, 1975, Correlation of receptive field properties of monkey LGN cells with the conduction velocity of retinal afferent input, *Brain Res.* **92:**137–144.

Marrocco, R. T., and R. H. Li, 1977, Monkey superior colliculus: Properties of single cells and their afferent inputs, *J. Neurophysiol.* **40:**844–860.

Marshall, W. H., C. N. Woolsey, and P. Bard, 1941, Observations on cortical somatic sensory mechanisms of cat and monkey, *J. Neurophysiol.* **4:**1–24.

Marshall, W. H., S. A. Talbot, and H. W. Ades, 1943, Cortical response of the anesthetized cat to gross photic and electrical afferent stimulation, *J. Neurophysiol.* **6:**1–15.

Marzi, C. A., 1978, Abnormal distribution of crossed and uncrossed ganglion cells in the Siamese cat's retina, *Neurosci. Lett. Suppl.* 1.

Marzi, C. A., 1980, Vision in Siamese Cats, *Trends Neurosci.* **3:**165–169.

Marzi, C. A., and M. Di Stefano, 1981, Hemiretinal differences in visual perception, *Doc. Ophthalmol. Proc. Series,* **30:**273–278.

Masland, R., K. L. Chow, and D. L. Stewart, 1971, Receptive field characteristics of superior colliculus neurons in the rabbit, *J. Neurophysiol.* **34**:148–156.

Mason, C. A., and J. A. Robson, 1979, Morphology of retino-geniculate axons in the cat, *Neuroscience* **4**:79–97.

Mason, C. A., N. Sparrow, and D. W. Lincoln, 1977, Structural features of the retinohypothalamic projection in the rat during normal development, *Brain Res.* **132**:141–148.

Mason, R., 1975, Cell properties in the medical interlaminar nucleus of the cat's lateral geniculate complex in relation to the sustained/transient classification, *Exp. Brain Res.* **22**:327–329.

Mason, R., 1979, The retino-recipient zone of the feline pulvinar: Should it be considered part of the lateral geniculate complex? *J. Physiol. (London)* **289**:19P.

Mason, R., 1981, Differential responsiveness of cells in the visual zones of the cat's LP–pulvinar complex to visual stimuli, *Exp. Brain Res.* **43**:25–33.

Mathers, L. H., 1971, Tectal projection to the posterior thalamus of the squirrel monkey, *Brain Res.* **35**:292–294.

Maturana, H., and S. Frenk, 1963, Directional movement and horizontal edge detectors in the pigeon retina, *Science* **142**:977–979.

Maturana, H. R., J. Y. Lettvin, W. S. McCulloch, and W. H. Pitts, 1960, Anatomy and physiology of vision in the frog (*Rana pipiens*), *J. Gen. Physiol.* **43**:(Suppl. 2):129–171.

Mayr, E., 1959, Typological vs. population thinking: Darwin on the evolutionary theory in biology, Anthropological Society of Washington, Washington, D.C., pp. 409–412.

Mayr, E., 1969, *Principles of Systematic Zoology,* McGraw-Hill, New York.

Mayr, E., E. G. Linsley, and R. S. Usinger, 1953, *Methods and Principles of Systematic Zoology,* McGraw-Hill, New York.

Melzak, R., and P. D. Wall, 1963, On the nature of cutaneous sensory mechanisms, *Brain* **85**:331–335.

Merzenich, M. M., and J. H. Kaas, 1980, Principles of organisation of sensory–perceptual systems in mammals, *Prog. Psychobiol. Psychol.* **9**:2–42.

Merzenich, M. M., J. H. Kaas, M. Sur, and C.-S. Lin, 1978, Double representation of the body surface within cytoarchitectonic areas 3b and 1 in "S1" in the owl monkey (*Aotus trivirgatus*), *J. Comp. Neurol.* **181**:41–74.

Meyer, G., and K. Albus, 1981, Spiny stellates as cells of origin of association fibres from area 17 to area 18 in the cat's neocortex, *Brain Res.* **210**:335–341.

Michael, C. R., 1969a, Receptive fields of single optic nerve fibers in a mammal with an all-cone retina. I. Contrast-sensitive units, *J. Neurophysiol.* **31**:249–256.

Michael, C. R. 1969b, Receptive fields in single optic nerve fibers in a mammal with an all-cone retina. II. Directionally selective units, *J. Neurophysiol.* **31**:257–267.

Mill, J. S. 1843, *A System of Logic,* Longmans, Green and Co., London (1961 edition).

Millar, J., 1979, Convergence of joint, cutaneous and muscle afferents onto cuneate neurones in the cat, *Brain Res.* **175**:347–350.

Millhouse, O. E., 1977, Optic chiasm collaterals afferent to the suprachiasmatic nucleus, *Brain Res.* **137**:351–355.

Minkowski, M., 1919, Uber den Verlauf, die Endigung und die zentrale Räpresentation von gekreuzten und ungekreuzten Sehnervfasern bei Saugetieren und bei Menschen, *Schweiz. Arch. Neurol. Psychiatr.* **6**:201–252.

Mitchell, D. E., R. D. Freeman, M. Millodot, and G. Haegerstrom, 1973, Meridional amblyopia: Evidence for modification of the human visual system by early visual experience, *Vision Res.* **13**:535–558.

Mitzdorf, U., and W. Singer, 1977, Laminar segregation of afferents to lateral geniculate nucleus of the cat: An analysis of current source density, *J. Neurophysiol.* **40**:1127–1244.

Molotchnikoff, S., P. LaChappelle, D. Richard, P. L'Archeveque, and I. Lessard, 1979, Latency distribution from orthodromic stimulation at the optic nerve, in the lateral geniculate nucleus and superior colliculus of rabbits, *Brain Res. Bull.* **4**:579–581.

Moore, R. Y., 1973, Retinohypothalamic projection in mammals: A comparative study, *Brain Res.* **49**:403–409.

Moore, R. Y., and V. B., Eichler, 1972, Loss of circadian adrenal cortico-sterone rhythm following suprachiasmatic lesions in the rat, *Brain Res.* **42**:201–206.

Moore, R. Y., and D. C. Klein, 1974, Visual pathways and central neural control of a circadian rhythm in pineal serotonin *N*-acetyltransferase activity, *Brain Res.* **71**:17–33.

Moore, R. Y., and N. J. Lenn, 1972, A retinohypothalamic projection in the rat, *J. Comp. Neurol.* **146**:1–14.

Morales, R., D. Duncan, and R. Rehmet, 1964, A distinctive laminated cytoplasmic body in the lateral geniculate body neurons of the cat, *J. Ultrastruct. Res.* **10**:116–123.

Moran, J., and B. Gordon, 1982, Long term visual deprivation in a human, *Vision Res.* **22**:27–36.

Motokawa, K., T. Oikawa, and K. Tasaki, 1957, Studies of neuronal processes in the retina by antidromic stimulation, *Jpn. J. Physiol.* **7**:119–131.

Mountcastle, V. B., 1957, Modality and topographic properties of single neurons of cat's somatic sensory cortex, *J. Neurophysiol.* **20**:408–434.

Montcastle, V. B., and E. Henneman, 1952, The representation of tactile in the thalamus of the monkey, *J. Comp. Neurol.* **97**:409–440.

Movshon, J. A., 1975, The velocity tuning of single units in cat striate cortex, *J. Physiol. (London)* **249**:445–468.

Movshon, J., and R. C. van Sluyters, 1981, Visual neural development, *Annu. Rev. Psychol.* **32**:477–522.

Movshon, J. A., I. D. Thompson, and D. J. Tolhurst, 1978*a*, Spatial summation in the receptive fields of simple cells in cat's striate cortex, *J. Physiol. (London)* **283**:53–77.

Movshon, J. A., I. D. Thompson, and D. J. Tolhurst, 1978*b*, Receptive field organization of complex cells in the cat's striate cortex, *J. Physiol. (London)* **283**:79–99.

Movshon, J. A., I. D. Thompson, and D. J. Tolhurst, 1978*c*, Spatial and temporal contrast sensitivity of neurones in areas 17 and 18 of the cat's visual cortex, *J. Physiol. (London)* **283**:101–120.

Muir, D. W., and D. E. Mitchell, 1973, Visual resolution and experience: Acuity deficits in cats following early selective visual deprivation, *Science* **180**:420–422.

Munk, O., 1970, On the occurrence and significance of horizontal band-shaped retinal areas in teleosts, *Vidensk. Medd. Dan. Naturhist. Foren. Khobenhavn* **133**:85–120.

Murakami, D. M., and P. D. Wilson, 1981, Monocular deprivation affects cell morphology in laminae C and C1 in cat lateral geniculate nucleus, *Soc. Neurosci. Abstr.* **269.1**.

Murakami, D. M., M. A. Sesma, and M. H. Rowe, 1982, Characteristics of nasal and temporal retina in Siamese and normally pigmented cats, *Brain Behav. Evol.* **21**:61–113.

Musgrave, A., 1976, Method or madness?, in: *Essays in Memory of Imre Lakatos (R. S. Cohen, eds.)* Reidel, Dordrecht.

Nelson, J. I., 1974, Motion sensitivity in peripheral vision, *Perception* **3**:151–152.

Nelson, R., E. V. Famiglietti, and H. Kolb, 1978, Intracellular staining reveals different levels of stratification of on- and off-center ganglion cells in cat retina, *J. Neurophysiol.* **41**:472–483.

Newton, I., 1730, *Opticks,* 1952 issue based on 4th edition (1730), Dover Publications, New York.

Niimi, S. K., and J. M. Sprague, 1970, Thalamo-cortical organisation of the visual system in the cat, *J. Comp. Neurol.* **138**:219–250.

Noda, H., and K. Iwama, 1967, Unitary analysis of retino-geniculate response time in rats, *Vision Res.* **7**:205–213.

Norton, T. T., V. A. Casagrande, and S. M. Sherman, 1977, Loss of Y-cells in the lateral geniculate nucleus of monocularly deprived tree shrews, *Science* **197**:784–786.

Ogden, T. E., and R. F. Miller, 1966, Studies of the optic nerve of the rhesus monkey: Nerve fiber spectrum and physiological properties, *Vision Res.* **6**:485–506.

Ogle, K., 1962, Spatial localization through binocular vision, in: *The Eye,* Vol. 4 (H. Davson, ed), pp. 350–406, Academic Press, New York.

Ogren, M. P. 1977. Evidence for a projection from pulvinar to striate cortex in the squirrel monkey (*Saimiri sciureus*), *Exp. Neurol.* **54**:622–625.

Ogren, M. P., and A. Hendrickson, 1976, Pathways between striate cortex and subcortical regions in *Macaca mulatta* and *Saimiri sciureus:* Evidence for a reciprocal pulvinar connection, *Exp. Neurol.* **53**:780–800.

Ogren, M. P., and A. E. Hendrickson, 1977, The distribution of pulvinar terminals in visual areas 17 and 18 of the monkey, *Brain Res.* **137**:343–350.

Ohno, T., T. Kiyohara, and J. I. Simpson, 1970, Postsynaptic potentials evoked in cells of area 19 and its lateral zone during stimulation of the optic pathway in the cat, *Brain Res.* **20**:453–456.

Ohno, T., U. Misgeld, S. T. Kitai, and A. Wagner, 1975, Organization of the visual afferents into the LGd and the pulvinar of the tree shrew *Tupaia glis, Brain Res.* **90**:153–158.

Olavarria, J., 1979, A horseradish peroxidase study of the projections from the latero-posterior nucleus to three lateral peristriate areas in the rat, *Brain Res.* **173**:137–141.

Oldroyd, D. R., 1980, *Darwinian Impacts,* New South Wales University Press, Sydney.

O'Leary, J. L., 1940, A structural analysis of the lateral geniculate nucleus of the cat, *J. Comp. Neurol.* **73**:405–430.

O'Leary, J. L., 1941, Structure of the area striata of the cat, *J. Comp. Neurol.* **75**:131–164.

Olson, C. R., and R. D. Freeman, 1978, Eye alignment in kittens, **41**:848–859.

Orban, G. A., and M. Callens, 1978, Influence of movement parameters on area 18 neurones in the cat, *Exp. Brain Res.* **30**:125–140.

Osmotherly, S., 1979, Retinal topography of the dog, *Proc. Aust. Physiol. Pharmocol. Soc.* **10**:43P.

Otsuka, R., and R. Hassler, 1962, Über die Aufbau und Gliederung der corticalen Sehsphäre der Katze, *Arch. Psychiatr. Nervenkr.* **203**:212–234.

Oyster, C. W., J. I. Simpson, E. S. Takahashi, and R. F. Soodak, 1980, Retinal ganglion cells projecting to the rabbit accessory optic system, *J. Comp. Neurol.* **190**:49–62.

Oyster, C., E. S. Takahashi, and D. C. Hurst, 1981, Density, soma size, and regional distribution of rabbit retinal ganglion cells, *J. Neurosci.* **1**:1331–1346.

Packwood, J., and B. Gordon, 1975, Stereopsis in the normal domestic cat, Siamese cat and cat raised with alternating monocular occlusion, *J. Neurophysiol.* **38**:1485–1499.

Palmer, L. A., and A. C. Rosenquist, 1974, Visual receptive fields of single striate cortical units projecting to the superior colliculus in the cat, *Brain Res.* **657**:27–42.

Partlow, G. D., M. Colonnier, and J. Szabo, 1977, Thalamic projections of the superior colliculus in the rhesus monkey, *Macaca mulatta:* A light and electron microscopic study, *J. Comp. Neurol.* **171**:285–318.

Partridge, L. D., and J. E. Brown, 1970, Receptive fields of rat retinal ganglion cells, *Vision Res.* **10**:455–460.

Pasik, T., and P. Pasik, 1971, The visual world of monkeys deprived of striate cortex: Effective stimulus parameters and the importance of the accessory optic system, in: *Visual Processes in Vertebrates* (T. Shipley and J. E. Dowling, eds.), *Vision Res. Suppl.* **3**:419–435.

Pearlman, A. L., and N. W. Daw, 1970, Opponent color cells in the cat lateral geniculate nucleus, *Science* **167**:84–86.

Pearson, L. J., K. J. Sanderson, and R. T. Wells, 1977, Retinal projections in the ringtailed possum *Pseudocheirus peregrinus, J. Comp. Neurol.* **170**:227–240.

Perenin, M. T., and M. Jeannerod, 1979, Subcortical vision in man, *Trends Neurosci.* **2**:205–207.

Perry, V. H., 1979, The ganglion cell layer of the retina of the rat: A Golgi study, *Proc. R. Soc. London Ser. B* **204**:363–375.

Perry, V. H., and M. Walker, 1980a, Morphology of cells in the ganglion cell layer during development of the rat retina, *Proc. R. Soc. London Ser. B* **208**:433–445.

Perry, V. H., and M. Walker, 1980b, Amacrine cells, displaced amacrine cells and interplexiform cells in the retina of the rat, *Proc. R. Soc. London Ser. B* **208**:415–431.

Peters, A., C. Proskauer, M. Feldman, and L. Kimerer, 1979, The projection of the lateral geniculate nucleus to area 17 of the rat cerebral cortex. V. Degenerating axon terminals synapsing with Golgi impregnated neurons, *J. Neurocytol.* **8**:331–357.

Pettigrew, J. D., 1974, The effect of visual experience on the development of stimulus specificity by kitten cortical neurones, *J. Physiol. (London)* **237**:49–74.

Pettigrew, J. D., 1978, The paradox of the critical period for striate cortex, in: *Neuronal Plasticity* (C. W. Cotman, ed.), Raven Press, New York.

Pettigrew, J. D., T. Nikara, and P. O. Bishop, 1968, Responses to moving slits by single units in cat striate cortex, *Exp. Brain Res.* **6**:373–390.

Pfaffmann, C., M. Frank, L. M. Bartoshuk, and T. C. Snell, 1976, Coding gustatory information in the squirrel monkey chorda tympani, *Prog. Psychobiol. Physiol. Psychol.* **6**:1–27.

Phillips, D. P., and D. R. F. Irvine, 1981, Parallel input to the cat's auditory cortex: Evidence on the properties of neurones in the anterior auditory field (AAF), *Proc. Aust. Physiol. Pharmacol. Soc.* **12**:32P.

Pickard, G., 1980, Morphological characteristics of retinal ganglion cells projecting to the suprachiasmatic nucleus: A horseradish peroxidase study, *Brain Res.* **183:**458–465.

Poggio, G. F., 1974, Central neural mechanisms in vision, in: *Medical Physiology* (V. B. Mountcastle, ed.), Mosby, St. Louis.

Polyak, S., 1941, *The Retina,* University of Chicago Press, Chicago.

Polyak, S., 1957, *The Vertebrate Visual System,* University of Chicago Press, Chicago.

Pomeranz, B., and S. H. Chung, 1970, Dendritic-tree anatomy codes form-vision physiology in tadpole retina, *Science* **170:**983–984.

Pöppel, E., R. Held, and D. Frost, 1973, Residual visual function after brain wounds involving the central visual pathways in man, *Nature (London)* **243:**451–454.

Popper, K. R., 1959, *The Logic of Scientific Discovery,* Hutchinson, London.

Popper, K. R., 1962, *The Open Society and its Enemies,* Vol. I, *Plato,* 4th ed., Routledge & Kegan Paul, London.

Popper, K. R., 1970, Normal science and its dangers, in: *Criticism and the Growth of Knowledge* (I. Lakatos and A. Musgrave, eds.), Cambridge University Press, London.

Popper, K. R., 1972, *Conjectures and Refutations: The Growth of Scientific Knowledge,* 4th ed., Harper Torchbooks, New York.

Pratt, V., 1972, Numerical taxonomy—A critique, *J. Theor. Biol.* **36:**581–592.

Pratt, V., 1977, Foucault and the history of classification theory, *Stud. Hist. Philos. Sci.* **8:**163–171.

Provis, J. M., 1979, The distribution and size of ganglion cells in the retina of the pigmented rabbit: A quantitative study, *J. Comp. Neurol.* **185:**121–139.

Provis, J. M., and C. R. R. Watson, 1981, The distributions of ipsilaterally and contralaterally projecting ganglion cells in the retina of the pigmented rabbit, *Exp. Brain Res.* **44:**82–92.

Raczkowski, D., and I. T. Diamond, 1978, Connections of the striate cortex in *Galago senegalensis, Brain Res.* **144:**383–388.

Rakic, P., 1977, Genesis of the dorsal lateral geniculate nucleus in the rhesus monkey: Site and time of origin, kinetics of proliferation, routes of migration and pattern of distribution of neurons, *J. Comp. Neurol.* **176:**23–52.

Ranson, S. W., 1921, Afferent paths for visceral reflexes, *Physiol. Rev.* **1:**477–522.

Rapaport, D. H., and P. D. Wilson, 1983, Retinal ganglion cell size groups projecting to the superior colliculus and dorsal lateral geniculate nucleus in the North American opossum, *J. Comp. Neurol.* **213:**74–85.

Rapaport, D. H., M. A. Sesma, and M. H. Rowe, 1979, Distribution and central projections of ganglion cells in the retina of the grey fox, *Soc. Neurosci. Abstr.* **5:**804.

Rapaport, D. H., J. M. Provis, and B. Dreher, 1981a, Rabbit retinal ganglion cell morphology, *Proc. Aust. Physiol. Pharmacol. Soc.* **12:**162P.

Rapaport, D. H., P. D. Wilson, and M. H. Rowe, 1981b, The distribution of ganglion cells in the retina of the North American opossum (*Didelphis virginiana*), *J. Comp. Neurol.* **199:**465–480.

Reiner, A., N. Brecha, and H. J. Karten, 1979, A specific projection of retinal displaced amacrine cells to the nucleus of the basal optic root in the chicken, *Neuroscience* **4:**1679–1688.

Reuter, J. H., and K.-P. Hoffmann, 1980, The conduction velocity in retino-geniculate afferent fibers in the rabbit, *Neurosci. Lett.* **16:**175–179.

Reuter, T., and K. Virtanen, 1972, Border and colour coding in the retina of the frog, *Nature (London)* **239:**260–263.

Rezak, M., and L. A. Benevento, 1979, A comparison of the organization of the projections of the dorsal lateral geniculate nucleus, the inferior pulvinar and adjacent lateral pulvinar to promary visual cortex (area 17) in the macaque monkey, *Brain Res.* **167:**19–41.

Rhoades, R. W., and L. M. Chalupa, 1979, Conduction velocity distribution of retinal input to the hamster's superior colliculus and a correlation with receptive field characteristics, *J. Comp. Neurol.* **184:**243–264.

Rigamonti, D. D., and M. B. Hancock, 1978, Viscerosomatic convergence in the dorsal column nuclei of the cat, *Exp. Neurol.* **61:**337–348.

Riley, J. N., J. P. Card, and R. Y. Moore, 1981, A retinal projection to the lateral hypothalamus in the rat, *Cell Tissue Res.* **214:**257–269.

Rioch, D. M., 1929, Studies on the diencephalon of Carnivora. I. The nuclear configuration of the thalamus, epithalamus and hypothalamus of the dog and cat, *J. Comp. Neurol.* **49:**1–119.

Riva Sanseverino, E., C. Galletti, and M. G. Maioli, 1973, Responses to moving stimuli of single cells in the cat visual areas 17 and 18, *Brain Res.* **55**:451–454.

Riva Sanseverino, E., C. Galletti, M. G. Maioli, and S. Squatrito, 1979, Single unit responses to visual stimuli in cat cortical areas 17 and 18. III. Responses to moving stimuli of variable velocity, *Arch. Ital, Biol.* **117**:248–267.

Rizzolatti, G., V. Tradardi, and R. Camarda, 1970, Unit responses to visual stimuli in the cat's superior colliculus after removal of the visual cortex, *Brain Res.* **24**:336–339.

Robson, J. G., 1966, Spatial and temporal contrast-sensitivity functions of the visual system, *J. Opt. Soc. Am.* **56**:1141–1142.

Rockel, A. J., C. J. Heath, and E. G. Jones, 1972, Afferent connections to the diencephalon in the marsupial phalanger and the question of sensory convergence in the "posterior group" of the thalamus, *J. Comp. Neurol.* **145**:105–130.

Rockland, K. S., and D. N. Pandya, 1979, Laminar origins of cortical connections of the occipital lobe in the rhesus monkey, *Brain Res.* **179**:3–20.

Rodieck, R. W., 1965, Quantitative analysis of cat retinal ganglion cell response to visual stimuli, *Vision Res.* **5**:583–601.

Rodieck, R. W., 1967, Receptive fields of the cat's retina: A new type, *Science* **157**:90–92.

Rodieck, R. W., 1973, *The Vertebrate Retina,* Freeman, San Francisco.

Rodieck, R. W., 1979, Visual pathways, *Annu. Rev. Neurosci.* **2**:193–226.

Rodieck, R. W., and B. Dreher, 1979, Visual suppression from nondominant eye in the lateral geniculate nucleus: A comparison of cat and monkey, *Exp. Brain Res.* **35**:465–477.

Rodieck, R. W., and J. Stone, 1965a, Response of cat retinal ganglion cells to moving visual patterns, *J. Neurophysiol.* **28**:819–832.

Rodieck, R. W., and J. Stone, 1965b, Analysis of receptive fields of cat retinal ganglion cells, *J. Neurophysiol.* **28**:833–849.

Rolls, E. T., and A. Cowey, 1970, Topography of the retina and striate cortex and its relationship to visual acuity in rhesus monkeys and squirrel monkeys, *Exp. Brain Res.* **10**:298–310.

Rosenquist, A. C., and L. A. Palmer, 1971, Visual receptive field properties of cells of the superior colliculus after cortical lesions in the cat, *Exp. Neurol.* **33**:629–652.

Rosenquist, A. C., S. B. Edwards, and L. A. Palmer, 1974, An autoradiographic study of the projections of the dorsal lateral geniculate nucleus and the posterior nucleus in the cat, *Brain Res.* **80**:71–93.

Rossignol, S., and M. Colonnier, 1971, A light microscope study of degeneration patterns in cat cortex after lesions of the lateral geniculate nucleus, *Vision Res. Suppl.* **3**:329–338.

Rowe, M. H., and B. Dreher, 1979, Perikaryal size of forebrain-projecting W-type ganglion cells of the cat's retina, *Proc. Aust. Physiol. Pharmacol. Soc.* **10**:116P.

Rowe, M. H., and B. Dreher, 1982a, Functional morphology of beta cells in the area centralis of the cat's retina: A model for the evolution of central retinal specializations, *Brain Behav. Evol.* **21**:1–23.

Rowe, M. H., and B. Dreher, 1982b, The W-cell projection to the medial interlaminar nucleus of the cat: Implications for ganglion cell classification, *J. Comp. Neurol.* **204**:117–133.

Rowe, M. H., and J. Stone, 1976a, Properties of ganglion cells in the visual streak of the cat's retina, *J. Comp. Neurol.* **167**:99–126.

Rowe, M. H., and J. Stone, 1976b, Conduction velocity groupings among axons of cat retinal ganglion cells and their relationship to retinal topography, *Exp. Brain Res.* **25**:339–357.

Rowe, M. H., and J. Stone, 1977, Naming of neurones: Classification and naming of cat retinal ganglion cells, *Brain Behav. Evol.* **14**:185–216.

Rowe, M. H., and J. Stone, 1979, The importance of knowning our own presuppositions, *Brain Behav. Evol.* **16**:65–80.

Rowe, M. H., and J. Stone, 1980a, Parametric and feature extraction analyses of the receptive fields of visual neurones: Two streams of thought in the study of a sensory pathway, *Brain Behav. Evol.* **17**:103–122.

Rowe, M. H., and J. Stone, 1980b, The interpretation of variation in the classification of nerve cells, *Brain Behav. Evol.* **17**:123–151.

Rowe, M. H., E. Tancred, and B. Freeman, 1977, Properties of ganglion cells in the retina of the brush-tailed possum (*Trichosurus vulpecula*), *Soc. Neurosci. Abstr.* **2**:1089.

Rowe, M. H., L. A. Benevento, and M. Rezak, 1978, Some observations on the patterns of segregated geniculate inputs to the visual cortex in New World primates: An autoradiographic study, *Brain Res.* **159**:371–378.

Rowe, M. H., D. Murakami, and M. A. Sesma, 1980, Abnormalities of temporal retina in the Siamese cat, *Neurosci. Abstr.* **6**:584.

Rowe, M. H., P. D. Wilson, and D. H. Rapaport, 1981, Conduction velocity groups in the optic nerve of the North American opossum (*Didelphis virginiana*): Retinal origins and central projections, *J. Comp. Neurol.* **199**:481–493.

Saini, K. D., and L. J. Garey, 1981, Morphology of neurons in the lateral geniculate nucleus of the monkey: A Golgi study, *Exp. Brain Res.* **42**:235–248.

Saito, H. T. Shimahara, and Y. Fukada, 1970, Four types of responses to light and dark spot stimuli in the cat optic nerve, *Tohoku J. Exp. Med.* **102**:127–133.

Salinger, W. L., P. E. Garraghty, M. G. Macavoy, and L. F. Hooker, 1980, Sensitivity of the mature lateral geniculate nucleus to components of monocular paralysis, *Brain Res.* **187**:307–320.

Sanders, M. D., E. K. Warrington, J. Marshal, and L. Weiskrantz, 1974, Blindsight: Vision in a field defect, *Lancet* **34**:707–708.

Sanderson, K. J., 1971, The projection of the visual field to lateral geniculate and medial interlaminar nuclei in the cat, *J. Comp. Neurol.* **143**:101–118.

Sanderson, K. J., and S. M. Sherman, 1971, Nasotemporal overlap in visual field projected to lateral geniculate nucleus in the cat, *J. Neurophysiol.* **34**:453–466.

Schiffman, S. S., and R. P. Erickson, 1980, The issue of primary tastes versus a taste continuum, *Neurosci. Biobehav. Rev.* **4**:109–117.

Schilder, P., T. Pasik, and P. Pasik, 1971, Extrageniculostriate vision in the monkey. II. Demonstration of brightness discrimination, *Brain Res.* **32**:383–398.

Schilder, P., T. Pasik, and P. Pasik, 1972, Extrageniculostriate vision in the monkey. III. Circle vs. triangle and red vs. green discrimination, *Exp. Brain Res.* **14**:436–448.

Schiller, P. H., and J. G. Malpeli, 1977a, Properties and tectal projections of monkey retinal ganglion cells, *J. Neurophysiol.* **40**:428–445.

Schiller, P., and J. G. Malpeli, 1977b, The effect of striate cortex cooling on area 18 cells in the monkey, *Brain Res.* **126**:366–369.

Schiller, P. H., and J. G. Malpeli, 1978, Functional specificity of lateral geniculate nucleus laminae of the rhesus monkey, *J. Neurophysiol.* **41**:788–797.

Schiller, P. H., J. G. Malpeli, and S. J. Schein, 1979, Composition of geniculatriate input to superior colliculus of the rhesus monkey, *J. Neurophysiol.* **42**:1124–1133.

Schmidt, M. L., and H. V. B. Hirsch, 1980, A quantitative study of the occurrence and distribution of cytoplasmic laminated bodies in the lateral geniculate nucleus of the normal adult cat, *J. Comp. Neurol.* **189**:235–247.

Schneider, G. E., 1969, Two visual systems, *Science* **163**:895–902.

Schoppmann, A., and K.-P. Hoffmann, 1979, A comparison of visual responses in two pretectal nuclei and in the superior colliculus of the cat, *Exp. Brain Res.* **35**:495–510.

Sefton, A. J., 1969, The innervation of the lateral geniculate nucleus and anterior colliculus in the rat, *Vision Res.* **8**:867–881.

Sefton, A. J., and M. Swinburn, 1964, Electrical activity of lateral geniculate nucleus and optic tract of the rat, *Vision Res.* **4**:315–328.

Semm, P., 1978, Antidromically activated direction selective ganglion cells of the rabbit, *Neurosci. Lett.* **9**:207–211.

Shapley, R. M., and J. Gordon, 1978, Ganglion cell classes and spatial mechanisms, *J. Gen. Physiol.* **71**:139–155.

Shapley, R. M., and Y. T. So, 1980, Is there an effect of monocular deprivation on the proportions of X and Y cells in the lateral geniculate nucleus? *Exp. Brain Res.* **39**:41–48.

Shapley, R., E. Kaplan, and R. Soodak, 1981, Spatial summation and contrast sensitivity of X- and Y-cells in the lateral geniculate nucleus of the macaque, *Nature (London)* **192**:543–545.

Sherk, H., 1978, Area 18 cell responses in cat during reversible inactivation of area 17, *J. Neurophysiol.* **41**:204–215.

Sherk, H., and M. P. Stryker, 1977, Quantitative study of cortical orientation selectivity in visually inexperienced kittens, *J. Neurophysiol.* **40**:260–283.

Sherman, S. M., 1973, Visual field defects in monocularly and binocularly deprived cat, *Brain Res.* **49:**25–45.

Sherman, S. M., 1977, The effect of superior colliculus lesions upon the visual fields of cats with cortical ablations, *J. Comp. Neurol.* **172:**211–229.

Sherman, S. M., 1979*a*, The functional significance of X- and Y-cells in normal and visually deprived cat, *Trends Neurosci.* **2:**192–195.

Sherman, S. M., 1979*b*, Development of the lateral geniculate nucleus in cats raised with monocular lid suture, in: *Developmental Neurobiology of Vision* (R. D. Freeman, ed.), pp. 79–97, Plenum Press, New York.

Sherman, S. M., and R. W. Guillery, 1976, Behavioral studies of binocular competition in cats, *Vision Res.* **16:**1479–1481.

Sherman, S. M., and J. Stone, 1973, Physiological normality of the retina in visually deprived cats, *Brain Res.* **40:**224–230.

Sherman, S. M., K.-P. Hoffmann, and J. Stone, 1972, Loss of a specific cell type from the dorsal lateral geniculate nucleus in visually deprived cats, *J. Neurophysiol.* **35:**532–541.

Sherman, S. M., T. T. Norton, and V. A. Casagrande, 1975*a*, X- and Y- cells in the dorsal lateral geniculate nucleus of the tree shrews, *Brain Res.* **93:**152–157.

Sherman, S. M., J. R. Wilson, and R. W. Guillery, 1975*b*, Evidence that binocular competition affects the postnatal development of Y-cells in the cat's lateral geniculate nucleus, *Brain Res.* **100:**441–444.

Sherman, S. M., R. W. Guillery, J. H. Kaas, and K. J. Sanderson, 1975*c*, Behavioral, electrophysiological and morphological studies of binocular competition in the development of the geniculo-cortical pathways of cats, *J. Comp. Neurol.* **158:**1–18.

Sherman, S. M., J. R. Wilson, J. H. Kaas, and S. V. Webb, 1976, X- and Y-cells in the lateral geniculate nucleus of the owl monkey (*Aotus trivirgatus*), *Science* **192:**475–476.

Shkolnik-Yarros, E. G., 1971, Neurones of the cat's retina, *Vision Res.* **11:**7–26.

Shoumura, K., 1973, Pathway from dorsal lateral geniculate nucleus to visual cortex in cats, *Brain Res.* **49:**277–290.

Simpson, G. G., 1961, *Principles of Animal Taxonomy*, Columbia University Press, New York.

Singer, W., 1976, Modification of orientation and direction selectivity of cortical cells in kittens with monocular vision, *Brain Res.* **118:**460–468.

Singer, W., and N. Bedworth, 1973, Inhibitory interaction between X and Y units in the cat lateral geniculate nucleus, *Brain Res.* **49:**291–307.

Singer, W., and N. Bedworth, 1974, Correlation between the effects of brain stem stimulation and saccadic eye movements on transmission in the cat lateral geniculate nucleus, *Brain Res.* **72:**185–202.

Singer, W., F. Tretter, and M. Cynader, 1975, Organization of cat striate cortex: A correlation of receptive field properties with afferent and efferent connections, *J. Neurophysiol.* **38:**1080–1098.

Singer, W., J. Zihl, and E. Pöppel, 1977, Subcortical control of visual thresholds in humans: Evidence of modality specific and retinotopically organized mechanisms of selective attention, *Exp. Brain Res.* **29:**173–190.

Singer, W., B. Freeman, and J. Rauschecker, 1981, Restriction of visual experience to a single orientation affects the orientation columns in cat visual cortex, *Exp. Brain Res.* **41:**199–215.

Sireteanu, R., and K.-P. Hoffmann, 1977, Interlaminar differences in the effects of early and late monocular deprivation on the visual acuity of cells in the lateral geniculate nuclues of the cat, *Neurosci. Lett.* **5:**171–175.

Sireteanu, R., and K.-P. Hoffmann, 1979, Relative frequency and visual resolution of X- and Y-cells in the LGN of normal and monocularly deprived cats: Interlaminar differences, *Exp. Brain Res.* **34:**591–603.

Slonaker, J. R., 1897, A comparative study of the area of acute vision in vertebrates, *J. Morphol.* **13:**445–500.

Smith, D. C., P. D. Spear, and K. E. Kratz, 1978, Role of visual experience in postcritical-period reversal of effects of monocular deprivation in cat striate cortex, *J. Comp. Neurol.* **178:**313–328.

Sneath, P. H. A., and R. R. Sokal, 1973, *Numerical Taxonomy: The Principles and Practice of Numerical Classification*, Freeman, San Francisco.

Solomon, S. J., T. Pasik, and P. Pasik, 1981, Extrageniculostriate vision in the monkey. VIII. Critical structures for spatial localization, *Exp. Brain Res.* **44:**259–270.

Somogyi, G., F. Hajdu, R. Hassler, and A. Wagner, 1981, An experimental electron microscopical study of a direct retino-pulvinar pathway in the tree shrew, *Exp. Brain Res.* **43**:447–450.

Spatz, W. B., 1978, The retino-geniculo-cortical pathway in *Callithrix*. I. Intraspecific variations in the lamination pattern of the lateral geniculate nucleus, *Exp. Brain Res.* **33**:551–563.

Spear, P. D., and J. J. Braun, 1969, Pattern discrimination following removal of visual neocortex in the cat, *Exp. Neurol.* **25**:331–348.

Spear, P. D., D. C. Smith, and L. L. Williams, 1977, Visual receptive field properties of single neurons in cat's ventral lateral geniculate nucleus, *J. Neurophysiol.* **40**:390–409.

Spehlmann, R., 1967, Compound action potentials of cat optic nerve produced by stimulation of optic tracts and of optic nerve, *Exp. Neurol.* **19**:156–165.

Spoendlin, H., 1969, Innervation patterns in the organ of Corti of the cat, *Acta Oto-Laryngol.* **67**:239–254.

Spoendlin, H., 1970, Structural basis of peripheral frequency analysis, in: *Frequency Analysis and Periodicity Detection in Hearing* (R. Plomp and G. F. Smoorenburg, eds.), pp. 1–40, Sitjhoff, Leiden.

Spoendlin, S. F., 1972, Innervation densities of the cochlea, *Acta Oto-Laryngol.* **73**:235–248.

Sprague, J. M., J. Levy, A. DiBerardino, and G. Berlucchi, 1977, Visual cortical areas mediating form discrimination in the cat, *J. Comp. Neurol.* **172**:441–448.

Sprague, J. M., H. C. Hughes, and G. Berlucchi, 1981, Cortical mechanisms in pattern and form perception, in: *Brain Mechanisms and Perceptual Awareness* (O. Pompeiano and C. Ajmone Marsan, eds.), Raven Press, New York.

Stanford, L. R., M. J. Friedlander, and S. M. Sherman, 1981, Morphology of physiologically identified W-cells in the C-laminae of the cat's lateral geniculate nucleus, *J. Neurosci.* **1**:578–584.

Steinberg, R. H., M. Reid, and P. L. Lacy, 1973, The distribution of rods and cones in the retina of the cat (*Felis domesticus*), *J. Comp. Neurol.* **148**:229–248.

Steinman, R. M., 1975, Oculomotor effects on vision, in: *Basic Mechanisms of Ocular Motility and Their Clinical Implications* (G. Lennerstrand and P. Bach-y-Rita, eds.), Pergamon Press, New York.

Sterling, P., and B. G. Wickelgren, 1970, Function of the projection from the visual cortex to the superior colliculus, *Brain Behav. Evol.* **3**:210–218.

Stone, J., 1965, A quantitative analysis of the distribution of ganglion cells in the cat's retina, *J. Comp. Neurol.* **124**:337–352.

Stone, J., 1966, The naso-temporal division of the cat's retina, *J. Comp. Neurol.* **126**:585–600.

Stone, J., 1972, Morphology and physiology of the geniculocortical synapse in the cat: The question of parallel input to the striate cortex, *Invest. Ophthalmol.* **11**:338–345.

Stone, J., 1973, Sampling properties of microelectrodes assessed in the cat's retina, *J. Neurophysiol.* **36**:1071–1080.

Stone, J., 1978, The number and distribution of ganglion cells in the cat's retina, *J. Comp. Neurol.* **180**:753–772.

Stone, J., and R. M. Clarke, 1980, Correlation between soma size and dendritic morphology in cat retinal ganglion cells: Evidence of further variation in the gamma-cell class, *J. Comp. Neurol.* **192**:211–218.

Stone, J., and B. Dreher, 1973, Projection of X- and Y-cells of the cat's lateral geniculate nucleus to areas 17 and 18 of visual cortex, *J. Neurophysiol.* **36**:551–567.

Stone, J., and M. Fabian, 1966, Specialized receptive fields of the cat's retina, *Science* **152**:1277–1279.

Stone, J., and R. B. Freeman, 1971, Conduction velocity groups in the cat's optic nerve classified according to their retinal origin, *Exp. Brain Res.* **13**:489–497.

Stone, J., and R. B. Freeman, 1973, Neurophysiological mechanisms in the visual discrimination of form, in: *Handbook of Sensory Physiology,* Vol. VII/3A (R. Jung, ed.), Springer-Verlag, Berlin.

Stone J., and Y. Fukuda, 1974*a*, Properties of cat retinal ganglion cells: A comparison of W-cells with X- and Y-cells, *J. Neurophysiol.* **37**:722–748.

Stone, J., and Y. Fukuda, 1974*b*, The naso-temporal division of the cat's retina re-examined in terms of W-, X-, and Y-cells, *J. Comp. Neurol.* **155**:377–394.

Stone, J., and K.-P. Hoffmann, 1971, Conduction velocity as a parameter in the organisation of the afferent relay in the cat's lateral geniculate nucleus, *Brain Res.* **32**:460–466.

Stone, J., and K.-P. Hoffmann, 1972, Very slow conducting ganglion cells in the cat's retina: A major new functional type? *Brain Res.* **43**:610–616.

Stone, J., and E. Johnston, 1981, The topography of primate retina: A study of the human, bushbaby and New- and Old-world monkeys, *J. Comp. Neurol.* **196**:205–223.

Stone, J., and J. S. Keens, 1978, The distribution of medium and small classes of ganglion cells in the cat's retina, *Proc. Aust. Physiol. Pharmacol. Soc.* **9**:187P.

Stone, J., and J. S. Keens, 1980, The distribution of small and medium-sized ganglion cells in the cat's retina, *J. Comp. Neurol.* **192**:235–245.

Stone, J., J. Leicester, and S. M. Sherman, 1973, The nasotemporal division of the monkey's retina, *J. Comp. Neurol.* **150**:333–348.

Stone, J., J. E. Campion, and J. Leicester, 1976, The nasotemporal division of retina in the Siamese cat, *Proc. Aust. Physiol. Pharmacol. Soc.* **7**:51P.

Stone, J. C. R. R. Watson, and H. Holländer, 1977, A temporal–nasal soma size gradient in the cat's retina, *Proc. Aust. Physiol. Pharmacol. Soc.* **8**:128P.

Stone, J., J. E. Campion, and J. Leicester, 1978a, The nasotemporal division of retina in the Siamese cat, *J. Comp. Neurol.* **180**:783–798.

Stone, J., M. H. Rowe, and J. E. Campion, 1978b, Retinal abnormalities in the Siamese cat, *J. Comp. Neurol.* **180**:753–772.

Stone, J., B. Dreher, and A. G. Leventhal, 1979, Hierarchical and parallel mechanisms in the organization of visual cortex, *Brain Res. Rev.* **1**:345–394.

Stone, J., A. G. Leventhal, C. R. R. Watson, J. S. Keens, and R. M. Clarke, 1980, Gradients between nasal and temporal areas of the cat retina in the properties of retinal ganglion cells, *J. Comp. Neurol.* **192**:219–235.

Stryker, M. P., D. H. Hubel, and T. N. Wiesel, 1977, Orientation columns in the cat's visual cortex, *Soc. Neurosci. Abstr.* **3**:1852.

Stryker, M. P., H. Sherk, A. G. Leventhal, and H. V. B. Hirsch, 1978, Physiological consequences for the cat's visual cortex of effectively retricting early visual experience with oriented contours, *J. Neurophysiol.* **41**:896–909.

Sumitomo, I., K. Ide, and K. Iwama, 1969a, Conduction velocity of rat optic nerve fibers, *Brain Res.* **12**:261–264.

Sumitomo, I., K. Ide, K. Iwama, and T. Arikuni, 1969b, Conduction velocity of optic nerve fibers innervating lateral geniculate body and superior colliculus in the rat, *Exp. Neurol.* **25**:378–392.

Sur, M., A. L. Humphrey and S. M. Sherman, 1982, Monocular deprivation affects retinogeniculate terminations in cats, *Nature* **300**:183–185.

Sur, M. and S. M. Sherman, 1982, Retinogeniculate terminations in cat: Morphological differences between X and Y-cell axons, *Science* **218**:389–391.

Sur, M., J. T. Wall, and J. H. Kaas, 1981, Modular segregation of functional cell classes within the postcentral somatosensory cortex of monkeys, *Science* **212**:1059–1061.

Swanson, l. W., W. M. Cowan, and E. G. Jones, 1974, An autoradiographic study of the afferent connections of the ventral lateral geniculate nucleus in the albino rat and cat, *J. Comp. Neurol.* **156**:143–164.

Szenthagothai, J., 1973, Neuronal and synaptic architecture of the lateral geniculate nucleus, in: *Handbook of Sensory Physiology*, Vol. VII/3B (R. Jung, ed.), Springer-Verlag, Berlin.

Takagi, S. F., 1979, Dual systems for sensory olfactory processing in higher primates, *Trends Neurosci.* **2**:313–315.

Takahashi, Y., T. Ogawa, T. Takimori, and H. Kato, 1977, Intracellular studies of rabbit's superior colliculus, *Brain Res.* **123**:170–175.

Talbot, S. A. 1942, A lateral localization in cat's visual cortex, *Fed. Proc.* **1**:84.

Tancred, E., 1981, The distribution and sizes of ganglion cells in the retinas of five marsupials, *J. Comp. Neurol.* **196**:585–604.

Tancred, E., and M. H. Rowe, 1979, The distribution of ipsilaterally and contralaterally projecting ganglion cells in the brush-tailed possum (*Trichosurus vulpecula*), *Proc. Aust. Physiol. Pharmacol. Soc.* **10**:118P.

Tanji, D. G., S. P. Wise, R. W. Dykes, and E. G. Jones, 1978, Cytoarchitecture and thalamic connectivity of the third somatosensory area of cat cerebral cortex, *J. Neurophysiol.* **41**:268–284.

ter Laak, H. J., and J. M. Thijssen, 1978, Receptive field properties of optic tract fibres from on-center sustained and transient cells in a tree shrew (*Tupaia chinensis*), *Vision Res.* **18**:1097–1111.

Teuber, H. L., W. S. Battersby, and M. B. Bender, 1960, *Visual Field Defects after Penetrating Missile Wounds of the Brain,* Harvard University Press, Cambridge, Mass.

Thibos, L. N., and W. R. Levick, 1982, Astigmatic visual deprivation in cat: Behavioral, optical and retinophysiological consequences, *Vision Res.* **22:**43–53.

Thorpe, S., and M. Glickstein, 1972, Translators and editors of *The Structure of the Vertebrate Retina* by S. R. Cajal, Thomas, Springfield, Ill.

Thuma, B. D., 1928, Studies on the diencephalon of the cat. I. The cyto-architecture of the corpus geniculatum laterale, *J. Comp. Neurol.* **46:**173–200.

Tiao, Y.-C., and C. Blakemore, 1976, Regional specialization in the golden hamster's retina, *J. Comp. Neurol.* **168:**439–458.

Tigges, J., M. Tigges, and A. A. Perachio, 1977, Complementary laminar terminations of afferents to area 17 originating in area 18 and in the lateral geniculate nucleus in squirrel monkey, *J. Comp. Neurol.* **176:**87–100.

Tolhurst, D. J., 1973, Separate channels for the analysis of the shape and the movement of a moving visual stimulus, *J. Physiol. (London)* **231:**385–402.

Tolhurst, D. J., 1975, Sustained and transient channels in human vision, *Vision Res.* **15:**1151–1155.

Tolhurst, D. J., and L. P. Barfield, 1978, Interactions between spatial frequency channels, *Vision Res.* **18:**951–958.

Toyama, K., and K. Matsunami, 1968, Synaptic action of specific visual impulses upon cat's parastriate cortex, *Brain Res.* **10:**473–476.

Toyama, K., K. Matsunami, and T. Ohno, 1969a, Antidromic identification of association, commissural and corticofugal efferent cells in cat visual cortex, *Brain Res.* **14:**513–517.

Toyama, K., S. Tokashiki, and K. Matsunami, 1969b, Antidromic identification of association, commissural and corticofugal efferent cells in cat visual cortex, *Brain Res.* **14:**513–517.

Toyama, K., K. Maekawa, and T. Takeda, 1973, An analysis of neuronal circuitry for two types of visual cortical neurones classified on the basis of their responses to photic stimuli, *Brain Res.* **61:**395–399.

Toyama, K., K. Maekawa, and T. Takeda, 1977a, Convergence of retinal inputs onto visual cortical cells. I. A study of the cells monosynaptically excited from the lateral geniculate body, *Brain Res.* **137:**207–220.

Toyama, K., M. Kimura, T. Takeda, and T. Shiida, 1977b, Convergence of retinal inputs onto visual cortical cells. II. A study of the cells disynaptically excited from the lateral geniculate body, *Brain Res.* **137:**221–231.

Tretter, F., M. Cynader, and W. Singer, 1975, Cat parastriate cortex: A primary or secondary visual area? *J. Neurophysiol.* **38:**1099–1113.

Trevarthen, C. B., 1968, Two mechanisms of vision in primates, *Psychol. Forsch.* **31:**299–337.

Tusa, R. J., A. C. Rosenquist, and L. A. Palmer, 1979, Retinotopic organization of areas 18 and 19 in the cat, *J. Comp. Neurol.* **185:**657–678.

Tuttle, J. R., and L. C. Scott, 1979, X-like and Y-like ganglion cells in the *Necturus* retina, *Invest. Ophthalmol.* **18:**524–527.

Tyner, C. F., 1975, The naming of neurones: Applications of taxonomic theory to the study of cellular populations, *Brain Behav. Evol.* **12:**75–96.

Ulinski, P. S., 1980, Functional morphology of the vertebrate visual system: An essay on the evolution of complex systems, *Am. Zool.* **20:**229–246.

Updyke, B. V., 1975, The patterns of projection of cortical areas 17, 18 and 19 onto the laminae of the dorsal lateral geniculate nucleus in the cat, *J. Comp. Neurol.* **163:**377–396.

Updyke, B. V., 1977, Topographic organisation of the projections from cortical areas 17, 18 and 19 onto the thalamus, pretectum and superior colliculus in the cat, *J. Comp. Neurol.* **173:**81–122.

Uttal, W. R., 1971, The psychobiological silly season—or—what happens when neurophysiological data become psychological theories, *J. Gen. Physiol.* **84:**151–166.

Uttal, W. R., 1977, Review of Szenthagothai and Arbib 1974, *Brain Behav. Evol.* **14:**238–240.

van Buren, J. M., 1963, *The Retinal Ganglion Cell Layer,* Thomas, Springfield, Ill.

van Dongen, P. A. M., H. J. ter Laak, J. M. Thijssen, and A. J. H. Vendrik, 1976, Functional classification of cells in the optic tract of the tree shrew (*Tupaia chinensis*), *Exp. Brain Res.* **24:**441–446.

van Essen, D., and S. Zeki, 1978, The topographic organization of rhesus monkey peristriate cortex, *J. Physiol. (London)* **277:**193–226.

Vaney, D. I., W. R. Levick, and L. N. Thibos, 1978, Axonal conduction latencies of rabbit retinal ganglion cells, *Proc. Aust. Physiol. Pharmacol. Soc.* **9:**55P.

Vaney, D. I., L. Peichl, H. Wässle, and R.-B. Illing, 1981, Almost all ganglion cells in the rabbit retina project to the superior colliculus, *Brain Res.* **212:**454–460.

van Hof, M. W., and C. Lagers-van Haselen, 1973, The retinal fixation area in the rabbit, *Exp. Neurol.* **41:**218–221.

van Hof-van Duin, J., 1977, Visual field measurements in monocularly deprived and normal cats, *Exp. Brain Res.* **30:**353–368.

van Nes, F. L., J. J. Koenderick, H. Nas, and M. A. Bouman, 1967, Spatiotemporal modulation transfer in the human eye, *J. Opt. Soc. Am.* **57:**1082–1088.

Vastola, E. F., 1961, A direct pathway from lateral geniculate body to association cortex, *J. Neurophysiol.* **24:**489–497.

Victor, J. D., and R. M. Shapley, 1979, Receptive field mechanisms of cat X and Y retinal ganglion cells, *J. Gen. Physiol.* **74:**275–298.

Vital-Durand, F., L. J. Garey, and C. Blakemore, 1978, Monocular and binocular deprivation in the monkey: Morphological effects and reversibility, *Brain Res.* **158:**45–64.

von Gudden, B., 1886, Demonstration der Sehfasern und Papillarfasern des nerv. opticus, *Sitzungsber. Ges. Morphol. Physiol. Muenchen* **1:**169–170.

von Noorden, G. K., 1973, Histological studies of the visual system in monkeys with experimental amblyopia, *Invest. Ophthalmol.* **12:**727–728.

von Noorden, G. K., and M. L. J. Crawford, 1978, Morphological and physiological changes in the monkey visual system after short-term lid suture, *Invest. Ophthalmol. Vision Sci.* **17:**762–768.

von Noorden, G. K., and J. E., Dowling, 1970, Experimental amblyopia in monkeys. II. Behavioral studies of strabismic amblyopia, *Arch, Ophthalmol.* **84:**215–220.

von Noorden, G. K., J. E. Dowling, and D. C. Ferguson, 1970, Experimental amblyopia in monkeys. I. Behavioral studies of stimulus deprivation amblyopia, *Arch. Ophthalmol.* **84:**206–214.

Wagner, H. G., E. F. MacNichol, and M. L. Wolbarsht, 1960, The response properties of single ganglion cells in the goldfish retina, *J. Gen. Physiol.* **43**(Suppl.):45–62.

Wagner, H. G., E. F. MacNichol. and M. L. Wolbarsht, 1963, Functional basis of 'on'-center and 'off'-center receptive fields in the retina, *J. Opt. Soc. Am.* **53:**66–70.

Walls, G. L. 1953, The lateral geniculate nucleus and visual histophysiology, *Univ. Calif. Berkeley Publ. Physiol.* **9:**1–100.

Wan, Y. K., and B. Cragg, 1976, Cell growth in the lateral geniculate nucleus of kittens following the opening or closing of one eye, *J. Comp. Neurol.* **166:**365–372.

Wässle, H., and R. B. Illing, 1980, The retinal projection to the superior colliculus in the cat: A quantitative study with HRP, *J. Comp. Neurol.* **190:**333–356.

Wässle, H., W. R. Levick, and B. G. Cleland, 1975, The distribution of the alpha type of ganglion cells in the cat's retina, *J. Comp. Neurol.* **159:**419–438.

Wässle, H., L. Peichl, and B. B. Boycott, 1978, Topography of horizontal cells in the retina of the domestic cat, *Proc. R. Soc. London Ser B* **203:**269–291.

Watanabe, S., M. Konishi, and O. Creutzfeldt, 1966, Postsynaptic potentials in the cat's visual cortex following electrical stimulation of afferent pathways, *Exp. Brain Res.* **1:**272–283.

Webb, S. V., and J. H. Kaas, 1976, The size and distribution of ganglion cells in the retina of the owl monkey, *Aotus trivirgatus, Vision Res.* **16:**1247–1254.

Webster, K., 1977, Somaesthetic pathways, *Brit. Med. Bull.* **33:**113–120.

Webster, W. G., 1973, Assumptions, conceptualizations, and the search for the functions of the brain, *Physiol. Psychol.* **1:**346–350.

Weiskrantz, L., 1972, Behavioural analysis of the monkey's visual nervous system, *Proc. R. Soc. London Ser. B* **182:**427–455.

Weiskrantz, L., 1978, Some aspects of visual capacity in monkeys and man following striate cortex lesions, *Arch. Ital. Biol.* **16:**318–323.

Weiskrantz, L., E. K. Warrington, M. D. Sanderson, and J. Marshall, 1974, Visual capacity in hemianopic field following a restricted occipital ablation, *Brain* **97:**709–728.

Weller, R. E., J. H. Kaas, and A. B. Wetzel, 1979, Evidence for the loss of X-cells of the retina after long-term ablation of visual cortex in monkeys, *Brain Res.* **160:**134–138.

White, E. L., 1978, Identified neurons in mouse SmI cortex which are postsynaptic to thalmocortical axon terminals: A combined Golgi-electron microscopic and degeneration study, *J. Comp. Neurol.* **181**:627–662.

White, E. L., 1979, Thalamocortical synaptic relations: A review with emphasis on the projections of specific thalamic nuclei to the primary areas of the neocortex, *Brain Res. Rev.* **1**:275–311.

Wicklegren, B., and P. Sterling, 1969a, Effect on the superior colliculus of cortical removal in visually deprived cats, *Science* **224**:1032–1033.

Wickelgren, B., and P. Sterling, 1969b, Influence of visual cortex on receptive fields in the superior colliculus of the cat, *J. Neurophysiol.* **32**:16–23.

Wiesel, T. N., 1960, Receptive fields of ganglion cells in the cat's retina, *J. Physiol. (London)* **153**:583–594.

Wiesel, T. N., and D. H. Hubel, 1963a, Effects of visual deprivation on morphology and physiology of cells in the cat's lateral geniculate body, *J. Neurophysiol.* **26**:978–993.

Wiesel, T. N., and D. H. Hubel, 1963b, Single-cell responses in striate cortex of kittens deprived of vision in one eye, *J. Neurophysiol.* **26**:1003–1017.

Wiesel, T. N., and D. H. Hubel, 1965, Comparison of the effects of unilateral and bilateral eye closure on cortical unit responses in kittens, *J. Neurophysiol.* **28**:1029–1040.

Wiesel, T. N., and D. H. Hubel, 1966, Spatial and chromatic interactions in the lateral geniculate body of the rhesus monkey, *J. Neurophysiol.* **29**:1115–1156.

Wiesel, T. N., and D. H. Hubel, 1974, Ordered arrangement of orientation columns in monkeys lacking visual experience, *J. Comp. Neurol.* **158**:307–318.

Wilson, H. R., 1978, Quantitative characterization of two types of line-speard function near the fovea, *Vision Res.* **18**:971–981.

Wilson, H. R., and J. R. Bergen, 1979, Four mechanism model for threshold spatial vision, *Vision Res.* **19**:19–32.

Wilson, J. R., and S. M. Sherman, 1976, Receptive-field characteristics of neurons in cat striate cortex: Changes with receptive field eccentricity, *J. Neurophysiol.* **39**:512–533.

Wilson, M. E., and B. G. Cragg, 1967, Projections from the lateral geniculate nucleus in the cat and monkey, *J. Anat.* **101**:677–692.

Wilson, P. D., and J. Stone, 1975, Evidence of W-cell input to the cat's visual cortex via the C laminae of the lateral geniculate nucleus, *Brain Res.* **92**:472–478.

Wilson, P. D., M. H., Rowe, and J. Stone, 1976, Properties of relay cells in the cat's lateral geniculate nucleus: A comparison of W-cells with X- and Y-cells, *J. Neurophysiol.* **39**:1193–1209.

Winans, S. S., 1967, Visual form discrimination after removal of the visual cortex in cats, *Science* **158**:944–946.

Winfield, D. A., K. C. Gatter, and T. P. S. Powell, 1975, Certain connections of the visual cortex of the monkey shown by the use of horseradish peroxidase, *Brain Res.* **92**:457–461.

Winfield, J. A., A. E. Hendrickson, and J. Kimm, 1978, Anatomical evidence that the medial terminal nucleus of the accessory optic tract in mammals provides a visual mossy fibre input to the flocculus, *Brain Res.* **151**:175–182.

Winkler, C., and A. Potter, 1914, *An Anatomical Guide to Experimental Researches on the Cat's Brain*, T. Versloys, Amsterdam.

Witpaard, J., and H. E. D. J. ter Keurs, 1975, A reclassification of retinal ganglion cells in the frog, based upon tectal endings and response properties, *Vision Res.* **15**:1333–1338.

Wong-Riley, M. T. T., 1972, Neuronal and synaptic organisation of the normal dorsal lateral geniculate nucleus of the squirrel monkey *Saimiri sciureus*, *J. Comp. Neurol.* **144**:25–60.

Wong-Riley, M. T. T., 1974, Demonstration of geniculocortical and callosal projection neurons in the squirrel monkey by means of retrograde axonal transport of horseradish peroxidase, *Brain Res.* **79**:267–272.

Wong-Riley, M. T. T., 1976, Projections from the dorsal lateral geniculate nucleus to prestriate cortex in the squirrel monkey as demonstrated by retrograde transport of horseradish peroxidase, *Brain Res.* **109**:595–600.

Wong-Riley, M. T. T., 1977, Connections between the pulvinar nucleus and the prestriate cortex in the squirrel monkey as revealed by peroxidase histochemistry and autoradiography, *Brain Res.* **134**:225–236.

Yoshida, K., and L. A. Benevento, 1981, The projection from the dorsal lateral geniculate nucleus of the thalamus to extrastriate visual association cortex in the macaque monkey, *Neurosci. Lett.* **22:**103–108.

Yukie, M., and E. Iwai, 1981, Direct projection from dorsal lateral geniculate nucleus to the prestriate cortex in macaque monkey, *J. Comp. Neurol.* **201:**81–98.

Yukie, M., Y. Umitsu, S. Kido, T. Niihara, and E. Iwai, 1979, A quantitative study of the cells projecting from the lateral geniculate nucleus to the prestriate cortex in the monkey with horseradish peroxidase, *Neurosci. Lett. Suppl.* **2:**44.

Zeki, S., 1969, The secondary visual areas of the monkey, *Brain Res.* **13:**197–226.

Zeki, S., 1978a, Functional specialisation in the visual cortex of the rhesus monkey, *Nature (London)* **274:**423–428.

Seki, S. M., 1978b, Uniformity and diversity of structure and function in rhesus monkey prestriate visual cortex, *J. Physiol. (London)* **277:**273–290.

Index